CROP PRODUCTION

CROP PRODUCTION

Principles and Practices

Stephen R. Chapman and Lark P. Carter
Montana State University

W. H. FREEMAN AND COMPANY
San Francisco

Library of Congress Cataloging in Publication Data

Chapman, Stephen R
   Crop production: principles and practices.

   Includes bibliographies and index.
   1. Field crops.   2. Field crops—North America.
I. Carter, Lark P., joint author.   II. Title.
SB185.C52   633   75-40318
ISBN 0-7167-0581-8

Copyright © 1976 by W. H. Freeman and Company

No part of this book may be reproduced by any mechanical, photographic, or electronic process, or in the form of a phonographic recording, nor may it be stored in a retrieval system, transmitted, or otherwise copied for public or private use, without written permission from the publisher.

Printed in the United States of America

9 8 7 6 5 4 3 2

Contents

Preface vii

Part One Principles 1

1. Plant Science and Human Welfare 3
2. Cell Structure and Function 15
3. Vegetative Growth and Development 39
4. Reproductive Growth and Development 63
5. Plant Classification and Nomenclature 86

Part Two Plants and Their Environment 95

6. Soils and Crop Production 97
7. Moisture and Crop Growth 122
8. Light and Temperature 146
9. Insects, Diseases, and Weeds 164
10. Cropping Systems and Resource Conservation 194

Part Three Production Practices 213

11. Marketing Crop Products 215
12. Seeds and Seeding 227

13	Cereals and Man	247
14	Corn and Sorghum	259
15	Rice	281
16	Wheat	291
17	Barley	311
18	Oats and Rye	325
19	Edible Legumes and Man	337
20	Soybeans	345
21	Peanuts	359
22	Field Beans and Peas	371
23	Cotton	383
24	Tobacco	397
25	Sugar Beets and Sugarcane	411
26	Potatoes	431
27	Safflower, Rapeseed, Flax, and Sunflower for Oil	443
28	Forages and Man	463
29	Forage Legumes	470
30	Forage Grasses	493
31	Forage Management and Utilization	508
32	Progress Through Research	523
	Glossary	539
	Index	558

Preface

Man is dependent on field crops for food, for fiber, and for feed for his domestic animals. As world population increases, the need to produce more from each acre and the need to preserve and protect the environment and natural resources that make such production a possibility have become apparent not only to the professional agriculturalist and other scientists, but to the general public as well. To meet the challenge of increased productivity, producers must understand the fundamental principles of crop growth and how to apply the most efficient cultural practices.

Successful crop production depends on understanding how crops develop and grow, how various factors affect crop growth and development, and how each factor can be modified or managed. In this book, we first discuss the basic anatomic, morphologic, and physiologic features of field crops. Next, the major factors that affect plant development and growth, and the nature of their effects are introduced. Finally, modern production practices for major crops and groups of crops are considered in terms of the underlying principles presented in the first two parts.

Because the book covers more material than would normally be covered in a single course in plant science or crop production, it can readily be adapted to meet the needs of different groups of students. The first five chapters (Part One) are suitable for a review of fundamental botany or, in association with selected in-depth studies of the various crops considered in Part Three, for a course in applied botany: practical examples can be drawn from any of the production chapters. Part Two, in which we consider those factors that affect how and where crop species grow, can be used in conjunction with either the first or the last part. Part Three consists of discussions of actual production practices, in terms of plant biology, for the major American and Canadian field crops: separate chapters cover single or closely related crops.

Selected references are given at the end of each chapter. These have been chosen to present the subject matter of the chapter in greater depth and usually from a different, but not necessarily opposite, point-of-view. The study questions have been designed to encourage independent thought, not memorized answers.

Our general philosophy has been that modern, highly productive farming practices depend on an understanding of the principles of plant growth. When the principles are grasped, the practices follow logically.

It is impossible to thank individually the many contributors to this book. We are deeply indebted to the entire staff of the Department of Plant and Soil Science, and to other colleagues at Montana State University for advice and criticism. In addition, we express our sincere appreciation to Ralph L. Obendorf, Cornell University, and Donald G. Woolley, Iowa State University, who twice reviewed the entire manuscript and made uncounted valuable additions and corrections. We are also indebted to K. W. Clark, The University of Manitoba, and Robert P. Patterson, North Carolina State University at Raleigh, who reviewed some of the production chapters. Mary Cline had the patience and stamina to type the full manuscript.

Finally, we express our deepest thanks to our families—Barbara Chapman and daughters Cynthia and Cheryl and Jean Carter and daughters Nancy and Anne—for their patience, understanding, and unceasing encouragement without which we would not have attempted, let alone completed, *Crop Production: Principles and Practices*.

Stephen R. Chapman
Lark P. Carter

August, 1975

CROP PRODUCTION

Part One   Principles

One

Plant Science and Human Welfare

Throughout human history, people have been divided into the "haves and the have-nots." Today, entire nations fall into one or the other of these categories with respect to food for an ever-increasing population. A "have-not" nation is one that does not produce enough plant products to supply a reasonable diet for its people. Animal products also play a critical role in human nutrition, but animals are dependent on plants. The dependency is in fact on a complex phenomenon called photosynthesis, which is found almost exclusively in green plants. With the exception of several algae and a few bacteria, no organism other than a green plant is autotrophic: no other organism can convert nonnutritious inorganic material into nutritious organic material. To alter its status, a "have-not" nation must make its land green by developing an effective system of crop production.

The significance of green plants is best explained in terms of the use, storage, and loss of energy by all forms of life. Energy is required to maintain organization, from the simple organization of the subatomic particles of an atom to the complex, multilevel organization of millions of cells in an adult human. Without energy, there can be no life.

Energy enters the ecosphere in a multitude of forms, but solar radiation (sunlight) is by far the most important, fundamental source. Of the total energy entering the ecosphere, more than 99.9 percent is solar in origin. Before humans can use solar energy to satisfy their physiological needs, it must be converted into usable forms through photosynthesis.

The amount of solar energy reaching the earth's atmosphere per year is staggering: 1.3×10^{23} gram calories, or a continuous

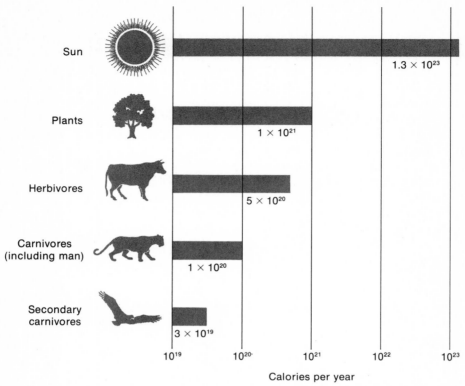

Figure 1-1
The utilization of solar energy decreases with each step in the food chain. Bars are plotted on a logarithmic scale, based on the fact that plants use only 0.08 percent of the solar energy that reaches the earth's atmosphere. [From The ecosphere by Lamont C. Cole. Copyright © 1958 by Scientific American, Inc. All rights reserved.]

daytime supply of two and a half billion billion horsepower. Of this quantity, less than 0.1 percent is used directly to produce food. Nearly all the energy used for food production for all forms of life enters the food chain through photosynthesis.

The conversion of solar energy into food (plant materials) is the foundation of the food chain. Plant materials are consumed by plant-eating animals (herbivores) and herbivores are consumed by meat-eating animals (carnivores). In each step, the amount of energy transferred is reduced (Figure 1-1). For example, green plants use approximately 16 percent of the energy that they trap from solar radiation for plant growth and reproduction; the rest is stored in plant tissue, part of which can be utilized directly by some kinds of animals. Green plants are primary producers of food and herbivores (e.g., cattle) are primary consumers. Carnivores, including man, which feed on herbivores, are secondary consumers.

To illustrate how the amount of solar energy available to the various consumers in the food chain is reduced with every step in the chain, let us consider the utilization of solar energy that is converted into grass, which is then eaten by a cow that in turn is consumed by man. The cow can utilize about 50 percent of the solar energy stored in the grass; the other 50 percent is "lost" in digestive processes. Of course, not all of the energy from the plant becomes beef on your table. Much of the energy that the cow obtains from its feed is used for life processes. Only 20 to 30 percent of the energy available is used to produce new tissue (meat). Thus, considering that only 50 percent of the plant's energy can be utilized by the cow (much of which must be used to sustain the animal), no more than 10 to 15 percent of the solar energy stored in the plant tissue is ultimately converted into the meat that is eaten by man. A similar energy "loss" takes place when a herbivore is consumed by a carnivore. A carnivore converts about 20 percent of the energy in the tissue of a herbivore into new tissue and uses as much as 70 percent of the energy stored in the herbivore for normal activities other than growth. The rest of the energy stored in the herbivore is not converted for any use by the carnivore.

The "loss" of energy in going from the primary producer to the primary and then to the secondary consumer takes place only in the chain under consideration—in this case from plant to cow to man. Energy is not lost: animal wastes, plant debris, and similar materials become parts of other food chains. If all life forms are considered, the food chain becomes part of a continuous energy cycle. Although a plant scientist usually considers the food chain to start with a terrestrial (crop) plant, minute, aquatic plants are also treated as the base of a food chain.

The maximum potential for food production is effectively impossible to estimate accurately. Losses in the food chain, discussed above, preclude a pure meat diet for humans. Such a diet is in no way realistic for other reasons, including man's need to consume certain plant products directly. On the other hand, if man were to change his eating habits and large quantities of aquatic plants were incorporated into the human diet, less than 5 percent of the earth's productivity would be required to feed its population. The current nutritional problem seems to be in feeding people a diet based on crop plants as primary producers.

In 1969, Paul Ehrlich predicted that all of the world's people would soon be "have-nots:" the population growth rate is so great that mass starvation is a real and immediate threat (Ehrlich 1972). The rate of world population growth is in fact a matter of grave concern. The pattern is logarithmic: $2 \rightarrow 4 \rightarrow 8 \rightarrow 16$, and so forth (Figure 1-2). The world's population approximately doubles every twenty-five years (a human generation, or the span of time between the birth of parents and the birth of their children), which means that each day there are nearly 100,000 more people to be fed. Currently, the world's population is nearly 3 billion. Although the rate of growth in the United States and elsewhere is below that projected ten years ago, by 1990 the population could approach 6 billion: 3 billion more people to be fed. It is reported that from three to four of every five deaths in the world can be traced in part or exclusively to malnutrition. About one-half of the world's people subsist on a diet of less than 1500 kilocalories per day; by FAO standards, a balanced diet of 2200 kilocalories per day is a suggested minimum diet required by a working man. A balanced diet is nearly as critical as the caloric content. Nearly 35 percent of the average North American diet is animal protein (meat, poultry, fish, milk, eggs, cheese, etc.), 40 percent is cereal (wheat, barley, and oats), and approximately 25 percent consists of vegetables, fruits, and other luxury items. In Asia, only about 8 percent of

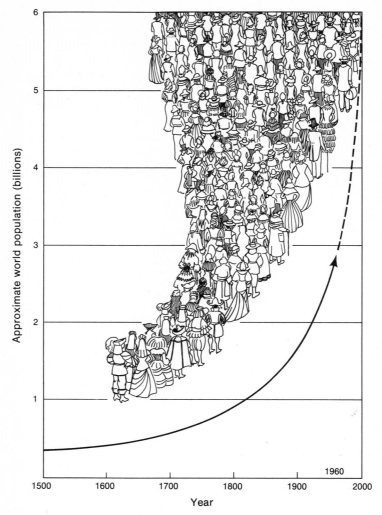

Figure 1-2
Pattern of population growth for five centuries, depicting the population explosion.

the diet is animal protein and more than 75 percent comes from cereals (mostly rice), as shown in Figure 1-3.

The inclusion of relatively large proportions of animal protein in the diet—common in North America but rare in Asia—reflects national wealth in terms of agricultural development of arable land. Land is available to produce feed for livestock, which is then consumed by man. Feeding plant materials to animals is an inefficient intermediate step in converting solar energy through photosynthesis into human nourishment. Only those nations having abundant agricultural re-

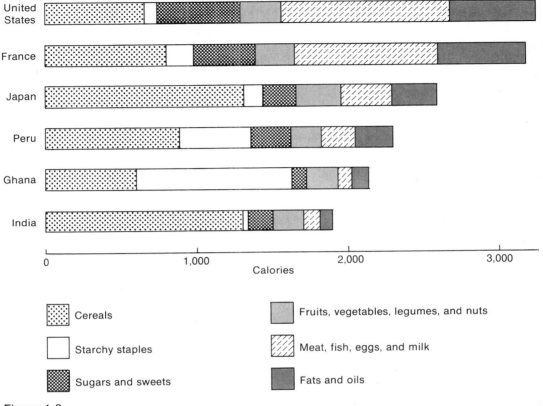

Figure 1-3
Caloric content and composition of diets in six countries. Note that all diets have essentially the same components, but the proportions vary. Less-developed, heavily populated countries are highly dependent on cereals. [From *Seed to civilization* by Charles B. Heiser, Jr. W. H. Freeman and Company. Copyright © 1973.]

sources in terms of arable land and sophisticated production technology can afford this.

The shortage of food will become greater in the future because there will be less room in which to produce it. Currently, in the United States there are about 2.8 acres (1.1 hectares) of land per person. According to some plant scientists, it takes about 2.5 acres (1.0 hectare) to produce a well-balanced diet of 2500 kilocalories by present methods of agriculture. With a U.S. population of 400 million in the year 2000, there would be about 1.2 acres (0.5 hectare) per person. The amount of land available is not the only criterion for judging agricultural resources. Compared with the United States, Canada is rich in available land: the population density in Canada is slightly more than 5 persons per square mile; in the United States, it is about 55 per square mile. Yet, a greater diversity of crops can be produced in the United States than in Canada because of more favorable climatic conditions in some regions—for example, in the corn and cotton belts. In much of Asia, however, the amount of land available (an acre or less per person) is the major limiting factor in food production.

Concern about the shortage of food is not new to man. In the late eighteenth century, Thomas Malthus, a British economist, made predictions quite similar to those currently being made about population growth and the world's food supply. Malthus noted that the rate of population growth was far greater than the increase in food production and suggested that these factors would soon lead to mass starvation. Malthus's predictions were not entirely correct. Food productivity has accelerated as a result of scientific discoveries and advancing technology; in addition, more land has been brought under cultivation. Unfortunately, today there is little additional land that can easily be brought under cultivation. If mass starvation is to be prevented and the current problems of human hunger and malnutrition are to be reduced, the plant scientist must continue to improve crop yields and quality through the development of new cultivars, production practices, and technology (including ecologically safe pesticides). At the same time, it seems apparent that man must also regulate and reduce the rate of population growth.

It would be simple, and not wholly inaccurate, to explain man's past success in developing food supplies in terms of his inventive genius. However, this disregards several factors that are of primary importance in increasing the world's agricultural productivity.

Research in archaeology and in many other fields indicates that farming (plant domestication) had its roots in the fertile valleys of the Tigris, Euphrates, and Nile rivers, where the earliest records of farming date back more than seven thousand years. There is good evidence that "farming" may have been started at the same time in China; the crop producer in ancient China was a man to be highly respected, inferior in class only to civic officials and priests. Primitive crop production was first practiced in Central and South America two thousand years ago.

Early man was a hunter and a gatherer. He was nomadic, following herds of wild animals, which provided much of his diet, and gathering fruits and berries when they were available. Early man must have experimented with the plants that he came across in his wanderings. By trial and error, he identified those plants of greatest value to him and found that the seed of such plants could be saved and planted to produce more plants, thereby insuring a food supply. With the ability to produce plants where desired, wandering in pursuit of game was replaced by primitive farming. With feed available, livestock domestication also became possible. Thus, man settled down, leading to the development of early societies and cultures (Figures 1-4 and 1-5).

Man used plants everywhere he wandered; yet farming seems to have originated in fairly isolated areas—a result of the natural availability of plants amenable to cultivation in various parts of the world. Although many of the steps in tracing plants to their origins are still missing, there is conclusive evidence that major crop species originated, or evolved, from wild ancestors in specific areas that are now known as centers of origin. The origin of wheat has been studied in great detail. Cultivated ancestral forms of what today is common wheat (*Triticum aestivum*) tracing back to 7500 B.C. have been identified in an area extending from eastern Turkey to Pakistan and Kashmir. Other cereal crops, including barley and oats, also originated in this area, the location of the earliest known farming communities. Evidence of the earliest settlements in the New World strongly suggests that corn originated in Central and South America (as did lima beans).

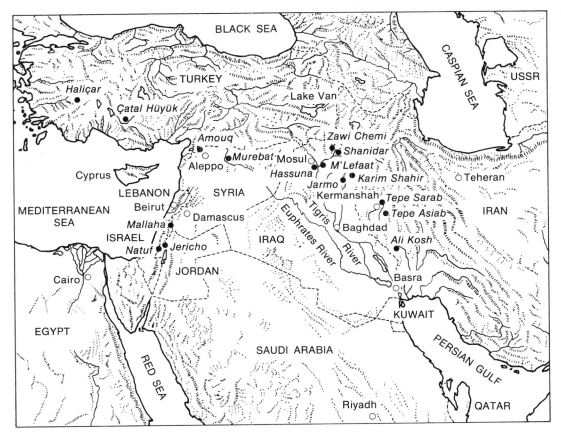

Figure 1-4
Sites of primitive farming and early farming communities (indicated by the solid dots) in the Near East. Note the large number of sites flanking the Fertile Crescent, which lies between the Tigris and Euphrates rivers. These sites are also known to be the centers of origin of cereal crops. [From *Seed to civilization* by Charles B. Heiser, Jr. W. H. Freeman and Company. Copyright © 1973.]

Centers of origin are characterized as having climatic conditions that favor the survival of the widest array of forms and types of crop plants. Early farmers unwittingly became plant breeders by preserving the seeds of plants most suitable for their needs in preference to those of less-suitable plant types. Even today, centers of origin are important to plant breeders because of the diversity of plant types that may possess valuable traits to be used in plant breeding.

International, national, and private organizations sponsor plant-exploration teams to continue the task of identifying and evaluating the centers of origin of cultivated crops.

In the earliest days of farming, crop production consisted of seven basic steps: (1) gathering and preserving seeds for planting, (2) destroying the original vegetation on fields that were to be planted, (3) stirring the soil to prepare seedbeds for planting, (4) planting seeds at the appropriate time of the year, (5)

Figure 1-5
Sites in Central and South America similar to those illustrated in Figure 1-4. These sites are known to be the centers of origin of corn and lima beans. [From *Seed to civilization* by Charles B. Heiser, Jr. W. H. Freeman and Company. Copyright © 1973.]

destroying weeds, (6) protecting crops from various predators, and (7) gathering, processing, and storing crops. These same steps are the basic steps in crop production today. Science and technology have made each of the steps easier, but have not entirely eliminated any of them.

Primitive agriculture evolved with the advent of other scientific and cultural events. The appearance of the plow in approximately 3000 B.C. was a landmark in the development of crop cultivation. The replacement of the man-drawn plow by an ox-drawn one was another dramatic advance. In the metal ages, "technology" greatly enhanced the development of farming implements—for example, the metal sickle replaced the stone one used in harvesting crops.

Records of agriculture in general are fairly complete for the era of the ancient Greeks, although from the eighth century B.C. through the first century A.D., little is found on agriculture in the writings of the Greek scholars and philosophers.

The Romans were basically agrarian. Although farming practices changed little from about 300 B.C. to 400 A.D., the Romans contributed significantly to agriculture through extensive writings on various aspects of plants and crop production. The precedent for many contemporary land laws and for price support can be traced to the Romans.

Modern agriculture had its beginning in England early in the eighteenth century. In 1701, the grain drill was invented by an English gentleman-farmer named Jethro Tull. In 1733, Tull published *Horse-hoing Husbandry*, which established the principles of row-crop cultivation. England remained a leader in improving crop production from 1700 to 1850. However, other nations contributed to the technology of crop production as well. In France, economist François Quesnay stressed the importance of agriculture to the well-being of the nation. Benjamin Franklin brought many of Quesnay's ideas to the United States where they were ultimately embodied in legislation that protected agriculture. Justus von Liebig, a German, published a guide to agricultural chemistry in 1840 entitled *Chemistry in its Application to Agriculture and Physiology*; a second volume, *The Natural Law of Husbandry*, was published in 1862.

In the early days of the United States, less than 10 percent of the population was urban. Leaders of the young nation either were gentlemen-farmers themselves or were closely associated with farming and its problems. Individual leaders like Benjamin Franklin spoke out loudly in support of agriculture.

During the American Revolution, when national unity was the prevailing mood, farm organizations sprang up throughout the growing nation. The first groups were formed to protest colonial regulations; later they protested low prices. Shortly after the war, in 1785, two organizations focusing on agriculture were founded: the Philadelphia Society for Promoting Agriculture and, in Charleston, South Carolina, the Society for Promoting and Improving Agriculture and Other Rural Concerns. These groups numbered among their members national leaders of the day and were forerunners of agricultural organizations in existence today.

Agricultural fairs had been held in many parts of the world for centuries. The first producer-centered fair to be held in the United States was organized in 1811 by Elkahana Watson, a Pittsfield, Massachusetts, farmer. State governments sponsored an increasing number of such fairs until about 1825, but from then until after the Civil War interest diminished. Then, shortly after the war, attention once again turned to the social and educational importance of the agricultural fair. At the same time, new agricultural groups were formed, including the Grange in 1867, the Farmers' Educational and Cooperative Union in 1902 (today one of the largest farm organizations), and the American Farm Bureau federation in 1919. Marketing and local special interest groups developed at this time as well.

The federal government was not standing still. In 1862, the U.S. Department of Agriculture was established. In the same year, two federal laws that were to have far-reaching effects on the growth of agriculture were

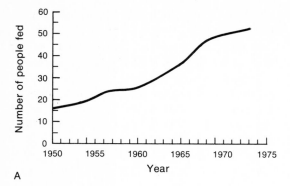

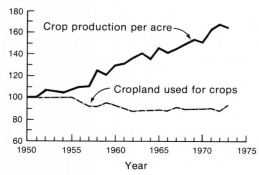

Figure 1-6
A. Number of people fed per farm worker from 1950 to 1973. B. Relationship between crop yield per acre and acres farmed for the same period. [USDA data.]

passed: the Morrill Act, which led to the establishment of land-grant colleges, and the Homestead Act. In 1914, the Cooperative Extension Service was authorized; in 1935, the Soil Conservation Service was founded.

In Canada, the impact of legislative action on crop production was felt in the early 1840s. Laws favoring wheat and corn produced or processed in Canada were passed by 1843. Today, Canadian agricultural research receives the support of local, provincial, and national governments, in addition to private support. Agricultural practices are also regulated by the government in Canada.

What has contributed to the progress that has been made in agriculture? Two hundred years ago, a farmer could produce only enough food for a family of four. As recently as a quarter of a century ago, a single farm worker produced enough for four families of four. Today, he produces enough food—of higher quality—for more than ten families of four (Figure 1-6). The doubling of corn yields in the past thirty years, or, for that matter, the substantial increase in rice production in India since 1967 (making India almost self-sufficient), can no more be attributed to mere chance than the development of the cotton gin or of the mechanical tomato harvester. It took more than trial and error to develop insecticides and weed killers. Years of dedicated effort have yielded superior fertilizers, better irrigation systems, more effective ways to store and process crops, and crop cultivars that are well suited to specific areas and resistant to diseases and pests. Since the beginning of the eighteenth century, science has played a leading role in the modernization of agriculture. Nearly every physical and biological science has contributed to its development, including botany, zoology, physiology, chemistry, and mathematics.

Merely to identify the scientific fields involved with progress in agriculture does not explain the progress. A simple definition of science, "a branch of knowledge which deals with systematically arranged facts which show the operation of natural laws," is at least a starting point, but what are facts and natural laws and how are they established?

Scientific research is a continuing process. As new facts are verified, they are added to a systematic arrangement of other related facts. The accumulation of facts can then be used to develop new facts. Agriculture's growth is a result of both the development of new facts and the application of known facts. The scientific method is an accepted approach to establishing facts, but not a guarantee. To establish new facts and new applications of known facts by means of the scientific method, the following steps must be taken: (1) recognize the problem; (2) develop one or more hypotheses for its solution (by using known facts); (3) test the hypotheses, (4) identify the best hypothesis based on the test results (it then becomes a theory), and (5) continue to test the theory under a wide range of conditions. If the theory is proven to be valid over a long period, it becomes the basis of a natural law.

Crop producers, like scientists, can use the scientific method as a guide to solving problems. Although the crop producer obviously omits several of the steps in practice, he should at least consider each one. For example, a farmer knows that his crop yields are too low; he wants to increase them in order to raise his standard of living or to meet market demands. He must now develop hypotheses for increasing yields and then, in some way, test these hypotheses. Some of the hypotheses he develops may not actually be tested; from his own experience or for other reasons, he knows they are impractical. For example, cultivating more land may be a suitable method of increasing farm productivity, but, if land is not available or if adequate equipment and labor are not available, the hypothesis to cultivate more land to increase yield is not worth testing. Other, more practical hypotheses might be (1) to plant a different cultivar, (2) to use a new fertilizer, or (3) to combine the two—planting a certain, new cultivar and applying a specific amount of a particular fertilizer.

These hypotheses can then be tested. The tests consist of field experimentation followed by moderately sophisticated statistical procedures, which may be essential for evaluating the hypothesis (see Chapter 32). Field experimentation includes planting plots of various new cultivars and applying several combinations of fertilizer to each cultivar. The yield of each plot is determined, and appropriate statistical procedures determine which combination or combinations of cultivar and fertilizer will result in the highest yield. On the basis of the tests, the producer can establish a theory: a particular combination of cultivar and fertilizer is superior to all others tested. Because plants react differently to environmental conditions, the results obtained in a given year are not necessarily the same the next year. The entire experiment may have to be repeated for several years before the hypothesis becomes a theory. Ultimately, the theory is put into practice. At this point, the theory may be the basis for "a natural law" for the producer, although only if it proves to have general application elsewhere can it be considered to be a natural law in the scientific sense.

The critical steps in applying the scientific method to a specific problem are (1) to identify reasonable hypotheses and (2) to devise experiments that will establish which ones are valid. Certainly experience and education are essential to both steps. The objectives of this text are to present the important principles of plant growth and current practices of crop production in North America so that producers may become better able to develop ways in which to obtain the maximum yield. The importance of increasing crop yields can be traced through the literature and history of mankind. More than two hundred years ago, Jonathan Swift expressed it clearly in *Gulliver's Travels*: "Whoever could make two ears of corn or two blades of grass to grow on a spot of ground where only one grew before would deserve better of mankind and do more essential service to his country than the whole race of politicians put together."

Selected References

Cole, LaMont C. 1958. The ecosphere. *Sci. Amer.* 198(4):83–92. Available as *Sci. Amer.* Offprint 144.

Ehrlich, Paul R. 1972. *The population bomb*, rev. ed. Ballantine.

Heiser, Charles B., Jr. 1973. *Seed to civilization*. W. H. Freeman and Company.

Commission on Population Growth and the American Future. 1972. *Population and the American future*. New American Library.

Study Questions

1. Until about 1950, what factors prevented predictions of mass starvation from being realized?
2. What are the major weapons currently being used in the war on hunger?
3. Would the production of high-calorie food eliminate the world's hunger problems?
4. What are the apparent relationships between the centers of origin of crops and early farming and society?

Two

Cell Structure and Function

No manufacturer would be so naive as to start operating an assembly line without a thorough understanding of how it works and what is required for its efficient operation. To do so would invite chaos and, more than likely, financial ruin. To a crop producer, the green plant is an "assembly line." To guard against possible ruin, he must understand how plants grow and what factors he can control to foster efficient growth. Thus, a background in crop botany is essential for modern crop producers. In recent years, botany, like so many other sciences, has gone through a period of rapid growth and specialization; crop botany comprises many specialized subjects of botany, including plant morphology, the study of the form of plants, with emphasis on their external parts; plant anatomy, the study of their organs; plant physiology, the study of the living chemistry of plants; plant taxonomy, the study of scientific classification or naming of plants; and plant pathology, the study of their diseases.

Living organisms have many important features in common: they all have a life cycle and certain basic requirements for completion of the cycle. The distinction between major groups of organisms such as plants and animals can be made according to how the organisms meet these requirements. The life cycle of a plant begins with a seed and ends when another seed is produced. Depending on the plant, this may take less than a year, two years, or several years. Many important crop species are of the type in which the same individual plant produces seed for many years. Plants that produce seed and die in a single year or less (such as wheat, oats, barley, and corn) are called annuals; those that require two years to produce seed (such as sugar beets) are biennials; and those that produce seed indefinitely (such as alfalfa and many other forage crops, as well as fruit trees) are perennials. Among the annual crops, a distinction of agricultural importance is the difference between winter annual and summer annual species: a winter annual, such as winter wheat, is planted in the fall and matures the following spring or summer; summer annuals are planted in the spring or early summer and mature the same summer or in the fall. Regardless of its growth period—one year, two years, or more—the life cycle of a plant can be traced from seed to seed.

Each type of crop has specific requirements at different stages of its life cycle. Recognizing the stages of growth and development and determining (and fulfilling) the needs of the plant at each stage enables a crop producer to obtain the highest yield in the most efficient manner. Consider the life cycle of any annual crop. Starting with the seed, it is essential to know how to determine its quality. If poor seed is planted, a poor crop can be expected. Given high-quality seed, the time at which it is planted, the method used, and the amount to be planted for maximum yield depend on specific environmental factors such as light, temperature, and moisture. An understanding of the effects of these factors enables a crop producer to make the right decisions in modifying them: for example, irrigation can supplement an inadequate supply of water or planting can be delayed until the weather becomes warmer if the soil temperature is too low.

As the seed germinates and the seedling emerges, the plant's requirements change. All plants require specific minerals in precise combinations for maximum growth. If the right kinds are not present in the soil in the right amounts, fertilizers must be added. Also, seedlings differ in their responses to stress caused by inadequate moisture and high or low temperatures. Knowing how plants respond to different factors and knowing the relative importance of these factors will help in making wise decisions in the field management of a crop. For example, spring wheat can be planted in relatively cool soil (4°C) and it will grow; corn requires much warmer soil for germination (13°C or above). Because wheat has smaller seed than corn, corn can be planted more deeply than wheat.

As the plant develops from a seedling to the juvenile or vegetative stage, its needs change. In the early stages of development, a plant needs both nitrogen and phosphorus, but the balance is critical: phosphorus is essential for root growth and a surplus of nitrogen may cause excessive leaf growth. As the plant reaches the vegetative stage, its demand for nitrogen increases. Once again, a producer must know what a plant requires and the effects of not meeting these requirements. Either deficiencies or excesses of water or minerals may affect plant growth and development unfavorably and reduce yields.

As the plant reaches sexual maturity, a producer must understand what affects the production and quality of its seed. For example, cereals must have the right amount of moisture both during and after flowering to insure plump kernels. If grain with a high protein content is desired, a large amount of nitrogen in the soil is necessary during seed development. During flower development, either excessively high or low temperatures are damaging to many crops. In protecting a crop from temperature extremes, a crop producer has to take into consideration the crop itself, the facilities available, and the potential damage.

When the plant reaches maturity, the product to be harvested—fruit, seed, foliage, or root—dictates the harvesting method to be used to obtain the maximum yield of the highest quality. Harvesting green, immature seeds reduces grain quality, but waiting too long may result in excessive losses due to shattering. In addition, equipment must be adjusted for the crop and local conditions.

If the plant is a biennial or perennial, the impact of current management on yields in subsequent years must be considered. A producer must understand what the plant requires not only to survive but to thrive from year to year. Overgrazing a range or pasture causes yields to be lower in subsequent years; however, yields can be maintained with irrigation, fertilizers, and reasonable grazing management. Even after harvesting, an understanding of crop biology is important. The storage and processing of crops depends on the crop and on local conditions. Hundreds of tons of hay are wasted annually owing to spoilage resulting from improper storage.

All crops must be protected from an alarmingly broad range of pests. The identification of insects, diseases, and weeds, an evaluation of the damage they might cause, and knowledge of the ways in which control measures might affect crop growth determine the methods to be used. For example, many chemical weed killers (herbicides) can be used to control broadleaf weeds, such as mustard or thistles, growing among cereals, because cereals are not adversely affected by the chemicals. On the other hand, safflower is severely harmed by herbicides such as 2, 4-D.

Cytology

Cells are the basic building blocks of plants. Many of the processes typical of entire plants are also typical of a single cell, where they can be more readily examined. The detailed study of cells is called cytology. Although plant scientists should have a thorough understanding of the complexity of cells, only the eight cell parts of most concern to agriculturalists will be considered here.

CELL WALL

If a plant cell is examined under relatively low magnification without the aid of special staining, the most evident structure is the cell wall. Although some scientists suggest that this wall is not actually a part of the living cell, for present purposes the cell wall will be considered to be a part of the living cell; it is typical of all plant cells. The physical appearance of the cell wall belies its structural complexity, which is diagrammatically illustrated in Figure 2-1. In describing its structure, it is convenient to consider the wall that separates two, adjacent cells. There are at least three distinct layers: The middle lamella is the layer in the center; it consists of a cementing material containing calcium pectate and serves to hold adjacent cells together. The primary cell wall of each adjacent cell lies on either side of the middle lamella and is composed of cellulose fibers. Cellulose is a complex carbohydrate consisting of polymers of glucose, built from carbon, hydrogen, and oxygen. As cells differentiate and specialize, their walls change markedly. Secondary cell-wall material may be deposited just inside the primary cell wall. Secondary cell-wall materials include lignin, which makes the wall harder and stronger, and waxes (suberin and cutin), which make it waterproof. The location and arrangement of lignin deposits are used to identify various types of tissues. Waxes are also found on the outer surfaces of many of the exterior cells of leaves and stems.

The major function of the cell wall is to support the cell. The size and shape of the mature cell are ultimately defined by the wall. The cell wall also supports the entire plant, cell by cell. The arrangement of the cellulose fibers of the primary cell wall and of lignin and other materials in the secondary wall

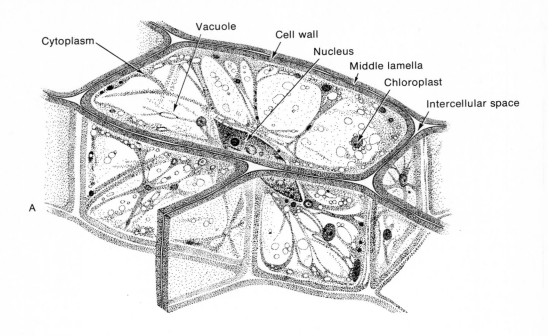
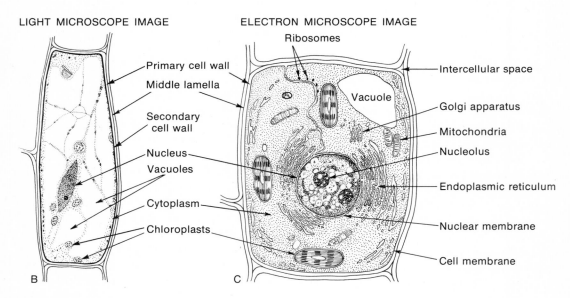

Figure 2-1
Three views of a general plant cell with major parts identified. [Parts A and B after *Plant anatomy,* by K. Esau. Wiley, 1953. Part C after The living cell by J. Brachet. Copyright © 1961 by Scientific American, Inc. All rights reserved.]

account for the remarkable strength of the cell wall.

The interface between cells is not smooth and continuous. Intercellular spaces, generally located at the "corners" of a cell, allow for the movement of solutions and gases.

In addition to being of botanical interest, the cell wall is of real agricultural interest. Cotton fiber, a leading cash crop in the United States, is cell-wall material. The digestibility and the palatability of many forages are affected by the substances in the cell walls; lignified material is difficult to digest. The presence of waxes in cell walls reduces water losses. Damage to cell walls allows disease-causing organisms to enter the cells.

CELL MEMBRANE

The cell membrane, or plasmalemma, lies immediately inside the cell wall. This membrane, which is generally present in both plant and animal cells, is frequently considered to be the outer limit of the living cell. (Cells have many other membranes that are quite similar to the plasmalemma in basic composition and structure). The cell membrane is extremely thin, 75 Å across,* and cannot be seen with the standard light microscope. It can be seen with the electron microscope, and both its structure and its chemical composition have been the subject of diligent research in the past two decades.

The core or backbone of the membrane consists of a lipid layer in association with proteins. (A protein is a group of linked amino acids, and all amino acids contain nitrogen.) The arrangement of the proteins and lipids in the membrane is not fully understood. One model suggests that the lipids are sandwiched between an upper and a lower layer of proteins. A more recent model suggests that the proteins are globular and are either on the surface of the lipid layer (extrinsic) or partly or fully embedded in it (intrinsic). Furthermore, the degree to which the intrinsic proteins penetrate the lipid layer depends on the activity or function of the membrane. Figure 2-2 shows the two types of structures indicated by these models.

The primary function of the cell membrane is to regulate the flow of materials into and out of the cell. Various elements and compounds are found in different concentrations in the cell. At times, these concentrations seem to be in violation of physical laws which say that they should be equal both inside and outside the cell. How the cell membrane allows the laws of diffusion and osmosis to be violated is not known. Rarely can a cell survive if the cell membrane is broken and not repaired. Other membranes are vitally involved with the functions of cellular organelles such as mitochondria and chloroplasts.

CYTOPLASM

The cell membrane surrounds and serves to contain a highly developed protein matrix, or gel, called the cytoplasm. The cytoplasm contains not only a myriad of cell parts whose functions are essential for cellular life, but also many enzymes that are equally essential for cellular life.

VACUOLE

The vacuole is the single largest internal body in maturing plant cells. It is not an empty space, as its name implies, but is a region surrounded by a membrane similar to the cell membrane, in which various substances can be stored for future use. In addition, certain water-soluble pigments—such as anthocyanins, which give the characteristic red color to tomatoes and some kinds of apples—ac-

---

*An angstrom, Å, is 1×10^{-7} mm or about 0.00000025 inch.

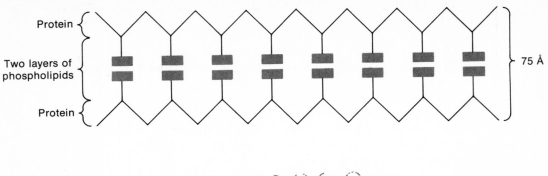

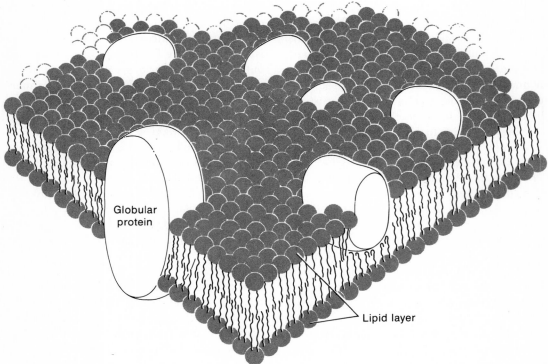

Figure 2-2
Models of the structure of cell membranes: (top) the unit membrane with static structure; (bottom) globular proteins on the surface of or inside the phospholipid core. [Bottom model after S. J. Singer and G. L. Nicolson. *Science* 175(1972):723. Copyright 1972 by the American Association for the Advancement of Science.]

cumulate in the vacuole. As osmotic forces cause the vacuole to increase in size, pressure is exerted on the entire cell, causing it to elongate or grow.

NUCLEUS

The nuclei of plant cells (and animal cells) can be seen with a light microscope if proper staining is used. During most of the cell cycle, the nucleus is surrounded by the nuclear envelope, a double membrane structure with "pores." Within the nuclear envelope is a watery substance called the nuclear sap, or karyolymph, which is to the nucleus what the cytoplasm is to the entire cell, and chromosomes. Within a chromosome is a very long linear molecule of deoxyribonucleic acid (DNA), segments of which are the genes or units of inheritance that direct the functions of the cell. The DNA contains information, written in a genetic code, for sequences of events by which proteins are made. The importance of protein becomes apparent when we consider that nearly all cellular functions and activities are regulated by enzymes; enzymes are proteins and their production is thus dependent on DNA. Chromosomes divide in a precise, highly regulated manner to insure a continuity of life. During certain stages of cell development, chromosomes can be clearly observed with a microscope. From such observations several significant facts have been discovered. With the exception of sex, or reproductive, cells, within any plant there is a specific, constant number of chromosomes in all nuclei; the number is the same for all plants of the same species. For example, barley has fourteen chromosomes; wheat, forty-two; and corn, twenty. The twenty chromosomes in corn can be grouped into ten pairs. The members of each pair are referred to as homologous chromosomes. In the absence of rare abnormalities, homologous chromosomes are structurally and morphologically identical. The significance of this and the behavior of chromosomes during cell division will be explored more thoroughly in the discussion of mitosis and meiosis.

MITOCHONDRIA AND CHLOROPLASTS
Two other cell structures are the mitochondrion and the chloroplast. Both play major roles in energy transformations in the cell.

Mitochondria, present in varying numbers in both plant and animal cells, are the sites of oxidative respiration. The mitochondrion is surrounded by two membranes: the inner one folds in and out, but the outer one is more or less smooth (Figure 2-3). Each mitochondrion is rich in a series of enzymes required to break foodstuff (carbohydrates, fats, etc.) down to carbon dioxide (CO_2) and water, releasing energy for maintaining the cell.

Chloroplasts are common only to green plants and are found mainly in certain cells of leaves and stems. They are surrounded by two membranes and have a complex interior structure, where chlorophyll is located (Figure 2-4). Chlorophyll is the substance that gives a green plant its color; it is capable of absorbing light energy, which is ultimately converted into foodstuff through a complex set of chemical reactions. These chemical reactions make up the process of photosynthesis, the only process through which raw nonfood material (inorganic nonnutritious compounds) can be converted into the food material that provides energy for all animal cells and for plant cells that do not have chloroplasts.

ENDOPLASMIC RETICULUM AND RIBOSOMES
The endoplasmic reticulum (ER) is an extensive network of membranes that permeates the cytoplasm in most cells. Ribosomes are often associated with ER, but they are also found in great numbers in many cells independent of ER. A ribosome comprises two roughly spherical subunits composed of ribonucleic acid (RNA) and protein. Whether associated with ER or not, ribosomes play a key role in chemically linking amino acids together to form different proteins, including enzymes. The sequence of amino acids in each protein is specified by the genetic information in chromosomes.

Nearly all of the instructions for cellular activities are carried in chromosomal DNA. Messages are coded in the DNA by specific sequences of the four nitrogenous bases of which DNA is composed. The instructions are transferred to a messenger RNA molecule, which then leaves the nucleus. The message generally carries instructions for the synthesis

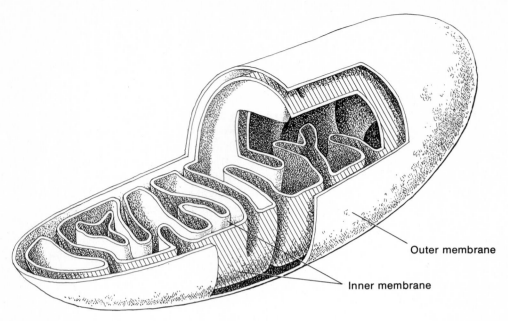

Figure 2-3
Diagram of a mitochondrion showing the two membranes: an unfolded outer membrane and a complexly folded inner one.

of a particular enzyme. Remember, nearly all cellular functions depend on enzymes. At the site of enzyme synthesis in the cytoplasm—that is, at the ribosomes—the messenger RNA associates with another type of RNA, called ribosomal RNA. The amino acids, of which enzymes (and all other proteins) are composed, are moved to the messenger RNA-ribosomal RNA complex by a third type of RNA, called transfer RNA. Transfer RNA picks up a particular amino acid and allows it to join in a chain of amino acids; the sequence of amino acids in this chain is spelled out on the messenger RNA that came from chromosomal DNA.

The concept of the gene has been subjected to a number of redefinitions as our understanding of the biochemical basis of inheritance has increased in the past several decades. For our purposes, a gene is the DNA in a very specific section (locus) of a chromosome, and the enzyme coded or specified by the gene controls the expression of one particular trait. A gene may take slightly different forms in terms of the trait as a result of slight differences in the DNA at its locus. These forms are called alleles of the gene. For example, the control of resistance of a wheat plant to the disease stem rust may be specified at one locus. At this locus two allelic forms may be possible: resistance to the disease (designated r) or susceptibility to the disease (designated R). If two homologous chromosomes both have R, or if they both have r, then the nucleus is referred to as being homozygous at this locus. If on the other hand one homolog has r and the other has R, the nucleus is referred to as being heterozygous at the locus. Of course, there are thousands of loci controlling all aspects of the growth and

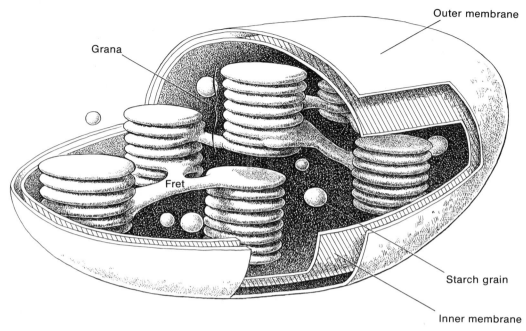

Figure 2-4
Diagram of a single chloroplast showing the double membranous envelope and the complex interior structure. Chlorophyll is associated with the granum.

development of a plant. It is likely that some loci would be homozygous and others heterozygous.

Although the concepts of homozygosity and heterozygosity have been introduced in terms of a single nucleus, they are completely appropriate for an individual plant. The reasons for this will become apparent in the following discussion of nuclear division.

Nuclear Division

The nucleus plays a major role in plant growth and development. Even though a single cell can carry out a wide array of important processes, it cannot carry out all of the complex processes of an entire plant (or of an entire animal). A plant, like an animal, starts its life not as a seed, but as a *single, fertilized egg*, a single cell. This cell grows, increasing in size and volume, and divides into two cells. Thus, plant growth is a result of two processes: one increases the size of cells and the other increases their number. Cell division is contingent on nuclear division, and nuclear division on chromosomal division.

MITOSIS

Cells increase in number by the process of nuclear division known as mitosis followed by cytokinesis, which is the actual division of a single cell into two cells. Cell division is a continuous process; it is divided into phases solely for purposes of discussion. The individual phases of the division cycle are not discrete, but merge one into another.

A cell in which the nucleus is not actively dividing is referred to as an interphase cell, or as being in the resting phase. It is resting only in the sense that the nucleus is not actively dividing. The interphase cell is carrying out the various functions typical of a living cell. The stimuli that cause the cell to proceed from the resting phase to the initial phase of mitosis are not clearly understood. Division seems to depend on cell volume, energy build-up in the cell, and the level of differentiation of the cell. When a cell reaches a certain volume and has built up a large enough reserve of energy, the division process is initiated. However, highly specialized cells tend to divide less than undifferentiated cells.

To understand mitosis, a thorough understanding of the basic parts of the nucleus is essential. Recall that the nucleus is surrounded by a double membrane structure, the nuclear envelope; inside this envelope are the nuclear sap (karyolymph) and the chromosomes. Each cell in an organism, except the highly specialized reproductive, or sex, cells, has the same number of chromosomes. The chromosomes exist as homologous pairs.

Mitosis is generally divided into four phases starting with prophase and continuing through metaphase, anaphase, and telophase (Figure 2-5). Cytokinesis follows telophase and is sometimes treated as part of the mitotic cycle. Although individual chromosomes cannot be clearly distinguished during interphase or in the early parts of prophase, a great deal is known about their structure and function. Late in interphase, a marvelous event takes place: the DNA in each separate chromosome replicates, or duplicates itself exactly. This complex event signals the onset of nuclear division.

During prophase the nuclear envelope starts to fragment, and the spindle or mitotic apparatus is formed. Also, because of twisting and coiling, the chromosomes shorten and thicken and can be clearly distinguished with proper staining under a light microscope. The fibers of the mitotic apparatus attach to a special, constricted region of each chromosome, the centromere. It can be readily seen that each chromosome consists of two chromatids that are identical with each other as a direct result of DNA duplication. The chromosomes, still composed of two chromatids, migrate to the center of the cell and reach the equatorial region, or metaphase plate.

When the chromosomes are aligned on the metaphase plate, the division cycle is in metaphase. The chromosomes continue to become shorter and thicker as a result of coiling as the dividing nucleus approaches metaphase. The forces that cause the migration of the chromosomes are not fully understood, but the contractile protein tubules that are a part of the mitotic apparatus appear to play a role in chromosome movement. Because of the attachment of the spindle fibers to the centromere, as each chromosome moves to the equatorial plate it takes on a characteristic shape based on the position of the centromere, and homologous chromosomes can be identified. However, the members of a pair of homologous chromosomes are independent throughout mitosis. They do not physically pair, or apparently establish any sort of relationship.

In anaphase, the centromere splits and the chromatids separate. The movement of the separated chromatids from the equatorial region is due to repulsion of the centromeres and may be due in part to contraction of the spindle fibers that are attached to the individual chromatids. Separation and movement of the two chromatids of each chromosome is precise: one chromatid goes to one pole and the other goes to the opposite pole. Once two chromatids are completely separated, they may be thought of as daughter chromosomes that are identical because of the precision of DNA duplication.

At telophase, daughter chromosomes reach opposite poles of the mitotic apparatus. At each pole there is one daughter chromosome

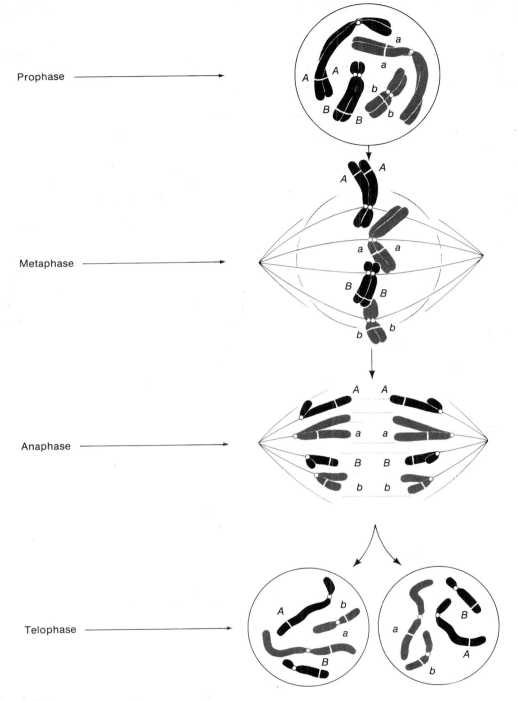

Figure 2-5
Mitosis. Letters designate genes on the chromosomes. If two homologous chromosomes have different alleles at a given locus, one may be given a capital, and the other a lowercase, letter.

descended from each chromosome of the original interphase nucleus. The chromosomes uncoil and individual chromosomes can no longer be distinguished under the microscope. Mitosis ends as a new nuclear envelope forms around each group of daughter chromosomes. As a result of mitosis the cell contains two nuclei that are identical to each other as well as to the parent, interphase nucleus.

During cytokinesis, the cytoplasm divides and new cell walls begin to form so that the nuclei established in telophase end up in separate cells. After the chromatids separate and start to migrate toward opposite poles, a cell plate forms along the plane of the equatorial plate. The middle lamella and new primary and secondary cell walls are ultimately formed from this region. Cytokinesis usually follows telophase immediately, but there are many cases in which it does not. Mitosis and cytokinesis occur in all somatic (nonsexual) tissue. These two closely related processes increase the size of an organism by increasing the number of its cells.

The genetic consequences of mitosis can be explained by considering a cell containing chromosomes that are heterozygous for two loci with the *A* locus being on one pair of homologous chromosomes and the *B* locus on a different pair. The genetic constitution, or genotype, is *Aa Bb*. After DNA duplication the genotype is *AA aa BB bb*. During mitosis, when the chromatids with *AA* separate, one *A* allele goes to one pole, and the other to the opposite pole. Separation is the same for *aa*, *BB*, and *bb*. Thus, at the end of mitosis and cytokinesis, two cells having *AaBb* genotypes have formed.

MEIOSIS

Mitosis precisely duplicates cells, and functions in growth by increasing the number of vegetative, or somatic, cells. A second type of division is required to produce reproductive cells, which are either gametes (eggs and sperms) or special spores that ultimately produce gametes. The production of reproductive cells is called meiosis (Figure 2-6). The entire meiotic cycle actually comprises two divisions. Like mitosis, for convenience it is divided into several phases, but in fact it is a continuous process. Consider once again the corn plant with twenty chromosomes in each cell. Remember these chromosomes can be grouped into ten homologous pairs. In plants meiosis occurs only in specialized cells in flower parts; in corn, as in most plants, it occurs in spore mother cells. Each division has the same four phases as mitosis—prophase, metaphase, anaphase, and telophase—but within these phases there are some differences from what happens in mitosis. Because there are two divisions in the full meiotic cycle, the phases must be further identified and this is done by adding a roman numeral; for example, prophase of the first cycle is called prophase I, and telophase of the second cycle, telophase II. Meiosis is preceded by premeiotic interphase, called interphase I, in which replication of the chromosomes takes place. After the first meiotic division, the chromosomes may return to an interphase state called interphase II.

In prophase I, a peculiar event takes place: the members of each homologous pair of chromosomes (each of which is composed of two chromatids as a result of the duplication during interphase I) physically pair and wind around each other in a very precise manner. Rather than having twenty individual chromosomes, there are ten pairs of chromosomes. Following this event, which is unique to meiosis, the nuclear envelope fragments and the spindle apparatus forms. In meiosis, apparently, filaments from one pole attach to one member of a pair of chromosomes at its centromere, and filaments from the opposite pole

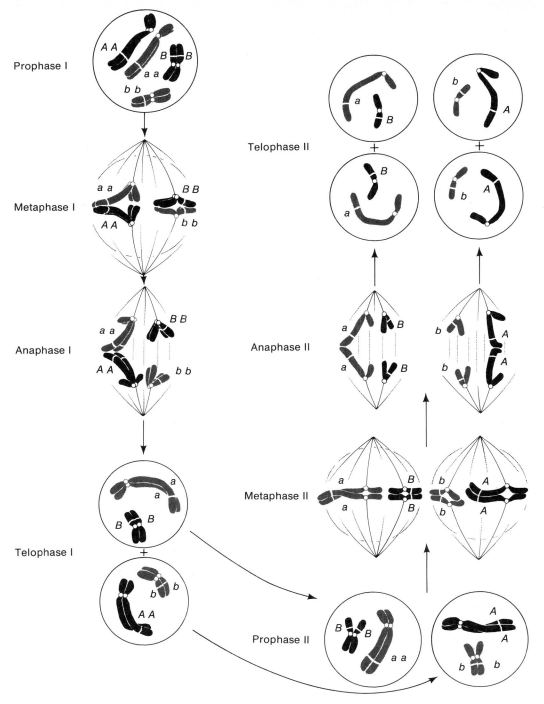

Figure 2-6
Meiosis. The composition of the 4 telophase II nuclei depends on the separation of independent chromosomes at metaphase I (*A* could move with *b*) as well as on complex events that take place in prophase I.

attach to the other member of the pair at its centromere. Such attachments are made to all pairs of chromosomes. In meiosis, as in mitosis, as a result of coiling or twisting, chromosomes become shorter and thicker, and thus easier to see.

Metaphase I is the stage in which the pairs of chromosomes have migrated to the center of the cell and are aligned on the metaphase plate. Migration is dependent in some way on the spindle apparatus, but, as in mitosis, how the chromosomes move is not fully understood. The members of each pair of chromosomes are held together at their centromeres and at points of contact, or chiasmata (singular chiasma), which form between chromatids of opposite members of the pair of chromosomes. Chiasmata occur as a result of chromosome pairing in prophase I. They are most visible late in prophase I and during all of metaphase I.

In anaphase I, the paired chromosomes separate, and the chromosomes move towards opposite poles. The separation of homologous chromosomes into different gametes accounts for a basic law of inheritance: segregation of genetic factors, or genes, which, for example, results in the production of both red- and white-flowered plants from the seeds of a single, red-flowered plant. Throughout anaphase I each chromosome is still composed of two chromatids. It has been possible to demonstrate, however, that there has been an exchange of chromosomal material between chromatids of two homologous chromosomes at the chiastmata. These exchanges occur in prophase I and become apparent in anaphase I. This phenomenon is termed crossing over and leads to genetic recombination. In anaphase I the members of any pair of chromosomes separate independently of the members of all other pairs. (In any individual of a sexually reproducing species, all of the maternally derived chromosomes do *not* move to one pole while all of the paternally derived ones move to opposite poles; instead, the group of chromosomes moving to either pole is a random mixture of both maternally and paternally derived chromosomes). However, exactly half of the chromosomes, regardless of their origin, move to each pole. The random assortment of members of pairs of homologous chromosomes has the observable effect of crossing the alternate expressions of one trait carried on one chromosome pair (such as flower color) with the alternate expressions of another trait carried on a different chromosome pair (such as fruit shape): different combinations of flower color and fruit shape may be seen in plants grown from the seeds of a single plant.

Telophase I is the stage at which the chromosomes, one member from each pair, arrive at opposite poles. Recall that at this stage in mitosis there were as many chromosomes at each pole as there were in the initial nucleus. At telophase I there are exactly one-half as many chromosomes as were present in the initial nucleus, but each of these chromosomes consists of two chromatids. By convention, the chromosome number in the initial nucleus is referred to as the 2N or diploid number (e.g., in corn 2N = 20). Although meiosis is not completed at telophase I, note that by the end of this stage each nucleus has one-half of the original number of chromosomes; that is, N chromosomes. This is termed the haploid number. At the end of telophase I, new nuclear envelopes form around each nucleus. In some species, cytokinesis does not occur; in others, it follows telophase I immediately, or it may be delayed. As a result of the first meiotic division, a diploid nucleus is divided into two haploid nuclei.

Interphase II follows telophase I. The duration of interphase II is variable; however, no critical events occur during this phase. In contrast to interphase I and mitotic interphase, there is no DNA duplication during interphase II.

Depending on the plant being studied, prophase II may follow telophase I almost immediately or be delayed hours or even days. The chromosomes, which have uncoiled and become nearly invisible during interphase II, coil and once again can be seen with the microscope if they are properly stained. Both cells that result from a first meiotic division go through a second meiotic division. During prophase II, the nuclear membranes of each telophase I nucleus fragment and a new spindle apparatus forms.

In metaphase II, the individual chromosomes, still composed of two chromatids from the initial DNA duplication of interphase I, move to the equatorial plate. When the chromosomes are aligned on the equatorial plate, metaphase II is completed.

The two chromatids of which each chromosome is composed as a result of DNA duplication in interphase I separate in anaphase II. It is probable that separation occurs as a result of the contraction of the spindle fibers that extend from each pole to opposite sides of each centromere. This results in the splitting of the centromeres and the separation of the chromatids into independent chromosomes. Two chromatids that had been joined go to opposite poles.

Telophase II is the final step of the second meiotic division. From each of the two cells formed in telophase I, two more cells have formed so that a total of four cells results from the original interphase I cell. Each of these cells has one-half of the chromosomal material present in the original cell. As a result of segregation, independent assortment, and recombination in the first meiotic division, these cells are not identical to each other. In plants, the four cells that result from the meiotic process are spores, not gametes.

The genetic consequences and implications of meiosis are more complex than those of mitosis, but, like mitosis, they can be described for the simple case of heterozygosity at each of two loci that are on different pairs of homologous chromosomes. Assume that a cell initially has the genotype $Aa\ Bb$; as a result of DNA duplication, just as in mitosis, this becomes $AAaa\ BBbb$. Now two distinct cases must be considered; remember, the behavior of nonhomologous chromosomes, such as A and B is independent.

In the first case, as a result of chance alignment of pairs of chromosomes on the metaphase plate, the chromosomes separate at anaphase so that AA and BB go to one pole, and aa and bb go to the opposite pole. At the conclusion of meiosis, the four reproductive cells would be AB, AB, ab, and ab (verify this by tracing the steps of meiosis).

In the second case, once again as a result of chance alignment of pairs of chromosomes on the metaphase plate, the chromosomes separate at anaphase I so that AA and bb go to the same pole, and aa and BB go to the opposite pole. The results of meiosis in this case are four cells that are Ab, Ab, aB, and aB. In plants, more than a single cell goes through meiosis so that several male and female gametes are ultimately produced. Because the two cases described above are equally likely to occur, in a plant of the genotype $Aa\ Bb$, the four types of spores are produced in equal numbers—$1AB:1Ab:1aB:1ab$. These spores produce gametes. If three independent loci are heterozygous, the maximum number of types of spores is 2^3 or $2 \times 2 \times 2 = 8$. If X loci are heterozygous, the maximum number of types of spores is 2^X. The role of spores and gamete production are treated in Chapter 4, where floral morphology is discussed.

COMPARING MITOSIS AND MEIOSIS

The major differences between mitosis and meiosis are that mitosis occurs in all somatic tissue, meiosis only in special tissue. In mitosis homologous chromosomes do not pair, in meiosis they do (this leads to genetic recombination). In mitosis at metaphase pairs of chromatids line up on the equatorial plate; in meiosis at metaphase I pairs of chromosomes line up. In mitosis in anaphase doubled centromeres split and chromatids separate; in meiosis in anaphase I doubled centromeres do not split, but chromosomes, composed of two chromatids each, separate.

The second meiotic division is quite similar to mitosis, but the cells going through the second meiotic division are haploid. The end products of mitosis are two cells that are identical to each other and to the original cell. The end products of meiosis are four cells, each having one-half of the chromosomal material of the original cell; these cells are generally not identical with each other.

Cellular Energetics

RESPIRATION

Energy is required for the cell to maintain its complex organization and to carry out the activities that, taken together, constitute life. This energy is made available within the cell through a complex series of chemical reactions referred to collectively as respiration. Chemically, respiration is oxidation. In oxidation, either oxygen is added to the material being oxidized or hydrogen is removed from it. Respiration converts food, commonly carbohydrates (CH_2O), into a form of energy that may be used to carry out work in the cell. The overall process of respiration can be expressed by the simple equation

$$CH_2O + O_2 \rightarrow CO_2 + H_2O + energy$$

In most situations, rapid oxidation is accompanied by a large increase in temperature. In the living cell, which would be damaged by high temperatures, the oxidation of food is controlled by enzymes that allow the process to take place in a series of steps, without dangerous temperature increases. Carbohydrate is broken down to carbon dioxide (CO_2) and water (H_2O). Oxygen is required, and energy is released and stored in forms that are readily available for cellular needs. Before many complex food materials can be respired —converted into useful energy—they must be converted into simpler substances. This conversion of complex substances into simpler ones is digestion; for example, starch, a complex carbohydrate is converted into glucose, a relatively simple carbohydrate. Energy is not released by digestion; in fact, digestion requires energy. Nearly all living cells are capable of both digestion and respiration. Energy is released in respiration because elements or atoms in a food molecule are held together with energy. When these elements or atoms are broken apart through respiration, the energy is not lost but is stored in forms that are readily available for cellular activity. Energy is most commonly stored in the cell in adenosine triphosphate (ATP). This is a specialized molecule in which an atom of phosphorus, through a complex series of steps, is added to adenosine diphosphate (ADP) to create the triphosphate form. The chemical bond by which this third atom of phosphorus is held is unusual. The bond is stable, but when ATP is hydrolized as a result of the rearrangement of electrons, it has the capacity to release energy. The free energy released when the terminal phosphate bond is hydrolized varies from 7 to 12 kcal per mole of ATP. Thus, this terminal phosphate bond is frequently referred to as a high-energy bond. Energy released in the respiration of food is transferred to specific bonds in molecules of ATP. In some instances it may be convenient to think of the ATP molecule as a source of stored energy.

As food is respired, not all of the energy is transferred to ATP. In some stages of respiration, after hydrogen is removed from food, it is bound to a hydrogen acceptor. Part of the energy that bound the hydrogen to the food is then transferred to the bond between the hydrogen and the acceptor. This energy can be used directly by the cell or to form ATP. A very common hydrogen acceptor is nicotinamide adenine dinucleotide (NAD). There are other hydrogen acceptors.

Energy can take many forms, among which are heat, magnetism, light, and motion. In biology, energy is usually expressed in units of heat or calories. A calorie (cal) is defined as the amount of energy required to raise the temperature of one gram of water one degree Celsius. A large calorie (kcal) equals 1000 cal.

Much of the foodstuff respired by plants is stored in the plant as a complex starch. Through various digestive processes, it is converted into glucose, a fairly simple type of sugar whose formula is $C_6H_{12}O_6$ (a chain of six carbon atoms, to which are attached twelve hydrogen atoms and six oxygen atoms). These atoms are held together with energy. When glucose is respired, the energy is released. If a mole of glucose is completely oxidized ($C_6H_{12}O_6 + 6 O_2 \rightarrow 6 CO_2 + 6 H_2O$), a total of 673 kcal of free energy is released. When glucose is respired in the cell, some, but not all, of the energy holding the molecule together is transferred to the terminal phosphate bond to produce ATP from ADP (ADP + P $\rightarrow$ ATP). Each mole of ATP stores from 7 to 12 kcal from each mole of glucose that is oxidized.

Glycolysis The complete respiration of glucose to carbon dioxide and water requires two different sets of reactions, each set comprising several steps. The first set, which consists of ten steps, is glycolysis, in which a molecule of glucose is broken down into two molecules of pyruvate (Figure 2-7); oxygen is not required for this process. Although four moles of ATP are produced from a mole of glucose, two moles of ATP are used in the initial steps. Thus, there is a net gain of only two moles of ATP. If the minimum of 7 kcal of energy is stored per mole of ATP produced, then the efficiency of this process, in terms of the percentage of total energy in a mole of glucose that is recovered for use by the cell, is 7/673, or about 2 percent—in other words, only 2 percent of the energy in glucose is recovered and transferred to ATP to do work in the cell. Obviously, this is too low and too inefficient to fulfill the cell's energy needs.

The apparent inefficiency is compensated for in part by the production of two reduced hydrogen acceptors:

$$2 \text{ NADH} + 2 \text{ H}^+ \rightarrow 2 \text{ NADH}_2$$

The two molecules of $NADH_2$ have the energy potential ultimately to produce a total of four ATPs from four ADPs; that is, 4 × (ADP + P).

The enzymes necessary to carry out all of the steps of glycolysis are in the cytoplasm. Glycolysis is thus a process associated with the cytoplasm of the cell, rather than with some specialized part of the cell.

Krebs cycle Glycolysis does not result in the complete oxidation of a mole of glucose: the end products are two moles of pyruvate, not carbon dioxide and water. Also, some of the energy that held the original glucose molecules together is tied up in holding the pyruvate molecules together. When adequate oxygen is available, the end product of glycolysis—pyruvate—is oxidized and energy is then stored in ATP through a second respiratory sequence of eighteen steps known collectively as aerobic respiration, which comprises the Krebs cycle and electron transport (Figure 2-8). The Krebs cycle is also referred to as the

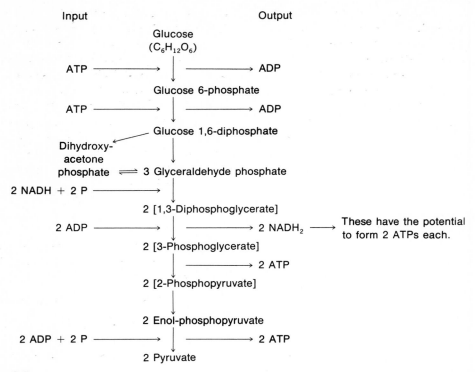

Figure 2-7
The steps and products of glycolysis, showing where ATP is required and where it is produced in the oxidation of glucose to pyruvate.

citric acid cycle or the tricarboxylic acid (TCA) cycle. The enzymes necessary for the Krebs cycle and for electron transport are located in the mitochondria.

Two molecules of $NADH_2$ ($NADH + H^+ \rightarrow NADH_2$) are formed in the first step, which converts a molecule of pyruvate into acetyl coenzyme A (acetyl CoA). Each of these two molecules has the potential to produce three ATPs. In addition, the $NADH_2$ formed during glycolysis may, by losing hydrogen, release electrons to pass through the transport system: two moles of ATP may be produced from each mole of $NADH_2$ coming from glycolysis. The acetyl CoA complex is fully oxidized to yield a total of twenty-four additional ATPs. As the process goes on, hydrogen is removed until only a molecule consisting of carbon and oxygen remains, which is ultimately given off as CO_2. The hydrogen is moved through the electron-transport system. At the last step, the final hydrogen acceptor is oxygen: hydrogen and oxygen unite to form water. If oxygen is not available, the entire system is blocked and aerobic respiration effectively ceases. Cellular life depends on a source of energy. Efficient energy production in green plants requires oxygen; thus, life that depends on green plants requires oxygen. Through the Krebs cycle and electron trans-

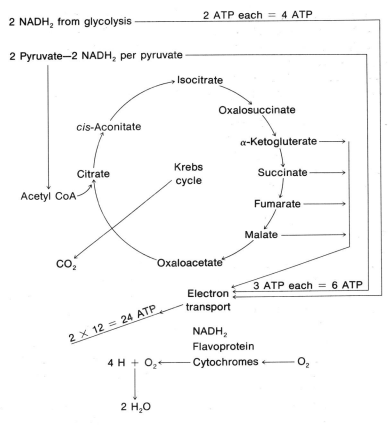

Figure 2-8
The steps and products of the Krebs cycle and electron transport through which pyruvate produced by glycolysis is oxidized to carbon dioxide and water, and the ATP potential of NADH$_2$ is realized. Note that oxygen is required in the final step.

port, thirty-four moles of ATP are produced; adding this to the end product of glycolysis gives thirty-six moles of ATP for each mole of glucose. Because each mole of ATP transfers 7 kcal, a total of 252 (7 × 36) kcal is transferred. The efficiency of respiration is 252/673, or about 37 percent. The other 63 percent of the energy is lost in various forms, such as heat.

The process through which ATP is produced in respiration is collectively called oxidative phosphorylation. Respiration is also an oxidative process; however, until the final step, which requires oxygen, oxidation is dehydrogenation (removal of a hydrogen atom and with it an electron). This is why enzymes that act as or regulate hydrogen acceptors are critical in respiration.

Through the Krebs cycle, fats and amino acids can be respired as well as pyruvate. The energy released and stored depends on the material being respired. If oxygen is not available to complete the Krebs cycle, in some instances alternative paths are taken and alcohol is among the possible end products.

The process giving alcohol is fermentation, which is of great industrial importance in, for example, manufacturing wine and liquors and making silage. Fermentation resulting in the accumulation of lactic acid is desired for the production of high-quality silage. Whenever the Krebs cycle is not completed, the amount of energy, in the form of ATP, recovered from the food being respired is reduced.

PHOTOSYNTHESIS

Recently, a nationwide survey of university biologists was made. From a collection of 114 concepts the one they regarded as the most important was: "with minor exceptions living things obtain their energy directly or indirectly from the sun through the process of photosynthesis." There are two simple reasons why it is the responsibility of plant scientists to obtain a thorough understanding of photosynthesis. First, it is the only significant process through which nonnutritious inorganic compounds are converted into essential foodstuffs for both plants and animals, including man. Second, it is a process carried out only in specific parts (chloroplasts) of certain cells of green plants. The very term photosynthesis, when broken down, well describes the process: *photo* means light and *synthesis* means building; thus, photosynthesis means building with light. Another significant, direct benefit of the photosynthetic process is the release of oxygen (O_2) from plants for use in respiration by both plants and animals. Air contains about 21 percent oxygen; green plants supply a significant part of this. The basic process of photosynthesis can be described in a simple equation:

$$CO_2 + H_2O \xrightarrow[\text{chlorophyll}]{\text{light energy}} CH_2O + O_2$$

in which inorganic nonnutritious carbon dioxide is combined with water to form a carbohydrate, using light energy in the presence of chlorophyll. Oxygen from the splitting of water molecules is given off as a by-product. Note that carbohydrates are composed of carbon, hydrogen, and oxygen, in the ratio of 1:2:1. Thus glucose, $C_6H_{12}O_6$, and pyruvate, $C_3H_6O_3$, are both carbohydrates.

Photosynthesis is functionally the reverse of respiration. In respiration carbohydrates are broken down to carbon dioxide and water and energy is released. Some of the energy is stored in ATP; the hydrogen from the glucose reduces and is associated with special hydrogen acceptors or carriers. In photosynthesis carbon dioxide and water are combined to form carbohydrates, and solar energy is stored in, or transferred to, the bonds that hold the carbohydrate molecules together. The breaking down of a substance with the concomitant release of energy, as in respiration, is a catabolic type of reaction. A reaction in which a substance is built up and energy is stored is anabolic.

All enzymes necessary for photosynthesis are located in chloroplasts. The key substance in the chloroplast is the pigment chlorophyll, which reacts to light energy and ultimately allows this energy to be converted into the energy stored in food—mainly carbohydrates. In most plants there are two types of chlorophyll—*a* and *b*—both of which are present in a single chloroplast. There are also other pigments that are active in photosynthesis. The total process of photosynthesis can be separated into two distinct phases: the light reactions and the dark reactions. The light reactions require light to proceed; they are also referred to as photochemical reactions. The dark reactions may go on in the presence of light, but do not require light; they are sometimes referred to as carbon dioxide fixation.

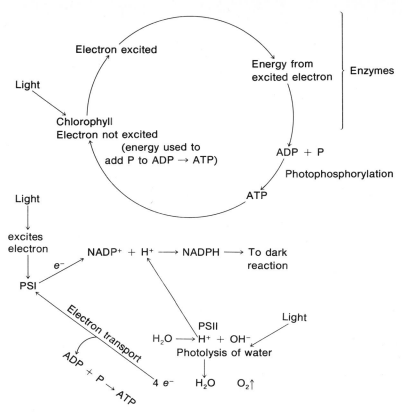

Figure 2-9
Light reactions of photosynthesis: (top) cyclic photophosphorylation in which energy from a light-excited electron is used to form ATP; (bottom) part of noncyclic photophosphorylation in which some light energy is used to produce ATP and some to bind hydrogen ions derived from water to carriers for use in dark reactions.

Light Reactions The light reactions through which light energy is converted into chemical energy can be subdivided into two sets: cyclic (or cyclic photophosphorylation) and noncyclic. The first step in both sets is the absorption of light energy, which activates the chlorophyll by exciting (raising the energy level of) one of its electrons. The high-energy electron leaves the chlorophyll molecule and is passed along through a series of electron-carrier molecules (Figure 2-9).

In the cyclic reactions, the excited electron is transferred first to an iron-containing protein, ferredoxin. From ferredoxin, the electron is transferred by means of a sequence of carrier molecules to plastocyanin, a copper-containing protein. The electron completes its cycle when it is transferred from plastocyanin to its original position in the chlorophyll molecule; it arrives at this starting point in its original, nonexcited state because, as the light-excited, high-energy electron is transferred from one electron carrier to another, its energy level is continuously lowered. The energy is not lost; some of the light energy that initially excited the electron is converted into the chemical energy of the terminal phosphate bond in ATP.

The noncyclic reactions are more complex; they are subdivided into two photo (or pigment) systems that are linked by an electron-transport system. The systems are designated

PSI and PSII. Both PSI and PSII use the longer wavelengths of light in the near-red to red part of the visible spectrum; however, light of a slightly shorter wavelength is used by PSII. This suggests that PSI and PSII occur in slightly different types of chlorophyll.

In PSI, light energy excites an electron, which is passed along through a series of enzymes until it ultimately binds hydrogen ions (produced in PSII) to a special hydrogen carrier, nicotinamide adenine dinucleotide phosphate ($NADP^+$): $NADP^+ + H^+ \rightarrow NADPH$. The high-energy electron is temporarily lost from the chlorophyll molecule of PSI, but it is replaced by an electron that comes from water as a result of the reactions associated with PSII.

Although the sequence of reactions is not yet understood, the key events in PSII center on the photolysis of water in which the water molecule is split with light energy into H^+ and OH^- and on the transfer of both the H^+ and the electrons from OH^-. The end results of PSII are:

$$4\ H_2O \rightarrow 4\ H^+ + 4\ OH^-$$

$$4\ OH^- \rightarrow 4\ OH^+ + 4\ e^-$$

$$4\ OH \rightarrow 2\ H_2O + O_2$$

These reactions account for the oxygen produced during photosynthesis. The hydrogen that is freed during PSII is bound to $NADP^+$ by the activated electron from PSI, and the four electrons released by the reactions of PSII are passed along through a series of enzymes to replace the electrons lost in PSI. As the electrons are transferred from PSII to PSI, ATP is formed.

Light energy is converted into chemical energy in both the cyclic and the noncyclic light reactions. The light energy excites electrons, which are then transported through a series of enzymes, and, as the electrons pass from enzyme to enzyme, the light energy that excited them is converted into chemical energy. Only chlorophyll and a very few other special pigments have the peculiar structure that permits this type of energy conversion.

Photosynthesis uses only part of the light that falls on a leaf. Sunlight, although it appears "white" to the unaided eye, is composed of light of different colors—this is what is seen in a rainbow. The color of light depends on its wavelength; light travels in waves, just as sound does. Also, light behaves as a physical particle, or photon: this concept is used in physics and in advanced plant physiology.

Photosynthesis uses light in the range of wavelengths normally visible to the human eye. This light ranges from violet (wavelength about 3800 Å) to red (wavelength about 6750 Å). The light used in photosynthesis is absorbed by chlorophyll. The light not used in photosynthesis is transmitted through the leaf or is reflected by the leaf. The pattern of light absorption by a photosynthesizing leaf is given in Figure 2-10. Note that light in the shorter wavelengths (blue light about 4500 Å) and in the longer wavelengths (red light about 6750 Å) is absorbed. The green plant appears green because chlorophyll absorbs mostly blue and red light so that mainly green light is transmitted or reflected.

Dark reactions Hydrogen and the electrons that move with it are carried by NADP (now, with two hydrogen atoms this carrier is designated $NADPH_2$) and are active in the dark reactions. In these reactions hydrogen is given up by $NADPH_2$ to combine with carbon dioxide, yielding carbohydrate. This is known as CO_2 fixation. There are several intermediate forms of carbohydrate, but the end product is starch.

In the past ten years, a great deal more detail about photosynthesis has been verified. Many of the new concepts can be of signifi-

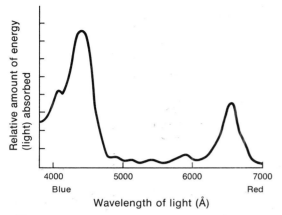

Figure 2-10
Absorption spectrum of chlorophyll. Generally, blue and red light are absorbed and green light is transmitted.

cance to crop producers, although they involve advanced biochemistry. Plants now can be classified as one or the other of two types according to the pathway of CO_2 fixation (that is, according to the first identifiable product incorporating CO_2 taken in from the environment). Plants whose first carbon compound has a basic chain of three carbon atoms are designated C_3 plants. The most common initial product is phosphoglyceric acid. Other plants form first carbon compounds having a basic chain of four carbon atoms—for example, oxaloacetic acid. These plants are designated C_4 plants. Generally speaking, plants that evolved in the temperate regions, such as winter annuals or cool-season crops including wheat, barley, and many forage grasses, are C_3. Many warm-season crops are C_4; corn and sugarcane are major examples.

The differences between C_3 and C_4 plants are also expressed in other ways. In C_4 plants there are two types of chloroplasts: one type is quite similar to the chloroplast typical of a C_3 plant (see Figure 2-4), but starch does not accumulate in it as it does in the C_3 chloroplast. In certain parts of the leaves of C_4 plants (see Chapter 3) a second type of chloroplast is evident. This chloroplast is larger than the C_3 chloroplast and has few or no grana.

As a result of normal respiratory activity, plants use some of the carbohydrates they produce by photosynthesis. In terms of plant growth, or ultimate yield, the crop producer's goal is a plant that achieves maximum net photosynthesis. Net photosynthesis, although extremely difficult to measure in the field, is defined as total carbohydrates produced (measured by CO_2 fixation) minus the carbohydrates respired. The net photosynthetic rate of C_4 plants is markedly higher than that of C_3 plants. This difference is due to the phenomenon of photorespiration. Normally, respiration rate increases as temperature increases. In C_3 plants, respiration in the chloroplasts (not in mitochondria, as might be expected) increases with light. From 20 to 50 percent of the carbohydrates produced can be lost as a result of photorespiration in C_3 plants. The phenomenon is not found in C_4 plants. Thus, under conditions of high temperatures and bright light, C_4 plants are potentially more productive than C_3 plants; under cool conditions with more limited light (see Chapter 8), C_3 plants may be more productive. Productivity depends on a complex interrelationship between plants and environment.

The processes of respiration and photosynthesis are functionally the opposite, but closely related in that the end products of one are the starting products of the other (Figure 2-11). In addition, although these processes take place in different, specialized parts of the cell, both require similar enzymes to regulate electron-transport systems. Although life depends on energy converted through respiration, respiration is, in fact, dependent on photosynthesis; both plants and animals ultimately depend on the products of photosynthesis for energy. Obviously, the goal of

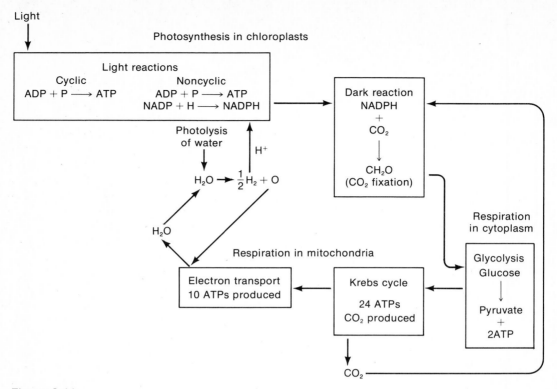

Figure 2-11
Relationships between photosynthesis and respiration. Both processes produce ATP. Water and carbon dioxide can be traced as end products of respiration and starting products of photosynthesis.

crop production is to maximize photosynthesis for the production of plants that are of value to man.

Lehninger, A. 1961. How cells transform energy. *Sci. Amer.* 205(3):62–73. Available as *Sci. Amer.* Offprint 91.

Selected References

Brachet, J. 1961. The living cell. *Sci. Amer.* 205(3):50–61. Available as *Sci. Amer.* Offprint 90.

Capaldi, Roderick A. 1974. A dynamic model of cell membranes. *Sci. Amer.* 230(3): 26–33. Available as *Sci. Amer.* Offprint 1292.

Cohn, Norman S. 1969. *Elements of cytology.* 2d ed. Harcourt, Brace, and World.

Greulach, Victor A. 1973. *Plant function and structure.* Macmillan.

Study Questions

1. Trace the flow of energy from CO_2 and H_2O to carbohydrates and back again to CO_2 and H_2O: include the key processes and where they take place. Does this explain why green plants support nearly all forms of life?
2. Why are both mitosis and meiosis necessary for life? How do these processes differ?
3. Describe the structure of a typical cell and note the functions of each part.

Three

Vegetative Growth and Development

Although individual cells are capable of carrying out an amazingly wide array of vital processes, they can become neither large enough nor specialized enough to meet all the needs of crop plants. Crop plants, like all higher organisms, are multicellular and highly differentiated. An understanding of the structures of plants is as important as an understanding of cellular functions for effective management of crop plants. It is convenient in describing plant structures to treat those below ground—the roots—separately from those above ground—the shoots, including stems and leaves and flowering or reproductive structures. In this chapter, only vegetative growth and structures are considered; floral growth and reproduction are discussed in Chapter 4.

Roots

FUNCTION

All crop plants have rather extensive root growth. The mass of roots formed by a plant may be referred to as its root system. The specific types of roots and the patterns of growth vary from crop to crop. In general, there are three major functions served by the roots: First, roots serve as the source of contact between the plant and the soil. Their major function in this regard is to absorb the water, oxygen, and minerals that are required for plant growth. Understanding the differences in patterns of root growth and development among various crops, and at various stages of development of a specific crop, is essential in planning when or how much to irrigate and how to place fertilizers in the soil. Crops that have shallow roots and crops that are young (seedlings) require less water per irrigation, but more frequent irrigations, than deeply rooted, older crops. Also, crops differ with respect to the breadth and depth of their roots. Understanding the distribution of roots in the soil is essential for efficient and effective application of fertilizers. Second, roots anchor the plant by branching throughout the soil. This property is important for two reasons: (1) if the plant is even partly uprooted, its roots may not be able to carry out the necessary absorption of minerals, water, and oxygen; and (2) if plants are uprooted, a crop will be difficult to harvest. Such uprooting would have serious economic consequences for wheat and barley producers. The selection of crop cultivars with extensive and strong root systems tends to lessen the possibility of severe losses from lodging. Plants with extensive root systems are also able to hold the soil, thus reducing erosion due to wind and rain—a characteristic of major importance in selecting plants for the purpose of soil conservation. The final major function of roots is storage. Carbohydrates are produced through photosynthesis. These provide food (material for respiration) for the entire plant. The cells in which photosynthesis takes place cannot store all of the carbohydrates produced. Many of the products of photosynthesis are translocated through the phloem and stored in the roots. Food storage in the roots of annual crops is not critical; however, in biennial or perennial crops it is critical, both to the plant and to the crop producer. The stored carbohydrates are the energy source for spring growth and regrowth of many forage species. A part of forage management consists of cultural practices that insure adequate carbohydrate reserves for the new growth. Many perennial grasses store carbohydrates in the lower parts of stems as well as in the roots. The particular form in which carbohydrates are stored in plants differs for different species: for example, fructosans are stored by many C_3 temperate or cool-season grasses; other species store starch, and sugar beets store sucrose.

Sugar beets are a biennial crop. During the first year, the plant grows vegetatively. The products of photosynthesis are stored in a single root. If the crop is not harvested at the end of the first growing season, in the second year the food stored in the root is used to provide the necessary energy for the plant to grow, produce flowers, and make seeds. Although, biologically, sugar beets are biennials, they are farmed as annuals. To maximize his profit, a producer must know how to force the plant to store the greatest amount of sugar possible in the roots and must know when to harvest them. Sugar storage by sugar beets is largely dependent on moisture and fertilizers. If the plant runs out of nitrogen late in the season, the starch stored in the roots will be converted into sugar, and it is the sugar content of the raw material that is of primary interest to beet processing companies.

Root storage is equally, if not more, important in perennial crops such as forage grasses (bromegrass, wheatgrass, orchardgrass), alfalfa, sainfoin, and other prime forages. These crops stop growing and become dormant (but do not die) with the onset of cold weather. The leaf tissue is killed by the cold. In the following spring when new growth begins, there is no green tissue to carry out photosynthesis and provide energy for growth; this energy comes from starch or other forms of carbohydrates stored in the roots and the base of the stems (the crown region) of each plant. If the crops are harvested or grazed too late in the season, food reserves stored in the crown are used to develop new shoots. If these shoots do not carry out photosynthesis for a long enough period before fall freezing to replace the food that has been used for new growth, there will be inadequate food to support growth in the subsequent spring. Stands of forage crops will then be weakened and reduced. The same consequences can be expected if forages are overgrazed. Leaf material is removed too frequently, and the plant exhausts its food reserves in the crown and ultimately is weakened and dies. The same type of problem exists in areas in which plants are dormant in the summer because of high temperatures or a lack of water, or both.

Figure 3-1
Major patterns of root growth; (left) taproot typical of legumes and (right) fibrous roots of a corn seedling.

In considering the storage function of roots, it may be helpful to think of the roots as a bank—the products of photosynthesis that are stored in the roots as deposits and the energy used for new growth as withdrawals. As everyone knows, there is serious trouble if the amount withdrawn from an account exceeds the amount deposited. Cultivation to control perennial weeds should be so scheduled as to force the weed to exhaust its stored carbohydrate reserves until it is weakened and ultimately dies.

ROOT MORPHOLOGY AND ANATOMY
There are only two basic patterns of root distribution—one predominantly vertical, the other horizontal. These yield, respectively, a taproot system or a fibrous root system (Figure 3-1). Taproots, such as sugar beets, garden beets, and carrots, are more generally harvested for the food stored in them. Some crops have taproots that are not harvested, however. Alfalfa has perhaps the deepest taproot of any of the common crop plants, which allows the plant to absorb water and minerals from great depths in the soil—in some cases, more than 20 feet (6 m). Some trees also have extremely deep taproots; walnut and oak trees are good examples.

Many annual crops have fibrous root systems: most of the cereal crops (wheat, barley, oats, rye, and rice) do. Perennial grasses also have predominantly fibrous roots, as do evergreens. This trait accounts for the ease with

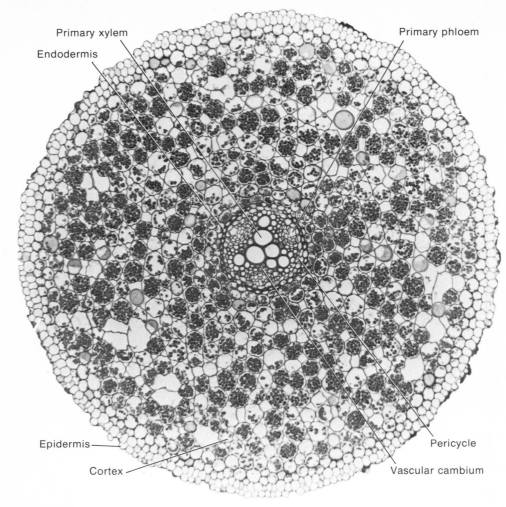

Figure 3-2
Cross section of a root of a perennial plant. Compare the arrangement of the vascular tissue with that in Figure 3-3. [Photomicrograph by Triarch, Inc.]

which many evergreens can be transplanted.

The internal morphology of all roots is fairly consistent. A cross section of the root of a perennial is shown in Figure 3-2. However, the roots of annuals have a core of pith (Figure 3-3), which is not found in the roots of perennials. The outer layer of cells (the "skin") is the epidermis. Root hairs are elongations of individual cells of the epidermis (Figure 3-4). Root hairs increase the area of contact between roots and soil and account for much of the absorption by the roots.

Immediately inside the epidermis is an area of cells with lighter walls called the cortex. This is the storage area. The innermost layer of cells of the cortex is the endodermis. Next to the endodermis is a ring of specialized cells called the pericycle. Branch roots arise from pericycle cells.

Inside the ring of pericycle cells lies the vascular cylinder, which consists of phloem cells on the outside and xylem cells on the inside. The phloem transports products of photosynthesis from the leaves of the plant to

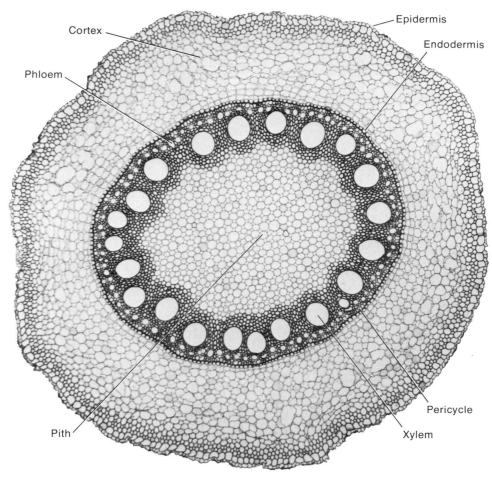

Figure 3-3
Cross section of a root of an annual plant. Compare the vascular tissue with that in Figures 3-2, 3-5, and 3-6. [Photomicrograph by Triarch, Inc.]

the roots as a source of food for respiration and for storage. The xylem conducts water and essential minerals up and throughout the entire plant. Cells in the vascular tissue generally have elaborate secondary cell walls. The phloem is made up primarily of sieve elements, which are long, tapered, and perforated at the ends. Sieve elements have no nuclei, but they are accompanied by companion cells that do. The nucleus of the companion cell apparently serves the minimal needs of the sieve element. The xylem in crop plants is made up primarily of vessel elements. Generally, vessel elements have well-lignified secondary walls, which are quite hard and afford a great deal of support to the entire plant. In perennial plants there is a special region between the phloem and xylem that can give rise each year (sometimes more frequently) to new vascular tissue, both xylem and phloem. This is the vascular cambium. Secondary phloem is formed on the outside of the vascular cambium and secondary xylem on the inside. In annual plants the center

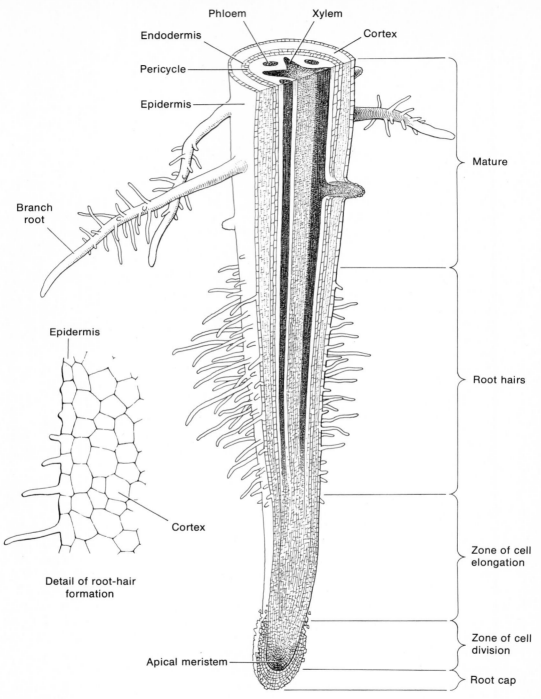

Figure 3-4
Longitudinal section of a root showing the location and development of root hairs.

of the roots is filled with a loosely arranged cell mass of pith material.

ORIGIN OF ROOTS

Roots have been classified according to their origin. Each seed contains an embryo, or a minute, partly developed plant. Microscopic examination of the embryo reveals the presence of both root and stem parts. In the embryos of such plants as sugar beets and carrots —which have taproot systems—a single root, formed from the radicle, can be distinguished. This structure is usually the first part of the embryo to grow as germination starts. It grows from its tip and becomes the primary root. Other roots branch from the pericycle of this primary root and become secondary roots. The primary root is usually the most evident root throughout the life of these plants.

In the embryos of plants that have fibrous root systems—such as corn, wheat, barley, and oats—the radicle, or primary root, plus two other pairs of roots that are not branches from the radicle but form near the base of the embryonic stem can be detected. These roots are referred to as seminal (or seed) roots. In most plants having fibrous roots, the seminal root system does not serve the plant for its entire life—not even in annuals. A third type of root system develops, consisting of adventitious roots, which is the permanent root system.

Adventitious roots arise from organs other than the primary root. Except for the radicle, the seminal roots are in fact adventitious roots. However, they are special in that they are differentiated in the embryo, and they do not generally become a major part of the permanent root system. In many plants, mainly grasses, a network of adventitious roots arises from the nodes at the bases of stems at or below the soil surface (the crown region). This network constitutes the fibrous root system. It can include the seminal roots, but rarely does it include the radicle. In some species, such as corn, adventitious roots arise from nodes well above the soil surface; they enter the soil at an angle, much like a guy wire on a power or telephone pole. In addition to serving the three major functions of roots, these roots, called brace roots or aerial roots, serve as supporting braces for the plant.

Whether a root is a taproot or a fibrous root, whether it is primary, secondary, seminal, or adventitious, it grows from meristematic cells located at the root tip. These meristematic cells are originally undifferentiated and can give rise to any of the internal parts of the root. The meristematic tissue in a root tip is called the apical meristem. This tissue is protected by a special layer of cells called the root cap. The apical meristem differentiates into:

> Protoderm, which gives rise to the epidermis
>
> Ground meristem, which gives rise to the cortex and endodermis
>
> Procambium, which gives rise to the pericycle (which yields lateral roots), primary phloem, vascular cambium, primary xylem, and pith (in some plants)

Several regions are identifiable in the lower part of a root (Figure 3-4). The apical meristem is protected by a dense root cap. Meristematic cells undergo mitosis; this, coupled with the cell elongation taking place in the region immediately above the apical meristem, forces the root tip downward through the soil. Immediately above the zone of elongation is the zone of maturation; the cells in this zone perform specific functions. Root hairs form in this region. As the root increases in diameter, many of these fine cellular extensions are torn off and are not replaced in the same cells. As the roots penetrate the soil, the zone of maturation becomes deeper; hence, the area in which new root hairs form is deeper.

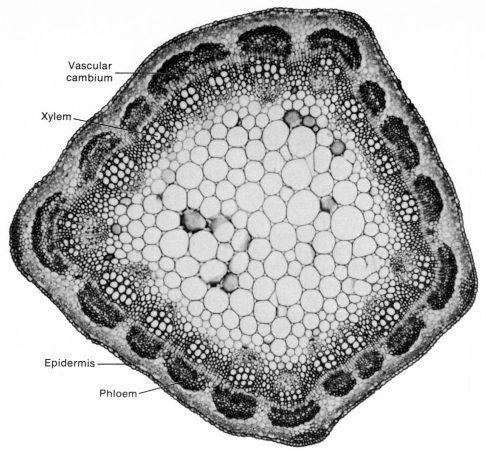

Figure 3-5
Cross section of a stem of an annual plant. Compare the vascular tissue with that in Figures 3-2, 3-3, and 3-6. [Photomicrograph by Triarch, Inc.]

This growth pattern allows the plant to reach ever deeper in the soil to obtain water and essential minerals.

Stems

FUNCTION AND ANATOMY

The shoot, that part of a plant that is above ground, can be divided into the stem, leaves, and flowers. There are two primary functions of stems: to support and display leaves, enabling them to obtain the needed light for photosynthesis, and to house the vascular system so that water may move up through the entire plant and the products of photosynthesis may move down. The vascular system, consisting of xylem and phloem, is continuous from roots through shoots. The arrangement of the vascular tissue in shoots is different from that in the root and varies between annuals and perennials (Figures 3-5 and 3-6). In annuals the vascular bundles (composed of phloem on the outside and xylem on the inside) are scattered throughout the pith area. In perennials the vascular tissue is arranged in a ring with phloem on the outside and xylem on the inside. The phloem and xylem are separated by vascular cambium, which yields secondary phloem and xylem. The wood of a tree is generally vascular tissue; heartwood and sap-

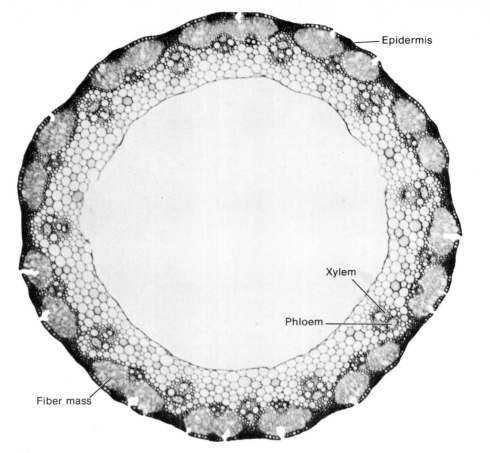

Figure 3-6
Cross section of a stem of a perennial plant. Compare the vascular tissue with that in Figures 3-2, 3-3, and 3-5. [Photomicrograph by Triarch, Inc.]

wood are secondary xylem. The annual rings of trees are a result of the development of secondary vascular tissue. As secondary phloem develops, primary phloem is forced to the outside where it becomes cork or bark tissue in older plants. The primary xylem is forced to the center of the stem.

The epidermis constitutes the outer surface of a stem; it surrounds a layer of cortex, which surrounds the vascular tissue. Neither endodermis nor pericycle is present in stems. Like root growth, stem growth comes from meristematic tissue. The differentiation of meristematic tissue in stems is slightly different from that in roots. It differentiates into:

> Protoderm, which gives rise to the epidermis
>
> Ground meristem, which gives rise to the cortex
>
> Procambium, which gives rise to primary phloem, vascular cambium (which yields secondary phloem and xylem), and primary xylem

In addition, ground meristem cells develop into pith and pith ray cells. The pith rays move materials laterally across stems.

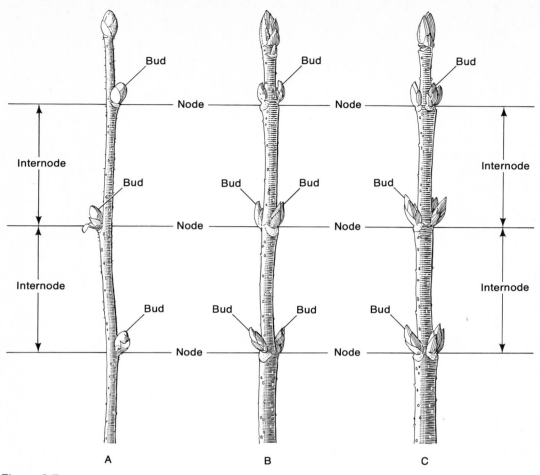

Figure 3-7
Classification of buds by arrangement on the stem; (A) alternate, (B) opposite, and (C) whorled.

MORPHOLOGY

The branches, leaves, and flowers of a plant arise from buds that are formed at nodes, which are slightly enlarged or swollen areas along each stem (see Figure 3-7). The space between two successive nodes is an internode. The length and the number of internodes determine plant height. In most woody perennials the number of nodes is indeterminate (growth in height is somewhat open-ended); perennial grasses, on the other hand, have a fixed number of internodes. Annuals such as barley, wheat, and corn also have a fixed number, whereas vine types of beans do not. With the exception of grasses, which will be discussed separately, stem growth is similar to root growth in that it is initiated in apical meristematic tissue.

BUDS

Buds are clusters of primary meristematic tissue and can be classified in at least three ways. The first is by their arrangement on the stem, which may be alternate, opposite, or whorled (Figure 3-7). Alternate leaf buds give rise to a single leaf at each node; each successive bud is on the opposite side of the stem from the one preceding it. The leaf-bud arrangement in alfalfa and walnuts, for example, is alternate. Opposite buds are arranged so that there are two buds on opposite sides of the stem at each node, each bud giving rise to a single leaf. This type of leaf-bud arrangement is characteristic of common beans (white, pinto, and kidney) and of lima beans. An arrangement in which more than two leaf buds are located at each node is a whorled arrangement. Whorled arrangements are not found in major crop plants and are in fact rare in other types of plants as well.

If a stem is viewed from the top, the leaves are seen to be in a spiral arrangement. This arrangement affords each leaf the best opportunity to receive sunlight for photosynthesis and prevents upper leaves from shading lower leaves.

The second way in which buds can be classified is according to their position on the stem (Figure 3-8). Buds located at the tips of stems are terminal buds. Lateral (or axillary) buds are found in the axils between leaves or branches and the stem. An axil is the angle between the upper surface of a branch or leaf and the main stem. Adventitious buds are neither terminal nor lateral; they frequently form on stems that have been injured.

A third way in which buds can be classified is according to the structures into which they develop: either leaves or flowers or mixed. A mixed bud may give rise to both leaves and flowers.

If a stem has a terminal flower bud, its growth is determinate; when the flower bud develops, the stem will not increase further in length. Wheat and barley are examples of crops having stems that terminate in floral buds. Other examples are red clover and some cultivars of dry beans and lima beans. Stems whose terminal buds are leaf (or stem) buds are indeterminate in growth. Alfalfa, vine types of beans, grapes, hop plants (the hops of which are an important ingredient in beer), and all so-called creeping plants are examples of plants whose growth is indeterminate. (For the moment, sod-forming grasses will not be considered.)

A knowledge of bud location and type is of practical importance. In many fruit trees the location of buds determines trimming or pruning practices. The location of floral buds, which ultimately yield fruit, determines where fruit trees should be trimmed or pruned. Floral buds on citrus trees form on both new and old branches; thus, citrus trees can be pruned to a specific shape for ease of picking fruit by hand without undue consideration of bud location. Such is not the case for cherries: flower buds form on two-year-old branches and care must be taken not to remove new growth so that future floral buds can form; thus, only a limited amount of older growth can be removed. Floral buds on peach and apricot trees form only on new branches; thus, these trees can be heavily pruned to remove older growth, thereby encouraging a great deal of new growth and many floral buds.

Leaves

SIMPLE LEAF

Photosynthesis is carried out in special cells in leaves; thus, leaves must be considered to be a critical part of plants. However, this essential process could not be carried out if any of the other plant parts failed to function. Leaves vary extensively in their external shape or morphology. A common leaf form is the simple leaf (Figure 3-9). Although the margins of simple leaves may vary (they may be smooth, toothed, or deeply lobed), the basic morphological structure is the same. They have a relatively large, flat surface called the leaf blade. At the base of the leaf blade is the stalk, called the petiole, by which the blade is attached to the stem. This type of leaf is found on many nongrass crop plants;

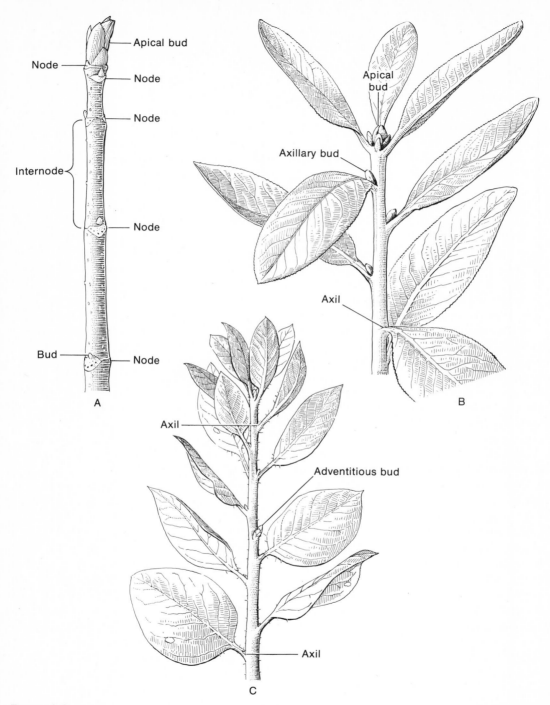

Figure 3-8
Stem morphology, and bud arrangement: (A) node and internode with apical bud, (B) leaf axils and axillary bud, and (C) adventitious bud.

cotton and sugar beets are common examples. In some crop species, hooklike, or minute leaflike appendages called stipules are found at the base of the petiole. In other species, the leaf is attached directly to the stem; the petiole is entirely absent. Such leaves are sessile, whereas those attached by petioles are petiolate. Sessile leaves are found in safflower and related plants in the sunflower family.

VENATION
Casual examination of a leaf blade reveals a fine network of veins throughout the blade (Figure 3-9). This network is a continuation of the vascular system in the roots and stems. It branches in one of two ways: if all the main branches come from the base of the leaf blade where the blade and the petiole join, the branching is palmate; if the branches of the vascular system do not arise from a common point but come from one main line, the branching is pinnate. Although these differences are of minor agronomic significance, they are useful in distinguishing different types of plants.

COMPOUND LEAVES
The leaves of several important crop plants differ markedly from the simple leaf described above; these are compound leaves (Figure 3-9). A compound leaf has a petiole or central axis, much like a simple leaf, but the blade is divided into individual leaflets. Employing the terms used for venation of simple leaves, a compound leaf may be palmately compound or pinnately compound. All the leaflets of a palmately compound leaf, such as red clover, are joined to the petiole at one point, whereas the leaflets of a pinnately compound leaf, such as crown vetch, arise at various points along a common axis (the equivalent of the midrib of a simple leaf). Leaflets of the compound leaf may be attached directly to their central axis in a sessile manner, as in clovers (palmately compound) or as in vetch, sainfoin, and locoweed (pinnately compound), or one or more of the leaflets may be on a short stalk. It is difficult to determine whether trifoliolate leaves (three leaflets per leaf) are pinnately compound or palmately compound in some plants. For example, the leaf of an alfalfa plant may seem to be palmately compound, but in fact it is pinnately compound; each of the three leaflets is borne on a short stalk. It is also sometimes difficult to distinguish the leaflets of compound leaves from simple leaves. Knowing several characteristics of compound leaves will help to identify them: When a compound leaf drops in the fall, the entire leaf drops, the individual leaflets remaining attached to the central axis of the leaf. If stipules are present, they are found at the base of the petiole of the compound leaf, not at the base of each leaflet. Buds are found at the base of the leaf, not at the bases of the leaflets.

LEAF ANATOMY
Leaf blades are composed of several specialized layers of cells (Figure 3-10). The upper and lower layers are the epidermal layers. Pairs of special cells, called the guard cells, are present in both the upper and lower epidermis. Between each pair of guard cells is an opening to the interior of the leaf. This opening is a stoma (plural stomata). Stomata serve two purposes: they are the point of entry for carbon dioxide, which must reach chloroplasts for photosynthesis; they also are the openings through which water taken in through the roots is lost to the atmosphere. The walls of the guard cells are highly specialized: those adjacent to the opening are thick and inelastic compared with those adjacent to other epidermal cells. Because of this difference, when the cells are turgid the stoma is fully opened. When water is not adequately available, the guard cells lose their turgidity,

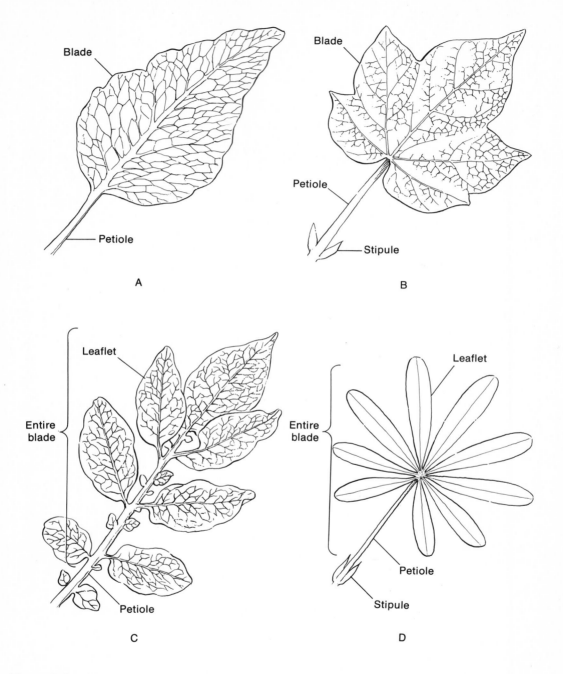

Figure 3-9
Simple and compound leaves and their venation: (A) a simple leaf that is pinnately venated; (B) a simple leaf palmately venated; (C) a pinnately compound leaf; and (D) a palmately compound leaf.

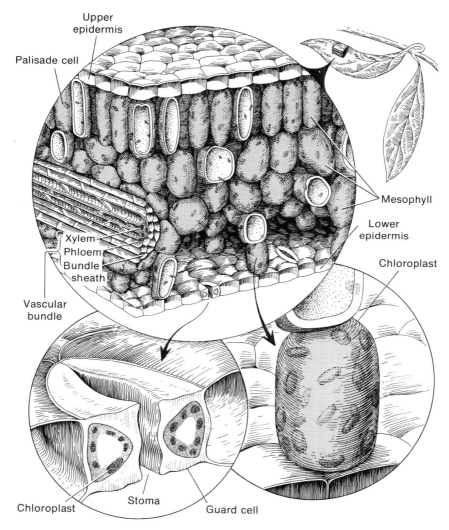

Figure 3-10
Cross section of a leaf with detail of stoma and guard cells of a chloroplast. [From *Principles of plant physiology* by James Bonner and Arthur W. Galston. W. H. Freeman and Company. Copyright © 1952.]

and the area of the stoma is reduced. This reduces water loss from the plant. Guard cells are the only cells in the epidermis that have chloroplasts and carry out photosynthesis. Their photosynthetic activity plays a role in the opening and closing of the stoma by changing the concentration of sugars in the guard cell, thereby altering osmotic relationships of the cell.

Crops differ in the amount of water they require to produce a pound of dry matter; one way in which they differ is in number of stomata. In general, the more stomata a leaf has, the more water can pass through it. For ex-

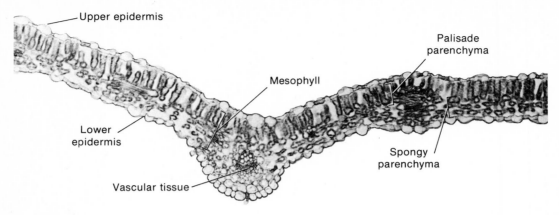

Figure 3-11
Cross section of a leaf of a dicot. Note the subdivision of the mesophyll layer and compare it with that in Figure 3-14. [Photomicrograph by Triarch Inc.]

ample, alfalfa uses a great deal of water to produce a pound of dry matter. There are nearly 17,000 stomata per square centimeter on the upper epidermis and 14,000 on the lower epidermis. By comparison, corn uses relatively little water; it has 5,200 stomata per square centimeter on the upper epidermis and 6,800 on the lower.

Nearly all of the photosynthetic activity in a leaf takes place in mesophyll cells, which are thin-walled and loosely packed and which comprise all of the tissue, except vascular tissue, between the upper epidermis and lower epidermis. In crop species other than grasses, the mesophyll is differentiated into two layers (Figure 3-11). Immediately below the upper epidermis are the cells that are very active in photosynthesis called the palisade parenchyma cells. They are long and narrow and contain chloroplasts. The arrangement of this layer and of the chloroplasts in each cell allows the best utilization of the sunlight that falls on the leaf surface. Also, there are rather large intercellular air spaces that allow the circulation of the carbon dioxide that enters through the stomata.

Below the palisade parenchyma cells is a mass of loosely packed, irregularly shaped cells, the spongy parenchyma. These cells are also active in photosynthesis and are the initial storage area for the products of photosynthesis.

The mesophyll layer of grasses is not generally differentiated into palisade and spongy parenchyma layers. Mesophyll cells in grasses are densely packed compared with those in various dicot species (see Chapters 4 and 5). In addition, in those grasses having the C_4 photosynthetic pathway rather than the C_3 pathway (see Chapter 2), carbohydrates are not generally stored in the mesophyll cells. Also, in C_4 plants, larger chloroplasts that have few if any grana are concentrated in groups of cells surrounding the vascular bundles in the leaf (see Chapter 2). The vascular bundles in the leaves of C_4 plants are closer together than in C_3 plants, and there are fewer cells between the vascular bundles. These differences may provide the C_4 plants with a more efficient method of transporting the products of photosynthesis from the mesophyll tissue and may also contribute to the higher net photosynthetic rates of C_4 plants.

Although a great deal is still unknown,

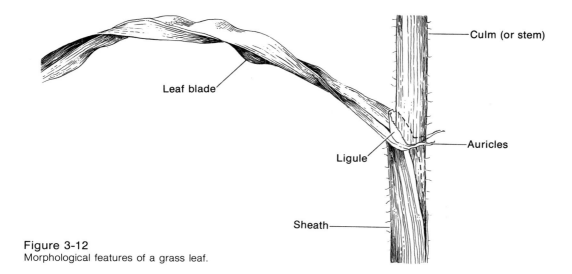

Figure 3-12
Morphological features of a grass leaf.

plant breeders, working with plant physiologists and other specialists, are trying to determine what accounts for the higher efficiency of C_4 plants, so that in the future such traits can be developed in or transferred to less efficient C_3 plants (see Chapter 32).

The lower epidermis is similar in structure to the upper epidermis but generally has more stomata, with the exception of a few species.

THE GRASS PLANT
The root, stem, and leaf structures and the cellular functions discussed so far are common to all crop plants. However, there are modifications in grass plants that merit separate consideration. (Many major food and feed crops are grasses; in addition to forage grasses —such as bromegrass, wheatgrass, and orchardgrass—the cereals—corn, barley, wheat, oats, rye, and rice—are grasses.)

The stem of a grass plant is termed a culm. Like all stems, it is divided into nodes and internodes. Leaves arise at the nodes; however, the grass leaf does not have a petiole, nor is it truly sessile. The grass leaf consists of a blade and a sheath; the sheath is wrapped around the culm (Figure 3-12). At the junction of the blade and the sheath are two structures: the auricles, which are clawlike appendages extending from the junction around the culm, and the ligule, which is a membranous structure between the blade and the culm that arises from this junction. The auricles and the ligule are useful in identifying several crops in the vegetative stage: among the common cereal crops—wheat, barley, and oats—wheat has neither conspicuous auricles nor a ligule; barley has large, pronounced auricles and no ligule; and oats have no auricles and a large, obvious ligule.

Growth of the grass culm is different from that of other stems. It is not apical—that is, growth is not due to elongation from the tip. Instead, there is a meristematic region at the base of each internode. New cells are produced sequentially, starting at the first or lowest internode, then at the second internode, and so on. The pattern of growth is analogous to extending a telescope or an aerial on a car, which are lengthened not by adding to the upper end but by extension at each of the joints (Figure 3-13). Leaves from all nodes are initiated prior to culm elongation and, as

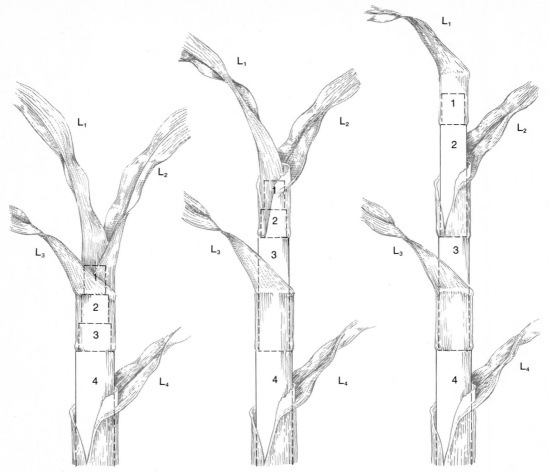

Figure 3-13
Pattern of elongation of a grass stem. Sheaths of leaves formed at the bases of unelongated internodes (i.e., internodes 1, 2, and 3 and leaves L_1, L_2, and L_3) enfold the culm. Internodes elongate from their bases, the lowest internode elongating first. A leaf is elevated as the internode below the node at which it is formed elongates.

a result of leaf initiation and growth, the culm is literally wrapped in leaf sheaths. The tip of the culm is ultimately pushed through the leaf sheaths as the culm elongates. This type of growth is termed intercalary growth. Leaf blades and sheaths also elongate as a result of growth from basal regions; thus, their growth pattern is also intercalary.

Buds form in the axils between leaves and stem (where the sheath and stem join). At the lower internodes that do not elongate, buds develop into new stems, which are called tillers or stools. New tillers generally form before culm elongation; thus, a young grass plant that is still quite short may have several stems. Rarely are tillers produced at the higher nodes. Thus, grasses do not appear to branch like many trees or like alfalfa, clovers, or sainfoin. The basal area of the plant—the first node or two and the roots—is called the crown. Each year new growth in perennial grasses (forage grasses) comes from buds at the crown,

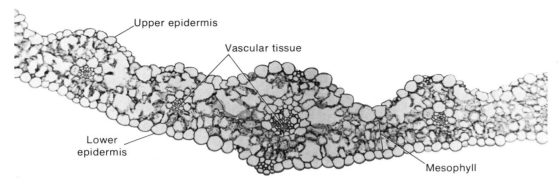

Figure 3-14
Cross section of the leaf of a monocot (grass) plant. Compare the mesophyll layer with that in Figure 3-11. [Photomicrograph by Triarch, Inc.]

and the crown is where the energy needed for this growth is stored.

Besides being morphologically different from the leaves of other crop plants, the leaves of grass plants are anatomically different. The major distinguishing feature is the absence of the palisade parenchyma layer in grass leaves (Figure 3-14).

VEGETATIVE VARIATIONS
There are several unusual forms of stems of significance to plant scientists in general and to crop producers in particular. One form is a stolon—a long, horizontal stem that runs along the soil surface. At each node adventitious roots and new shoots may arise. If the internode on either side is cut, the newly rooted nodular piece becomes another plant, which is just like the original plant that developed from seed. Stolons are common to a number of plants; bermudagrass is a typical example. Rhizomes are like stolons except that they are underground stems. Many of the creeping or sod-forming grasses have rhizomes—for example, bromegrass and quackgrass. Because any piece of a rhizome (or stolon) that includes a node may produce a new plant, plants that have either of these can be spread easily by cultivation equipment and may become serious weed pests.

Tubers are modified stems; they are enlarged tips of undergound stems and have nodes and internodes. The common (Irish) potato is a tuber. The "eyes" on a potato are in fact nodes. If a potato is cut into pieces for planting (seed potato), a new plant can develop from each node, just like the nodes of a rhizome or stolon. Corms, although not common to major crops, are enlarged underground stems in which food can be stored and from which new plants can arise. Corms provide the basis for new growth annually in perennial forage grasses such as timothy. Bulbs are not solely stems; they are short segments of stem surrounded by a mass of highly modified, fleshy leaves. The common onion is an example. Bulbs, like corms and tubers, are storage organs.

Plant Nutrition

The basic equation for photosynthesis

$$CO_2 + H_2O \xrightarrow[\text{chlorophyll}]{\text{light}} CH_2O + O_2$$

is an over simplification. A photosynthesizing plant requires a number of specific minerals for life and growth. Based on the relative amounts of each required, these minerals are classified as macronutrients (relatively large amounts required) or micronutrients (relatively small amounts required).

MACRONUTRIENTS
Carbon, hydrogen, and oxygen are macronutrients and are required for the synthesis of carbohydrates through photosynthesis. Carbon enters the plant through stomata in the form of carbon dioxide. Normally, the air is only 0.03 percent carbon dioxide, and at times the amount present in the air may limit the rate of photosynthesis. Little can be done in the field to increase the concentration of carbon dioxide, but it can be increased in greenhouses, where flowers or tomatoes may be grown, by pumping the gas in. Carbon dioxide also supplies some of the oxygen required for carbohydrate synthesis, as does water.

Hydrogen is obtained from molecules of water, which are split during the noncyclic part of the light reactions in photosynthesis (see Chapter 2). Oxygen comes from several sources. As a result of the photolysis of water, oxygen is available for respiration in leaves and stems; it also enters the leaves through the stomata. Much of the oxygen required for normal respiratory activity in the roots comes from air in the soil. If the soil is flooded, adequate oxygen is not available for respiration in the roots and normal root functions may be impeded. Ultimately, should flooded conditions persist, this may cause the entire plant to be weakened or killed (see Chapters 6 and 7). In a sense, then, because oxygen is not available to the plant in adequate amounts from soil that is flooded, the plant "drowns."

Nitrogen, phosphorus, and potassium are also macronutrients. They are termed the primary plant nutrients because they are most frequently applied as fertilizers. Nitrogen is the most important and most frequently limiting plant nutrient. It generally enters the plant in the form of nitrate ion (NO_3^-) from the soil. About 2 percent of the dry weight of a plant is nitrogen; however, nitrogen-containing compounds make up nearly 25 percent of the total dry matter of most plants. Nitrogen is a constituent of amino acids and, thus, of proteins. Proteins coupled with fats form the various membranes in cells and all enzymes are proteins. The regulation of cell functions—respiration and photosynthesis—depends on enzymes and therefore on nitrogen. In addition, many of the high-energy intermediates in photosynthesis contain nitrogen. Plants that are deficient in nitrogen are generally weak and light green, or chlorotic. Chlorosis usually appears first on the lower leaves and later on new growth. Nitrogen deficiency leads to a reduction in yield in all crops.

One group of plants, the legumes (such as peas, beans, and alfalfa), can convert atmospheric nitrogen (N_2) into nitrates (NO_3^-), which can be either used by the plant itself or deposited in the soil for use by other plants. This beneficial conversion is carried out in symbiotic association with bacteria (*Rhizobium*) that live only in the roots of legumes. The bacteria receive energy (carbohydrates) from the plant and in turn supply available nitrogen to the plant. A specific strain of *Rhizobium* is required for a specific crop.

Phosphorus is taken in through the roots as phosphate ion (PO_4^-). Only about 0.2 percent of the dry weight of a plant is phosphorus; however, phosphorus plays a major role in plant metabolism in the formation of ATP. The fats associated with proteins to form membranes contain phosphorus and are called phospholipids. Phosphorus-deficient plants have characteristic symptoms: they take longer to mature; their leaves turn a deep reddish purple and may appear to be fired or burned; and their fruit may drop before it is mature.

Potassium is taken in through the roots as potassium oxide (K_2O). The exact localization and function of potassium in plants are not known, but it is required in mitosis. Although it is not a part of carbohydrates, it is needed for the plant to produce sugars and starches. It affects the straw strength of grasses. Adequate amounts tend to increase winter hardiness in many crops. Plants deficient in potassium initially have characteristically mottled leaves (leaves with white or light spots). The leaves tend to appear burned along the margin and they curl up. These symptoms also appear first on the lower leaves and later on new growth. Potassium deficiency also results in a weakened root system, and the plant may be uprooted if it is lodged by wind.

Four other macronutrients are classified as secondary plant foods: calcium, magnesium, sulfur, and iron. Calcium (Ca) is necessary for the formation of the cell wall (as calcium pectate in the middle lamella). Plants deficient in calcium typically have poor bud growth and a breakdown of meristematic tissue because of a failure in cell-wall formation following mitosis.

Magnesium (Mg) is also taken up through the roots. It is a key element in the chlorophyll molecule and is essential for the formation of some amino acids and vitamins. Plants deficient in magnesium are very chlorotic (light green to white); the chlorosis is evident first between the leaf veins of older leaves.

Sulfur (S) is a constituent of two essential amino acids—methionine and cysteine. Although it is not part of the chlorophyll molecule, it is required for chlorophyll formation. In an unknown manner, it affects the flavor of certain foods such as cabbage and onions. Symptoms of sulfur deficiency are quite similar to those of nitrogen.

The precise function and localization of iron (Fe) in plants are unknown. Like sulfur, it is required in chlorophyll formation. Symptoms of iron deficiency generally appear first on new growth; the veins of leaves are unusually dark green and the interveinal areas are extremely chlorotic. Iron is sometimes considered to be a micronutrient.

MICRONUTRIENTS

There are six known micronutrients: boron, manganese, zinc, molybdenum, copper, and chlorine. All are required in relatively small quantities. The exact function of boron (B) is unknown. It plays a role in flowering and pollen germination on the stigma; it may also play a role in sugar transport from leaves to roots. It is apparently associated with enzymatic activity. Plants deficient in boron have a wide array of symptoms: terminal buds die, leaves become thick and brittle, foliage may develop in a rosette, and fruit tends to crack and be discolored. Plants that have excess boron show signs of toxicity. For example, excess boron in cereal crops (barley, wheat, and rice) may cause small black flecks or spots to appear on the leaves.

Manganese (Mn) is required for chlorophyll development. It also participates in oxidation-reduction reactions in cells. Unless masked by deficiency symptoms of other elements, manganese deficiency can be identified by the dark green veins and light green interveinal areas on leaves. Severe or prolonged deficiency causes these areas to turn white. Manganese deficiency can also cause leaves to be lost prematurely.

Zinc (Zn) is required in very small amounts.

Like those of several of the other micronutrients, its localization and function are unknown. It plays a part in the development and function of plant growth regulators (plant hormones), in internode elongation, in chloroplast development, in starch development, and in seed development in legumes. The leaves of citrus trees that are deficient in zinc are irregularly chlorotic or mottled. In field crops there is a general interveinal chlorosis.

Molybdenum (Mo) is required in minute amounts, but its role is critical. This element is essential in reducing nitrates. The reduced forms are incorporated in amino acids. Plants deficient in molybdenum show the symptoms of nitrogen deficiency. In addition, they may have rolled leaves and poor fruit or grain development.

Copper (Cu) participates in nitrogen metabolism. Its exact role is unclear, but it may act in various enzyme systems in respiration and photosynthesis. Deficiency symptoms are vague; chlorosis, twig dieback, and dead spots on leaves are a few. In some cases, excess copper is toxic to crop plants.

The element most recently discovered to be essential to plant growth is chlorine (Cl). Its role in the plant has not been determined, but it seems to be active in the light reactions in photosynthesis and possibly in translocating starches. Deficiency symptoms include wilting, stubby roots with excessive branching, and some chlorosis.

All of the micronutrients are usually taken in through the roots. They are generally plentiful in the soil; they are also included as impurities in fertilizers. In some cases, these elements are applied to the leaves directly, but with questionable effectiveness. However, in some trees zinc deficiency has been remedied simply by driving a galvanized (zinc coated) nail into the tree trunk.

In summary, there are four principal ways in which essential minerals are utilized by plants. First, they may be an integral part of the organic complexes of which plants are built. Second, they may play key roles in assisting in the function of enzymes; most of the micronutrients seem to function in the binding of enzymes to substrates. Third, they may participate in maintaining the correct ionic balance for normal cellular function. Fourth, they seem to play key roles in oxidation-reduction reactions in both respiration and photosynthesis.

A deficiency of any macronutrient, except carbon, hydrogen, and oxygen, may lead to some degree of chlorosis and, thus, to reduced photosynthesis. The most severe chlorosis results from deficiencies of nitrogen, magnesium, and iron. Chlorosis may also result from deficiencies of sulfur and potassium. A calcium deficiency may result in chlorosis if potassium is also deficient. Similarily, a phosphorus deficiency may lead to chlorosis as a result of an unfavorable ratio of iron to phosphorus.

In addition to the essential elements, plants take up other elements not required for their growth and development. Among these are cobalt, which is required by animals (and obtained from plants) to aid digestion; vanadium, which is required by algae and some bacteria; sodium, which in some cases might serve in place of potassium; iodine, which is required by animals; fluorine, which is also required by animals for proper skeletal development; and silicon and aluminum, which are taken up in large quantities and seem to serve no useful purpose to plants or animals.

Before any element can be considered to be essential, plants must be grown in a solution completely lacking it. If these plants are in any way abnormal and if only the addition of the element that is lacking alleviates the abnormality, then the element may be considered to be essential. A major problem in studying micronutrients has been the presence of essential elements as impurities in chemicals used in laboratory plant-nutrition studies. Only in recent years has it been possible to eliminate such impurities; in the 1960s, the means for removing impurities containing chlorine was devised. In future years, as more impurities can be removed, additional elements may be found to be essential for plant growth.

Plant Growth Regulators

Plant growth and development from the fertilized egg to the mature sporophyte are controlled by different groups of growth regulators or hormones. One group, the auxins, is simple and rather nonspecific in its action. Auxins are active in controlling cell elongation and cell division. They are frequently localized in shoot tips or the shoot apex, and their presence there inhibits the development of lateral (axillary) buds down the stem. This inhibition is termed apical dominance. When the tip is removed, the source of auxin is also removed, and lateral buds may develop. This phenomenon accounts for the development of new branches when limbs are pruned in fruit and ornamental trees and increased tillering in some grasses after grazing. An understanding of the phenomenon of apical dominance and of bud location allows a skillful nurseryman to control the shape of many kinds of trees.

As a result of their effect on cell elongation, auxins control the phototropic response of plants, the tendency of plants to bend toward light (see Chapter 8). Similarily, changes in the concentration of auxins are responsible for geotropic responses; that is, stems growing up and roots down.

Because of the role played by auxins in stimulating growth, they can be effective in weed control. A common herbicide, 2,4-dichlorophenoxyacetic acid (2,4-D) is an artificial auxinlike growth regulator. In the right concentration it can enter a plant through its leaves and disrupt metabolism to the point of killing the plant.

Nine or more closely related compounds are classified as gibberellins. A type of gibberellin was originally isolated from a fungus found growing on rice. Crop plants are known to synthesize gibberellins. The gibberellins, like auxins, direct cell elongation and cell division. They also may control the flowering response of plants in terms of the plant's response to environmental conditions: the number of hours of daylight required for a plant to initiate flowers (the photoperiodic response, which is discussed in more detail in Chapter 8), and the temperature required to break dormancy. When sprayed on grape crops, gibberellins are known to increase the berry size. Malting barley may be treated with gibberellins to increase its enzymatic activity for brewing beer. When this group of hormones was first discovered, it was believed by a few dreamers that the world's food problems could be solved by treating all crops with some form of gibberellic acid, thereby inducing the production of giant seeds or fruits. These people have obviously been disappointed.

Another group of hormones, the kinins or cytokinins, also stimulates cell division. These hormones are active in the development of scar tissue at wounds. In addition, they seem to be very active in differentiation in connection with hormones in the other groups. The destiny (pattern of differentiation) of various meristematic cells depends on the relative concentrations of the hormones, although in what way is still unclear.

All deciduous plants (perennial plants that lose their leaves each fall) form a special cell layer, called the abscission layer, at the union of the petiole and stem. Leaf abscission, bud dormancy, and physiological seed dormancy (rather than the "dormancy" created by impermeable seed coats, which inhibit water uptake) are regulated by another hormone—abscisic acid. These three phenomena are so similar physiologically that they are regulated at least in part by the same hormone.

Many of a plant's activities are controlled by some type of hormone. Dormancy of both seeds and buds is under hormonal control; even when conditions are just right, some seeds will not germinate. For example, alfalfa ceases active growth and hibernates for the winter because of hormonal action.

Another plant activity that is under hormonal control is floral development. It is well known that crop plants vary markedly in their temperature and light-duration requirements for the initiation of floral-bud development. Many of the responses of an organism to specific environmental conditions are governed by hormones.

Not every hormone or group of hormones acts independently; each has a fairly broad spectrum of effects, but together they affect cell elongation. Although the joint effects and interactions of the hormones at the cellular level have not been explained, there is little doubt that such effects are extremely important in plant growth and development and in how plants respond to various environmental conditions.

Selected References

Crookston, R. Kent, and Dale N. Moss. 1974. Interveinal distance for carbohydrate transport in leaves of C_3 and C_4 grasses. *Crop Sci.* 14:123–125.

Esau, K. 1960. *Anatomy of seed plants.* Wiley.

Greulach, Victor A. 1973. *Plant function and structure.* Macmillan.

Robbins, W. W., T. E. Wier, and C. R. Stocking. 1966. *Botany.* 3d ed. Wiley.

Study Questions

1. Why is a knowledge of plant morphology and anatomy important in crop production?
2. Why might meiosis be of greater importance in annual plants than in perennial plants?
3. Describe several anatomical differences between annuals and perennials.
4. What minerals are essential for plant growth? What are the results or symptoms of mineral deficiency?
5. Why is a knowledge of bud function and location of practical importance?

Four

Reproductive Growth and Development

All crop plants, even perennials, must reproduce at some time. The most common form of reproduction is sexual (through seed), although plants can reproduce asexually (through stolons, rhizomes, and other vegetative bodies, such as pieces of tuber). Plants that reproduce asexually also produce seed. Sexual reproduction begins in the flower and follows a pattern (gamete formation and fertilization) similar to that of animals. Most crops are planted as seed (except for a few such as potatoes and sugarcane). Moreover, most food crops are grown for their seed or fruit. Thus, a crop producer must have a knowledge of floral structures and seed and fruit formation.

Floral Morphology

Reproductive organs in plants, just as in animals, are classified as male or female (Figure 4-1). The male organ is the stamen, which is composed of a long filament and an anther. Inside the anther is special tissue that undergoes meiosis and eventually produces pollen (microspores). Each pollen cell develops and finally produces the male reproductive nucleus, or gamete, the equivalent of the sperm in animals. The female reproductive organ is the pistil, which is composed of a stigma, a style, and an ovary. The stigma is at the top of the pistil and provides a suitable site for pollination, which starts the process that finally ends in fertilization. The style serves as a link between the stigma and the ovary. As in animals, the ovary in plants is the protective body in which the female reproductive cell (or cells), the ovum, is housed. The pistil may be composed of one or more carpels: in plants with a simple pistil, the pistil comes from a single carpel; compound pistils are composed of two or more fused carpels. The carpel is a modified leaf (or leaflike structure). Along the margins of each carpel there is a layer of specialized tissue that undergoes

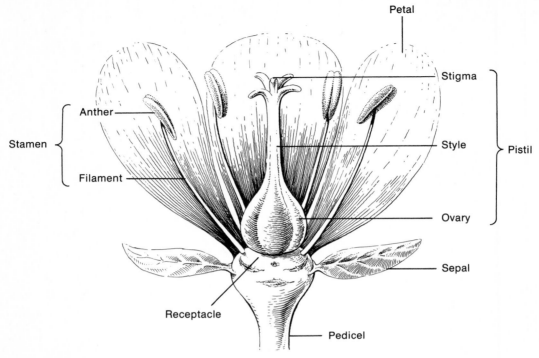

Figure 4-1
A complete and perfect flower showing the major parts. In some flowers some of the parts are fused.

[From *Lehrbuch der allgemeinen Botanik,* 2d ed., by H. von Guttenberg. Akademie-Verlag, 1952.]

meiosis to produce megaspores. Depending on the type of plant, one or more megaspores develops and produces the egg nucleus, or female reproductive cell, called the ovum. Although not always apparent, the ovary wall is composed of three layers: the exterior layer is the exocarp; the middle layer, the mesocarp; and the exterior layer, the endocarp.

Nearly all flowers have parts in addition to the pistil and stamen (Figure 4-1). Usually, the colorful parts of a flower are the petals. The entire group of petals is called the corolla. In many plants there is a group of green, leaf- or bractlike structures below the corolla called the calyx; each leaflike structure is called a sepal. The calyx and corolla together constitute the perianth. The position of the ovary in relation to the receptacle (Figure 4-2) is a characteristic that has been useful in tracing the evolution of plants and in their scientific classification (see Chapter 5). The receptacle is the enlarged upper end of the stem to which the "whole" flower is attached. The stalk below the receptacle is called the pedicel or peduncle, depending on the plant. If the ovary is on top of the receptacle, it is designated a superior ovary; in this case, all other parts of the flower originate below the base of the ovary. If, on the other hand, the ovary is imbedded in the receptacle and the other flower parts originate above the base of the ovary or are fused to the ovary, it is desig-

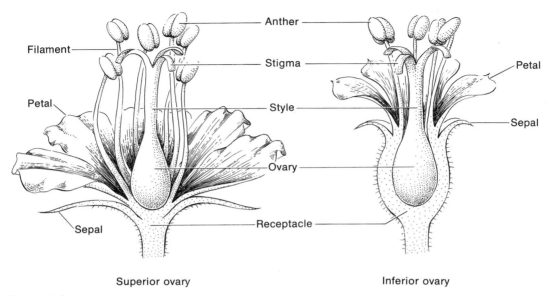

Figure 4-2
The position of the ovary in relation to the receptacle determines whether the ovary is classified as superior or inferior. The receptacle and ovary wall are fused in inferior ovaries; thus, other floral parts seem to originate above the base of the ovary.

nated an inferior ovary. Many fleshy fruits, such as apples, include part of the receptacle or other floral tissues.

Flowers can be classified in three ways according to the presence or absence of various parts. A complete and perfect flower has a calyx, a corolla, stamens, and one or more pistils. The flowers of alfalfa, clover, and beans are examples of complete and perfect flowers. An incomplete but perfect flower is both pistil-bearing and stamen-bearing, but lacks either a calyx, or a corolla, or both. In agriculture the most frequently encountered incomplete flower is the grass flower common to wheat, barley, oats, and many forage grasses, which will be discussed in detail in the next section. Some flowers are not only incomplete, but also imperfect. Imperfect flowers have either stamens or pistils in a single flower, but not both. Corn is an example of a plant that has imperfect flowers. The tassels of a corn plant are inflorescences that contain male, or staminate, flowers in which pollen is produced. Numerous pistils are found on the ear. The "silk" of the corn ear is really an elongated style with a sticky, stigmatic surface, one from each ovary on the ear. Each ovary develops into a corn kernel. Plants such as corn that have both staminate and pistillate flowers on the same plant are termed monoecious. Plants that have male flowers on one plant and female flowers on another are dioecious. Examples are figs, asparagus, buffalo berry, Canada thistle, and marijuana.

The Grass Flower

The common grass flower is incomplete but perfect (Figure 4-3). It is lacking both a calyx

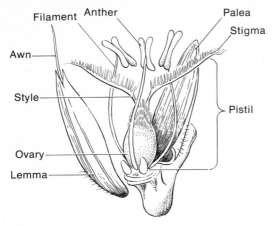

Figure 4-3
A typical grass floret, which is a part of the flowering unit (called the spikelet) shown in Figure 4-4.

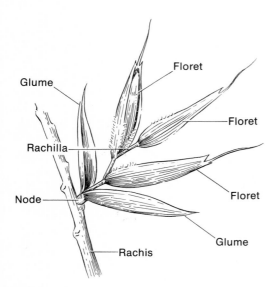

Figure 4-4
A grass spikelet with three florets (a single floret is shown in detail in Figure 4-3). Spikelets of some species have a single floret (e.g., barley), whereas others have several (e.g., wheat).

and a corolla but has both a pistil and stamens. The individual flower is termed a floret. The outer parts of the floret are leaflike structures called the palea and the lemma. In some grasses—for example, in most barley cultivars, in some wheat cultivars, and in many forage grasses—the central vein of the lemma extends into a long, thin spine called an awn. The basic floral unit of a grass plant is a spikelet, not a floret (Figure 4-4). A single spikelet may consist of one or more florets. Each spikelet has a pair of glumes, which are its outer limits, and a central axis, called the rachilla.

Inflorescences

The flowers of many crops are borne in groups, rather than singly; such a group is an inflorescence. There are six common types of inflorescences found in different species (Figure 4-5).

The spike is common to such crops as wheat, barley, and rye, as well as various types of wheatgrass, and the ear of corn is a modified spike. (The inflorescence of wheat and barley is sometimes erroneously referred to as a head type of inflorescence.) The central axis of the spike is the rachis, which forms at the tip (receptaclelike point) of each reproductive culm (not all culms bear spikes or reproductive parts). The rachis is divided into nodes and internodes. At each node one or more sessile spikelets may form. In barley, for example, there are three spikelets per rachis node and only one floret per spikelet; thus, there are three pairs of glumes at each node. Each pair of glumes surrounds a palea and a lemma, which enclose a pistil and three stamens. In six-row barley, all three spikelets have a fertile floret and form seed. In two-row barley only the middle or central spikelet is fertile; the two outer ones are sterile. In wheat there is a single spikelet per rachis node, each

Figure 4-5
Common types of inflorescences and species illustrated are (A) spike (wheat); (B) raceme (mustard); (C) panicle (oats); (D) cyme (white cockle); (E) umbel (poison hemlock); and (F) head (sunflower). Among field crop species, the spike, panicle, and raceme types of inflorescences are most common.

spikelet having two or more fertile florets; thus, at each node there is a single pair of glumes and several florets.

The panicle type of inflorescence is found in rice, oats (including wild oats), bromegrass, sorghum, and many other crops. The tassel of a corn plant is a panicle with only male florets. The central axis of a panicle is branched, and spikelets composed of one or more florets are borne on short stalks, or pedicels, that are attached to these branches.

The raceme type of inflorescence is like a single branch of a panicle (and a panicle is like a branched raceme). Single flowers are borne on pedicels that arise from the central axis of the raceme. Mustard, radishes, alfalfa, and sainfoin have raceme inflorescences. A raceme is different from a spike in that sessile spikelets are borne at the nodes of the rachis of a spike and flowers are borne on pedicels attached to the central axis of a raceme inflorescence.

The head type of inflorescence is typical of crops such as sunflower and safflower, as well as weeds such as thistles and dandelions. The base of the head is an enlarged, flattened receptacle to which individual sessile flowers are attached. Generally, there are two types of flowers on a head: ray flowers around the edge, which have petals but are usually sterile; and disc flowers in the center, which are incomplete (lack colorful petals) but are fertile and produce seed. The "sunflower" is not a flower; rather, it is a group of flowers (an inflorescence).

The umbel type of inflorescence is typical of birdsfoot trefoil and of carrots and such weeds as leafy spurge and poison hemlock. Single flowers are borne on pedicels of equal length that arise from the base of the peduncle. The umbel may have two leaflike bracts attached at the base of the peduncle.

The cyme type of inflorescence is found in flax, an important oil crop. It is also found in many weeds—chickweed and false flax, for example. The vegetative stem terminates in a meristematic mass from which a single flower develops. Below this point, branches arise that in turn branch and form terminal flowers.

Floral Initiation

There are important differences between vegetative growth and reproductive or floral growth. Although a number of factors that affect the flowering activities of many plant species are known, the complexities of the entire process are not well understood. At least three general conditions are essential for a plant to initiate reproductive growth: first, it must reach a stage of "ripeness to flower," called the stage of floral induction. Nearly all plants must produce a minimum amount of vegetative growth before floral initiation occurs. (For example, a rye plant—discussed in Chapter 18—must have at least seven leaves before the culm initiates inflorescence development.) Second, many species must be exposed to quite specific periods of light and dark to induce floral initiation. This phenomenon, called photoperiodism, is discussed in Chapter 8. Finally, some species require specific temperature conditions to induce flowering. (Winter wheat, for example, must be exposed to subfreezing temperatures for several weeks before floral development is initiated; this explains, in part, why winter wheat cannot be planted in the spring.) A discussion of plant response to temperature is included in Chapter 8; a discussion of vernalization in Chapter 16; and a discussion of bolting in Chapter 25.

In conjunction with light, temperature, and biochemical changes, other factors under the management of the crop producer affect floral development. For example, large amounts of nitrogen fertilizer can delay flowering in cereals, pasture grasses, alfalfa, and clovers and can hasten flowering in corn, sorghum, cotton, tobacco, and safflower.

Floral initiation is known to be regulated by hormones (see Chapter 3). Auxins play a minor role as do cytokinins; the gibberellins have a major effect. The application of gibberellins can offset the requirement for long light periods in some long-day types of plants (see Chapter 8). Another hormone—florigen—plays a direct role: as a plant grows and matures, floral initiation is closely associated with changes in the concentration of florigen. This hormone is apparently translocated from leaf tissue to shoot apical meristematic regions where floral primordia form.

The anatomical and morphological changes associated with floral development are most apparent in shoot apical meristematic tissue. During vegetative growth, shoot apical meristematic tissue differentiates to produce buds that become stems and leaves. When the appropriate conditions are satisfied, there is a dramatic change: this same apical meristem starts to produce the primordia of buds that ultimately develop into flowers or inflorescences. The plant allocates more products of photosynthesis to reproductive parts than to vegetative parts; thus, the total needs of the growing plant for light, water, and essential minerals must be met at this stage. For some crops it is desirable to prevent or limit flowering; for example, the removal of flower buds from tobacco plants causes the plants to produce larger leaves of higher quality (see Chapter 24). In general, understanding the development of reproductive parts, including seed and fruit formation, is of practical importance in determining how and when crop plants respond to manageable environmental factors such as the rate and timing of fertilizer application and of irrigation. Efficient field management of crop plants at any stage of development requires an understanding of how the plants develop and grow.

Mating Systems in Plants

If a flower is perfect, it may be self-pollinated; that is, pollen is transferred from an anther to the stigma of the same flower or of another flower on the same plant. Even monoecious plants like corn can be self-pollinated, although they generally are not. Cereal crops such as wheat and barley usually are.

Many plants with perfect flowers are naturally cross-pollinated; that is, pollen is transferred from a flower on one plant to a flower on another plant.* Alfalfa and many forage grasses (such as bromegrass and some of the wheatgrasses) are cross-pollinated. In cross-pollination, pollen must be carried by some means from one plant to another. Wind carries the pollen of cross-pollinated grasses from one plant to another. Bees are the pollen carriers (or pollen vectors) of alfalfa. Differences in flower morphology of various plants determine the type of vector required.

There are several variations in flower development that force cross-pollination; the most obvious is in species in which individual plants bear either male (pistillate) or female (staminate) flowers, but not both (i.e., dioecious plants). Monoecious plants can be forced to self-pollinate, but generally they are cross-pollinated. The most common monoecious plant, corn, is cross-pollinated by wind.

Male (or pollen) sterility can occur in plants whose flowers are morphologically perfect, but whose pollen is nonfunctional. Seeds that form on such sterile plants are necessarily the result of cross-pollination.

---

*If the two plants are genetically the same, however, the effect is the same as self-pollination.

Male sterility is an inherited trait that has been used extensively by plant breeders to develop new, superior cultivars of several crops. The production of hybrid seed corn for planting depended on the use of male-sterile types. At one time, tassels of seed-parent plants were removed by hand before the pollen was shed to insure pollination only from specific plants. Even today, valuable hybrids are maintained by hand-detasseling and careful hand-pollination. To develop hybrid cultivars that are resistant to a serious foliar disease, commercial seed producers have been forced temporarily to abandon the use of male sterility and to depend on hand emasculation. The economic feasibility of developing both hybrid wheat and hybrid barley depends to a great extent on achieving the same degree of success in using male sterility for these crops as for corn. The breeding of hybrid corn is discussed in detail in Chapter 32.

Another phenomenon that leads to cross-pollination is incompatibility. In some crops, the flowers may be perfect, but they will not normally self-fertilize. The pollen from one plant will function only on a few other plants. It is rather like "selective" male sterility and, like true male sterility, it is inherited. Incompatibility has been more of a nuisance to plant scientists than a benefit. In crops like cherries, figs, and tobacco this phenomenon requires fruit-bearing plants and pollinator plants.

The natural mating system of a plant may have tremendous genetic consequences. Consider, for example, plants that are 100 percent cross-fertilized and that are heterozygous at two independent loci: plants whose genotype is $AaBb$. (Recall that spore mother cells divide in the process of meiosis to yield spores that ultimately yield gametes, both male and female.) Each plant would produce four types of pollen (sperms) and four types of eggs: AB, Ab, aB, and ab. These would be produced in

Table 4-1
All possible combinations of four types of female and male gametes.

Female gamete	Male gamete	Embryo
AB	AB	AABB
Ab	AB	AABb
aB	AB	AaBB
ab	AB	AaBb
AB	Ab	AABb
Ab	Ab	AAbb
aB	Ab	AaBb
ab	Ab	Aabb
AB	aB	AaBB
Ab	aB	AaBb
aB	aB	aaBB
ab	aB	aaBb
AB	ab	AaBb
Ab	ab	Aabb
aB	ab	aaBb
ab	ab	aabb

Note: By convention, capital letters representing genes are given first regardless of whether they come from the female or male parent.

equal numbers so that each combination would be found in 25 percent of all pollen and eggs. Assume that in cross-pollination each of the four types of pollen has an equal chance of landing on a stigma and ultimately fertilizing any of the four types of eggs to produce an embryo. Given these conditions, embryos (or seeds) of nine different genotypes can be expected. For example, if AB pollen fertilizes an Ab egg, the resulting embryo is $AABb$. The possible genotypic combinations are summarized in Table 4-1.

A Punnett square provides a simple method for determining all possible combinations of male and female gametes. The square shown in Figure 4-6 is divided into sixteen smaller squares, with the four male gametes lined up at the left and the four female gametes across the top. Each small square represents the genotype resulting from fertilization, the union of egg and sperm. The Punnett square can be used to determine all possible combinations of any number of gametes, but,

	Female gamete			
Male gamete	AB	Ab	aB	ab
AB	1 AABB	5 AABb	9 AaBB	13 AaBb
Ab	2 AABb	6 AAbb	10 AaBb	14 Aabb
aB	3 AaBB	7 AaBb	11 aaBB	15 aaBb
ab	4 AaBb	8 Aabb	12 aaBb	16 aabb

Figure 4-6
Punnet square showing all possible combinations of four male and four female gametes. The order in which the squares are numbered is arbitrary.

if more than eight types of gametes are produced, this method becomes unwieldy.

Although all four types of both eggs and sperms are produced in equal numbers, chance combinations yield nine genotypes, which do not occur in equal numbers. The production of a given type of sperm is independent of the production of a given type of egg. The probability that a given sperm will fertilize a given egg equals the product of the probability of finding that sperm and that egg. In this example, there is one chance in four of finding a given type of sperm (e.g., *AB*). Similarly, the probability of finding an *AB* egg is 1/4. The odds that these two gametes will unite to produce an *AABB* embryo are 1/4 × 1/4 = 1/16. The probability of fertilization yielding an *AABb* embryo is 1/8. This embryo can result from an *AB* sperm fertilizing an *Ab* egg; the probability that this would happen is 1/16. An *AABb* embryo can also result from an *Ab* sperm fertilizing an *AB* egg; the probability of this happening is also 1/16. Because either combination results in an *AABb* embryo, the probability of obtaining that embryo is 1/16 + 1/16 = 1/8. For convenience, each cell in Figure 4-6 has been numbered. Note that, of the sixteen possible combinations, there are only nine possible types of embryos. The *AABB* embryo is found only in cell 1; thus, the probability of producing this embryo is 1/16. The *AABb* embryo is found in cells 2 and 5 or in 2/16 (1/8) of the cells. The *AaBb* type is found in cells 4, 7, 10, and 13, or in 4/16 (1/4) of the cells. Verify that the embryo genotypes and their frequencies are those given in Table 4-1. As long as no force disrupts the population of plants, in every generation the frequency of the various genotypes will remain the same when plants are randomly cross-pollinated.

The results of continued self-fertilization are quite different (Figure 4-7). To understand them, the frequencies of all genotypes must be traced through several generations. We will start with a plant of the first-filial (F_1) generation. It is the product of a cross between homozygous parental plants. Seeds produced by self-fertilization of the F_1 plant become the F_2 generation; those produced by self-fertilization of the F_2 plant become the F_3 generation; and so on. The heterozygous F_1

Parents				AABB × aabb					
Gametes				AB	ab				
F₁ embryo				AaBb					
Gametes (male and female)				AB Ab aB ab					
Frequencies and genotypes of F₂ embryos from self-pollination of F₁	1 AABB	2 AABb	1 AAbb	2 AaBB	4 AaBb	2 Aabb	1 aaBB	2 aaBb	1 aabb
Gametes produced by each F₂ plant	AB	AB Ab	Ab	AB aB	AB Ab aB ab	Ab ab	aB	aB ab	ab
F₃ embryos from self-pollination of F₂ plants	AABB	AABB AABb AAbb	AAbb	AABB AaBB aaBB	AABB AABb AAbb AaBB AaBb Aabb aaBB aaBb aabb	AAbb Aabb aabb	aaBB	aaBB aaBb aabb	aabb
Summary of F₃ genotypes and their frequencies:									
Genotype	AABB	AABb	AAbb	AaBB	AaBb	Aabb	aaBB	aaBb	aabb
Frequency	9/64	6/64	9/64	6/64	4/64	6/64	9/64	6/64	9/64

Figure 4-7
The result of self-pollination of an F₁ plant to obtain F₂ plants and of self-pollination of each F₂ plant to obtain F₃ plants. Gametes from each plant are shown and the genotype frequencies in the F₃ generation are summarized. Note that there are more homozygous genotypes than would be expected from random cross-pollination.

plant produces four types of both male and female gametes: *AB, Ab, aB,* and *ab,* which occur in equal frequency. All possible combinations of male and female gametes occur in the F₂ generation in different, but specific, frequencies as follows: 1/16 *AABB,* 1/8 *AABb,* 1/16 *AAbb,* 1/8 *AaBB,* 1/4 *AaBb,* 1/8 *Aabb,* 1/16 *aaBB,* 1/8 *aaBb,* and 1/16 *aabb.* In self-fertilized plants, each homozygous genotype (*AABB, AAbb, aaBB,* and *aabb*) produces only one type of gamete; that is *AB, Ab, aB,* and *ab.* Consequently, the homozygous geno- types perpetuate themselves. In addition, heterozygous genotypes produce various combinations of homozygotes and heterozygotes in each generation, but once a plant is homozygous, it produces only homozygous offspring. In our example, 1/8 of the plants produced are *AABb.* When plants of this genotype are self-fertilized, a second generation will consist of three genotypes. Because the *AABb* plant can produce two types of gametes, *AB* and *Ab,* all possible combinations will be 1/4 *AABB,* 1/2 *AABb,* and 1/4

AAbb. As a result of self-fertilization, one half of the embryos formed by the original *AABb* heterozygous plant are homozygous: 1/4 *AABB* and 1/4 *AAbb*. The results of self-fertilization for two generations are summarized in Figure 4-7.

Unlike most animals, many plants are polyploid; they have more than the 2N number of chromosomes in their vegetative cells. This condition complicates the process of meiosis and may markedly alter the ratios discussed above. Polyploidy is a phenomenon of significant interest in the study of evolution, in genetics, and in plant breeding. However, because of its complexity, it will not be treated further here.

Asexual Reproduction

There are at least three forms of reproduction that do not require the complex processes of meiosis, spore formation, pollination, and fertilization. All three are asexual forms of reproduction. Plants reproduced asexually are identical with the female plant from which they came. On the other hand, plants developed from seed formed by means of the sexual process may be quite different from either parent as a result of genetic mixing or recombination, which takes place during meiosis.

VEGETATIVE

When plants are reproduced from parts other than seed, the mode of reproduction is vegetative. Different plant parts can be sown from which entire plants will develop. These plant parts include bulbs and pieces of stolons, rhizomes, and tubers. Among the major crops of the world, the most common examples of the use of vegetative reproductive structures are tubers for potatoes and pieces of stem that include nodes and buds for sugar cane. Although seed is not used directly in the production of these crops, both can and do produce seed in a manner similar to other crop plants. Many weedy species also reproduce vegetatively and can be spread widely when rhizomes, stolons, or similar structures are transported from field to field on equipment.

GRAFTAGE

Another form of asexual reproduction is graftage; it is important in orchard crops and grapes. Graftage includes both grafting and budding. Both processes require a stock (the trunk of a tree), which produces roots, and a scion (the cutting to be attached to the trunk), which grows and produces the upper, fruit-bearing part of the tree or vine. Budding is done while the tree is actively growing (that is, during the growing season) and grafting is done while it is dormant.

Before the scion and stock can grow together, contact must be made between the meristematic tissue of both the stock and the scion. All scions from a common source are genetically the same; thus, "trees" from these scions are uniform in such traits as fruit quality and flowering time. Another advantage of graftage is that disease-resistant rootstock can be used with desirable scions, a common practice in the production of grapes. Graftage allows the plant scientist to produce a plant that has a rootstock of one genotype and a top of another. Some rootstocks are more winter-hardy than others, and some have a dwarfing effect on the scion, yielding a tree more desirable in size and shape. Apples, grapes, and pears are propagated by grafting; cherries and peaches by budding. Grafting is not used in field-crop production.

APOMIXIS

The seeds and fruits of some types of plants are developed without fertilization. This type of asexual seed (or fruit) production is termed apomixis. Some plants require pollin-

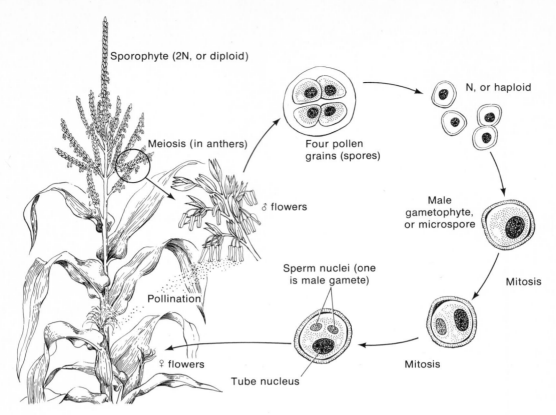

Figure 4-8
Steps in the production of the male gamete: the microspore (or male gametophyte) becomes a free living plant; the sporophyte is the crop plant.

ation, but not fertilization, for seed or fruit production. There are two types of apomixis: parthenogenic and parthenocarpic. Parthenogenesis is a form of reproduction in which seeds are produced without fertilization; parthenocarpy is a form in which fruit is produced without fertilization. Some grasses (bluegrass being a good example) are parthenogenic, and many of the valuable seedless fruits, such as grapes and oranges, are the products of parthenocarpy.

Alternation of Generations

The life cycle of a crop plant (and of any other higher plant) from seed to seed actually involves two discrete generations and, in a sense, three plants. One generation in the life cycle is greatly reduced and not readily observed.

In the anther of a mature plant there is specialized tissue that undergoes meiotic cell division, each spore mother cell dividing into four cells, called microspores. Each microspore has a single, haploid (N) nucleus; after mitotic division, it becomes a pollen grain. Each pollen grain is a free living plant (Figure 4-8). Before the pollen grain germinates, it cannot effect fertilization. The plant that produces pollen is the spore producer, or the sporophytic plant. The spore that is produced is the gametophyte, or the gametophytic plant.

After the pollen is shed from the anther

and lands on the stigma of a flower, it goes through a developmental sequence, including mitosis but not cytokinesis. The original haploid nucleus divides into two nuclei. One of these becomes the tube nucleus and undergoes no further division. The other is the generative or reproductive nucleus. This nucleus divides again to yield a sperm nucleus (the male reproductive nucleus, which is equivalent to the sperm in animals) and an endosperm nucleus. Thus, the original pollen grain, which is a free living plant, develops and divides to produce a male gamete.

The same sequence of events takes place in the female reproductive organs (Figure 4-9). However, the events are far less evident because they take place within the confines of the ovary. In the ovary there are spore mother cells that undergo meiotic cell division to produce female spores. At the conclusion of meiosis, there are four cells, each having a single, haploid (N) nucleus. These cells are surrounded by two layers of cells, called the integuments, which also arise from the ovary. Three of the four cells resulting from meiosis degenerate and are lost in most plants. The remaining cell is the female spore (or the female gametophyte), which then undergoes three mitotic divisions, resulting in an embryo sac containing eight haploid nuclei. One of these nuclei is the egg nucleus or the female gamete. On either side of the egg nucleus are the synergid nuclei. They play an indirect role in fertilization and need not be considered further now. In the center of the embryo sac there are two more nuclei, the polar nuclei. At the end opposite the egg nucleus are three nuclei, the antipodal nuclei, which disintegrate. The embryo sac and the integuments surrounding it constitute the ovule. The ovule develops into a seed, following fertilization. The mature seed is a miniature plant in an arrested stage of development. The integuments become the seed coat or testa. They surround the embry-

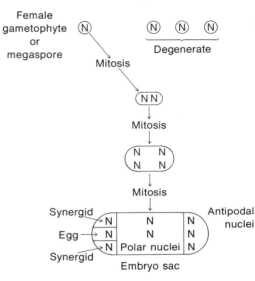

Figure 4-9
Steps in the production of the female gamete. The megaspore is confined in the ovary of crop species and other higher plants; it ultimately develops into the embryo sac, which includes the female gamete or egg.

onic plant, which consists of a stem, cotyledons at the first node, and, in some plants, several other leaves and roots. This minute plant, under the right conditions, will grow into a complete plant and reproduce.

The mature crop plant that is seen in the field is a sporophyte; as a result of meiosis it produces spores. For most species, the same plant produces both male (micro) and female (mega) spores. The spores are the gametophytes, which are free living but greatly reduced plants.

Because sexual reproduction in plants consists of spore production by the sporophyte followed by gamete production by the gametophyte, there is an alternation of generations: sporophyte → gametophyte → sporophyte, and so forth.

When the anther is mature, it ruptures and the pollen grains are released. Ultimately, the pollen grain lands on a stigma—an event called pollination. The tube nucleus of the pollen grain penetrates the stigma and grows down the style. The two other nuclei, the sperm and endosperm nuclei, follow down the tube. The sperm nucleus and the egg nucleus unite to form the zygote; this is fertilization. (Remember that the genetic makeup of the zygote, and of the plant into which it ultimately develops, is the sum of the genetic makeup of the gametes that unite at fertilization to form it.) The endosperm nucleus joins the two polar nuclei in the embryo sac; this is a second fertilization and it eventually yields endosperm. The two events together are referred to as double fertilization, which occurs in crop plants as well as many other higher plants.

The endosperm develops into stored food used ultimately by the embryo as an energy source for germination. Following double fertilization, the zygote grows and develops into an embryo or young plant. The endosperm increases in volume also, through mitosis. Cells of the endosperm are triploid: as a result of the haploid endosperm nucleus fertilizing the two haploid polar nuclei, there are three homologs of each chromosome rather than two as there are in each embryo cell. The entire life cycle is diagrammed in Figure 4-10. The integuments, which protected the embryo sac, develop and serve as the protective seed coat or testa. In the ovary, then, there is a seed comprising the embryo, the endosperm, and the seed coat that arises from the integuments. In some plants the endosperm cannot be found; the reserve food material that results from double fertilization is used to develop the very heavy, fleshy leaves that form at the first node of the stem in the embryo. These leaves are called cotyledons, and plants that have two of them at the first node are called dicotyledons or dicots. Alfalfa, beans, peas, and fruit trees, as well as many weeds, are dicots. Close examination of a bean, not the pod, after it has been soaked in water and split open reveals that the halves of the bean are in fact cotyledons (Figure 4-11).

Plants that have only a single leaf (called a scutellum rather than a cotyledon) at the first node of the stem in their embryos are monocotyledons, or monocots (Figure 4-11). The scutellum is small, but sugar and important proteins are stored in it. However, the main storage area for monocots is the endosperm region. Grasses are the most important monocots.

Double fertilization occurs in both monocots and dicots. In monocots, the result is triploid endosperm. In dicots, in which endosperm is generally not found, energy for the germinating seed is stored in the cotyledons. Castor beans are an exception: inside of the seed coat there is a ring of true endosperm; in addition, true fleshy cotyledons are also formed.

Fruits

Seeds are the mature ovules that are enclosed within the ovary. The mature seed plus the ovary (and sometimes part of the receptacle of the flower) constitute a fruit.

CLASSIFICATION

Fruit classification is based on four factors: the number of ovaries constituting the fruit; the number of flowers; the properties of the mature ovary wall (pericarp); and whether the fruit is dehiscent or indehiscent. The three major kinds of fruits—simple, aggregate, and multiple—are distinguished by the number of ovaries and/or flowers from which they are formed.

Simple fruits develop from a single ovary

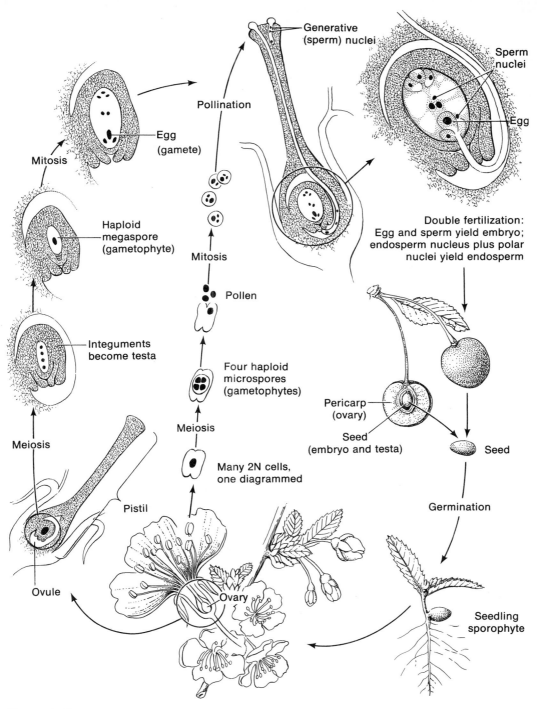

Figure 4-10
Summary of male and female life cycles showing spore production and fertilization.

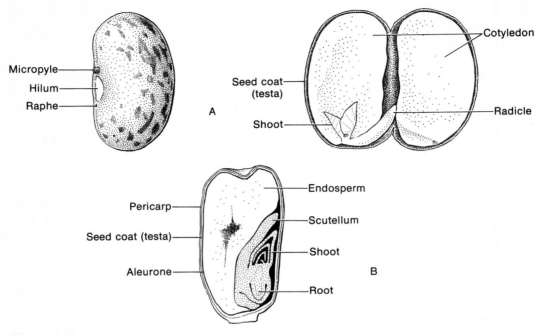

Figure 4-11
Details of (A) a dicot seed and (B) a grass fruit or caryopsis.

and have one or more seeds; they may be formed from a compound pistil—a pistil developed from two or more fused carpels. Simple fruits are classified as either dry or fleshy (having either a dry or a fleshy pericarp at maturity). The pericarp consists of three layers: the outermost is the exocarp; the middle, the mesocarp; and the innermost, the endocarp. In dry fruits, which are further classified as being dehiscent or indehiscent, these layers are fused and not evident. In fleshy fruits, one or more layer is prominent: the meat of a cherry is mesocarp and the hard outer layer of the pit, endocarp. Aggregate fruits form from a number of ovaries that come from a single flower on a single receptacle. Strawberries are aggregate fruits; individual ovaries become simple fruits. Multiple fruits, typical of sugar beets, are formed from the ovaries of two or more fused flowers.

Diagrams of various fruits are given in Figures 4-12 and 4-13, and the classification of fruits is summarized, with examples, in Table 4-2. Fruits and seeds are the plant parts most commonly consumed by humans; leaves and stems are more frequently used as livestock feed. Understanding how fruits and seed are formed is essential to knowing how to manage a crop to obtain the greatest possible yields of the highest quality. In addition, both fruit and seed characteristics are used in the scientific identification and classification of plants. This may seem to be of minor importance for familiar, cultivated plants, but it may be of practical significance in surveys of range plants and in identifying weeds.

What is generally referred to as wheat or corn seed (or the "seed" of any grass) is in fact a fruit. It is a simple, dry indehiscent fruit called a caryopsis or, commonly, a grain. It contains the true seed, comprising the integuments, the embryo, and the endosperm. The

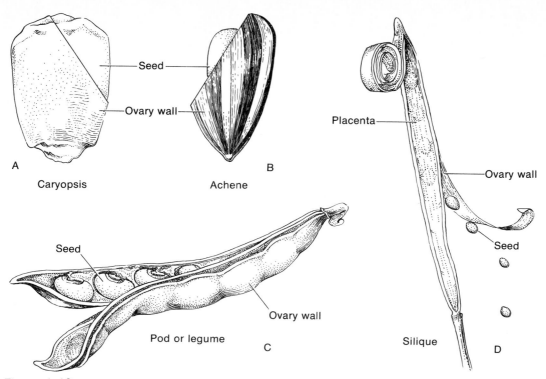

Figure 4–12
Dry fruits: (A and B) indehiscent and (C and D) dehiscent. The caryopsis, the achene, and the pod form from a simple pistil; the silique from a compound pistil.

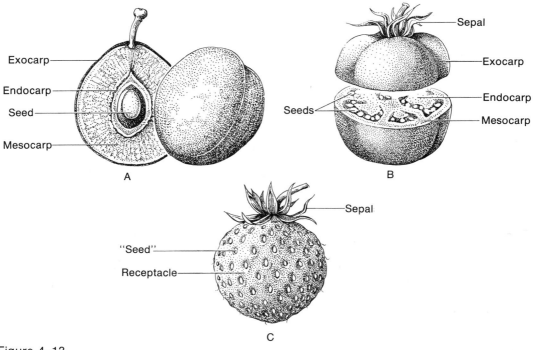

Figure 4–13
Examples of fleshy fruits (A and B) and an aggregate fruit (C).

Table 4-2
Classification of some common types of fruits.

Type	Name	Example
Simple, dry, indehiscent	Achene	Safflower, sunflower
	Caryopsis or grain	Wheat, barley, corn
	Nut	Walnut, hickory
Simple, dry, dehiscent	Legume or pod	Peanut, bean, pea
	Follicle	Many weeds
Simple, dry, dehiscent (formed from compound pistil)	Silique	Mustard family
	Schizocarp	Carrot, celery
	Capsule or boll	Jimsonweed, poppy, flax
Fleshy	Drupe	Plum, cherry, peach
	Berry	Tomato, cucumber
	Pome	Apple
Aggregate	Individual fruits are achenes	Strawberry
Multiple		Sugar beet

common bean or pea pod is a simple, dry (generally) dehiscent fruit. The pod is the mature ovary wall; within it are several true seeds.

Crop Botany and Crop Management

The optimum field management of any crop, from selecting seed to planting, harvesting, storing, and processing, requires an understanding of the plant. Economic considerations must also be taken into account, but understanding how a plant develops and grows provides a basis on which to make economic decisions regarding fertilizing or irrigating or methods and frequency of weed control.

From a biological point of view, the goal of crop production is to manage a field so that maximum products of photosynthesis are harvested (although economic considerations can modify this goal). For effective management, the crop producer must understand not only the growth and development of individual plants, but also how groups or populations of plants interact in the field.

In a broad sense, growth may be defined as an increase in dry matter, including increases in the number of cells and in cell size. The pattern of growth for nearly any organism takes the form of a sigmoidal, or S-shaped, curve (Figure 4-14). Growth can be divided into three stages: (1) initial growth, which is slow; (2) a period of rapid growth; and (3) a plateau in which little additional growth occurs. As plants mature and die, as do annuals, or become dormant, as do biennials and perennials, the plateau stage ends and there is a marked decrease in dry matter. This phenomenon is not usually considered part of the period of active growth. In some cases there may be an apparently premature loss of dry matter early in the plateau stage. This phenomenon is fairly typical of soybeans, which suddenly lose leaves about three weeks before the crop is mature. For annual species, this curve represents growth from planting through maturity and harvest. For many perennials, it represents growth from its initiation in the spring until dormancy in the fall. For species that grow actively in the fall, winter, and spring, it represents growth from the

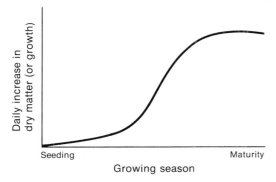

Figure 4-14
Pattern of daily growth, or dry-matter accumulation, from seeding to maturity.

end of a dry season to the start of the next dry season.

The stage of development in which the most rapid growth occurs differs for different crop species and cultivars. For example, in a study of several different cultivars of corn, the time it took to reach the stage of development at which ten leaves were fully visible on each plant was the same for each cultivar. However, the time it took to grow from this stage of development to the silking stage was markedly different for different cultivars. The time required for seed weight to increase depended on the cultivars, and in some cultivars dry matter that had been stored was translocated to support seed (fruit) development. During periods of rapid growth, a plant requires the maximum amount of moisture, minerals, and other essential nutrients, and a producer must, as far as possible, insure that they are available in adequate quantities (see, for example, the relationship between growth rate and the use of nitrogen in sugarcane, Chapter 25).

Not only does rapid growth occur at different stages of development for different species and cultivars, but different parts of plants grow more rapidly at one stage than another; thus, the timing of such management practices as the application of fertilizers or irrigation may affect one plant part more than another, as may the time at which stress occurs, such as high temperatures, drought, or attack by diseases or insects.

In cotton, total dry matter increases in the usual sigmoidal pattern (Figure 4-15). Vegetative material (leaves and stems) contributes uniformly throughout the season. Lint production takes place in forty to eighty days, starting in mid-August. The growth pattern of spring wheat, on the other hand, clearly demonstrates the different stages at which different plant parts grow most rapidly (Figure 4-16). From midseason to maturity, stem dry matter is apparently translocated to developing kernels. Total dry matter increases, and the distribution changes.

In cereals, seed yield is a function of three components: the number of inflorescences (spikes or panicles) or reproductive tillers per plant, the number of kernels per inflorescence, and kernel weight. Early in vegetative development, environmental factors have the greatest effect on the number of tillers; they have

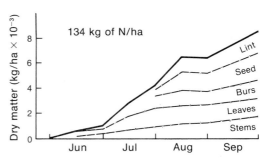

Figure 4-15
Pattern of dry-matter accumulation of a cotton plant and the contributions made by different parts of the plant throughout the growing season. [Adapted from D. M. Bassett, W. D. Anderson, and C. H. E. Werkhoven. Dry matter production and nutrient uptake in irrigated cotton (*Gossypium hirsutum*). Argon. J. 62(1970): 299–303.]

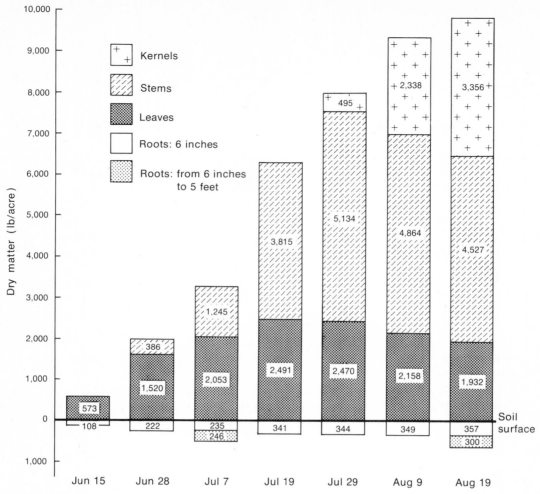

Figure 4-16
Relative contribution to total dry-matter accumulation by different plant parts of spring wheat throughout the growing season. [From F. H. McNeal, M. A. Berg, and C. A. Watson. Nitrogen and dry matter in five spring wheat varieties at successive stages of development. *Argon. J.* 58(1966):605–608.]

less effect on the number of kernels; and the least on kernel weight. Uniform yields can be obtained from different plants by adjusting any of these components: if a plant has many tillers, the number of kernels and their weight may be reduced, whereas fewer tillers may result in an increase in the number of kernels and in their weight (see Chapters 13 and 16). Usually the crop producer considers yield from a given area, an acre or a hectare, not from a single plant; however, these same principles are as valid for, say, tillers per acre as for tillers per plant.

Rates and patterns of development become more complex when considered for an entire field rather than for a single plant. Nearly all field management practices are based on the developmental pattern of single plants and how the patterns of groups of plants can be integrated for maximum yield per acre (or hectare). Decisions start with seeding and continue through harvesting.

Seeding rate and planting pattern are important in determining yield per acre. Seeding rate is expressed either as the number of seeds sown per acre (or hectare) or as pounds of seed per acre (or kilograms per hectare). Planting patterns consist of the amount of space between rows of plants, the amount of space between plants within a row, and other factors such as seeding in beds and transplanting.

To achieve the goal of maximum production from photosynthesis requires the optimum use of all factors essential for plant growth. Like a chain, plant growth is no stronger than its weakest link, but the link that is weakest may change as the plant grows and develops. For example, the distribution of roots increases rapidly as a seed germinates and as a seedling grows. Water requirements also increase. Seeding rates are determined by the pattern of root distribution and water requirements to the extent that excessively high seeding rates may lead to competition among plants for available water and may reduce the total yield per acre. Seeding rates can be adjusted so that competition for essential materials is minimized: if minimum moisture is available, seeding rates should be low (e.g., 18,000 corn plants per acre or 45,000 per hectare, and 31 pounds of winter-wheat seed per acre or 35 kilograms per hectare); if optimum moisture is available, rates may be increased two or threefold. The same adjustments can be made for the availability of essential minerals.

Under certain conditions, moisture and minerals can be essentially eliminated as factors that limit crop growth through irrigation and the application of fertilizers. In that case, planting patterns and seeding rates may be determined by light and its availability to the plants. Plants differ in the amount of light they require or can utilize. For a variety of reasons, not all the leaves on a single plant are equally active in photosynthesis. The lower leaves of a soybean plant, for example, contribute relatively little to productivity compared with the upper leaves, which receive more light. The crop producer can do little about intraplant shading or competition for light, but plant breeders are developing shade-tolerant cultivars that can be seeded at relatively higher rates (see Chapter 14). The crop producer can do a great deal by selecting seeding rates and planting patterns that will minimize shading between plants. For each crop there is an optimum total leaf surface area per area of land for optimum photosynthesis. The ratio of leaf surface area to land area, the leaf area index (LAI), is significant in determining appropriate seeding rates and planting patterns, as is the distribution of leaves from the upper part of plants to the base of the plant.

Because of intra- and interplant shading, all leaves do not receive the same quantity of light. Seeding rates and planting patterns for various crops are designed so that each crop will receive adequate light for optimum photosynthesis. In pastures in which different species are mixed, this is critical because competition for light may reduce yields. Weeds are also serious competitors for light (see Chapter 9). Weed management requires an understanding of the growth and development patterns of both the crop species and the weed species so that yield-reducing competition from the weeds can be eliminated. For example, in northern areas spring weeds can be a serious problem for warm-season crops (those planted in summer) because they have a head start in growth and may therefore use water that is needed by the crop, or they may shade young, later-growing seedlings.

Successful crop production starts with quality seed. The seed must be plump and free of injuries or damage. Something as simple as a split seed coat in a bean, for example, may indicate that the fleshy cotyledons have been separated from the rest of the embryo; if that were the case, the food supply would not be available to the seedling. Remember, too, that seed is alive, and the manner in which it is stored is critical: oxygen must be available for respiration and the temperature must be controlled so that essential enzymes are not altered or destroyed by excessive heat, in which case the seed would not germinate, or the seedling would be weakened.

When the seed is planted, it becomes solely dependent on the seedbed for materials and conditions required for germination. Soil temperature is critical for germination and temperature requirements vary from crop to crop. If conditions do not favor germination, the seed may rot, resulting in a total crop failure. Similarly, adequate moisture is essential, but the amount varies. Too much water forces air from the soil, and seeds or seedlings cannot respire normally and may appear to "drown." For some crops, planting dates must be adjusted to meet soil-temperature requirements. They may also be arranged to take advantage of expected rainfall if irrigation facilities are unavailable.

When the seed finally germinates, the distribution of the seedling roots in the soil dictates fertilizer placement and the frequency and amount of irrigation. As the seedling grows, it has specific requirements with respect to minerals and moisture, which must be met at the appropriate time in the correct amount for highest yields. Corn, for example, uses less water than alfalfa, but corn requires heavy applications of nitrogen fertilizer and alfalfa rarely, if ever, requires nitrogen fertilizer.

As the plant approaches maturity and its growth changes from vegetative to reproductive, its needs change too. In cereals (e.g., wheat), moisture stress during and after flowering frequently leads to light, shrunken grain. High temperatures at flowering cause many types of beans to shed their flowers. Field management must be designed to prevent, or at least minimize, damage. Also, large amounts of nitrogen generally lead to high-protein grain. This is frequently desirable for wheat, but it is undesirable for malting barley.

Sugar beets are biennial plants, but they are farmed as annuals. Proper fertilizer management is essential for maximum sugar yields. The management of all types of perennial forages depends on a knowledge of the plant. Bud location and function and sources of energy for plant growth after harvesting must be determined so that the best grazing management can be practiced. Hay and silage quality depends on the quality of the plant material to be processed, which in turn depends on fertilizers, irrigation, seed quality, and planting patterns.

Harvesting methods, although highly mechanized, must be adapted to the crop. Using improper or poorly adjusted equipment, or harvesting too early or too late may seriously reduce the quality of a prime crop in the field.

All crops are subject to a number of pests and diseases. A crop producer must be able to identify the weeds, insects, or diseases that threaten his crop; then he must determine the potential damage to the crop and take measures to protect it accordingly. Many broadleaf weeds that grow in cereals can be controlled with 2,4-D, but this chemical will kill beans and potatoes and many other dicots. The effectiveness of the chemical depends on the stage of growth of both the crop plant and the weed and is modified by climatic factors.

Insects are not always seen and therefore must be identified by the damage they cause. This requires a knowledge of various plant parts and being able to distinguish normal from damaged parts. It may also require recognizing specific types of damage. Plant diseases, too, are identified by recognizing their symptoms. Knowing the plant and the pest or disease and their relationships is the fundamental step to control. Many pests of all types can affect the plant at different stages of its development, so that management practices change as the crop grows.

Selected References

Bassett, D. M., W. D. Anderson, and C. H. E. Werkhoven. 1970. Dry matter production and nutrient uptake in irrigated cotton (*Gossypium hirsutum*). Agron. J. 62:299–303.

Greulach, Victor A. 1973. *Plant function and structure.* Macmillan.

McNeal, F. H., M. A. Berg, and C. A. Watson. 1966. Nitrogen and dry matter in five spring wheat varieties at successive stages of development. Agron. J. 58:605–608.

Robbins, W. W., T. E. Weir, and C. R. Stocking. 1966. *Botany.* 3d ed. Wiley.

Study Questions

1. What alternatives to annual sexual reproduction do various plants have?
2. What are the differences between a seed and a fruit? Between a dry fruit and a fleshy one?
3. How can a single flower form a compound fruit?
4. What are the distinguishing features of the major types of inflorescences?
5. What are the differences between a spike, a spikelet, and a floret?
6. What are the genetic consequences of self-fertilization? of cross fertilization?
7. What are the major differences between monocots and dicots?

Five

Plant Classification and Nomenclature

Plants have always played a major role in the life of man, as a source of food, feed, medicine, and even shelter. When civilization was confined to small, isolated centers, local names for important plants were adequate. With the advent of the Crusades about 1000 A.D., communication between civilized centers increased as did knowledge of new cultivars of plants from widespread sources. By the early Renaissance, as scholarly work was spread among expanding civilized centers, the use of local plant names created confusion, and the need for a coherent system of plant nomenclature and classification became apparent.

The Scientific System

Although many botanists considered this problem before the middle of the eighteenth century, scientific nomenclature dates back to two works by Carolus Linnaeus: *Species plantarum* published in 1753 and *Systema naturae,* first published in 1735. Many species described and named in these books still bear the names Linnaeus gave them.* Linnaeus developed a binomial (two-name) system in which each plant is identified by both a generic name and a specific epithet of Latin derivation. This system of nomenclature, which is universally accepted in the scientific community, forms the basis for the science of plant classification known as plant taxonomy.

Plants are classified according to seven categories: kingdom, division, class, order, family, genus, and species. Subclassifications may be recognized in each category, for example subkingdom, subdivision, subclass. Today the plant kingdom is divided into two subkingdoms and fifteen divisions; each division is divided into several classes that are further divided into subclasses. Within each subclass there are several orders; these orders are subdivided into families; each family is composed of several genera (singular, genus), and there are usually several species within each genus. Each species has a specific name of its own. It is often called a binomial be-

---

*In scientific journals, the initials of the scientist who identified and classified a species follows its scientific name.

cause it always consists of two parts: the first, the name of the genus to which it belongs, is called the generic name, and the second is called the specific epithet. For example, *Triticum aestivum* is the specific name for wheat.

A species may be further divided into subspecies that characterize distinctions within the species. Such distinctions may be in flower color (like those separating subspecies of sainfoin or clover), in winterhardiness, or in the number of chromosome sets (i.e., diploid, tetraploid, etc.).

Crop producers, seedsmen, and processors must be familiar with cultivars, or varieties, of crop species or subspecies. Cultivars of most crops have been developed by plant breeders and their names are not selected in accord with the scientific system of classification; the cultivar name may reflect certain characteristics of the plant itself or it may reflect the area in which it was developed. For example, the forage legume sainfoin is classified as *Onobrachis viciafolia*. A cultivar of this species is *Onobrachis viciafolia* 'Remont'. It is known for its regrowth after harvest (hence "Re" in the cultivar name) and was developed by plant breeders at the Montana Agricultural Experiment Station (hence "mont").

Cultivars of crop species usually differ in traits of agronomic importance. For example, various cultivars of wheat (*Triticum aestivum*) differ in grain quality, length of growing season, plant height, or resistance to disease, such as rusts. A thorough knowledge of the characteristics of different cultivars allows the crop producer to choose the cultivar that is best suited to a specific location, climate, or market (see Chapter 12).

The scientific system of plant classification, which attempts to group plants according to their evolutionary relationships, is based on the identification of ancestral plant forms. For example, paleobotanical evidence, compiled from the study of fossil plants, indicates that angiosperms (class Angiospermae—flowering plants) were derived from gymnosperms (class Gymnospermae—evergreens or cone plants that do not have an enclosed seed) and that the grass floret (a perfect but incomplete flower) was derived from a complete and perfect flower. Although paleobotanists have clarified how plants evolved from their predecessors as a result of natural selection and the artificial selection imposed by early farmers, there are still some missing phylogenetic links. Researchers are working to fill in these evolutionary steps through detailed studies in plant biochemistry, anatomy, and morphology.

The post-Darwinian phylogenetic classification of living plants is based largely on the morphological and anatomical similarities and differences of two kinds of traits: floral (structures associated with reproduction) and vegetative. Floral morphology is an important trait because it is not greatly affected by environmental conditions. For example, all members of the grass family (Gramineae) lack the typical calyx and corolla, but have three stamens and one functional pistil. Members of the legume family (Leguminosae) have complete and perfect flowers with five petals (two of which are fused) and ten stamens (nine of which are fused). The vegetative traits, used mainly in identifying species, include leaf shape, the presence or absence of hairs on leaves and stems, and the presence or absence of stolons and rhizomes (used extensively in identifying grass plants).

With the development of sophisticated biochemical and genetic techniques of analysis, other traits have become important to the plant taxonomist: the presence or absence of specific enzymes (or in some cases amino acids) or variations in chromosome number, morphology, and behavior in the offspring of artificial hybrids can be used as bases for plant classification.

When all the available data have been analyzed, plants may be accurately classified according to their phylogenetic relationships. For example, the full classification of cultivated wheat is:

Kingdom	*Plantae*
Division	*Tracheophyta*, vascular plants that have specialized conducting cells (phloem and xylem)
Subdivision	*Pteropsida*, plants that have macrophylls
Class	*Angiospermae*, flowering plants that have enclosed seeds
Subclass	*Monocotyledonae*, plants that have a seed containing a single cotyledon
Order	*Graminales*, plants that have elongated leaves and bisexual flowers
Family	*Gramineae*, grasses
Genus	*Triticum*, wheat
Species	*Triticum aestivum*, summer-growing wheat

The binomial name for cultivated wheat, *Triticum aestivum*, consists of the name of the genus followed by an epithet denoting the species; it does not include the subspecies or cultivar name.

Compare the classification of wheat with that of the forage grass crested wheatgrass (*Agropyron desertorum*):

Kingdom	*Plantae*
Division	*Tracheophyta*
Subdivision	*Pteropsida*
Class	*Angiospermae*
Subclass	*Monocotyledonae*
Order	*Graminales*
Family	*Gramineae*
Genus	*Agropyron*, wheatgrass
Species	*Agropyron desertorum*, wheatgrass that grows well under dry conditions

The classification of crested wheatgrass is identical with that of wheat down through the family. The two classifications diverge at the level of the genus. The fact that the taxonomic separation of wheat from crested wheatgrass occurs at a very refined level of classification indicates that these plants are closely related.

Now compare either wheat or crested wheatgrass with alfalfa (*Medicago sativa*):

Kingdom	*Plantae*
Division	*Tracheophyta*
Subdivision	*Pteropsida*
Class	*Angiospermae*
Subclass	*Dicotyledonae*, plants that have a seed containing two cotyledons
Order	*Rosales*, plants that have simple or compound leaves and variable fruits
Family	*Leguminosae*, fruit has a single carpel that splits along two lines at maturity
Genus	*Medicago*, forage
Species	*Medicago sativa*, alfalfa, perennial, long taproot, grows from 1 to 3 feet tall

The taxonomic separation of alfalfa from both grasses occurs at the subclass level. Alfalfa is therefore not closely related to either wheat or crested wheatgrass.

Although close to 300,000 different species of plants have been classified under the binomial system, only 30,000 are of any real value in food and feed production, and of these only 15 species make a significant contribution to the food supply of the world. These 15 species are all in the subdivision Tracheophyta (the vascular plants) and in the class Angiospermae (the seed and true flowering plants). The major food plants are in the Gramineae—grass—family (see Chapter 13), including such crops as corn (*Zea mays*), wheat (*Triticum aestivum*), rice (*Oryza sativa*), barley (*Hordeum vulgare*), and sorghum (*Sorghum bicolor*). Several other food crops are classified in the Leguminosae—legume—family (see Chapter 19), including soybeans (*Glycine max*), field beans (*Phaseolus vulgaris*), peanuts (*Arachis hypogaea*), and lentils (*Lens esculenta*). These edible legumes are an increasingly important source of protein. Sugar plants such as beets (*Beta vulgaris*) and cane (*Saccharum officinarum*) are another significant source of food, but they can only be grown in restricted areas. Although the importance of each species of food plant varies from region to region, most of the major species can adapt to a range of climatic conditions and are grown on a worldwide scale. This explains why there are relatively few of them. When species used for forage (hay, pasture, and range) are added to those used as a source of food, the total number of plants valuable to man skyrockets. Finally, there are many plants of industrial or medical value that further extend the list of plants useful to man.

The binomial system of plant nomenclature is used universally among scientists because it is accurate and minimizes the possibility of multiple names for the same plant. Although new taxonomic discoveries generate revision within the system, the basic taxonomic categories do not change. For example, until 1959, the accepted name for cultivated wheat was *Triticum vulgare*; the name was then changed to *Triticum aestivum*. (The English translation of *vulgare* is common or ordinary, and, as the example on the facing page indicates, *aestivum* is translated to "growing in the summer.") Linnaeus apparently intended to assign the specific epithet *aestivum,* but *vulgare* appeared in his classification of wheat by mistake. The restoration was made in accord with the rules of priority accepted by the International Botanical Congress, which require that the name first assigned by Linnaeus be followed. Appropriate binomial designations of newly identified or reclassified species (or other taxonomic units) are recommended by International Botanical Congresses. When a species is first described, a preserved specimen of a typical plant (the type specimen) is filed in an herbarium for reference purposes; identification of other plants may be based on comparisons with the type specimen. Many universities, museums, and other research organizations have extensive herbaria staffed by skilled taxonomists who provide a service to other plant scientists by identifying unknown plants (Figure 5-1).

Figure 5-1
Herbarium specimens (right) are used as references to aid in the identification of unknown species (left). Many of the morphological features used to separate closely related species require microscopic examination.

Agronomic Classification

The agronomic classification of plants, a system of nomenclature that identifies a plant's agricultural use, is a business convention and not an attempt to classify plants in a precise scientific way. This system of classification specifies how a crop will be used, that is, for fiber, oil, or feed.

Cereals and forages are two major groups in the agronomic classification system. A cereal is defined as a grass grown for its edible seed. Wheat, rice, rye, sorghum, barley, oats, and corn are all cereals; they belong to seven genera within the Gramineae family: *Triticum, Oryza, Secale, Sorghum, Hordeum, Avena,* and *Zea*. Cereals are among the world's leading food and feed crops.

Forages include all plants whose products (sometimes referred to as roughages) are used for feed; they may be grazed, dried for hay, or processed to make pellets or wafers. However, cereal grains used as livestock feed are not considered forage; they fall under the broader heading of "feed crops." The forages constitute an extremely broad classification that includes members of both subclasses of the Angiospermae: Monocotyledonae (grasses) and Dicotyledonae (legumes, such as alfalfa).

Other agronomic groups are: seed legumes (beans, peas, and soybeans), bulb crops (onions and garlic), range crops (grasses, forbs, etc), sugar crops (beets, cane, and sorgos), oil crops (flax, soybeans, cotton, castor beans), fiber crops (flax, cotton, hemp), and the very broad classification, vegetable crops. In several instances, crops may be listed in two or more groups, For example, flax is an important oil crop (the source of linseed oil); it is also a fiber crop (the source of fiber for fine linens). Similarly, cotton is both an oil and a fiber crop. Many range crops are also considered forages. Because the agronomic classification of a crop may vary from region to region, crop producers must understand local usage for accurate crop identification.

Common Names

Common names for plants, which are used extensively in agriculture, are descriptive and bear no relationship to binomial nomenclature. Because common names vary from retion to region, crop producers must know which specific plants are designated by the common names of each region. For example, corn (*Zea mays*), the leading crop by acreage in the United States, is referred to as maize in some regions. In addition, the relationship between corn (*Zea mays*) and various types of sorghum (*Sorghum bicolor*) is unclear: in some regions, sorghum is called maize, in others, milo or milo maize. The common names of many forage grasses, particularly wheatgrasses, also vary greatly from location to location. The confusion reaches its peak with respect to weeds. Field bindweed (*Convolvulus arvensis*) may be called creeping Jenny or wild morningglory. In addition, the common name creeping Jenny refers to a completely different genus and species (*Polygonum convolvulus*), which in turn is also known as wild buckwheat or field bindweed.

Government Classification and Grade

The governments of many countries have established standards (or grades) for crops that are to be sold. A crop producer must have a knowledge of such standards in developing production practices and techniques that will insure a high-grade crop product. The United States government has established grades for many crops. It divides each crop (e.g., corn) into market classes and assigns grades within each class. Although the factors used in determining grades vary from crop to crop, as do the grade designations themselves, the evaluation procedure is similar for most crops and can be readily illustrated with wheat.

Wheat produced in the United States is divided into seven market classes: Hard Red Spring, Durum, Red Durum, Hard Red Winter,

Soft Red Wheat, White Wheat, and Mixed Wheat. For grading purposes, a given lot can be assigned to only one class and grade; however, a producer may grow lots of different classes or grades. A market class is determined by the type of kernel (color and hardness), the type of growing season (either winter or spring—see Chapter 2), and the area in which the crop is produced. Each class may be divided into two or more subclasses. Most of the domestic wheat used in bread is classified as Hard Red Spring (HRS). It has a hard, nearly brittle, reddish kernel and is a spring type of wheat that is planted in April or May and harvested the following autumn. This wheat yields flour with desirable properties for baking bread. HRS is divided into three subclasses: Dark Northern Spring, Northern Spring, and Red Spring. Wheat that is milled for pastas (macaroni and spaghetti) is in the Durum class. This market class is also divided into three subclasses: Hard Amber Durum, Amber Durum, and Durum. The market class Red Durum has no subclasses. Because it has an undesirable flour color after milling, Red Durum wheat is rarely grown or consumed in the United States. Wheat in the market class Hard Red Winter (HRW) is often blended with HRS wheat for use in baking bread. HRW is divided into three subclasses: Dark Hard Red Winter, Hard Red Winter, and Yellow Hard Winter. These wheats have reddish kernels that range from hard to brittle. They are planted in the fall and harvested in midsummer. Wheat in the market class Soft Red Wheat, which has no subclasses, is milled to produce flour for cakes and other pastries. Soft Red Wheat has reddish kernels that are softer than those of HRS and HRW wheat. The market class White Wheat is divided into four subclasses: Hard White Wheat, Soft White Wheat, White Club Wheat, and Western Wheat. Flour from White Wheat is also used for pastries. The final market class, Mixed Wheat, is divided into three subclasses that are determined by the type and percentage of the specific grains used in each mixture. Wheat in this market class is generally used as feed.

A grade from 1 to 5 is assigned to grain in each market class (Table 5-1); 1 is the best grade and 5 the poorest. Grade is based on three factors: test weight, defects, and amount of mixtures. The test weight is the weight in pounds of a bushel of grain. It is a good measure of the plumpness or fullness of the grain. For grading purposes, a bushel of wheat is defined as a volume of four pecks or 35.24 liters (about 1.25 cubic feet). Defects include heat-damaged kernels, damaged kernels, foreign material, and shrunken and broken kernels. A maximum of each of these defects is allowed for each grade, and the total of all defects may not exceed a specified amount. The final criterion of grade is the amount of mixture: there is an upper limit on the total amount of contrasting grain that may be mixed with each market class; this is known as the total mixture. Within the total mixture, the amount of grain from *each* contrasting market class (e.g., White Wheat mixed with Hard Red Winter) may not exceed a specified maximum.

It is essential to know the factors considered in grading in order to produce a high-grade crop (see Table 5-1). With wheat, all of the factors on which grade is based can be controlled to some extent by the crop producer. Good filling of the grain to achieve a high test weight requires adequate moisture at and after flowering; fallow, irrigation, and weed control all are important practices to follow in this regard. Selection of a cultivar that matures before water stress becomes critical may also help. Fertilization and seeding rates affect test weight: overfertilization may cause excessive vegetative growth and lighter grain; heavy seeding rates may lead to competition among wheat plants and thus to lighter grain. Harvesting and storing damp grain

Table 5-1
Standards used in determining U.S. grades of wheat.

Grade	Minimum test weight in pounds per bushel		Maximum limits in percentages					Wheat of other classes	
			Defects						
	Hard Red Spring Wheat or White Club Wheat	All other classes and sub-classes	Heat-damaged kernels	Damaged kernels (total)	Foreign material	Shrunken and broken kernels	Defects (total)	Contrasting classes	Wheat of other classes (total)
1	58.0	60.0	0.1	2.0	0.5	3.0	3.0	1.0	3.0
2	57.0	58.0	.2	4.0	1.0	5.0	5.0	2.0	5.0
3	55.0	56.0	.5	7.0	2.0	8.0	8.0	3.0	10.0
4	53.0	54.0	1.0	10.0	3.0	12.0	12.0	10.0	10.0
5	50.0	51.0	3.0	15.0	5.0	20.0	20.0	10.0	10.0

Source: Adapted from *The official United States standards for grain.* 1974. USDA, AMS.
*Red Durum Wheat of any grade may contain not more than 10.0 percent of wheat of other classes.

results in heat-damaged kernels. The problem can be avoided by harvesting truly mature, dry grain (i.e., grain with a moisture content of 14 percent or less) and storing it properly.

An excessive amount of damaged kernels is caused by the use of improperly adjusted harvesting equipment or equipment operating at too high a speed. Proper cylinder speed and concave adjustments can greatly reduce the percentage of damaged kernels.

The percentage of foreign material (i.e., material that cannot be screened out) can be reduced by using clean equipment, by properly adjusting fans and screens, by carefully adjusting header height, and by storing the grain in a clean granary before grading.

The number of shrunken and broken kernels can be minimized by supplying adequate moisture at and after flowering and by using properly adjusted harvesting equipment. In addition, because diseased plants produce more shrunken kernels than healthy plants, planting a wheat cultivar that is resistant to local diseases is good insurance against an excessive number of shrunken kernels.

Finally, the problem of mixtures of any type can be controlled by starting with a "clean field." Planting patterns should be selected from year to year so as to prevent the growth of undesirable volunteer plants. Fallow also helps to control the problem. It is essential to know what you sow; planting clean, high-quality seed that has been certified is well worth the cost. Clean planting and harvesting equipment is essential if mixtures are to be avoided, as is clean storage.

Federal grade classification is the job of a highly skilled, licensed technician. Grading may be started in the field or at the elevator (for grain), but some steps of the procedure require specialized laboratory equipment (Figure 5-2). Laboratory studies are also necessary for the evaluation of characteristics, such as protein content, that are not formally considered in federal grade classification.

Crop producers are under increasing economic pressure to produce larger quantities of high-grade crops. To survive, they must be aware of both the problems and their solutions. In addition to the factors that determine

Figure 5-2
Seed testing and grain grading require special laboratory facilities. Seed is inspected for damage, mixtures, and other characteristics that affect seed quality.

U.S. grade, other grain characteristics affect market value. Currently, high-protein grains command premium prices. Thus, crop producers face a double challenge: they must produce high-grade crops that also meet other market requirements.

Selected References

Beetle, Alan A. 1970. Recommended plant names. *Wyoming, Univ., Agr. Exp. Sta. Res. J.* 31.

Benson, Lyman. 1962. *Plant taxonomy.* Ronald Press.

Leonard, W. H., R. M. Love, and M. E. Heath. 1968. Crop terminology today. *Crop Sci.* 8:257–261.

Study Questions

1. Why are floral characteristics more important than vegetative traits in classifying plants?
2. What advantages and disadvantages do you see in the use of scientific names? of common names? of government standards?
3. What is the basis of the scientific system of plant classification?

Part Two Plants and Their Environment

Six

Soils and Crop Production

In considering the growth and development of any plant, it is essential to keep in mind that both the functions and the structures of plants can be significantly affected by a wide array of external factors. Because plants depend on soil for water and essential minerals, soil contributes to plant growth in a very fundamental way.

The soil in which crops are grown can be defined as the upper, weathered surface of the earth's crust that is capable of supporting plant life. Indeed, soil supports plant life in two ways: it supplies moisture and essential minerals and it provides anchorage for the roots of the plant. Soil is the basic natural resource for crop production. It is possible to grow plants in mineral solutions (hydroponics) or in artificial media, but the high cost and low yields limit such alternatives to a very few, highly valued crops, such as very early spring tomatoes in cold areas, several kinds of berries, and a few ornamental plants. For world food production, soil is essential.

Soil is not uniform, not even in a small area (i.e., less than an acre). Like nearly all other naturally formed bodies, soil exists in various forms or types. The type of soil dictates how it can best be managed by a crop producer to obtain the highest yield and what type of crop can be grown.

Although accurate data on the total amount of land available in the world for crop production would be difficult to assemble, several generalizations about land use for crop production can be made.

Only about 30 percent of the earth's surface is land (70.1 percent is water). Of the total land (about 15.3 billion hectares), only about 10 percent (or 1.5 billion hectares) can be cultivated, and less than half of that is

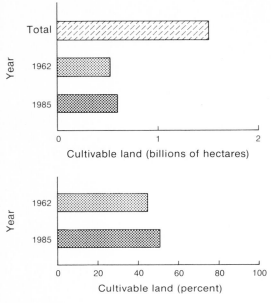

Figure 6-1
The amount of cultivable land in the world in use in 1962 and the percentage expected to be in use by 1985. [From A world agricultural plan by Addeke H. Boerma. Copyright © by Scientific American, Inc. All rights reserved.

being used to produce crops now (Figure 6-1). By 1985, nearly all of the potentially arable land in southern Asia and northwestern Africa will be in production; that of many countries in the Near East will also be almost completely utilized. In Central America and other parts of Africa, more than 75 percent will be utilized; in North America, almost all. In other words, almost all land that can be easily cultivated will be in crop production by 1985. If world crop production is then to continue to be increased by increasing the amount of land farmed, dense jungle in parts of Africa south of the Sahara desert and in South America will have to be cleared. Converting jungle into cultivable land will require immense expenditures not only for land clearing, but also for continual large applications of commercial fertilizers to obtain high yields. However, in the face of a worldwide food shortage, it may be imperative for all disciplines in the agricultural sciences to join forces to bring these jungle areas into production.

Data for land use in the United States, including Alaska and Hawaii, indicate that about 22 percent of the total land is used for crop production and 27 percent for grassland (Figure 6-2). Inasmuch as grasslands are a major feed source in livestock production, they must be considered a very important agricultural resource. Some grasslands can be diverted to the production of cultivated crops, but extreme caution must be exercised to prevent another dust-bowl disaster like those in the Midwest in the 1920s and 1930s. Some forest lands are used part time for grazing. The multiple use of such land—for timber, grazing, and recreation—is becoming acceptable to more people in a resource-conscious society.

Excluding Hawaii and Alaska, the total land area of the United States is about 1,853,000,000 acres (800 million ha). More than 70 percent, or 1.33 billion acres (0.6 billion ha), is used for grazing and field-crop production. Field crops include the major food and feed crops—wheat, corn, sorghum, barley, and rice—as well as cotton, the leading fiber crop. Dry beans are also considered field crops; most vegetables (which are luxury items in many diets) are not. More than 380 million acres (154 million ha) of field crops have been harvested annually in recent years. In spite of the growing population in the United States, less than 10 percent of its total land area is used for nonagricultural purposes: for cities, roads, parks, airports, and "waste" areas.

Although the percentage of land consumed by urbanization is fairly low, the expansion of many cities takes up prime agricultural

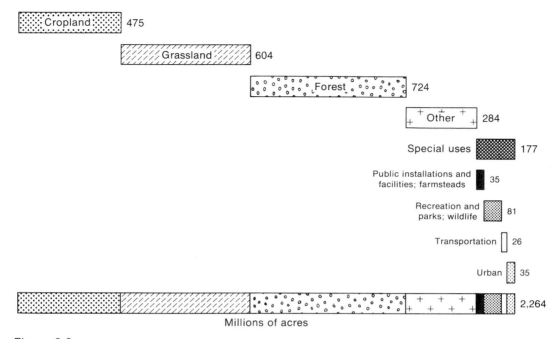

Figure 6-2
Major uses of land in all fifty states of the United States in 1969. (Grassland includes land in federal ownership.) Specific land use may change slightly from year to year, but the general pattern remains quite constant. As the population increases, more land may be diverted to urban use at the expense of food and feed production. [USDA data.]

lands—a consequence that is particularly evident in some parts of Florida and in California. Suburban development increases the value of land adjacent to cities, with a concomitant increase in property taxes, making the production of crops on such land uneconomical. Even if growing cities do not spread out to agricultural areas, a by-product of urbanization—namely, pollution— reduces the value of surrounding cropland.

The greatest potential for the development of new land on which to grow crops in the United States lies in the desert areas of the Southwest. However, in many parts of this region, extensive irrigation systems, which include getting water from the Pacific Northwest and the northern intermountain region to the Southwest, would have to be developed. In addition, problems of soil fertility would have to be solved and suitable crops selected. The conversion of desert into cultivable land is an enormous task, but not impossible.

Properties of Soil

TEXTURE

The mineral part of a soil is composed of different proportions of three types of particles: sand, silt, and clay. These particles vary in size: sand is the largest (ranging in diameter from 0.20 to 0.05 mm); silt is intermediate (from 0.050 to 0.002 mm), clay particles are the smallest (0.002 mm or less). The texture of soil is determined by the relative proportions of these particles in the surface layer of the soil. For example, a sandy loam contains from 50 to 70 percent sand, from 0 to 50 percent silt, and from 0 to 20 percent clay. The soil triangle

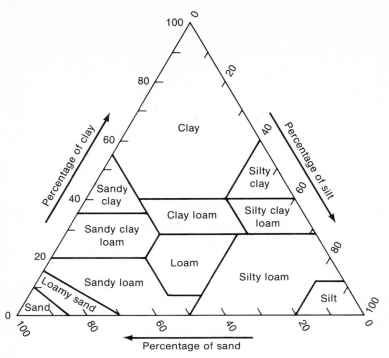

Figure 6-3
The soil triangle shows the proportions of clay, silt, and sand in each soil texture. [USDA data.]

in Figure 6-3 shows the proportions of sand, silt, and clay in various soil textures. Given the relative amounts of the three particle types in a soil, its texture can be determined.

Texture is important agriculturally for several reasons. So-called light soils are high in sand, and tillage is relatively easy. Less energy is required to cultivate a light, sandy soil than a heavy, clay soil, if all other factors are equal.

The pore space in any given volume of soil is largely dependent on soil texture. Both water and air are present in soil pores. In general, sandy soils have fewer but larger pores (because of the larger particles) than clay soils (whose particles are small). A clay soil holds more water than a sandy soil (Figure 6-4). If a soil is thoroughly wetted, then drained by gravity, it is considered to be at field capacity. The pores are not filled with water; a sheath of water coats the surface of each soil particle and is physically held to the soil. At field capacity, a very sandy soil can hold about 1 inch (2.5 cm) of water per foot (30 cm) of soil, a loam about 1.75 inches (4.4 cm), and a clay soil about 2.25 inches (5.7 cm). The water-holding capacity of a soil is greatly affected by the amount of organic matter in the soil, regardless of soil texture. Water added to the soil usually moves through it like a shade that is being drawn down over a window, the bottom of the shade comparing to the "bottom edge" of the water. This edge will not move down until all of the soil above it has been wetted to field capacity, just as the shade being drawn covers all of the window above its bottom edge.

As a crop grows, it removes water from the soil through transpiration; moisture is also lost directly from the soil through evaporation. Excessive transpiration can cause a plant to appear wilted; the plant loses more water

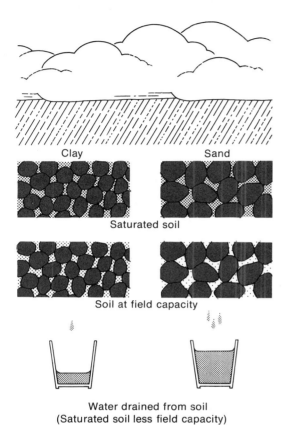

Figure 6-4
Differences in pore space and moisture-holding capacity between a finely textured (clay) soil and a coarse soil.

through its leaves (and stems) than it can take in through its roots. If the plant's appearance returns to normal when the transpiration rate is reduced (e.g., at night or with lower temperatures), the wilting is temporary and may not be too harmful to the plant. Temporary wilting may also be due to heat, wind, low humidity, or other similar conditions that cause high transpiration rates even if there is adequate moisture in the soil (see Chapter 7). If the plant remains wilted even though the transpiration rate is known to have been reduced, the wilting is permanent. Although the soil is not completely dry, the plant cannot remove water from it. The water content of a soil (as a percentage of dry soil weight) at which plants can no longer obtain adequate water is called the permanent wilting percent (PWP). In general, the PWP is higher in a clay soil than in a sandy soil. At the permanent wilting point, a clay soil may contain from 15 to 20 percent of the water present at field capacity. Permanent wilting can be prevented by adding water to the soil. If water is not added, plants will be severely damaged or killed.

The role of soil texture in determining soil porosity is of major importance in soil aeration—the exchange of gases, mainly oxygen and carbon dioxide, between the soil and the air above the soil and between the soil and plant roots. The importance of aeration to plant growth and development is clearly evident from studies of the responses of different types of plants to flooded or saturated soil, or to reduced soil porosity. As pores become filled with water, air is forced out of them.

For normal plant growth, the minimum soil porosity (the percentage of the volume of the soil not occupied by solids) is from 10 to 12 percent. The optimum porosity varies with crop species and with the developmental stage of the plant. Optimum porosity for sudangrass, for example, ranges from 6 to 10 percent, for wheat and oats from 10 to 15 percent, and for barley from 15 to 25 percent. In general, plants are more sensitive to reduced porosity at flowering than before. Environmental conditions such as light, temperature, and mineral availability also affect a crop's response to flooding (or to reduced or restricted porosity or aeration).

Reduced pore space can cause a reduction in crop yield in four ways: (1) total growth is reduced, (2) mineral uptake is reduced, (3) water uptake is reduced, and (4) toxic substances are formed in the plants and in the soil. If soil is flooded, adequate oxygen is not available for respiration to take place in a plant's roots, and the effects of poor aeration are first seen in the pattern and rate of root growth and development. The effects are most pronounced in root crops, such as sugar beets, in the form of smaller, misshaped roots. However, total root growth is also reduced in other crops. Because an entire plant is dependent on minerals and water taken up by the roots, the factors that reduce root growth ultimately affect the entire plant.

If the supply of oxygen is inadequate, certain by-products of respiration may be formed that are toxic to the plant. For example, on flooded soils, seedlings of both wheat and rice produce ethyl alcohol through respiration (see Chapter 2). Rice seedlings can tolerate this respiratory by-product better than wheat, which partly explains why rice can be grown on flooded soils. Relationships between plants and water are discussed in greater detail in Chapter 7.

Flooding adversely affects the soil and its relation to a crop plant in other ways. Flooded soils warm more slowly than soils at field capacity or drier. Thus, seeds and seedlings may be exposed to undesirably low temperatures for excessively long periods. In addition, the decay of organic matter and the release of essential minerals for plant growth are delayed in flooded soils; this is due in part to inadequate oxygen for respiration in bacteria and other microorganisms that promote decay. Certain plant pathogens are favored by flooded conditions; thus, flooding may contribute to disease (see Chapter 9). Oxygen is also required for certain minerals, such as iron and phosphorus, to be available in forms that plants can use. Finally, essential minerals (especially nitrogen) may be leached from the soil during drainage.

Soil structure is determined by the natural pattern in which the soil particles group to form aggregates. Aggregates are common to the upper layer (A horizon) of the soil. In lower layers, aggregates may group to form a fundamental soil unit, a ped. Soil structure, like texture, affects cultivability. It also affects soil porosity; when soil structure is damaged or destroyed, porosity may be reduced. Of more importance, structure determines in part how readily water, either natural precipitation or irrigation, infiltrates or enters the soil. Soil having poor structure may tend to puddle, and water will run off, rather than infiltrate, the soil. Well-aggregated soils accept water readily. Soil structure can be destroyed by excessive cultivation or by working wet soil, causing the soil to puddle and to crust. Crusting leads to poor emergence and spotty stands of many crops, which in turn are reflected in lower yields. Good farming practices preserve or improve soil structure.

An understanding of soil texture and structure is of practical importance in the field management of crops. In determining how frequently to irrigate and how much water to apply, the plant's use of water must be matched with the soil's field capacity and PWP. Although precise amounts cannot be determined, at least gross management errors in irrigation can be avoided. Similarly, a knowledge of the field capacities of soils having different textures can help in evaluating the effect of natural precipitation. For example, a light rain of, say, 0.4 or 0.5 inch (1–1.3 cm) may wet a sandy soil to a depth of 4 to 5 inches (10–13 cm), but it may wet only 2 or 3 inches (5–8 cm) of a clay soil, which may not be enough to reach the growing roots. The distribution of roots in the soil must also be considered in evaluating the benefit of a rain shower or in planning an irrigation program.

Although little can be done in the field to alter soil texture, soil structure can at least be protected and to some extent improved through a variety of management practices. The addition of organic matter by including green manure crops in a crop-rotation program or by spreading manure may improve soil structure. If problems associated with soil structure or texture cannot be remedied, planting a species that will tolerate adverse conditions may prevent losses. For example, in areas that are known to be subject to periodic flooding, the problems associated with poor aeration can be avoided by selecting a flood-tolerant species.

CATION EXCHANGE CAPACITY

The texture and structure of soil, and the characteristics associated with them, such as water-holding capacity, infiltration rate, and aertion, are physical properties. Chemical properties must also be considered in determining the agricultural potential of a soil and in developing crop-management programs. Cation exchange capacity (CEC), which is also called total exchange capacity or base exchange capacity, is both a physical and a chemical property of soil. It is defined as the capacity of the soil to exchange cations (positively charged atoms or molecules), such as H^+, Ca^{2+}, or NH_4^+, for an equivalent amount of other cations without undergoing any change in structure. Cation exchange takes place in the organic and clay components of the soil; the exchange process is an example of a more general phenomenon called ion exchange, in which ions are adsorbed (held on the surface of a soil particle) rather than absorbed (drawn into the particle). Cation exchange capacity is a rough indicator of the potential fertility of a soil and the extent to which fertilizers and lime will react with it.

The CEC of a soil for any cation is expressed in terms of milliequivalents of cation per 100 grams of soil. A milliequivalent is defined in grams as 1/1000 (0.001) of the atomic weight of the cation divided by its valence. Thus, a milliequivalent of hydrogen (H^+) is (1/1) × 0.001 g = 0.001 g of H^+; a milliequivalent of calcium (Ca^{2+}) is (40/2) × 0.001 = 0.020 g of Ca^{2+}.

The relationship of CEC to soil fertility is best explained by starting with a soil that has a CEC of 1; that is, a soil that can exchange 0.001 g of H^+ per 100 g of soil. This is a ratio of ten parts per million, or ten pounds of hydrogen per million pounds of soil (the approximate weight of 0.25 of an acre foot of soil). To exchange the hydrogen for calcium would require a minimum of 200 pounds of Ca^{2+}: 1 milliequivalent of Ca^{2+} = 0.020 g. This soil would exchange 0.020 g of Ca^{2+} per 100 g of soil or 200 parts per million. For NH_4^+, 180 pounds would be exchanged. If the CEC of a soil is known, then the exchangeable amount of any cation can be determined for any mass of that soil if the atomic or molecular weight and the valence of the cation are known.

As a rule, exhangeable cations are available to plants, but they are not readily leached from the soil. Although different cations are bonded to the clay and organic components (colloidal particles) of soil with different strengths, a surplus of any cation will foster the exchange of that cation at the collodial surface. Thus, when calcium, or a mineral in a fertilizer, is added to the soil in sufficient quantity, it replaces other cations (many of which may be hydrogen ions) on the collodial particles. The strength of the soil-cation bond for four important cations is: $Ca^{2+} > Mg^{2+} > K^+ > Na^+$.

SOIL pH

Another chemical property of soils is the degree to which they are acidic or basic, which is expressed in terms of pH—the logarithm of the reciprocal of the hydrogen ion concentration. A pH value of 7 is neutral; values lower than 7 are acidic; values greater

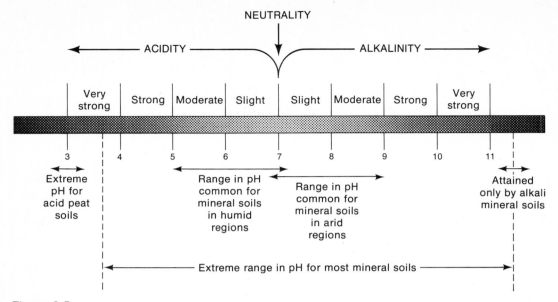

Figure 6-5
The extreme range in pH for most mineral soils and the ranges commonly found in humid regions and arid regions. [From *The nature and properties of soils,* 8th ed., by H. Buckman and N. C. Brady. Macmillan, 1974.]

than 7 are basic (alkaline). Because pH is a function of logarithms, pH = 5 is ten times as acidic as pH = 6.

The extreme range of soil pH is from 3 to about 11. For most agriculturally important soils, pH ranges from 5 to 9 (Figure 6-5). Many of the forage grasses tolerate soils that have a high pH; some types of sorghum (the sorgos), sugar beets, and barley are fairly tolerant of alkaline soils. Crops such as millets, buckwheat, rye, and alsike clover can be grown in quite acidic soils, those having a pH between 4 and 5. Corn, wheat, and beans can be produced in moderately acidic soils, those having a pH between 5 and 6. Although alfalfa can be grown in this same range, yields are reduced when pH falls below 6.8–6.5.

Soils in areas of heavy rainfall are generally acidic. Large amounts of moisture tend to wash out (leach) cations, such as Ca^{2+}, Mg^{2+}, and so forth, allowing an accumulation of H^+ on the clay particles.

The direct effects of hydrogen ions are minimal (except at the extremities of the pH range). More important are the effects of pH on the availability to plants of essential minerals in the soil (Figure 6-6). Phosphorus availability is adversely affected by slight acidity as is that of potassium, calcium, and magnesium to a lesser extent. The availability of micronutrients, including iron but not boron and molybdenum, is favored by a low pH; however, these elements are much less frequently found to be deficient. The relationship between pH and the availability of essential minerals to plants is an important consideration in choosing effective, efficient fertilizers for field crops.

Some degree of soil acidity may limit plant growth in parts of Canada and in the humid areas of the United States. The highly organic muck and peat soils in southwestern Quebec are strongly acid and require treatment to obtain maximum yields, as do soils in the area extending from the southeastern United States to Canada.

LIMING
As early as 200 B.C., liming in some form was known to be beneficial to plant growth. Now that the effects of acidity in soils have been determined, we know that a part of the bene-

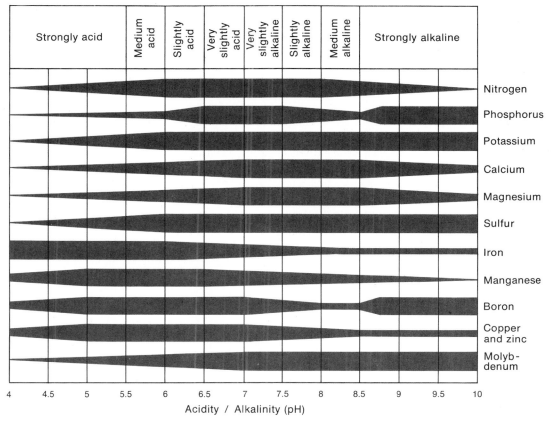

Figure 6-6
Relationship between soil pH and the availability of minerals that are essential for plant growth. The varying thickness of each band indicates the availability of the minerals. [From Chemical fertilizers by Christopher J. Pratt. Copyright © 1965 by Scientific American, Inc. All rights reserved.]

fit of liming may be that it reduces acidity. A major benefit is that it increases the availability of calcium and magnesium.

Currently, limestone—either calcite ($CaCO_3$) or dolomite [$CaMg(CO_3)_2$], or a combination of these—is most commonly used. Both calcite and dolomite have the capacity to neutralize acidity because the calcium and magnesium cations, if present in large quantities, can replace hydrogen ions, the cause of the acidity. The lime requirement depends on the total number of hydrogen ions in the soil: those in solution plus exchangeable hydrogen ions. The effectiveness of either source of cations largely depends on the size of the limestone particle applied to the soil: small particles dissolve quickly, and therefore neutralize acidity rapidly. Particle size is expressed as the percentage of the material that passes through a screen or sieve of a specified mesh or fineness. The amount of limestone to be applied also depends on particle size: a greater quantity of coarse material (one in which less than 30 percent of the particles passes through a 60-mesh sieve) is needed to match the acid-controlling capacity of a more finely ground type (one in which 80 percent passes through the sieve).

Lime requirements are often expressed as calcium oxide equivalents, or the number of equivalents of *pure* calcium oxide (CaO) that would be required to neutralize the soil. The neutralizing strength of commercial lime is then a function of its calcium oxide equivalence number. When lime is added to the soil, three sets of chemical reactions take place: (1) the lime dissolves, with hydrolysis, (2) the dissolved lime (calcium hydroxide, or "slaked lime") dissociates, and (3) the calcium ion reacts with exchangeable hydrogen ions. Consider a soil particle that has two hydrogen ions:

$$CaO \text{ (lime)} + H_2O \text{ (water in the soil)} \rightarrow Ca(OH)_2$$

$$Ca(OH)_2 \rightarrow Ca^{2+} + 2\ OH^-$$

Now the Ca^{2+} replaces the 2 H^+ on the soil particle:

$$Ca^{2+} + \text{soil particle (having 2 } H^+) \rightarrow 2\ H^+ + \text{soil particle (having } Ca^{2+})$$

The hydrogen ions that are liberated react with the hydroxide ions to form water:

$$2\ H^+ + 2\ OH^- \rightarrow 2\ H_2O$$

Soil conditions normally do not permit the persistence of calcium hydroxide [$Ca(OH)_2$]; usually calcium bicarbonate [$Ca(HCO_3)_2$] is formed and carbon dioxide is given off freely. Magnesium behaves in a similar manner. As the number of exchangeable hydrogen ions is reduced, the base saturation of the colloidal component of the soil is raised and the pH of the soil solution is increased correspondingly.

Unlike application rates of fertilizers, those of limestone are expressed in tons per acre, rather than pounds. Specific rates vary from place to place, depending on the chemical properties of the soil, the crop, weather conditions, and the type of limestone used. However, as much as six tons per acre have been shown to increase crop yields.

The effects of liming are complex: lime has a direct effect on soil pH; yet, doubling the amount of lime applied does not reduce acidity proportionately. Nor are the effects of liming immediate. As calcium and magnesium cations become available and replace hydrogen ions, the pH of the soil increases. Depending on particle size and how rapidly cations are exchanged, the reduction in acidity may continue for two or more years. However, this reduction may not be permanent: the natural conditions that cause the soil to be acidic cannot be eliminated by liming; thus, periodic treatment may be required. Altering the soil pH may also change the entire fertility status of the soil, and of the crop growing on it. Iron may become less available, and, if the pH is greatly increased, phosphorus availability may also be reduced.

Toxic quantities of certain minerals can accumulate in soils. Aluminum toxicity has long been known as a phenomenon associated with various acidic soils. In such soils, large quantities of aluminum become soluble and are adsorbed in preference to hydrogen. Aluminum ions (Al^{3+}) effectively block adsorption of cations that are essential for normal plant growth. Manganese, although required by plants in small amounts, may become overabundant in acidic soils and reach toxic levels. Correct liming reduces toxicity, but is beneficial only on acidic soils that are formed in humid and subhumid regions. Many agricultural soils are not acidic and liming is neither necessary nor beneficial.

Under dry (semiarid to arid) conditions, soils are usually basic. Some basic or alkaline soils can be classified as sodic (sometimes called alkali). Sodic soils have a high concentration of exchangeable sodium and also generally have a pH of 8.4 or higher. Basic soils have a pH above 7.0, regardless of the specific ions present, whereas any soil having a pH below 7.0 is an acidic soil. Saline soils may be acidic or basic, but always have excessive amounts of dissolved salts such as sulfates of sodium and magnesium in the soil water.

Like acidity, alkalinity affects the availability of essential minerals, which in turn has an impact on plant growth and development. Furthermore, excessive salts can significantly alter the soil moisture balance such that, because of osmotic conditions, a seed cannot absorb adequate moisture to germinate, or plants cannot obtain adequate moisture for normal growth and development.

Crop species vary markedly in their tolerance to salinity. Barley is rather tolerant of high concentrations of salt; field beans are not. Other field crops are more or less tolerant. Among the more common forages, salt grass, western wheatgrass, and birdsfoot trefoil tolerate salty conditions quite well; the clovers—alsike, red, and 'Ladino'—do not. Alfalfa and many other forages can tolerate intermediate concentrations of salt.

The fundamental source of all salts in sodic and saline soils is the weathering of the parent material or rock from which the soil is formed. There are few, if any, soils for which this source alone accounts for the presence of salt. Saline soils develop at sites where salts transported from other areas tend to build up—most being carried in the water that passes over or through the soil. Dissolved salts accumulate in water as it moves across materials such as ancient marine deposits, which are naturally high in soluble salts. In areas of poor drainage where the rate of movement of the water is reduced, the accumulated salts are concentrated and deposited. This is accelerated by the removal of water from the soil by plant transpiration or by evaporation, which leave the salts behind. Ocean mists and sprays on coastal lands and the presence of sea water in both rivers and wells that are used for irrigation add to the accumulation of salt. Increasingly heavy applications of fertilizers together with poor drainage may result in significant accumulations of salty, undesirable fertilizer residues.

The prevention of or remedies for saline conditions are not as straightforward as liming for pH control of acid soils. Saline conditions generally arise under dry conditions in areas that have poor drainage. The problem of salt accumulation can be alleviated or prevented by improving drainage and keeping water tables lower than the normal rooting zones of crop plants so that salts will not accumulate as water is transpired or evaporated. Improving drainage may require such measures as the installation of tile-lined drainage ditches, which can be very expensive. Even if drainage is economically and physically feasible, disposing of the salty water is still a problem. It cannot merely be drained onto crop land or into a stream to become a problem elsewhere.

Efforts are now being made to find plants that root very deeply, that can tolerate salty conditions, and that use large quantities of water, which will keep the water table at a safe level and minimize salt accumulation. Crop-rotation practices—notably, the crop-fallow system common to cereal production in many semiarid areas—are being revised to improve control of high water tables.

Prevention may be difficult and costly, but remedial measures are even more so. In addition to being adequately drained, the soil must be washed (or leached or flushed) with nonsalty water to remove the unwanted salt. In the past ten years, the growing number of acres of agricultural land in which the soil has become saline, and the consequent loss of such land for crop production in semiarid areas of both Canada and the United States, has become a major concern in resource conservation.

ORGANIC MATTER

Soil is not exclusively mineral material. Organic matter, composed of living or previously living material, is an important component that affects both its physical and its chemical properties.

Raw organic matter, which includes crop residue (stubble, beet tops, and roots left in the ground) and animal waste, is more important in its effect on soil tilth than on plant nutrition. It tends to loosen the soil, which enhances the structure, thereby increasing its water-holding capacity and the water infiltration rate.

Decomposed organic matter (humus) serves some of the functions of raw organic matter. In addition, minerals bound in the raw organic matter are released and become available for plant nutrition as the organic matter decays. Decomposed organic matter becomes part of the soil colloidal system, a system made up of very small particles (less than 0.001 mm in diameter). Many of the minerals essential for plant growth are held on colloidal particles. The greater the colloidal part of the soil, the more minerals can be held and the higher the potential fertility of the soil.

Soils also have a living component composed of nematodes, insects, worms, small rodents, and microorganisms such as bacteria, fungi, and algae. Their activity in the soil may be either beneficial or detrimental to crop production. Burrowing animals, including worms and insects, can improve soil aeration and the rate of water infiltration, but they also eat plants, thereby reducing crop yield or quality. Microorganisms play a major role in the decay of raw organic material and the release of essential minerals. Certain bacteria that live in the roots of legumes fix atmospheric nitrogen in a form available for plant use, but other types can harm crops by causing seeds to rot before they have been able to germinate and by causing serious diseases in crop plants.

DEPTH OF SOIL AND TOPOGRAPHY

Soil depth is the distance from the soil surface to a material in which roots cannot grow. This material may be gravel, a heavy, hard pan or clayey layer, bed rock, or water. Crops differ with respect to how deeply their roots penetrate the soil. Of the common crops, alfalfa roots penetrate the deepest; they may extend more than twenty feet (6 m) into the soil. At the opposite extreme, 'Ladino' clover roots are confined to about the top two feet. Grasses, including the cereal crops, have most of their roots in the surface foot of soil, but some extend to a depth of more than three feet (1 m) and, in some cases, to eight feet. In addition to rooting depth, consideration must be given to the lateral spread of plant roots. A crop that has a dense fibrous root system may extract more water and essential minerals from a shallow depth of soil than a deep taprooted species because the former permeates the soil more fully than the latter. Like grasses, legumes have roots that spread laterally; however, they do not spread as much and a larger proportion is found at a greater depth. Most of the roots of a potato plant are in the upper two feet of the soil (see Chapter 26). It should be noted that roots will not penetrate dry soil. The idea of withholding water (irrigation) to force a crop to develop deep roots is invalid. Likewise, frequent but shallow irrigations will prevent deep root penetration into the dry soil below and may cause the plant to develop a shallow root system.

Root penetration can be inhibited by the

presence of what may for agricultural purposes be classified as a hard pan in the soil at a depth beyond which crops normally extend their roots. The inhibition may be mechanical; hard pan presents a barrier through which roots cannot grow. It also retards soil drainage and may allow the water table to rise to a level at which the soil in the normal rooting zone becomes saturated and cannot supply oxygen to plants. Recall that, when water that contains salts rises, saline soils may be formed under arid or semiarid conditions. Under more humid conditions, the potential exists for the accumulation of materials other than salts that might be toxic to crop plants. Such materials may include excessive amounts of essential minerals—such as boron, copper, and manganeses—and, in some cases, of nitrates. Accumulations of iron oxides, silica, and various carbonates are also characteristic of many soils having underlying hard pans. In addition, toxic amounts of chemical pesticides or their residues may accumulate, a consequence that has become of increasing concern to environmentalists in recent years because of the possible widespread pollution of rivers and lakes. Few, if any, documented instances of significant pollution from these sources have been related to the application of agrichemicals, if properly used.

Hard pans that foster poor drainage can be attributed to a number of natural causes. In some instances, they are the result of the presence of heavy clay layers, of varying thicknesses, at different depths in the soil. In others, they may be due to the presence of a shale layer. A well-studied, specific type of hard pan, the fragipan, is found in some soils in all states east of the Mississippi River, as well as in Arkansas, Oklahoma, Louisiana, Missouri, and eastern Texas, extending northward to Minnesota. It is a subsurface layer of soil that is brittle and rigid when moist. It resists root penetration and is a barrier to good drainage. Thus, like other hard pans, it may be the cause of high water tables. Fragipans contain little clay (35-60 percent) and have low water-holding capacity. Thus, the addition of relatively small amounts of water to a soil with an underlying fragipan layer may result in a rapid elevation of the water table. In naturally wet periods, as in spring and fall, they either support crop growth or become saturated and restrict growth. In drier periods, on the other hand, plants growing on fragipan may suffer from moisture stress because the soil cannot retain adequate moisture to support growth. Because fragipans have very specific, characteristic features, they have been used in classifying soils.

Hard pans can also be created by farming activities. Plowing to the same depth repeatedly for a number of years produces a compacted soil layer below the plowing depth. This is caused by mechanical destruction of the soil structure and can be prevented by plowing at different depths and by plowing deeply periodically. Compaction of the soil can be hastened by working the soil (by any type of cultivation, or even by light traffic across a field) when it is wet.

Not all hard pans are undesirable. For example, in the rice-growing areas of the Sacramento Valley in California, water is more efficiently retained in paddies by the presence of a highly impervious layer of clay hard pan below the crop-rooting zone.

The topography of a soil is the physical configuration of its surface, including ridges, gullies, slopes, mounds, and so forth. It is an important consideration in the design of irrigation systems and in determining the direction in which to plow and plant. For example, it is advisable to plow and plant along the contours of slopes rather than up one side and down the other. This is a sound conservation practice in that it tends to retard erosion.

Figure 6-7
Soil profile showing surface vegetation, organic matter (dark layer) of A horizon, and relatively uniform lower layers.

Many other physical and chemical properties could be considered in a discussion of soils and their relation to crops. Readers who wish to pursue the subject further will find several good references at the end of this chapter.

The Soil Profile

There are many different kinds of soil; more than 10,000 have been classified for the United States. Much of the detailed description of specific soils is based on the soil profile. A soil profile is a two-dimensional view of the soil from its surface to the bedrock at its base (Figure 6-7). The basic indivisible three-dimensional unit of soil (a "cube" of soil) is called a pedon.

Soil profiles are divided into three major layers, or horizons. Depending on the soil, these horizons may be quite distinct or they may be very indistinct. Older soils have well-defined or differentiated horizons; younger soils have poorly developed horizons.

The surface horizon is designated the A horizon. It is referred to as the leached horizon because the minerals in it are leached downward. The A horizon is usually dark because of the organic matter it contains. The B horizon is immediately below the A horizon. It is referred to as the zone of accumulation because the minerals that have been leached from the A horizon are held there. Jointly, the A and B horizons (which are of primary importance in crop production) are referred to as the solum.

The C horizon is the parent material from which soil is formed by weathering. The nature of the parent material, of which there are several common types, can play a major role in the character of the soil formed from it. Many of the shallow, relatively unfertile mountain soils arose from granite parent material. In the Snake River areas of Idaho, and in Hawaii, the parent material is volcanic in origin. Some of the most productive agricultural soils in the world are alluvial soils, which result from repeated flooding and the deposition of silts. Many alluvial soils are relatively young and have poorly developed profiles. River, lake, and ocean sediments are other common forms of parent material.

Soil Formation

Generally, five factors—climate, parent material, organic material, topography, and time—are considered to be critical in forming a soil and determining its properties. First, a major role is played by climate, especially precipitation and temperature: water is the universal solvent required in chemically breaking down the rocky parent material, and the rate at which this takes place depends in part on temperature. Water also breaks the parent material down through erosion. Second, the chemical composition of a soil and, in part, its structural and textural character-

Figure 6-8
Soil monoliths prepared by soil scientists for detailed study of soil profile characteristics: (left) relatively immature soil with poorly developed horizons; (right) mature soil with well-developed horizons.

istics can be traced to the parent material from which it is formed. Soils formed from certain saline sedimentary deposits, for example, are naturally more salty than soils formed from granite. Third, both plants and animals affect the development of a soil:

through the root growth of plants and the boring of various animals, including worms and small rodents, parent material is broken into smaller pieces. These same activities facilitate the movement of air and water, which in turn affect soil formation. Both plants and animals are the source of organic matter in a soil. Fourth, topography determines the extent of erosion by wind and precipitation. Finally, time, measured in thousands of years, affects the character of a soil. The longer an event has been taking place—for example, the deposition of silt from rivers on flood plains—the more pronounced the effects of that event. This explains why most older soils have well-developed horizons due to climatic and biological factors and most younger ones have poorly developed profiles (Figure 6-8).

Soil Classification

A new system of soil taxonomy was adopted by the U.S. Department of Agriculture in 1967, but an older system of classification, the 1938 system, is still in use in some areas. Many excellent references, research reports, and studies incorporating the old system were published before 1967; thus, an agriculturalist should be familiar with both systems.

THE NEW SYSTEM

The new system of soil taxonomy, which took twenty years to formulate, classifies soils more accurately than the old system and is less descriptive than earlier systems of classification. There are two distinct advantages to this new system. First, the classification of a soil is based on specific, measurable physical and chemical properties of the soil itself. One of the important measurable traits used is cation exchange capacity. Location, parent material, and climate—criteria used in the 1938 system—are not required to identify a soil.

Second, the nomenclature itself provides a means for identification. In this new system there are six levels of classification: order, suborder, great group, subgroup, family, and series. There are ten orders; each has from two to six suborders. Suborders are divided into great soil groups, and these are subdivided into subgroups. Each subgroup is further divided into families and the families into series. With each subdivision, identification of a soil becomes more refined, and specific. More than 10,000 soil series have been identified in the United States. The soil series is based on specific characteristics of the soil below normal plowing depth. It is the narrowest, or most restrictive, classification in the new system and can be used to identify or describe the soil in quite restricted areas.

The identification of a soil from its name is based on the meanings of key parts or roots of words in each of the six levels of classification. The names of the ten orders, their roots, and their characteristics are given in Table 6-1.

The root of the name of a soil order constitutes the last three letters of the names of its suborders, which also have roots that identify particular characteristics. For example, the suborder Aquoll belongs to the order Mollisol (the oll indicates the order Mollisol); aqu indicates wetness and so the suborder name identifies a wet soil that is rich in organic matter and exchangeable bases. On the other hand, the suborder name Xeroll identifies a dry soil of the Mollisol order (xer indicating dryness, or a soil in an area with an annual dry season).

The names of the great groups have been constructed by adding prefixes (which are actually roots that identify the groups) to suborder names. Subgroup names have been constructed by preceding the great group names with roots that identify the subgroups. For example, the name Typic cryoboroll tells

Table 6-1
Soil orders in the new system of classification.

Order	Root	Characteristics
Entisol	ent	Young soils with indistinct profile
Vertisol	ert	Cracking clay soils
Inseptisol	ept	Young soils with poorly developed profiles
Aridisol	id	Soils in dry areas; low in organic matter
Mollisol*	oll	Soils rich in organic matter and in exchangeable bases.
Spodisol	od	Soils with an accumulation of amorphous material in the subhorizons
Alfisol	alf	Gray to brown soils with an accumulation of clay in the subsurface soil
Ultisol	ult	Soils with a clay horizon that are low in exchangeable bases
Oxisol	ox	Soils that contain quarts and (or) hydrated oxides
Histosol	ist	Soils rich in organic matter

*Mollisols are of great agricultural value.

us that soils in this subgroup are in the Mollisol order; the Boroll suborder [bor indicating soils in cool, northern regions that have a mean annual soil temperature of less than 47°F (8.3°C)]; the Cryoboroll great group [cryo indicating cold regions that have a mean summer soil temperature at a 20-inch (51-cm) depth of less than 59°F (15°C)]; and the Typic cryoboroll subgroup (typic indicating soils that are typical of the Cryoboroll soils). Taken overall, this subgroup name identifies soils that are typical of those that form in cold, northern regions and that are rich in organic matter and high in exchangeable bases. Typic cryoboroll is an agriculturally important soil, but not all crops can be grown in it because of the cool soil temperature.

Finally, a name identifying a family is added to the subgroup name, and a series name is added to the family name. The family and series names describe soil texture and other precisely measurable characteristics. The series is the lowest level of soil classification. However, a soil type is added as an ultimate identification of the soil. Soil type describes the texture of the surface soil.

THE 1938 SYSTEM

In the 1930s, the soils of the United States (and other parts of the world to a lesser extent) were divided into Great Soil Groups—a system of classification that was accepted by many soil scientists in 1938. Under this system, soils in the United States were divided into two broad categories: the soils in the eastern half of the United States being pedalfers and those in the western half pedocals. Pedalfers are rather acidic in the surface soil with an accumulation of aluminum (Al) and iron (Fe) in the B horizon. They are found in areas of high rainfall. Pedocals range from neutral to alkaline in the surface layer and are formed in regions of lower rainfall. Although this broad classification is of little practical value in identifying a specific soil, it does provide a starting point. The major divisions of the 1938 system classify soils into three orders: Zonal Soils, Azonal Soils, and Intrazonal Soils. These orders are then subdivided into Great Soil Groups.

Zonal Soils Zonal Soils cover fairly large areas; their boundaries are well-defined. The various types of Zonal Soils reflect the effects of climate and vegetation on their development. Although Zonal Soils are a broad classification, they identify a soil more precisely than the categories pedalfer or pedocal. Several of the very important Great Soil Groups are in the zonal soil order.

The soils of the Podzol Great Soil Group are commonly found in the Great Lakes region and in northern New England. They are formed under moist conditions and moderate to cool temperatures. Because these soils are not naturally very fertile, they are frequently used for growing trees or hay and for pasture. If they are used for crops, large quantities of fertilizer must be added to them. The soils of the Latosol Great Soil Group form under warm, moist conditions and are highly weathered. They are found in Hawaii and other tropical areas. Because of extensive leaching of plant nutrients, large quantities of fertilizers must be added to these soils to make the production of crops economically feasible.

Many soils in the Chernozem Great Soil Group are highly fertile and of great agricultural importance. Perhaps the most fertile of these are in the Red River Valley.

Azonal Soils Azonal soils differ from Zonal Soils in that they are relatively young and do not have well-differentiated profiles. The characteristics of Azonal Soils are greatly affected by their parent material. A very important Great Soil Group in the azonal order is Alluvial Soils; these are deep and highly fertile soils that have developed as a result of mineral deposits from periodic flooding. Alluvial Soils are found in the flood plains of the Nile and the Mississippi rivers, in the Great Valley in California, and in most other areas along rivers, including, for example, some very fertile land along the Yellowstone River in Montana. Other Great Soil Groups in the azonal soil order are Lithosol Soils—shallow soils that formed over bedrock—and Regosol Soils, which form over soft mineral deposits (sands).

Intrazonal Soils The characteristics of Intrazonal Soils are determined by specific local factors such as parent material or topography. There are three important Great Soil Groups in this order: the Hydromorphic Soils, the Halomorphic Soils, and the Calcimorphic Soils. Hydromorphic Soils are formed under conditions of poor soil drainage. These soils are recognized as bogs or swamps. Halomorphic Soils are formed in arid areas that have poor drainage and are characteristically very salty. (There are Halomorphic Soils along parts of the Yellowstone River in Montana.) Calcimorphic Soils are excessively high in calcium.

The Great Soil Groups cannot be used to describe precisely the soil at a specific farm or ranch (or, a potential highway site or housing development). Like the new system, the 1938 system classifies soils more specifically by series and type: the soil series is a classification based on characteristics of a soil profile; soil types are subsequently determined by the texture of the surface soil.

Soils and Plant Nutrition

MINERALS IN THE SOIL

Except for carbon dioxide (CO_2), the nutrients required for plant growth come directly or indirectly from the soil. An analysis of a soil for its mineral content may be misleading because many of the essential elements it contains can be bound in a form that cannot be used by plants. For example, the nitrogen in undecomposed protein cannot be used by plants; nor can the sulfur in certain amino acids. Many other elements are tied up in clays or rocks as parts of complex minerals. Cations (positively charged ions) may be chemically bound to soil colloids or clay particles, whereas anions (negatively charged ions such as nitrate, NO_3^-) generally are not. They dissolve in the free water in the pores of a soil and may be leached from the rooting zone as water moves down through the soil.

The minerals that are essential for plant

growth must be present in specific forms in order to be taken up and utilized by plants. Nitrogen is absorbed by plants as either ammonium cation (NH_4^+), which participates in the cation exchange reactions of the soil, or nitrate anion (NO_3^-). An organic form of nitrogen, urea, is also taken up by plants. Three major sources of nitrogen in a form available for plant growth are: (1) decomposed organic material, (2) fertilizers, and (3) symbiotic bacteria (*Rhizobium*) that live in the roots of legumes. Symbiosis is the relationship between two organisms that live together for their mutual benefit. *Rhizobium* bacteria obtain energy (sugar from photosynthesis) and water from the legume; in turn, they convert (or fix) atmospheric nitrogen (N_2, a form not available for plant use) into forms that ultimately may become available for plant growth (see Chapter 12). Nitrogen follows a path, called the nitrogen cycle, from the atmosphere through the soil into plants and animals and then back to the soil. The details of this cycle are shown in Figure 6-9.

Phosphorus is taken up as phosphate ion ($H_2PO_4^-$ or HPO_4^{2-}). In basic soils it is present as HPO_4^{2-} and in acidic soils as $H_2PO_4^-$. Phosphorus tied up in organic matter is unavailable for plant growth and must go through mineralization (a series of chemical changes) before it becomes available to plants. Phosphorus is present in organic matter and in the parent material of soils in the form of calcium, iron, and aluminum phosphates. Many soils are deficient in phosphorus and significant increases in crop yields can be expected from the use of fertilizers that contain this nutrient.

Sulfur is taken up by plants as sulfate ion SO_4^{2-}. Sulfur comes from organic matter as well as a number of types of soil parent material.

Other elements are taken up in their ionic form as well—for example, Ca^{2+}, Mg^{2+}, K^+, Zn^{2+}, and Cl^-. These elements are present in such mineral components of parent materials as feldspars and dolomite.

FERTILIZERS

The most common way to maintain or restore soil fertility or to replace minerals that are essential for plant growth is through the application of commercial fertilizers. A fertilizer can be defined as any material added to the soil that supplies one or more required minerals for plants. A soil amendment, on the other hand, is used to alter physical and chemical properties of the soil, such as pH.

Man has used natural fertilizers, such as mulches, marl (a soft limestone material), and animal droppings, since before the dawn of written history. In medieval times in Europe, legumes were included in crop rotations to revitalize fields. Now, of course, we know that atmospheric nitrogen can be converted into a form available to plants because of the symbiotic relationship between leguminous plants and certain bacteria. The first English settlers in North America reported that the practice used by Indians of placing a fish in each hill of corn at the time of seeding increased yields. Again, we now know that, as the fish decay, minerals are released for plant growth.

The use of various natural fertilizers was a step in the right direction. However, in terms of today's food requirements, these materials cannot supply adequate amounts of essential minerals to promote maximum yields. The development and use of artificial fertilizers in the past sixty to seventy-five years has had a staggering impact on crop yields. Modern commercial fertilizers allow a crop producer to apply specific essential minerals or groups of minerals in precise concentrations to attain maximum yields.

The manufacture of fertilizers is a major agri-industry. The elements in a commercial fertilizer are combined through a series of regulated chemical reactions. Regardless of

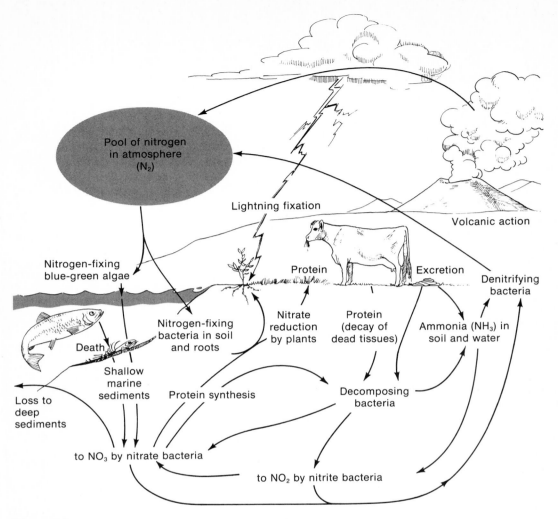

Figure 6-9
The nitrogen cycle. [From *Population, Resources, Environment*, 2d ed., by Paul R. Ehrlich and Anne H. Ehrlich. W. H. Freeman and Company. Copyright © 1972.]

the fertilizer, the basic source of nitrogen for fertilizers is synthetic ammonia. Ammonia is synthesized by combining nitrogen, which can be extracted from air, and hydrogen, which can be obtained from methane and from petroleum products. In areas where natural gas and petroleum products are not plentiful, solid fuels, such as coal, are used. If petroleum and coal become scarce, this source of hydrogen could be affected and the current supply of nitrogen fertilizers could be reduced.

The primary source of phosphorus for fertilizers is mineral deposits. Major deposits are found in Florida, throughout the western United States, in northern Africa, and in parts of the USSR. The rock phosphate that is mined is high in phosphorus, but the mineral is not readily available for plant use. Treatment of various types of rock phosphate with sulfuric acid and/or phosphoric acid yields a form of phosphorus—triple superphosphate—that is much more available to plants.

Large quantities of water-soluble potassium chloride are found throughout the world. Canada has extraordinary potassium resources in Saskatchewan, but the deposits are buried quite deeply in the earth. Problems in mining these deposits have been solved in recent years, and the large quantities of potassium obtained from them will be of worldwide benefit to agriculture, as will the potash obtained by Canadian "solution" mining techniques. The various forms of potassium are processed and marketed as muriate of potash (KCl), which is up to the equivalent of 60 percent potassium oxide (K_2O). Potassium is combined with other primary plant foods in commercial fertilizers. For example, rock phosphate is treated with nitric acid and ammonia to yield materials with both nitrogen and phosphorus. To these materials, sources of potassium are added to yield ultimately a fertilizer that may be as much as 20 percent nitrogen, 20 percent phosphorus pentoxide (P_2O_5), and 20 percent potassium oxide (K_2O). Common sources of mineral nutrients, manufacturing processes for producing commercial fertilizers, and the fertilizer product are summarized in Table 6-2.

Fertilizers can be applied in many forms; the most common are dry and granular. Nitrogen can be injected into the soil in the form of gaseous ammonia, or it can be added to irrigation water in ammonia or nitrate forms that have been dissolved in water. Some micronutrients can be applied to foliage, but this is of negligible importance in field crop production.

The percentages of the various minerals in a commercial fertilizer are given on the label in a group of three numbers—for example, "0-20-0" or "10-15-5." These numbers indicate in order what percentage of the fertilizer is nitrogen (N), what percentage is phosphorus pentoxide (P_2O_5), and what percentage is potassium oxide (K_2O). Thus, a 50-pound bag labeled 10-20-10 would be 10 percent nitrogen, 20 percent phosphorus pentoxide, and 10 percent potassium oxide and would contain $0.10 \times 50 = 5$ pounds of nitrogen, 10 pounds of phosphorus pentoxide, and 5 pounds of potassium oxide. The remaining 30 pounds of material is inert filler.

To find the number of pounds of the elements phosphorus and potassium in the bag requires a simple calculation, which will yield the percentages of these elements in their oxides. The atomic weights of phosphorus, potassium, and oxygen are approximately 31.0, 39.1, and 16.0, respectively. The molecular weight of phosphorus pentoxide (P_2O_5) is $2(31.0) + 5(16.0) = 62.0 + 80.0 = 142.0$, and the phosphorus content is 62.0/142.0, or 44 percent. Forty-four percent of 10 pounds of phosphorus pentoxide is $0.44 \times 10.0 = 4.40$; thus, the bag contains 4.40 pounds of phosphorus. Similarly, the molecular weight of potassium oxide (K_2O) is $2(39.1) + 16.0 = 78.2 + 16.0 = 94.2$, and the potassium content is 78.2/94.2, or 83 percent. Eighty-three percent of 5 pounds of potassium oxide is $0.83 \times 5.0 = 4.15$; the bag contains 4.15 pounds of potassium.

Some commercial fertilizers have two rows of numbers on their labels. The large values indicate the percentages of N, P_2O_5, and K_2O, the smaller values the percentages of N, P, and K. In determining the real cost of fertilizers, a grower must consider not only the actual cost of a given weight of the fertilizer, but also the concentration of the essential nutrients in it.

REMOVAL OF MINERALS BY PLANTS

When a crop producer harvests any plant product, from grain to forage, he is harvesting soil minerals. The supply of these minerals is not inexhaustible and sooner or later they must be replaced. The replacement is generally done with commercial fertilizers. The magnitude of mineral extraction in crops is surprisingly high. For example, if 100 bushels of corn per acre is harvested (a moderate yield

Table 6-2
Raw materials, processes, and products in the manufacture of commercial fertilizers.

Sources	Process	Product
Phosphate rock	Grind to about 0.1 ml	Ground phosphate rock (35% P_2O_5)
Sulfur, Air, Water	Catalytic oxidation and hydration	Sulfuric acid
Phosphate rock, Sulfuric acid	React and cure	Single superphosphate (20% P_2O_5)
Phosphate rock, Sulfuric acid	Dissolve and filter	Phosphoric acid; Gypsum
Phosphate rock, Phosphoric acid	React and cure	Triple superphosphate (48% P_2O_5)
Hydrocarbons, Steam, Air	Reform to hydrogen; Nitrogen from air; Synthesize	Ammonia (82% N)
Ammonia, Sulfuric acid	React and crystallize	Ammonium sulfate (21% N)
Ammonia, Phosphoric acid	React and crystallize or granulate	Ammonium phosphate (18% N, 46% P_2O_5)
Ammonia, Air, Water	Catalytic oxidation; Absorption	Nitric acid
Phosphate rock, Nitric acid, Ammonia	React; Ammoniate; Filter	Nitrophosphates (20% N, 20% P_2O_5); Calcium nitrate (15% N)
Ammonia, Nitric acid	React and crystallize or make into pellets	Ammonium nitrate (33% N)
Ammonia, Carbon dioxide	React and crystallize or make into pellets	Urea (46% N)
Ammonia, Carbon dioxide, Salt	Carbonate and filter; Ammoniate	Sodium carbonate; Ammonium chloride (23% N)
Coal, Limestone, Nitrogen	Fuse in arc furnace; Nitrify	Calcium cyanamide (24% N)

Source: Chemical fertilizers by Christopher J. Pratt. Copyright © 1965 by Scientific American, Inc. All rights reserved.

in the U.S. corn belt), more than 100 pounds of nitrogen, about 96 pounds of phosphorus pentoxide, and 18 pounds of potassium oxide are removed from the soil. To replace these minerals, a grower would have to apply 290 pounds of ammonium nitrate (NH_4NO_3), 200 pounds of superphosphate $Ca(H_2PO_4)_2$ and 40 pounds of potassium sulfate (K_2SO_4). If the stubble is also harvested, even more minerals are removed. If 40 bushels of wheat per acre is harvested, together with the straw, 73 pounds of nitrogen, 32 pounds of phosphorus pentoxide, and 98 pounds of potassium oxide are taken from the soil. If only the grain is removed and all of the stubble is returned to the soil, then 53 pounds of nitrogen, 11 pounds of phosphorus (or 25 pounds of P_2O_5), and 12 pounds of potassium (or 16 pounds of K_2O) are removed. To replace these would require nearly 160 pounds of ammonium nitrate, more than 100 pounds of super phosphate, and 29 pounds of potassium sulfate. As a general rule, 1.0 to 1.5 pounds of nitrogen is removed from the soil for each bushel of wheat harvested. The soil is like a savings account: withdrawals cannot be made indefinitely; some deposits must be made or the operation is in the red.

ORGANIC MATTER IN A SOIL AND THE NITROGEN CYCLE

The organic matter in a soil affects many of its physical properties and its fertility. In evaluating a soil's potential for producing crops, several facts about organic matter should be kept in mind. Raw organic matter contributes little directly to plant nutrition; it must decay and release minerals for plant growth. The mechanism of decomposition is critical to nutrients. Like nitrogen (Figure 6-9), other nutrients go through cycles. Carbon and water, for example, have cycles in which atmospheric carbon (CO_2) or water enters the soil through plants and is either returned to the atmosphere or added to the soil. The process of decomposition and the ultimate release of nutrients in a form available for plant growth consists of several steps. For example, nitrogen in raw organic matter is bound in complex proteins. Through a sequence of steps, a group of bacteria converts proteinaceous nitrogen into ammonia (NH_4^+), a process called ammonification. In nitrification, the ammonia is oxidized to a nitrite (NO_2^{2-}) and to nitrate (NO_3^-) forms by other, nitrifying bacteria. Denitrifying bacteria work to convert nitrates into nitrogen gas (N_2), a process called denitrification. Also, the bacteria in the roots of legumes (*Rhizobium*) are fixing or converting nitrogen gas into nitrates. The bacteria that fix nitrogen in this symbiotic relationship with legumes are highly specific for various legume crops (see Chapter 12). One strain of bacteria functions with alfalfa, another with beans, and a third with clovers.

The potential nutritional value of organic matter in a soil to plants depends on environmental conditions. The amount of nitrogen, for example, that becomes available to plants depends on temperature and moisture: in cold areas, little becomes available; in warm areas that have adequate rainfall, more becomes available. Consider an acre of soil in a meadow high in the mountains. The surface six inches would weigh about 2,000,000 pounds. If 5 percent of this soil was organic matter, there would be about 100,000 pounds of it in the soil. An acceptable value for the amount of nitrogen present in organic matter is 5 percent; thus, the 100,000 pounds of organic matter would contain 5,000 pounds of nitrogen. Because of low temperatures and reduced microbial activity, about 3 percent of the 5,000 pounds of nitrogen becomes available for plant growth every year. Thus, in the 2,000,000 pounds of soil, only 150 pounds of nitrogen is released per year under good conditions. In other areas, the rate of release, due to moisture and temperature conditions, may be much less. That 5 percent of a soil is organic matter and that 5 percent of the organic matter is nitrogen is a generalization. Some soils contain a much smaller percentage of organic matter and the organic matter in soils may be much less than 5 percent nitrogen. Rarely are agricultural soils more than 8 percent organic matter.

In the past (and possibly as an antipollution measure in the future), manure has been used to supply organic matter to soils. The nutritive value of manure is variable, but, as an approximation, one ton of large-animal manure (2,000 pounds) is about 0.32 percent nitrogen (6 pounds), 0.21 percent phosphorus pentoxide (4 pounds), and 0.16 percent potassium oxide (3 pounds), which indicates that manure is not a rich source of plant nutrients.

Because soils differ greatly in their innate fertility and as a result of past cropping patterns, the use of fertilizers, and climate, it is impossible to make meaningful generalizations about how much of any fertilizer should be added to a given kind of soil for optimum plant growth, or about the amount of lime to be added to an acidic soil. Valid recommendations must be based on soil tests and analyses.

Rarely if ever is a farmer equipped to do a soil analysis. Most analyses are done in special laboratories in which appropriate, rather sophisticated equipment is operated by specially trained technicians. Many state agricultural agencies provide soil-testing services for a nominal fee. In addition, many large fertilizer supply companies and private laboratories are equipped to do soil tests. An analysis of a soil includes a description of its texture and its pH. It also indicates the quantities of essential minerals present and the cation exchange capacity of the soil. It is necessary to know a soil's cation exchange capacity to determine the effectiveness of liming in reducing acidity and to estimate fertility potential. The information obtained from the soil analysis, the requirements of the crop in question for essential minerals, the availability of water, and a knowledge of other factors that might limit crop yields help in determining the amount and type of fertilizer to be applied.

Although a farmer may not do the actual test, or even interpret the test results, the validity of the test may depend on him because he takes the field sample that will be tested. If the samples of soil to be tested are not representative of the entire field, test results and subsequent recommendations may be inaccurate. Samples should be taken at various locations and precise records should be kept of the location of each sample. It is quite possible that cropland on hillsides, in low areas, on ridges, or in flat areas will require different fertilizers. In addition, samples should be taken from the rooting zone of the crop to be grown; for example, at one spot samples might be taken at the surface 6 inches (15 cm), at 12 inches (30 cm), and at 18 or 24 inches (45 or 60 cm). Usually, a shovel is adequate for taking all soil samples, but if more than a few deep samples are to be taken, a soil tube may be desirable.

If a field appears to be uniform and there are no obvious differences in slope, soil structure, or soil type, representative samples can be obtained by sampling in a grid pattern. If, on the other hand, a field is obviously not uniform—if it has low spots, sandy areas, heavy clay areas, or any of a variety of other obvious differences—then it is necessary to sample within each discrete area of the field. If careful records are maintained of the original soil tests, the fertilizer applications, and the crop yields, tests need not be made annually, and the amount of residual fertilizer can be estimated from year to year so that wasteful overfertilization can be minimized.

Before any field soil samples are taken, the laboratory that will do the testing should be consulted about how the samples should be handled. Depending on the methods used by the laboratory, samples either can be placed directly in paper or plastic sacks and sent to the laboratory or must be air dried first. Care must be taken not to mix samples. The benefits in the form of higher yields resulting from proper fertilizer applications more than offset the costs in time and money of testing soils.

Soil Conservation

As the population of the world expands and as greater quantities of food are required, it will become increasingly important to protect the soil—the basic resource in crop production. Several measures can be taken to do so. Land-use planning and zoning ordinances can help to conserve valuable agricultural lands. Soil surveys and the accurate classification of soils aid in determining appropriate uses.

In crop production a great deal can be done to reduce erosion. Fallow land, including trashy fallows in which all the stubble is not plowed into the soil, can be carefully managed not only to reduce erosion, but also to increase moisture infiltration. Excessive cultivation may seriously damage soil structure: minimizing the number of cultivations to accomplish a given job and avoiding the cultivation of excessively wet soil will not only save time and money, but also protect the soil. In even moderately hilly areas, plowing across a slope rather than over it reduces erosion due to wind and flowing water. Carefully designed irrigation systems and appropriate irrigation rates also aid in reducing erosion. Today, soil losses due to erosion must be considered an area of prime concern in soil conservation, and farming practices that minimize such losses should be employed wherever feasible.

A knowledge of fertilizers and of crop use of soil minerals will aid in preventing undue depletion of minerals in the soils. The best protection for a soil is a healthy crop. At the same time the overuse of fertilizers and chemical pesticides must be avoided to guard against pollution as well as economic losses.

Selected References

Buckman, Harry O., and Nyle C. Brady. 1969. *The nature and properties of soils.* 7th ed. Macmillan.

Donahue, Roy L., John C. Shickluna, and Lynn S. Robertson. 1971. *Soils.* 3d ed. Prentice-Hall.

Grable, Albert R. 1966. Soil aeration and plant growth. *Advan. Agron.* 18:57-106.

Pratt, Christopher J. 1965. Chemical fertilizers. *Sci. Amer.* 212(6):62-72. Available as *Sci. Amer.* Offprint 328.

Study Questions

1. Define the major physical properties of soils that are of major interest to crop producers.
2. Of what significance is soil texture to a crop producer? What are the major features of sandy soils? Of clay soils? Of what agricultural significance are these features?
3. What is cation exchange capacity? What conditions affect CEC? Why is CEC of interest to a crop producer?
4. Why are soils limed? How does lime react with the soil? What are the effects of liming?
5. Discuss the new system of soil taxonomy, compare it with the 1938 system, and consider why soil classification is of interest and importance to crop producers.
6. What are some of the effects of flooded soils on plant growth and plant-soil relationships?

Seven

Moisture and Crop Growth

Water is absolutely essential for plant life. Plants use more water than any other substance they absorb. The importance of water to the living plant can be summarized in terms of five general functions it serves:

1. Water is the major constituent of the living cell; between 85 and 95 percent of the live weight of most plant tissues is water.
2. Water in the living cell is the universal solvent that allows critical chemical reactions to occur.
3. Water is the solvent that carries essential nutrients through the plant.
4. Water, through its complex relations with osmotic subtances (such as salts) in the cell, is essential for cell turgidity and for cell elongation.
5. Water, through photolysis, provides electrons for carbon dioxide fixation, a key step in photosynthesis. Hydrogen for reducing NADP to $NADPH_2$ also comes from water.

Plants take up much more water than they actually use in the chemical reactions taking place in them. More than 95 percent of the water that enters the roots moves up through the xylem and is lost in the form of water vapor through transpiration. As water vapor is given off, water is drawn through the plant, causing more water from the soil to be absorbed by the roots. Additional water is removed from the soil because of evaporation; water lost by evaporation, of course, is not available for plant growth.

The Use of Water by Plants

TRANSPIRATION RATIO

Plants have been classified according to the amounts of water they require for normal growth. Xerophytes require little water, mesophytes require moderate amounts of water, and hydrophytes (or aquatics) require large amounts of water and may in fact grow submerged in water. Nearly all desert plants are xerophytes, most crop plants are mesophytes, and rice is the only major hydrophyte. Upland rice, although the same species as that grown in aquatic environments, is not grown in standing water; thus, rice can also be considered a mesophyte.

The amount of water in pounds required to produce a pound of dry plant tissue is expressed as the transpiration ratio. The amount of water required to sustain a plant throughout a growing season is determined by many factors. Obviously, xerophytes, mesophytes, and hyrophytes require different amounts. In addition, within any of these groups, species differ markedly in their water requirements. Climate—especially temperature, relative humidity, and wind—greatly affects the amount of water transpired by a plant. Water transpiration rates are high under hot, windy, and dry conditions and low under cool, calm, and moist conditions. Consequently, the amount of water required by a particular type of crop varies from location to location; for example, a wheat crop may require from 12 to 14 acre-inches of moisture per acre in cool intermountain regions and from 18 to 20 acre-inches in warm southern areas.

Although the water requirements of a species may vary greatly because of geographic differences in weather, the relative amounts of water used by different species are fairly constant, as indicated in Table 7-1. For example, alfalfa uses large amounts of water; in Colorado it may require up to 850 units (lb or kg) of water per unit of dry matter produced. Barley also uses large amounts—more than 800 units. Sugar beets (tops and roots) and corn, on the other hand, require markedly less water, about 370 units of water per unit of dry matter. Weeds, too, differ in their requirements: some major weeds thrive on little water (pigweed uses about 300 units of water per unit of dry matter, and purslane uses even less); some really steal water (lambs-quarters uses nearly 700 units per unit of dry matter). Water used by weeds is not available to crop plants.

Table 7-1
The relative water requirements of ten crops grown at Akron, Colorado.

Crop	Relative water requirement
Proso millet	1.00
Common millet	1.07
Sorghum	1.14
Corn	1.31
Barley	1.94
Wheat	2.09
Oats	2.18
Rye	2.37
Legumes	2.81
Grasses	3.10

Source: *Crop adaptation and distribution* by C. P. Wilsie. W. H. Freeman and Company. Copyright © 1962.

Disease generally increases the water requirements of crops drastically. For example, the water requirement of winter wheat that has leaf rust increases from 30 to 100 percent. In some crops, diseases may increase water requirements up to 400 percent, and, even if the additional water has been supplied, yields are reduced.

Plant water use is also related to soil fertility and fertilizer applications. When adequate moisture is available, yields increase with the application of appropriate fertilizers (mainly nitrogen), and the water required to produce a pound of dry matter may decrease (Table 7-2). This favorable response illustrates the necessity of balancing all plant requirements to obtain maximum yields. If adequate moisture is not available, the benefits of fertilizer applications are reduced; excessive applications may damage a crop because fertilizer salts can chemically burn root tissue if moisture is inadequate. Excessive applications, especially of nitrogen, also result in lush vegetative growth. Water is required to support such growth, which drains the soil of avail-

Table 7-2
Transpiration ratios for six grasses in a wet season and a dry season, with different amounts of nitrogen fertilizer.

Grass	Pounds of water used per pound of dry matter produced			
	1953 (rainfall: 39.7 in)		1954 (rainfall: 13.7 in)	
	Nitrogen added		Nitrogen added	
	50 lb/acre	200 lb/acre	50 lb/acre	200 lb/acre
Coastal bermudagrass	2,478	803	1,547	641
Suwanee bermudagrass	1,923	692	1,107	452
Common bermudagrass	6,812	1,546	9,738	4,336
Pensacola bahiagrass	2,200	870	3,103	1,239
Pangolagrass	2,249	2,240	2,843	3,016
Average	3,132	1,230	3,667	1,936

Source: *Crop adaptation and distribution* by C. P. Wilsie. H. Freeman and Company. Copyright © 1962.
Note: In either year, nitrogen fertilizer reduced the number of pounds of water required to produce a pound of dry matter and the average amount of water required in the dry year was greater than in the wet year.

able moisture before the crop is mature, thereby reducing yields.

WATER-USE EFFICIENCY
The transpiration ratio allows comparisons among species, but it does not indicate how efficiently a species uses water in terms of actual crop yield. It does not indicate how much of the potentially available water a crop uses for a given yield. Where water is in short supply, a crop producer should strive to obtain full use of available soil moisture, and of moisture added by irrigation, to realize maximum yields. Water-use efficiency is determined by the relation of yield (grain, forage, fiber, oil, etc.), to the amount of soil moisture used. (The amount of water used is expressed as a volume, frequently in acre-inches.) Water-use efficiency is a relative term and is generally used in comparing crops or crop management schemes. Crop species, and even cultivars of the same species, differ in their relative water-use efficiencies. By and large, C_4 types of plants have higher water-use efficiencies than C_3 types. This is attributable in part to the absence of photorespiration in C_4 types. If yields are comparable, species or cultivars with higher water-use efficiencies require less water than those with lower efficiencies.

CONDITIONS THAT AFFECT WATER USE AND WATER-USE EFFICIENCY
Regardless of water-use efficiency, the amount of water used is directly related to yield in all crops. As yields increase, total water use increases because more water is needed for the increased plant growth. The yields of many crops are related to stand density, or the number of plants per acre. The planting rates for corn, for example, are frequently expressed in plants per acre, rather than in pounds (or bushels) of seed per acre. Within the limits of available moisture, nutrients, and other variables, as stand densities increase, yields increase and total water use also increases. At the same time, water-use efficiency may also increase because the soil is permeated with roots so that the maximum amount of moisture that enters the soil is extracted from it and transpired by the crop.

For example, a plant population of 25,000 plants per acre may allow roots to explore 40 percent of the soil volume to a depth of three feet and extract 60 percent of the water in this volume of soil. If the planting density were increased to 50,000 plants per acre, 80 percent of the soil volume would be explored by roots and perhaps 75 percent of the water stored in it would be used by the crop. Both yield and water-use efficiency would increase, though they would not double. Before increasing stand densities, a grower should be certain that essential minerals are available in adequate amounts to support the increased number of plants. If water-use efficiency is improved, it may be necessary to increase applications of fertilizers, which may improve water-use efficiency even more. For example, nitrogen fertilizers increase the water-use efficiency of crested wheatgrass more than twofold: the additional nitrogen causes more vigorous root growth, resulting in a more thorough exploration of the soil by roots, which then extract more stored water. At the same time, total water use may be increased by increased vegetative growth.

The amount of available moisture present in the soil affects water-use efficiency. Generally, the closer a soil is to field capacity, the more water a plant will use. Plants can extract moisture more efficiently from a soil if it is held near field capacity than if it is close to the permanent wilting percentage. Crop plants use less energy in extracting water from a soil at field capacity, and this saving is converted into greater yields. This explains in part why winter wheat has a higher water-use efficiency if it is grown in a soil that had been fallow the previous year than it would under continuous cropping.

The stage of development of a plant plays a major role in relative water-use efficiency. Seedlings generally have a lower water-use efficiency than mature plants because their root systems do not fully permeate a large volume of soil; therefore, a great deal of potentially available water cannot be utilized. In addition, seedlings do not shade the soil surface, and significant amounts of water are lost from the soil by evaporation. As a plant grows, its roots develop and extract water more thoroughly from a given volume of soil: in the active vegetative stage of development, growth is so rapid that water-use efficiency is comparatively high.

In perennial forages, which may be harvested several times in a single season, water-use efficiency differs for different crop species and for different harvests. The water-use efficiency for the initial harvest is generally higher than that of regrowth. For example, for the initial harvest of sainfoin, water-use efficiency is fairly high; however, regrowth of most cultivars of sainfoin is slow and requires large amounts of water, thereby reducing water-use efficiency. On the other hand, regrowth of alfalfa is more rapid, and the differences in water-use efficiencies for initial harvests and for regrowth of this crop plant are not very great.

The overall health of a plant greatly affects its water consumption and water-use efficiency. Plants that are infected by any of a wide variety of pathogens (see Chapter 9) have greatly reduced water-use efficiencies, high transpiration ratios, and reduced yields. Many diseased plants display symptoms typical of moisture stress or drought: reduced total growth, few reproductive stems, light seed or fruit, and generally lower yields.

Water use and water-use efficiency are intimately involved with the fundamental physiological process of photosynthesis. Both are high when the photosynthetic rate is high. Because the photosynthetic rate is a function of light intensity, water use is high during daylight hours and low during hours of darkness (Figure 7-1). Water use is also high when

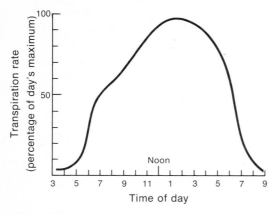

Figure 7-1
Daily pattern of water use by crop plants. Actual amounts of water used depend on the crop species, its stage of development, and climate. Water use is associated with both temperature and sunlight, and peak water use may lag behind the period of highest temperatures. [Adapted from L. J. Briggs and H. L. Shantz. *J. Agr. Res.* 5(1916):583.]

100 percent as water moves from the leaf to the air; transpiration is reduced because there is only a slight gradient for moisture between the air and the leaf. Wind moves this moist air away from the surface of the leaf and replaces it with drier air. As the moisture gradient between the air and the leaf surface increases, transpiration increases. Because relative humidity is a direct function of air temperature, the effects of temperature, wind, and relative humidity are interdependent, and all are tied to light and the process of photosynthesis.

In practice, a crop producer must consider not only the water use of an entire plant, but also that of many plants in a field. The effects of planting rates or densities on water-use efficiency have already been discussed. Because water use and water-use efficiency are closely associated with photosynthetic activity, factors that affect photosynthetic activity or efficiency may be related to water use.

Leaves intercept light and most photosynthesis occurs in leaf tissue; most of the water that is lost by a plant is lost through stomata on the surfaces of leaves. Although the relationships between light and leaves are treated in a later chapter on light and crop growth (Chapter 8), certain aspects are directly related to water use and merit consideration here.

Neither a single leaf nor all the leaves on a single plant can adequately represent the transpiration and photosynthesis taking place in a field of crop plants. All the leaves on all the plants must be considered. There are a number of methods that can be used to describe this mass of leaves; a common one is the leaf area index (LAI). The leaf area index is the ratio of the surface area of leaves—the area of any leaf being that of its upper or lower surface, but not both—to a given surface area of land, normally an acre (or hectare). An

temperatures are high, as is photosynthetic activity because high temperatures are largely associated with daylight hours.

The reasons for higher water use during photosynthesis are linked to the function and regulation of the opening and closing of stomata. More than 80 percent, and in some cases closer to 95 percent, of the water transpired by a plant passes to the atmosphere through stomata. Stomatal regulation depends on the turgidity of the guard cells, which in turn depends in part on the concentration of sugars, or other products of photosynthesis, in the guard cells (see Chapter 3).

Both relative humidity and wind affect water use by plants. If the relative humidity is 100 percent, there is no net movement (transpiration) of water from the plant to the atmosphere. The effect of wind on water use is related to the relative humidity of the air at the surface of each leaf: in calm air, the relative humidity at the leaf surface approaches

LAI value of 1 indicates one acre of leaf surface per acre of land, a value of 2 indicates two acres of leaf surface per acre of land. Maximum LAI values vary for different types of crops; for example, 4 is a reasonable value for wheat, and values of 3 to 5 are common for many crops. LAI values also vary with the developmental stages of a crop and with planting rates.

Leaves are not arranged on plants so that they are equally exposed to the sun. New leaves shade the older leaves on a plant, and the leaves on one plant shade those on another. Thus, in computing leaf area—for determining rates of either photosynthesis or transpiration—their three-dimensional arrangement (constituting a foliar canopy) must be taken into consideration. The amount of light that reaches a leaf depends on the location of the leaf in the canopy (see Chapter 8). In turn, the rates of photosynthesis and transpiration depend on the location of the leaf in the canopy. The canopy modifies the climate near the leaf surface: there is less wind movement in the lower part of the canopy than in the upper part, and the relative humidity is higher in the lower part than in the upper part. Thus, more transpiration takes place in the upper part of the canopy than in the lower part. A crop producer can control both the LAI and the arrangement of the canopy to some extent. Both seeding rate and planting pattern (the distance between rows and between plants in a row) can be managed to produce optimum results. In mixtures of species, which are common in pastures (see Chapter 31), the selection of compatible species will result in an LAI that will minimize shading so that optimum yields can be obtained. In addition, the timing of applications of fertilizers and irrigations can affect the LAI and canopy; late or excessive applications of nitrogen prolong the vegetative stage of growth in some types of plants, as does excessive water. For example, a late application of nitrogen prolongs the life of the flag leaf and awns of wheat, both of which through photosynthesis contribute significantly to grain yield.

Weed control is an important aspect of canopy management. One of the ways in which weeds reduce yield is through competition for light (see Chapter 9). Early, tall-growing weeds, such as Canada thistle, can tower over wheat and barley seedlings, robbing them of essential light as well as competing for water and minerals.

Unfortunately, all too frequently, only the minimal amount of moisture required per season is considered in the production of crops. For example, successful wheat production requires from 12 to 15 inches (30–38 cm) of water per year; for a growing season beginning in May and ending in September, corn requires about 20 inches (50 cm); alfalfa, a perennial crop, requires about 26 inches (66 cm) of water for a growing season from May through September with three harvests. Less water is required for a shorter growing season, fewer harvests, and lower yields. Water requirements also vary from location to location and from year to year at a given location. In areas with long, bright days, low humidity, and warm temperatures, water use is high; shorter days, higher humdity, and cooler temperatures reduce the total seasonal water requirement. However, to consider only the total seasonal water requirement for a crop can be misleading. As mentioned earlier, water use varies with the time of day; it also varies for different stages in the production of a crop. (Figure 7-2 shows a specific seasonal pattern of water use for barley.) Rarely, if ever, is enough moisture retained in the soil to meet a crop's changing needs. Thus, it is essential to consider not only how much water is needed during the entire season, but how much is needed at different times or

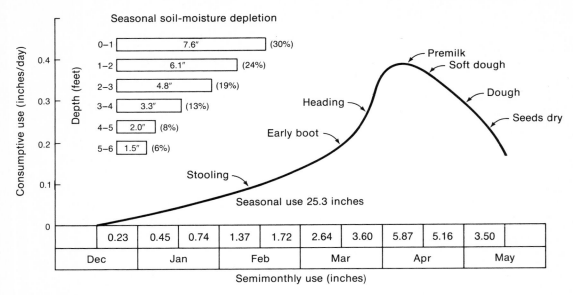

Figure 7-2
Consumptive use of water for barley grown at Mesa, Arizona. Note that the daily water requirements increase rapidly from the post-tillering through the flowering stages of development. Amounts of moisture required and period of highest use vary for different crops. [From L. J. Erie, O. F. French, and K. Harris. Consumptive use of water by crops in Arizona. *Ariz., Univ., Agr. Exp. Sta., Tech. Bull.* 169(September 1965).]

stages of plant growth. If moisture stored in the soil and precipitation do not satisfy a crop's needs during a growing season, irrigation is essential. A crop producer must manage his crops so that adequate moisture is present when and where it is needed.

Responses to moisture stress Of the major food crops, sorghum requires the least amount of water (and is the most drought tolerant) and rice requires the greatest amount. However, regardless of differences in water requirements, all plants are harmed to some degree by inadequate moisture. Under conditions of drought or of moisture stress, total plant growth, or dry-matter production, is reduced; therefore, yields are reduced.

The way in which a plant responds to moisture stress depends on the type of plant and its stage of development. The responses of wheat exemplify the general pattern of responses for all crop plants. Water is essential for germination: if adequate moisture is unavailable, seeds will not germinate; if seeds have been partially wetted and then allowed to dry (and germination takes place), the seedlings will be weak and the stands poor. Roots grow rapidly immediately after germination. At this stage, water-use efficiency is low because the seedling root system does not explore a large volume of soil. As the roots grow, they penetrate the soil more deeply, requiring water as they go for continued root growth. If moisture limits root growth, then the total development of the plant is restricted. As a rule, a plant does not develop more vegetative parts than its roots can support. However, conditions may change: for example, if adequate moisture had allowed vigorous root development and prolific tillering and then the plant was subjected to moisture stress, it would not be able to shed tillers, but fewer reproductive structures would form. Depending on the developmental stage of the plant, the decrease in reproductive structures could be due to fewer tillers that bear spikes, or, if inflorescence development has been initiated, to fewer fertile florets per spikelet or fewer spikelets per spike. The number of florets that will ultimately form grain may be determined at the time of flowering. If the moisture stress follows flowering, then kernel weight or plumpness will be reduced. For all crops that are grown for seed or fruit, moisture stress immediately before, during, and immediately after flowering seems to have the greatest effect on reducing yield.

Plants can survive conditions of moisture stress or drought in three ways. Some types of plants are drought tolerant. As the amount of available water is reduced, their total growth is reduced proportionately. This phenomenon is most common in forage grasses native to arid and semiarid areas. Examples include crested wheatgrass, Russian wildrye, and some fescues. Plants that are drought tolerant have rather extensive root systems that develop rapidly compared with those of less-tolerant plants. Plant species and cultivars of several crops differ with respect to drought tolerance and root growth.

Other types of plants avoid drought or moisture stress. This "avoidance" usually requires some type of internal adjustment in the plant. It may be an accumulation of water, as in cactus. Among cultivated crops, grain sorghum is a "drought avoider." When moisture becomes scarce, the leaves of a sorghum plant roll inward from their margins to form a loose, partial tube. Stomata are then protected from wind and the transpiration rate is reduced; therefore, less water is used. Some plants become dormant when subjected to moisture stress.

A final method of dealing with drought is escape or evasion. Crops that respond in this way hasten their entire life cycle as moisture becomes limited. They go from seed to seed before they run out of water.

Responses to excessive moisture Like moisture stress, excessive moisture can reduce crop yields. In general, yield reductions due to excessive moisture are related to poor aeration of the soil and reduced oxygen supply for the plant's respiratory needs. Poor soil aeration due to restricted soil drainage affects the growth and function of roots by reducing total root growth and decreasing the capacity of roots to absorb water efficiently. If root growth is restricted, total plant growth (and yield) is reduced. Excessive moisture late in the growing season can delay flowering and lead to poor seed set and lower quality seed: late precipitation can be a serious problem in northern parts of the corn belt because of its effect on the uniformity of both corn and soybeans; in other areas, late-season rains may disrupt the harvest of grain sorghum (in humid southeastern areas), delay threshing of beans, and cause losses of windrowed hay.

Excessive moisture also has indirect, long-term effects on plant growth. In time, essential minerals are leached from the soil, toxic substances may accumulate, and soil pH may be reduced (see Chapter 6). Flooding also reduces microbial activity, which in turn may reduce nitrogen availability. Many plants grown on flooded soil show symptoms typical of nitrogen deficiency. Flooded conditions also facilitate the development of several diseases.

Crop species differ in their tolerance for spring flooding or standing water. By planting at the right time and controlling irrigation, a crop producer can minimize losses due to flooding in many annual crops. Perennial forage species, however, are subject to potential floods every year in some regions. Choosing the right species for the expected duration of flooding is an important aspect of crop management (see Table 28-1 on page 467). However, the development of drainage systems may eliminate, or greatly reduce, losses due to flooding.

Sources of Moisture

RAIN

Natural precipitation, primarily in the form of rain, greatly affects the distribution of both cultivated and native plants. North America (and the entire world for that matter) can be divided into four major precipitation regions (Figure 7-3).

The arid region has less than 10 inches (25 cm) of annual precipitation. Parts of Nevada, Oregon, Washington, Wyoming, Utah, Colorado, and New Mexico are classified as arid. Crop production in arid areas is dependent on supplemental water supplied through irrigation. Agriculturally productive arid areas are found in southern California (the Imperial Valley), western Arizona, and southern Nevada.

The semiarid region has from 10 to 20 inches (25-50 cm) of annual precipitation. Much of the western half of the United States is semiarid, including Montana, North and South Dakota, most of Nebraska and Oklahoma, and the western half of Texas. The rich, grain-producing Canadian prairie provinces of Alberta and Saskatchewan are also semiarid. Crop production in these areas requires either farming practices that conserve water (such as fallow) or irrigation. Because annual precipitation in semiarid areas varies considerably, in some years there is adequate moisture to produce a profitable crop without any form of moisture conservation, and in other years, successful crop production may require irrigation. Both the short-grass and the tall-grass prairies are in the semiarid region: the short-grass areas have less annual precipitation (approximately 10 in., or 25 cm) and the tall-grass areas have more (closer to 20 in. or 50 cm). Parts of both the winter and spring wheat belts are found in the eastern half of the semiarid region, which has more rainfall than the western half.

Natural precipitation does not limit crop production in the semihumid region, which has from 30 to 40 inches (75-100 cm) of annual precipitation. In the United States, the semihumid region is east of the semiarid region and includes Minnesota and Wisconsin, where hay and pasture lands support a major dairy industry, and a large part of the corn belt (Illinois, Iowa, and Nebraska), extending to the northern edge of New England (Connecticut, Vermont, and Maine). It is a large area of diverse crop production.

The humid region has more than 40 inches (100 cm) of annual precipitation; it includes the Pacific northwest (northern California, Oregon, and Washington), the Gulf Coast, and the Atlantic seaboard. The cotton-belt states (Texas, Oklahoma, Arkansas, Louisiana, Mississippi, Alabama, and North and South Carolina) are located in the heart of the humid region.

A region with a high annual rainfall (more than 40 in., or 100 cm), high average annual temperatures (70–75°F, or 21°–24°C), and long frost-free periods is classified in a fifth category, the subtropical. Florida, Hawaii, and a narrow strip along the Gulf and Atlantic seaboard are subtropical areas in the United States. Citrus, sugarcane, and other specialty crops are grown in these areas.

Effectiveness of rainfall In the humid and semihumid regions, annual precipitation does not limit crop production, but, in the arid and semiarid regions, the relation of rainfall to

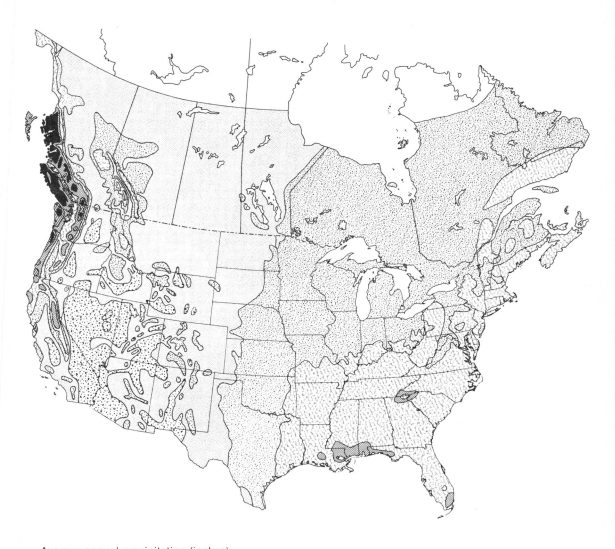

Average annual precipitation (inches)

- 0–10
- 10–20
- 20–40
- 40–60
- 60–80
- 80–100
- 100–150

Figure 7-3
Average annual precipitation (in inches) in North America. [USDA data.]

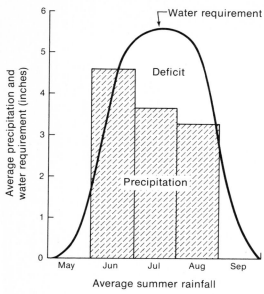

Figure 7-4
The relationship between crop water use and summer rainfall in five corn-belt states. [From L. M. Thompson. Regional weather relations. In *Advances in corn production*, ed. W. H. Pierre, S. A. Aldrich, and W. P. Martin. © 1966 by Iowa State University Press.]

crop growth varies. The effectiveness of precipitation in promoting crop growth depends on three main factors: distribution of moisture (in time), amount of moisture, and soil texture. These three factors play an important role in determining where a crop can be grown. Rainfall during the growing season (roughly from April through September) is much more beneficial to crops than rainfall during periods when crops are not growing rapidly. For example, the distribution of rain has a variable effect on corn yields. Corn requires about 8 inches (20 cm) of water from June through August. In the corn belt, the average precipitation is slightly excessive in June, adequate in July, and often inadequate in August (Figure 7-4). The crops survive because some moisture is stored in the soil. However, the excessive moisture in June reduces yields whereas increased rain in July and August would augment them (Figure 7-5).

Light rainfall (less than 0.25 in., or 0.6 cm) is of little direct value to most crops because it does not wet the soil deeply; nevertheless, because the effectivenss of light precipitation is directly related to the texture of the soil, the type of crop, the distribution of roots, and the requirements of the plants for water at different developmental stages, a light rainfall can be useful, for example, in establishing a stand of shallowly rooted forage grass. Heavy rainfall may be no more effective in promoting crop growth than light showers: if the rate of precipitation exceeds the rate at which the soil can absorb moisture (the infiltration rate), water will be lost either by evaporation or runoff. In addition, heavy precipitation can destroy soil structure; excessive runoff is a leading cause of soil erosion. Finally, if the soil is frozen or at field capacity, any rainfall is of little value because the soil cannot store the moisture for future use. Field management practices that protect soil structure and minimize erosion prepare the soil for maximum water infiltration and thus provide for maximum moisture storage.

SNOW

Snow is another source of moisture for crop production; it not only increases the volume of springs and creeks but also adds water to the soil that can be stored for use by crops. The effectiveness of snow in supplying water directly for plant use depends on the water content of the snow and the conditions under which it melts. The water content of snow varies from 5 percent (0.05 inch of water per inch of snow) in cold, dry snow to 10 percent in warmer, wet snow. If the snow melts slowly as the ground beneath it thaws, much of the moisture contained in the snow enters the soil and can be used by crop plants. If, on the

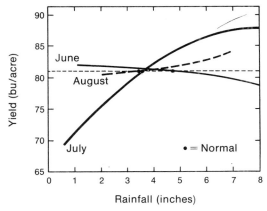

Figure 7-5
How variations in the amounts of precipitation in June, July, and August affect corn yields in the corn belt. [From W. L. Colville. Environment and maximum yield of corn. In *Maximum crop yield*. American Society of Agronomy Special Publication 9, 1967.]

other hand, there is a sudden warm, windy period before the ground has had a chance to thaw, the snow may melt and evaporate, or moisture may be lost through surface runoff.

Various methods are used to hold snow uniformly on a field so that the moisture is not lost. On fallow fields uniform snow accumulation is favored by a cloddy soil surface and trashy fallow (fallow in which crop residue—straw—is not fully turned under). Standing stubble is another excellent snow trap; or snow can be trapped in furrows where it acts as an insulating blanket covering winter wheat seedlings and later becomes a source of moisture.

In regions where saline seep has become a problem, snow accumulation is not desired. In these areas, cropping systems are being developed to utilize maximum amounts of water and at the same time prevent excessive water accumulation from snow and other sources (see Chapter 10).

The Fallow System of Moisture Conservation

The fallow system of moisture conservation, whereby land is cultivated but not seeded, creates optimum conditions for storing water from any source in the soil and enables one year's precipitation to be held in reserve for use by the next year's crop. The term "fallow" can refer to either the practice of cultivating without seeding or the bare land itself. Throughout much of the semiarid region of the United States and Canada, cereal production is based on a crop-fallow system in which the rotation of crop and fallow follows this pattern:

1. Current year, harvest wheat in early fall.
2. After harvest, cultivate to hasten decomposition of crop stubble.
3. Next spring, do not plant, but cultivate periodically to control weeds: this is the fallow period during which precipitation is stored in the soil.
4. Prepare seedbed and plant after full fallow year, (i.e., from eleven to thirteen months after harvest).
5. Repeat cycle.

Fallow may conserve from 10 to 30 percent of the precipitation that reaches the soil during the noncrop period. If 12 inches (30 cm) of rain falls during this period, from 1.2 to 4.0 inches (3–10 cm) will be stored in the soil. Winter wheat requires a minimum of 15 inches (38 cm) of precipitation to produce good yields. If the previous year's fallow has allowed the storage of 2.5 inches (7 cm) of water, then only 12.5 inches (31 cm) of precipitation will be required during the growing period to produce a substantial winter wheat crop. Besides the primary benefit of moisture storage, there are two other benefits from

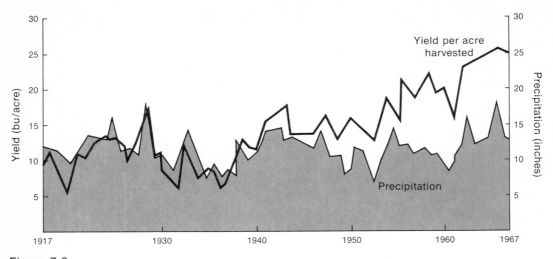

Figure 7-6
The relationship between precipitation and yield per acre of wheat harvested in eastern Montana. [From Don Bosley. Bill Reed: the man who got summer-fallowing off the ground. *Montana Farmer-Stockman* 56(8)(1969):17–18.]

fallow: weed control and the return of nitrates and other essential minerals to the soil through the decomposition of crop residue. It is essential to cultivate after the fall harvest as well as during the following spring and summer to prevent weeds from using the stored water, but cultivation during the fallow period requires care; if the ground is left completely bare, serious damage may result from wind and water erosion. Therefore it is often desirable to prepare a trashy fallow in which crop residues are not fully plowed into the soil: the residue left on the soil surface acts as a buffer to the destructive effects of wind and running water.

The fallow system has served to stabilize grain yields and reduce the extreme yearly differences in yield that used to lead to financial disaster for many growers. For example, in Montana before 1937, crop yields were closely tied to the amount of natural precipitation (Figure 7-6); note that the lower yields in 1919 were associated with lower rainfall. By 1937 fallow–crop rotation was widely accepted as a beneficial system for winter wheat so that, after 1937, the association between annual precipitation and yield is not as obvious. However, the fallow system does not create moisture. A period of several dry years (such as that which occurred in the late 1940s and early 1950s) causes a reduction in yields even if fallow is practiced. Moreover, the overall increase in yield illustrated in Figure 7-6 cannot be attributed exclusively to fallow. New disease- and insect-resistant cultivars, increased use of fertilizers, and chemical weed control contributed to higher yields. Nevertheless, the assurance of adequate moisture as a result of fallowing did encourage farmers to invest more in fertilizer and heavier seeding rates in pursuit of higher yields and greater profits.

In recent years the advantages of the fallow system have been weighed against the disadvantages. Fallow is costly and time consuming; it may lead to serious losses due to wind

7 / MOISTURE AND CROP GROWTH

and rain erosion. The development of saline seep areas has been associated with the practice of fallowing so that this cropping system should not be used in some areas. Furthermore, over long periods, average yields from continuously cropped land equal those from land on which crops are rotated with fallow. Yields from annual cropping tend to be comparable to those from crop-fallow rotations if the acreage of the entire farm is included in the calculation: for example, the average annual yield for two years of annual cropping on 50 acres might be 20 bushels per acre, whereas, in a crop-fallow system in which half the land is fallow and half is planted, the average annual yield from the 25 acres planted might be 40 bushels per acre. Even so, the average yield from the 50-acre farm would be 20 bushels per acre. The major reason to practice fallow is to avoid short-term, season-to-season variations in yield due to yearly differences in precipitation: fallowing can soften the boom-or-bust aspect of crop production.

Irrigation

If natural annual precipitation is inadequate for crop production, irrigation becomes essential. The object of irrigation is to supply the soil with the amount of water required by crop plants to produce optimum yields. Because irrigation is costly per unit of water (compared with natural precipitation), a knowledge of crop water requirements must be combined with a knowledge of how irrigation water is measured and how it can be applied most effectively for efficient irrigation management.

MEASUREMENT
OF WATER VOLUME AND FLOW

Water use by crop plants can be measured either in pounds of water per pound of dry matter produced (see p. 123), or in acre-inches of water. For irrigation, crop water use is usually expressed in acre-inches. An acre-inch is the amount of water required to cover one acre of land (43,560 square feet) with one inch of water. Twelve acre-inches equals one acre-foot. Many water rights, which specify the total volume of water that may be legally removed from a public stream or river, are also given in acre-inches. In buying land or planning programs for crop production, a thorough understanding of water rights is essential.

Water drawn from wells and rivers is measured in terms of flow. Well water is most frequently measured in gallons per minute (Figure 7-7). Flow in gallons per minute (GPM) is related to acre-feet as follows: GPM $\div$ 225 = the number of acre-feet in 24 hours. For example, if a pump plant delivers 585 GPM, then in 24 hours it will deliver 585/225 = 2.6 acre-feet. At this rate the pump would have to operate for about 46 hours to apply 6 inches of water to a 10-acre field (6 inches over 10 acres equals 60 acre-inches or 5 acre-feet; to deliver 5 acre-feet of water at the rate of 2.6 acre-feet every 24 hours the pump must operate 5.0/2.6 = 1.9 $\times$ 24 hours = 45.6 hours).

River water is commonly measured in cubic feet per second (CFS). A rate of flow of 1 CFS equals 450 GPM; this is the equivalent of 1 acre-inch per hour or 2 acre-feet in 24 hours. There is one additional and important measure of flow, the miner's inch; it is a measure of rate of flow and must not be confused with the acre-inch, which is a measure of volume. The flow of a miner's inch varies from state to state: in Montana, a miner's inch equals 11.25 GPM, and 40 miner's inches equal 1 CFS; in Idaho and Arizona a miner's inch equals 9 GPM. The miner's inch is used in determining stream and river flow mainly in areas that have a history of mining activity-

Figure 7-7
Well water delivered under low pressure is used for a variety of forms of surface irrigation.

such as Montana and the Sierra-Nevada region of California. In these areas, the miner's inch is also used in specifying water rights.

IRRIGATION METHODS
Irrigation methods are classified as either aerial or surface. The choice of how to apply water to a crop depends on the crop to be irrigated, the topography of the land (slope, hills, and gullies), the type of soil, the amount of water available, the applicability of special equipment, and cost. To conserve both water and money, accurate control of water application is essential. It does not make sense to apply more water than the soil can hold; besides wasting water, overirrigation results in minerals being washed from the soil, salt buildup in the soil, and soil erosion.

Aerial irrigation Aerial irrigation generally refers to a sprinkler irrigation system. Sprinklers can be used on most crops and are effective for a wide range of topography (flat or rolling fields), as well as for a wide range of soil types. In addition, the amount of water applied through sprinkler irrigation can be closely measured and controlled.

The major drawback of sprinkler irrigation is the initial cost of equipment. Distribution pipes, risers on which individual sprinklers are mounted, and the sprinkler heads are expensive. Moreover, the cost of pumping water is increased because of the high pressure (40 to 120 pounds per square inch) required for effective irrigation. Another drawback of sprinkler irrigation is the effect of wind on uniform water application. On calm days, with properly designed sprinkler arrangements, extremely uniform irrigation is possible, but even light winds disrupt this uniformity.

There is a variety of equipment available for sprinkler irrigation.

Permanent pipe. One system uses permanent pipes, risers, and sprinklers. This type of installation is generally limited to small permanent pastures or to very high income specialty crops, such as hops. All the benefits of sprinkler irrigation are realized by using a permanent system, but there is little flexibility in the use of such equipment.

Portable pipe. This type of system, which is quite flexible, is the most common form of sprinkler irrigation (Figure 7-8). It is suitable for all crops because risers and sprinkler heads can be changed to irrigate crops of different heights and the rate of flow can be controlled. The hard work required to move individual lengths of portable pipe can be reduced by using a "drag line" in which each pipe section is equipped with a sledlike device that can be attached to a small tractor or truck; several lengths of pipe can be moved from one location to the next at the same time. The only drawback to this arrangement is that lines can be moved only when an area is quite dry.

High-volume equipment. The most recent additions to sprinkler irrigation equipment are systems that irrigate large areas at one time and therefore deliver large quantities of water (Figure 7-9). These systems include a wide array of self-propelled or tractor-drawn lines and permanent high-volume individual units, such as center-pivot sprinklers. These systems are used to irrigate pastures and other forages grown on all but the most uneven terrain, as well as an increasing number of cereal crops. In spite of high initial costs, these new systems are becoming increasingly popular because of their efficient water delivery and minimum labor requirements.

Regardless of the amount of water to be applied through sprinkler irrigation, the rate of application should be between 0.33 and 0.50 acre-inch per hour. Rates higher than 0.50

Figure 7-8
Sprinkler system using portable pipe to irrigate alfalfa on rolling hills. [Photograph courtesy of Harold E. Jacobs.]

Figure 7-9
A self-propelled sprinkler line—one of a wide variety of new, high-volume systems.

acre-inch per hour result in runoff rather than infiltration. Runoff wastes water and erodes the soil. The hourly rate of application—the precipitation rate—is expressed by the equation

$$\text{Precipitation} = \frac{\text{GPM} \times 96.3}{\text{C} \times \text{S}}$$

GPM is the flow from any sprinkler head at the appropriate pressure (depending on the type of sprinkler head used, flow may vary from 2.5 GPM for small heads under low pressure to more than 200 GPM for large heads at very high pressure); the constant 96.3 accounts for the difference between the amount of water pumped and the amount of water received by the soil; C is the distance between adjacent pipe lines; S is the distance in feet between heads on the line. If GPM = 5.5, C = 40 feet, and S = 30 feet, the precipitation rate is $(5.5 \times 96.3)/(40 \times 30)$ = 529.65/1200 = 0.44 inches per hour. This is a reasonable rate of precipitation. For example, alfalfa requires 3–6 inches of irrigation water every four weeks. If a crop producer decides to supply 4 inches of irrigation water per month and the precipitation rate of his sprinkler system is 0.44 inch per hour, his pump will have to operate 9 hours before the pipes can be moved.

Surface irrigation Surface irrigation systems move water directly from a source (stream, ditch, or well) to an area in which crops are growing.

Flood irrigation. This is the simplest form of surface irrigation. Its object is to cover a field uniformly with just enough water to bring the soil to field capacity. In practice this goal is rarely achieved. There are three methods of flood irrigation: wild flooding, border strip irrigation, and basin irrigation. In wild flooding, water is allowed to run uncontrolled over a piece of land (Figure 7-10). This requires either opening a levee or building a dam in a ditch or stream. Wild flooding is undesirable because it is impossible to control the amount and flow of water: low points in a field may be flooded while high spots remain dry. Because crops and soils may be damaged by extensive wild flooding, the only crop lands suited to this method of irrigation are wild hay fields and uncultivated pastures.

Border strip irrigation is a common form of flood irrigation in which fields are divided into strips that are bounded by a low ridge or levee (Figure 7-11). The slope down the length of a strip varies with the type of soil and the crop being grown; it generally should not exceed 3 percent (3 feet of drop per 100 feet), but in established pastures it can be as much as 6 percent. The slope across a strip should not exceed 1 percent. The object of border strip irrigation is to allow a uniform sheet of water to move slowly across each strip. If the slope down the field is too great, water flows too rapidly and the high end of the field may remain dry while the low end is flooded. If the cross slope is too great, one side of the strip stays wet while the other remains dry. Field size and the cost of leveling land are the major factors considered in determining the dimensions of a border strip, which is usually a maximum of 800 feet long and 50 feet wide; the rate of water application and the type of soil must also be considered in determining the ideal dimensions of a strip. The ridges between the strips should be low enough and wide enough to be planted and to allow equipment to be driven over them. Border strip irrigation is suitable for nearly all hay and pasture crops and for closely seeded crops such as barley, wheat, and oats.

Basin irrigation is a more specialized form of flood irrigation (Figure 7-12). The areas that

Figure 7-10
Wild flooding on pasture land near Bozeman, Montana. Note that some areas are flooded and others are dry.

Figure 7-11
Border strip irrigation with water being delivered from a head ditch to individual strips. [USDA photograph.]

Figure 7-12
Basin irrigation of an orchard. [USDA photograph.]

are set off and flooded in basin irrigation are irregularly shaped because they follow the topography of the land; the levees that separate these irregular basins are high and more permanent than the ridges separating the strips in border strip irrigation. A basin must meet the same slope requirements as a border strip. The rice paddies of Asia and China are examples of basins. In the United States basin irrigation is used with rice crops and in orchards. Because land must be surveyed and permanent levees must be built, the initial cost in money, time, and energy of establishing suitable basins is quite high; it is offset by the permanency and ease of operation of a basin irrigation system.

Furrow irrigation. For many row crops (sugar beets, potatoes, vegetables, and, in some areas, corn) furrow irrigation, in which water is run down furrows between rows of plants, is a more suitable method of surface irrigation than flooding. The furrows are cut between rows of plants and water is introduced into them in one of several ways. The simplest technique is the use of "gated pipe" (Figure 7-13). Gated pipe is similar to sprinkler pipe, but instead of having openings for risers at specific intervals along the pipe, it has adjustable gates from which the flow of water into the furrows can be regulated. Gated pipe must be used in association with some type of low-pressure pumping system. Irrigation

Figure 7-13
Gated pipe used to deliver water for furrow irrigation of corn. [Photograph courtesy of University of California Agricultural Extension Service.]

Figure 7-14
Furrow irrigation of sugar beets with siphon tubes carrying water from a delivery ditch to individual furrows. Note the canvas check dam (lower right), which is designed to maintain the water in the ditch at a proper level for siphons to function.

by running water between elevated beds is quite similar to furrow irrigation. A common method of moving water from an open ditch to the furrows is through siphon tubes (Figure 7-14). The rate of flow partly depends on the diameter of the siphon tube used. This method is time consuming but effective. In some cases, water is discharged into flat basins and then allowed to flow at random down furrows. This method requires less work, but the resultant irrigation may be uneven, with some wet spots and some dry ones.

Regardless of how water is introduced into the furrows, furrow irrigation is time consuming. The furrows must be kept free of weeds and other plants, and each furrow must be watched to make sure water flows through it at the proper rate. Furrows must be carefully planned: length and slope depend on the type of soil and the proposed rate of water application. The cross slope of a field is relatively unimportant in furrow irrigation, but the slope down the length of the field is critical. If this slope is too steep, water runs over the soil but does not enter it. Excessive slope may cause severe erosion. The more level the land, the greater the allowable rate of water application. With a slope of 0.25 percent (3 inches in 100 feet), 40 GPM can be used to irrigate 1,300 feet of furrow in a clay soil. On the same soil, if the slope is 5 percent, only 2 GPM can be used to irrigate a maximum of 400 feet of furrow.

Corrugation irrigation. A final method of irrigation uses very small furrows called corrugations that are allowed to overflow and flood the land. These small furrows, which run the length of a field and follow natural contours, are not used to carry water, as does a normal furrow, but rather to guide its flow. The space between corrugations, as well as the length of each one, is highly variable and depends on the type of soil, the degree of slope, the amount of water to be applied, and the crop. Soils that take water slowly benefit from corrugation irrigation. It is also effective on sloping, uneven land that is planted in permanent pasture or hay crops; it has limited usefulness for cereal crops.

IRRIGATION MANAGEMENT

In truly arid regions (such as those in the southwestern United States), where natural precipitation is so scant and undependable that it may be more of a nuisance than a benefit in crop production, all water needs of crops must be met with irrigation. In these areas, all phases of crop production are adjusted for irrigation. Regardless of the method used to deliver water, irrigation must supply moisture to that part of the soil from which roots absorb it. The timing and amount of irrigation should be regulated to assure optimum crop yields.

As a general rule, as moisture increases either through natural precipitation or through irrigation, plant populations per acre (or seeding rates) increase and the spacing between rows decreases. For example, under semiarid conditions the seeding rate for winter wheat may be between 35 and 40 pounds per acre. With irrigation, the seeding rate for the same crop increases to more than 160 pounds per acre. However, it is essential to realize that cultivars may differ in their tolerance of heavier seeding rates. If a crop producer decides to irrigate, he may have to choose a cultivar that tolerates the heavier seeding rates made possible by the increase in moisture. In crops like corn, soybeans, and grain sorghum, the spacing between rows may be decreased to achieve higher plant populations per acre.

The composition of pastures, with respect to species, changes with variations in the relative seeding rates of different cultivars, which in turn vary according to the availability of moisture. Although local conditions determine the species, cultivars, and exact seeding rates, the correlation between average

Table 7-3
Seeding rates recommended for pasture mixtures in Montana for dry areas, moderately moist areas, and irrigated sites.

Pasture mixture	Seeding rate (lb/acre)		
	Annual precipitation		Irrigated
	Less than 18 in	More than 18 in	
'Ladak' alfalfa	0.5	1.0	1.5
Mixed with one of the following grasses:			
Crested wheatgrass	4.2	—	—
Smooth bromegrass	7.4	7.4	7.4
Intermediate wheatgrass	7.4	—	7.5
Green needlegrass	6.3	—	—
Russian wildrye	6.3	—	7.4
Pubescent wheatgrass	—	—	9.5
Orchardgrass	7.4	4.2	—

Source: Data from *Montana Agr. Exp. Sta. Circ.* 242.
Note: A dash indicates that the mixture is not recommended for that moisture classification.

annual precipitation and the seeding rates for various species and mixtures is illustrated in the recommendations for nonirrigated and irrigated pastures in Montana (Table 7-3).

The response of a crop to fertilizers is closely tied to the availability of adequate moisture. As moisture increases, more fertilizer may be required to produce optimum yields for two reasons: first, under conditions of increased moisture the soil can support larger plant populations; and, second, under such conditions plants can use fertilizers more efficiently. With irrigation, the fertilizer requirements of a crop may increase 50 percent or more; yields will increase proportionately. Increasing applications of fertilizer in the absence of adequate moisture can harm crops either by encouraging excessive, early vegetative growth and water consumption or by actually "burning" plants because of a buildup of toxic levels of fertilizer salts.

The adage that a chain is no stronger than its weakest link can be applied to crop production: each link is any factor that might limit yield, such as soil fertility, or available moisture, or length of growing season. Irrigation insures that available moisture is not the weakest link in the chain. Successful crop producers must also insure that no other link is weak compared with the "available moisture" link; at the same time they must avoid the inefficiency of making one link (e.g., fertilizers) unnecessarily strong. All factors must be balanced for maximum yields. By eliminating moisture as a limiting factor, irrigation allows more control over other potentially limiting factors.

Selected References

Isrelsen, O. W., and V. E. Hansen. 1962. *Irrigation principles and practices*. 3d ed. Wiley.

Salisbury, F. B., and C. Ross, 1969. *Plant physiology.* Wadsworth.

Water. 1955. *Yearbook Agr. USDA.*

Study Questions

1. Discuss the relationship between transpiration rate, water-use efficiency, and climate.
2. How does water use ultimately relate to factors that affect photosynthesis? How is stomatal regulation associated with water use and photosynthesis?
3. What factors would you consider in determining method, frequency, and amount of irrigation?
4. What are the common ways in which water is measured? How and why are they used? How are they interrelated?

Eight

Light and Temperature

Both light and temperature play major roles in determining how and where crops can be grown. Efficient crop management requires an understanding of how these factors affect plant growth, how they vary under field conditions, and how they can be managed.

Light

Light is a form of electromagnetic radiation that travels in waves; the wavelength determines the color as well as the energy level of the light. The shorter wavelengths, which form the blue part of the spectrum, are higher in energy than the longer wavelengths, which compose the red end of the spectrum. Only wavelengths from 350 to 700 Å are visible (those between 350 and 400 Å are violet to blue and those between 690 and 700 Å are red). This is a small part of the total spectrum of radiant energy, which also includes ultraviolet and infrared light. Sunlight appears as "white light" because it is a mixture of wavelengths.

Quality, quantity, and duration are three characteristics of light that greatly affect crop production. Although crop producers can do little to control light in the field, they must understand how plants react to critical variations in the quality, quantity, and duration of light in order to make decisions that will result in effective crop management.

QUALITY

The quality of light is determined by the relative proportion of different wavelengths (colors) contained in the sunlight that reaches the earth's surface. Light quality is not constant; it varies with the season and the geographic location, as well as with changes in the composition of the atmosphere surrounding the earth. Only some of the sunlight that reaches the upper atmosphere is transmitted to earth; the rest is either filtered ("screened out") by various components of the atmos-

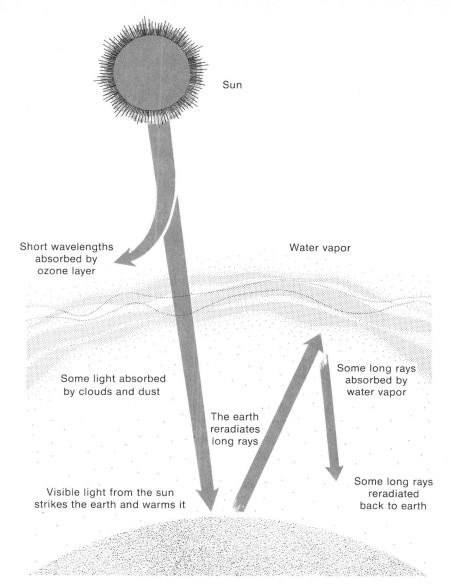

Figure 8-1
The fate of solar radiation as it passes from the sun to the earth's surface. Various layers of the atmosphere absorb and/or scatter a significant part of the radiation that comes from the sun. [Adapted from *Life of the green plant* by A. Galston. Prentice-Hall, 1961.]

phere or reflected back toward outer space (Figure 8-1). Light is filtered in various ways: light of short wavelengths is absorbed by atmospheric ozone (O_3); light of longer wavelengths is screened out by atmospheric moisture belts; light of other wavelengths is diverted through scattering and reflection by smoke and dust particles in the air near the earth's surface. In addition, some of the light that does reach the earth's surface is reflected back to the atmosphere. Filtering directly affects light quality, and geographic location, season, and the amount and types of pollutants in the air have an effect on the extent to which light is filtered. At the spring and fall equinoxes (when days and nights are of equal length), the mass of atmosphere through which sunlight must pass to reach the earth's surface is forty-five times as great at the poles as at the equator. At the poles, the sunlight traverses the atmosphere at a much greater angle and is filtered proportionately more. Because of the angle of the earth's axis in relation to its orbit around the sun, sunlight travels a shorter distance to the earth's surface and is relatively less filtered in summer than in winter. This may, unfortunately, be offset by more dust and other air pollutants, which increase filtering in summer.

Light quality is important in photosynthesis. Remember that light of shorter wavelengths (at the blue end of the visible spectrum —about 350–400 Å) and light of longer wavelengths (at the red end of the spectrum—about 650–700 Å) is absorbed by chloroplasts, which indicates that light of these wavelengths is the main source of energy used in photosynthesis. Most green light is transmitted or reflected by the plant and is not used as much in photosynthesis. Light of longer wavelengths and to some extent of middle wavelengths determines a plant's photoperiodic response; that is, how it responds to the daily duration of light. (The concept of photoperiod will be discussed in detail later in the chapter.) Red light is critical in promoting germination of certain seeds; for example, lettuce.

QUANTITY

Light quantity is frequently equated to intensity. In the United States it is most commonly measured in footcandles (FC); the light produced by a "standard candle" on a surface one foot from that candle equals 1 FC.

The international system of measurement uses the metric equivalent of the footcandle— the lux (plural luces). One lux, which equals approximately 0.093 FC, measures the illumination produced by one candle on a surface one meter from the light source. A full moon radiates about 0.05 FC of light and a comfortable reading light has an intensity of 20 FC. Most plants require a minimum light intensity ranging between 100 and 200 FC or more to maintain enough photosynthetic activity to support growth. It must be noted, however, that neither the footcandle nor the lux measures light energy. Because plants utilize light energy, data based on these measures of intensity can only furnish information on the responses of plants to light or the variation of light conditions in the field; such data cannot indicate how much light energy each plant receives. However, because a direct measure of light energy is difficult to obtain in the field, these measures of intensity are considered adequate for comparing differences in light requirements among plants. Responses to light intensity vary from crop to crop (and to a lesser extent from cultivar to cultivar of the same crop); moreover, differences in the rate of photosynthetic activity result in variable requirements for light among leaves of a single plant.

Any light of an appropriate wavelength will produce some photosynthetic activity in a living plant; the plant uses the oxygen produced during photosynthesis for respiration, and the carbon dioxide released during respiration for photosynthesis. The metabolic point at which the rates of photosynthesis and respiration are exactly balanced in terms of the production and use of photosynthate is called the compensation point. Plants functioning at the compensation point cannot survive for long: at night respiration relies on stored food because the light reactions of photosynthesis take place only during the day. In analyzing photosynthesis that occurs under field conditions, crop producers generally consider net photosynthesis—the photosynthesis that results in the production of carbohydrates above and beyond those necessary for respiration—in order to determine field practices and calculate potential yield. If no other factor is limiting photosynthesis, a plant is said to be light saturated when its photosynthetic rate (measured by the rate at which CO_2 is fixed) does not increase with increasing light intensity. When a plant reaches the point of light saturation, its entire photosynthetic "machine" is operating at maximum capacity. The maximum intensity of light that plants can use in photosynthesis under average field conditions is about 10,000 FC. If temperature, moisture, and other factors critical for photosynthesis are not limiting, the maximum light intensity that a plant can utilize in photosynthesis may greatly exceed 10,000 FC. It is possible to divide plants into two broad groups according to the light intensity required for saturation: sun plants reach saturation at light intensities of 5,000 FC or more; shade plants reach saturation at about 500 FC. Most crop plants and weeds that are serious pests are sun plants. Tobacco grown for certain commercial purposes (see Chapter 24) and native vegetation on the floors of forests are shade plants. Within each broad group, plants differ with respect to the minimum, optimum, and maximum light requirements. For example, C_4 plants utilize more light than C_3 plants, but even C_4 plants differ in this regard. The light saturation point for sun plants varies among and within species. Because plants with lower light saturation points tolerate shading relatively well, they can also tolerate heavy seeding rates (heavier seeding rates result in shading among plants). Thus, if available moisture and minerals can support heavy seeding rates, a shade-tolerant cultivar should be selected to produce maximum yields.

In addition to differences among species and among cultivars of the same species, there are significant differences among leaves on a single plant in the amount of light they receive to fuel photosynthetic activity. The amount of light reaching a leaf decreases as sunlight passes downward through the foliar canopy. Leaves higher up on the canopy shade the lower leaves; the upper leaves also reflect light away from the lower leaves. Thus the upper leaves may receive more light than they can utilize (light in excess of their saturation point), whereas the lower leaves may receive less light than they can use (light below their saturation point). Some leaves receive only enough light to remain at the compensation point: they neither contribute to nor detract from higher yields. However, if a leaf falls below the compensation point, that is, if its respiration exceeds its photosynthesis, it becomes a liability to the plant and causes reduced yields. The liability results from two facts: first, the leaf is using more products in respiration than it is producing through photosynthesis; second, although the photosynthetic rate is low, the leaf still transpires. Thus, the leaf causes the plant to lose water but does not contribute vital resources to the plant through photosynthesis. Of course, the rate of transpiration decreases from the top to the bottom of the foliar canopy (see Chapter 7).

Foliar canopies differ in their basic structures. In some canopies, leaves are held in a nearly horizontal position and shading is minimized by a whorling arrangement of leaves along the stem. In other canopies, leaves are held in a more vertical position, which results in minimum shading of lower leaves. If a leaf is held at an angle, it has more surface exposed per unit of ground surface than a leaf that is held in a horizontal position.

Corn cultivars that tolerate heavy seeding rates have more vertically held leaves than cultivars that are less tolerant of heavy seeding rates. This minimizes intraplant shading and allows plants to be placed closer together without excessive shading. These shade-tolerant cultivars frequently have lower light compensation points. Foliar canopies also differ in their densities. The growth habit of a plant affects the density of its foliar canopy. For example, plants with indeterminate vegetative growth (vines) may develop a denser foliar canopy than do plants with determinate (bush) growth. In determining planting rates and row spacings, a crop producer must consider the structure and density of the foliar canopy, among other factors, to minimize shading and maximize yield. If two or more species are mixed, as is frequently the case in hay fields and pastures, it is necessary to plant cultivars that have compatible foliar canopy structures and light saturation points. Compatible canopy structures minimize serious shading. For example, if two species of different heights are grown at the same time, the shorter species will be shaded by the taller species; to assure maximum yields, the shorter species must be tolerant of shading (have a lower light saturation point). Compatible species must also minimize other forms of competition.

If the amount of light received by a plant falls below the range of 100 to 200 FC, photosynthesis is markedly reduced or ceases and the plant falls below the compensation point. As photosynthesis is reduced because of inadequate light, a plant becomes etiolated: it develops white, spindly stems, elongated internodes, leaves that are not fully expanded, and a stunted root system. Etiolation commonly occurs in house plants grown in poorly lighted corners. Another example is the etiolation of maturing cereal crops caused by the rank growth of some weeds as they grow up through the crops. Excessive seeding rates in cereals, which cause shading among plants, may produce etiolation that results in lodging (Figure 8-2). The growth pattern produced by etiolation is governed by auxin.

Phototropism is another auxin-governed response to differences in the intensity of light, mainly of wavelengths in the red part of the spectrum, which result from the direction from which the light comes. This phenomenon is frequently observed in ornamental house plants that bend toward a window or bright light. Auxins on the lighted side of the stem are either destroyed or moved to the darker side of the stem where the relatively higher concentration of auxins causes cells to elongate more rapidly than those on the lighted side of the stem. The more rapidly growing cells force the stem to bend toward the light (Figure 8-3).

Although crop producers cannot alter natural light intensity to control how much light enters a field at any given time, they can control shading, the other main factor that affects the quantity of light a plant receives. They can also control the relationship between crop and light intensity by proper cultivar selection. Shading is a direct function of the planting rate (plants per acre), the distance between rows, and the number of plants per foot of row. It is unusual for a crop producer to have to foster shading, but, of the economically important crop plants, some types of tobacco require low light intensity; in some cases, entire fields are artificially shaded to enhance leaf quality, even at the expense of yield. It is more common to employ practices that minimize shading. Crops such as corn, soybeans, and cereals utilize higher light intensities. Cultivars of corn that tolerate little shading must be seeded at low planting rates (only 14,000 plants per acre); other cultivars, which tolerate much more shading, can be seeded at higher plant-

Figure 8-2
Lodged barley near Bozeman, Montana. Heavy seeding rates and high rates of nitrogen fertilizer produce tall plants with weak stems; if these plants are exposed to wind, driving rains, or hail, severe lodging results. It is more difficult to harvest lodged grain and field losses are greater than with normal, standing grain.

ing rates (up to 40,000 plants per acre). In mixed pastures, compatible cultivars can be planted to minimize the detrimental effects of shading on low-growing plants such as clovers by taller-growing grasses. In dense forests, new seedlings have difficulty becoming established, partly because of shading from mature trees. The forest manager must determine which trees to cut in order to reduce shading.

Effective weed control also helps to control shading. Part of the loss in yield (and thus in dollars) caused by weeds growing in crops may be attributed to competition for light. For example, early, tall-growing weeds inhibit the normal growth and development of crop plants by shading them, which results in less-than-maximum photosynthesis and possible etiolation. Aggressively growing weeds with an indeterminate type of growth are more serious pests in terms of competition for light than determinate, bush-type plants. Of course, competition by weeds is not limited to light; weeds also compete for moisture, minerals, and, in some cases, space.

Besides controlling shading, crop producers can regulate the crop-light-intensity relationship by selecting the right crop for a specific location. Light quantity and climate are related in that geographical areas differ in the number of foggy or cloudy days in a normal growing season. Cotton, which requires high light intensity, cannot be produced in

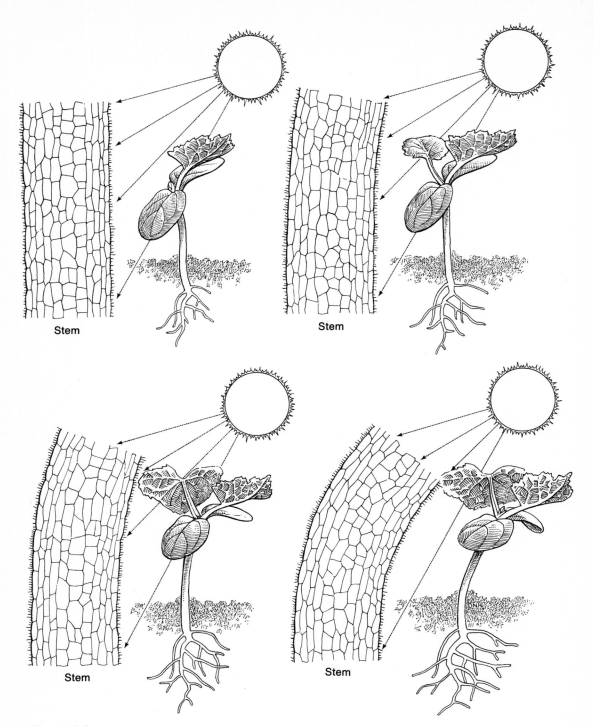

Figure 8-3
Light directed at only one side of a stem (or plant) causes excessive cell elongation on its shaded or dark side, which forces the stem to bend toward the light. The excessive elongation results from a relatively higher concentration of auxins on the shaded side of the stem.

cloudy regions, even if other conditions are suitable. In addition to the restriction of light due to clouds and fog, the filtering effect of smog limits crop production in some areas.

DURATION
The duration of light is measured by the hours of sunlight in 24 hours; light duration (commonly referred to as day length) varies with the location and the season. At the equator, the daily duration of sunlight is relatively constant at 12 hours. Twenty-five degrees north of the equator, at the tip of Florida, the duration of light varies from a minimum of 10.5 hours per day at the winter solstice to a maximum of 13.75 hours per day at the summer solstice. Further north, at the border between the United States and Canada (49° north latitude), there is an even greater seasonal variation in the number of hours of daylight: the shortest day has a little more than 8 hours of sunlight, whereas the longest day has slightly more than 16 hours. At the North Pole seasonal extremes produce 24 hours of darkness in winter and 24 hours of sunlight in summer. The same pattern of change in light duration occurs moving south from the equator, but the seasons are reversed.

It is easy to assume that the duration of light is important to plants: the longer the sun shines, the more photosynthesis occurs. Unfortunately, it is not that simple. Many crop plants require light of a specific duration to initiate flowering. The response of a plant to day length is its photoperiodic response. Plants can be classified three ways with respect to their photoperiodic responses: Long-day plants (barley, wheat, oats, rye, alfalfa, and red clover, for example) generally require a minimum of 13 hours of daylight before they will fully develop flower parts; in addition, the days must get progressively longer, not shorter. Short-day plants (rice, soybeans, and some types of field beans, for example) require a maximum of about 12 hours of daylight for the initiation and development of flowers, and the days must get progressively shorter. The longest day of the year in the Northern Hemisphere is June 21. The initiation of flowering in cereals is caused by the lengthening days of spring. For southern and tropical cultivars of corn, the shortening days of mid-to-late summer cause the initiation of flowering. Some plants are classified as day-neutral. They do not respond markedly to day length in terms of floral initiation. Although the southern and tropical cultivars of corn are short-day, the majority of the commercial, hybrid corn cultivars grown in the United States and Canada are more nearly day-neutral. Cotton and tobacco are two economically important day-neutral crops. Some classes and cultivars of field beans and lima beans are also day-neutral; others are short-day. A crop producer can do nothing to change the duration of light in the field at any given location, but he must nevertheless be aware of the photoperiodic response of plants. Because cultivars differ, even within a crop, in their precise photoperiodic requirements, he must select the appropriate cultivar for the location. For example, in California, because of the differences in day length between the northern and southern parts of the state, different cultivars of soybeans (generally a short-day plant) must be grown near the Oregon border than are grown in the Central Valley or near the Mexican border. Also, if rapidly growing (short-season), dwarf hybrid corn cultivars (which are suited to Utah, Colorado, and New Mexico) are moved north, they will grow unusually tall and will be delayed in flowering. The shortening day length in mid- or late summer in northern latitudes is too long for plants to initiate flowering, and by the time the days are short enough to cause floral induction, the crop may be killed by frost.

As a general rule, if other climatic factors such as temperature are disregarded, when short-day plants are moved northward from a natural southern location, they tend to stay in the vegetative stage longer and to produce excessive vegetative growth. In forage crops this may be desirable, but, in crops that are grown for seed or fruit, it is undesirable.

Temperature

MEASUREMENTS AND PLANT CLASSIFICATIONS

The importance of temperature in crop production is indicated by the fact that the major crops are classified as either "cool season" or "warm season" plants. Cool-season crops grow in temperatures as low as 40°F. Warm-season crops will not grow actively in temperatures below 50°F. In most popular writings in the United States, temperature is reported in degrees Fahrenheit (°F). Scientific reports use the Celsius scale (°C). It is important to be able to convert Celsius into Fahrenheit and vice versa. This conversion is quite simple: °C = (°F − 32) × 5/9, and °F = (°C × 9/5) + 32. The boiling point of water at sea level is 212°F; to convert this into °C:

$$\begin{aligned}°C &= (212 - 32) \times 5/9 \\ &= 180 \times 5/9 = 900/9 \\ &= 100°C\end{aligned}$$

The freezing point of water is 0°C; to convert this into °F:

$$\begin{aligned}°F &= (0° \times 9/5) + 32 \\ &= 0 + 32 \\ &= 32°F\end{aligned}$$

(Remember that 0 × anything = 0.)

In many physical and biological systems, the rate of various reactions depends on the temperature. As temperature increases, so does the rate of the reaction. The Q_{10} value is used to describe rate of change in reaction velocities in relation to changes in temperature. The rate of a chemical reaction usually doubles with each increase of 10°C, but the rate of a reaction cannot increase indefinitely. There are three key temperatures for each reaction—the cardinal temperatures: the minimum is the lowest temperature at which the reaction will occur, the maximum is the highest, and the optimum is that temperature at which the reaction will occur most readily or efficiently.

Cardinal temperatures vary markedly for different crops and for different stages of growth. However, there are a couple of generalizations that can be made by referring to a map of the United States and Canada that shows average summer temperatures (Figure 8-4). Although crops can grow actively in temperatures that range from near freezing (32°F or 0°C) to about 130°F (55°C), crops are not generally produced in areas that have an average summer temperature below 50°F (10°C). In fact, most of the areas in which major crops are grown have average temperatures that range from 50° to 68°F (10°–20°C) over a four-to-twelve-month period. Plants survive lower temperatures by ceasing active growth (going dormant).

Cool-season crops include wheat, rye, oats, barley, many forage grasses, and many vegetables. In general, the minimum temperature at which these crops grow actively is 40°F (4.5°C). Of all of the important cool-season crops, rye grows at the lowest temperatures; it is known to grow actively at temperatures slightly below freezing. Cool-season crops grow at maximum temperatures that range from 90° to 110°F (32°–44°C). If temperatures exceed this range, cool-season plants die. The optimum temperature for cool-season crops ranges from 60° to 90°F (16°–32°C).

Warm-season crops include corn, sorghum, sudangrass, rice, cotton, soybeans, and some

Figure 8-4
Mean summer temperatures (°F) in North America.

forages (switchgrass, gramagrass, and some clovers). Most of these crops will not grow actively at temperatures below 50°F (10°C). Nearly all will tolerate temperatures as high as 110°F (44°C), but their optimum is between 85° and 100°F (30°–38°C).

DAMAGE FROM TEMPERATURE EXTREMES

Both warm- and cool-season crops are damaged by excessively high temperatures. Heat damage may be attributed to desiccation, but frequently it is also associated with changes in enzyme structure and function (recall that enzymes are proteins). Heat may coagulate proteins (causing egg whites to solidify, for example); heat-induced changes cause enzymes to lose their ability to regulate or direct cellular activity.

During the flowering period of cool-season crops (in northern latitudes, approximately June and early July), high temperatures can cause pollen sterility: the pollen tube fails to grow or the pollen malfunctions in some other manner. This is due in part to the excessive drying of the pollen by the heat. High temperatures may also cause flower and fruit "blasting"—the premature dropping of the flower or fruit. Low temperatures at flowering of warm-season crops also result in pollen malfunctions and other floral abnormalities.

Low temperatures may cause various other forms of cold damage. If the temperature drops rapidly to below freezing, cytoplasm freezes and cellular organization is disrupted; enzymes may be precipitated, and cells die. If it takes eight hours or more for the temperature to fall below freezing, damage results from changes in water balance and relationships within the cell: water in the cell is warmed by metabolic activities (photosynthesis and respiration) and tends to move to the intercellular spaces. As a result of this water movement, the cytoplasm becomes dehydrated, and the cell dies. In some cases, the water in the intercellular spaces freezes, and the ice crystals that form actually rupture the cells, thereby killing them.

If a temperature decline is prolonged (lasting several weeks or more) and is coupled with decreasing day length, most perennial plants adapt by hardening. Hardening involves specific changes that lower the freezing point of the cytoplasm and increase the permeability of cytoplasmic membranes to water and certain charged molecules. The lower freezing point reduces the potential hazard of intracellular ice-crystal formation; the increased permeability allows water to leave the cell and reduces the potential of intracellular freezing. When cells harden, the structural viscosity of the cytoplasm is reduced; this is a third condition that limits freezing damage. Three cellular characteristics result from the cytoplasmic changes associated with hardening: cells have relatively low water content; cells have relatively high sugar content; and, cells are relatively high in osmotic pressure (i.e., the concentration of materials dissolved in the cytoplasm is increased). All plant tissues are not equally cold hardy. Tissues with smaller cells tend to harden more readily than tissues with larger cells. The greater hardening ability of smaller cells may be associated, in part, with their relatively smaller vacuoles. Changes that affect the physical and biochemical nature of the cytoplasm also affect the proteins (both structural and enzymatic) in the cell. With hardening, proteins are more protected from freezing as a result of both cytoplasmic changes and changes in their own molecular configurations and water relationships. Cold damage frequently produces symptoms similar to those that result from drought or moisture stress. It is worth noting that many of the cellular changes made as a plant hardens are also made as a plant adjusts to drought conditions.

In addition to hardening, many perennial plants become dormant with the onset of cold weather. Their life processes are reduced to a minimum, and they survive by respiring material stored in roots or crown tissue. Within various crops, the degree of hardening and the conditions required to induce dormancy vary widely. For example, there are alfalfa cultivars that are neither hardy nor dormant; these cultivars are not suitable for production in areas where winters are severe (such as in Montana). Other cultivars harden well and become dormant ('Ladak 65' is outstanding in its cold hardiness; 'Lahontan' is less cold hardy and only semidormant).

Frost can cause serious damage to plants. Frost occurs if the temperature falls to 32°F (0°C) or below after dew has formed. Dew formation depends on the relative humidity, which is the percentage of moisture in the air at any given temperature compared with the maximum moisture the air could hold at that temperature. Remember that, as air temperature drops, the absolute amount of moisture the air can hold decreases. When relative humidity reaches 100 percent, dew is formed. If this dew freezes, white frost forms. Sometimes the air is so dry that dew is not formed at 32°F (0°C). However, plants still suffer frost damage (usually burning of leaves or flower damage); this type of dry "frost damage" is referred to as "black frost."

Winter burn is another form of cold damage. It occurs if the aboveground parts of plants start active growth (break dormancy) while the ground is still frozen. The new growth requires water, but, because the plants cannot remove water from frozen soil, a droughtlike condition is established that causes leaf burn, bud death, and death of the entire plant if the conditions persist for a long enough period. Even a short warm period during the normally cold months may result in winter burn by causing the aboveground parts of plants to break dormancy. Winter burn can be a serious problem in ornamentals. In 1968, peculiar climatic conditions and a narrow temperature inversion layer in December caused severe burn of the evergreens along a narrow band in the Bridger Mountains north and east of Bozeman, Montana. Under normal conditions, as air temperature drops, a crop becomes dormant as the soil slowly cools to below freezing. Soils cool and warm more slowly than the air. Their rate of warming and cooling depends to some extent on their texture. The temperature of a sandy soil changes more rapidly than that of a heavier clay soil.

TEMPERATURE AND CROP MANAGEMENT

Temperature affects plants throughout their entire life cycles. Although the crop producer can do little to alter temperature range between germination and flowering, the temperature at germination can be controlled by the selection of a suitable planting date. For example, of the warm-season crops, corn as a rule should not be planted until the soil temperature approaches 55°F (13°C); cotton requires a minimum of 62°F (17°C), and tobacco even warmer temperatures. At the other end of the scale, a cool-season crop such as spring wheat should be seeded when the soil temperature rises to 37°F (3°C); oats require a slightly warmer temperature (about 43°F, or 6°C); and winter wheat should not be planted until the soil temperature drops to between 50° and 55°F (10°–13°C). At this temperature, germination and early fall seedling growth can occur, but a fungus that kills young seedlings is relatively inactive.

Although a crop producer cannot significantly alter temperatures during or after flowering, he can circumvent temperature-related problems by carefully selecting cultivars of sensitive crops as well as by controlling planting dates. If there is a danger of early hot weather, selection of a rapidly maturing cultivar of a cool-season crop is advantageous. If there is no real danger of early hot weather, but the possibility exists of cool, late-spring weather, then a later-maturing cultivar (one that stays in the vegetative growth stage for a longer period) is desirable. Similarly, to avoid cold damage to warm-season crops such as corn, rapidly maturing cultivars should be planted in areas that have a short growing season.

There are three practices a crop producer can rely on to control temperatures within a limited range in order to reduce damage from frost. Sometimes the operation of a sprinkler irrigation system will warm the air near plants enough to prevent frosting. This warming, which is due to the physical heat-exchange capacity of the sprinkled water and the air, will generally raise temperatures only 3° to 4°F (1.6°–2.2°C), from about 29°F (−1° to −2°C) to just above freezing. Of course, the use of this technique depends on the availability of sprinklers; it is most frequently used in lower growing row crops and in vineyards and orchards. A second technique that warms the air slightly near crops is the use of large fans. These fans are installed permanently on poles above orchards and vineyards. By stirring the air, they force higher, warmer air down into the crop's atmosphere. The higher air can be as much as 5°F (2.8°C) warmer than the air surrounding the crop. As the fans force this air to move down, the temperature of the air in contact with the surface of the plants increases from 2° to 5°F (1°–3°C). The use of fans is most effective where a thermal inversion traps warm air immediately above the trees or vines so that fans can move it down into the atmosphere immediately surrounding the plants. In areas of rolling foothills, local conditions often favor the use of fans.

Smudging is a third technique that can be used, especially to warm the air in orchards. Smudging requires some type of petroleum burner to serve as a heater; the equipment is not extremely expensive. Care must be exercised in placing the heaters so as to obtain efficient heating. The heaters are lighted as the temperature approaches freezing (32°F or 0°C); this means that someone must be up in the middle of the night to check temperatures and light the heaters if necessary. The dense canopy of branches and leaves in citrus orchards (where smudging has been used extensively) traps the burner-generated heat. Smudging can raise the temperature in an orchard more than 7°F (4°C). Of course, wind can alter this desired effect drastically. In some areas, the burning of oil-soaked rags and old tires has been used in place of commercial heaters. However, this practice has been declared illegal because of the air pollution it creates.

In addition to temperature range, the number of days from the last killing frost in the spring to the first killing frost in the fall—the frost-free period—also affects crop production. Frost-free periods differ markedly from area to area (Figure 8-5). In general, crops are not produced in regions where the frost-free period is less than 125 days. The production of winter and spring wheat is an exception. Most winter wheat cultivars are hardy enough to tolerate frost in the seedling stage. In general, if winter wheat is frost damaged, it is because of spring frosts, but, during the winter, it becomes relatively dormant and is protected from extreme cold by an insulating layer of snow. The depth of the insulating layer depends on the amount of snow and on field practices: both planting in furrows and allowing some stubble to stand in the field increase the depth of the insulating layer. In addition to protecting a crop during

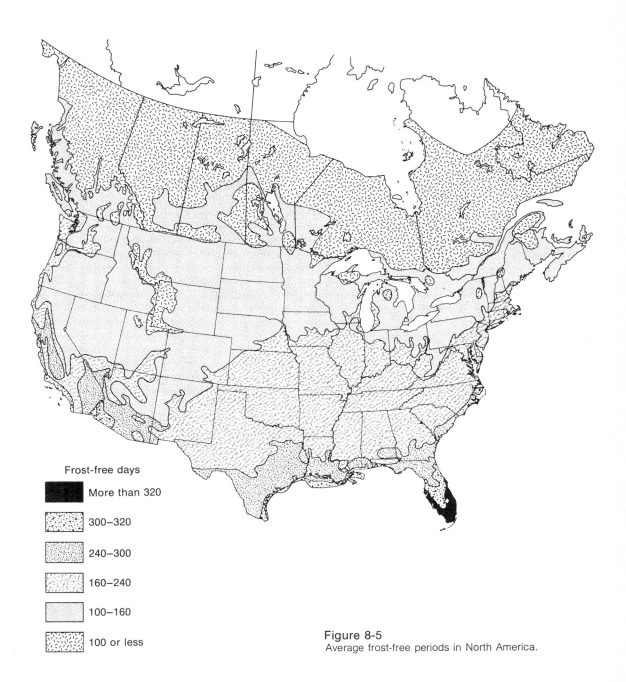

Figure 8-5
Average frost-free periods in North America.

the winter, the accumulated snow also provides moisture in the spring (see Chapter 7). With the coming of spring, winter wheat plants resume active growth and complete their life cycle. Spring wheat cultivars are less cold hardy than winter wheat. Because spring wheats mature later in the summer or early fall, they are subject to early-fall cold damage. Winter wheat is produced in regions with a frost-free period of as little as 100 to 110 days and spring wheat can grow in regions that have as little as 90 frost-free days. Many of the perennial forage species, including grasses and some legumes, are also exceptions to the rule that crops need a 125-day frost-free period. However, forage yields are reduced in regions that have fewer than 125 frost-free days, compared with yields in regions that have a longer growing season.

In contrast to winter wheat, crops such as cotton require a frost-free period of 200 days or more, plus warm temperatures. Such crops are grown in fairly well defined geographic areas (Figure 8-6).

In addition to a favorable temperature range and a suitable frost-free period, crops must receive a specific amount of energy to mature. This energy may be measured in degree days. The starting point in the calculation of degree days is the minimum temperature below which a crop will not grow significantly. The number of degrees by which the temperature exceeds this minimum in a 24-hour period represents the amount of energy accumulated by the crop in that span of time, or the number of degree days. For example, corn, sorghum, and sudangrass do not show significant growth below temperatures of 50°F; therefore, a 24-hour period with an average temperature of 55°F would be 5 degree days: $55° - 50° = 5 \times 1$ day $= 5$ degree days. If the average temperature is 65°F in a 24-hour period, there would be 15 degree days: $65° - 50° = 15 \times 1$ day $= 15$ degree days on the Fahrenheit temperature scale. The concept of degree days is useful in determining the potential growing season of a crop—the number of days from planting to harvest. Crops that mature in a short span of time generally tolerate warmer temperatures and require fewer accumulated degree days than those which require a longer growing season. Although each species has a minimum temperature at which growth occurs, cultivars within some crops (peas, hybrid corn, and sorghum) differ markedly in the number of degree days they require to mature. Crops that are grown to be canned or frozen (peas, some beans, and sweet corn) must be timed to provide processing plants with a constant supply of produce for a prolonged period. Weather data permit a grower to predict the number of degree days in any specific period so that he can stagger planting dates in relation to the number of degree days within a given period. He can thus fulfill the requirement of a constant supply of produce to the processor. For example, the pea cultivar 'Alaska' requires about 1,250 degree days to mature. If a first crop is planted in mid-May, it will ripen in mid-July; if a second crop is planted in early June, it will accumulate the required degree days in fewer calendar days and be ready for harvest soon after the first crop.

In biennial root crops such as sugar beets, cold temperatures may cause bolting. If there is a cool period (not necessarily freezing) during the first year of growth, the plant may bolt; it develops flower parts, but not the desired root. Normally sugar-beet growth in the first year is vegetative, and a major storage organ, the root, is formed. If temperatures are low, primordial floral or reproductive buds are induced to develop and grow in the second year; the plant uses the energy stored in the root to support this growth. In some instances, cold temperatures induce the development

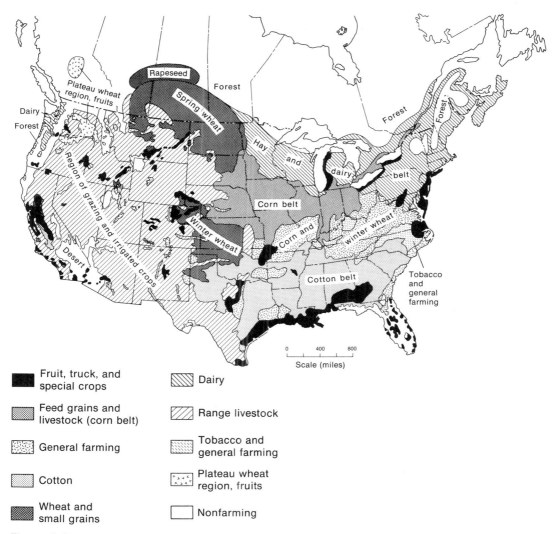

Figure 8-6
Major crop production regions in the United States. Although the specific crops and exact borders may vary for each area, well-known regions such as the cotton and corn belts are delimited to a large extent by the length of the frost-free period. [USDA data.]

of reproductive structures in the first year; that is, they cause bolting. Cultivars that tend to bolt with moderately cool temperatures (i.e., without a hard freeze) should not be planted in the intermountain west. They are more suitable for areas in California and Arizona.

Some crops require cold temperatures after planting. After winter wheat germinates in the early fall, several months of low temperature (near 28°F or −2°C) are required to vernalize the plants. Vernalization is a process that forces winter wheat plants to develop flowering parts (spikes) when exposed to appropriate, long photoperiods. If winter wheat is not vernalized, it will stay in the vegetative stage of growth and yield no seed because reproductive parts are not formed. Low temperatures after germination also stimulate tillering in fall-planted cereals. Many fruit trees require cold temperatures during the dormant period. If trees, such as cherry trees, are not subjected to several months of subfreezing temperatures during their dormant period, flower buds will not develop.

Climate and Crop Production

The distribution of both cultivated plants and native vegetation depends to a large extent on climatic conditions. The major factors of climate—light, temperature, and moisture—combine to form two broad classifications of climate: oceanic and continental. An oceanic climate is that typical of areas adjacent to large bodies of water. It is characterized by relatively mild temperatures, little change between summer and winter seasons, and small differences between day- and night-time temperatures. The body of water stabilizes temperatures seasonally by absorbing heat during the summer months and releasing it during the winter months. It also stabilizes temperatures daily by absorbing heat during the day and releasing it during the night. Areas that have a continental climate have cold winters and hot summers; there are also marked extremes in temperature from day to night. Even relatively small bodies of water may effect significant thermal buffering and account for a local "mini-oceanic" climate. For example, the Flathead Valley in northwestern Montana is subject to severe winters and the local climate would generally be classified as continental. However, adjacent to Flathead Lake, the climate is more oceanic. The lake keeps its surrounding shore areas colder in the early spring, which in turn keeps some plants dormant longer until the danger of late spring frost has passed; in the fall, as heat is radiated from the lake, adjacent areas are protected from early fall frosts. This thermal buffering allows sweet cherries to be grown in a relatively narrow band around the perimeter of the lake.

Climate is affected by a number of geographic factors. Among the more important ones are latitude (distance from the equator toward a pole), altitude (distance above sea level), distance from large bodies of water, oceanic currents, and the direction and intensity of prevailing winds. For agricultural purposes, three geographic factors have been correlated in Hopkins Bioclimatic Law, which states that each event in a growing season (planting, flowering, or harvesting) is delayed four days for every increase of 1 degree in north latitude in temperate North America (a distance of about 65 miles or 108 kilometers), or for every increase of 400 feet (120 m) in elevation, or for every change of 5 degrees longitude moving from west to east.

Efficient crop management requires a basic understanding of climatic factors, which combine in a variety of ways to engender local conditions. In evaluating light, temperature, moisture, and soil, a qualified crop manager recognizes that the slope and exposure of a field, the presence of trees and shrubs along its edges, the direction of prevailing winds, and a host of other seemingly minor factors can significantly affect the growth of crop plants.

Selected References

Meyer, B. S., and D. B. Anderson. 1959. *Plant physiology*. 2d ed. Van Nostrand.

Mitchell, Roger L. 1970. *Crop growth and culture*. The Iowa State University Press.

Salisbury, Frank B., and Clion Ross. 1969. *Plant physiology*. Wadsworth.

Zelitch, Israel, 1971. *Photosynthesis, photorespiration, and plant productivity*. Academic Press.

Study Questions

1. What aspects of light affect plant growth and development? Using examples, describe how they affect growth and development and how they can be managed under field conditions.
2. Discuss (with examples) the meaning and importance of the following terms in relation to the light requirements and field management of crops: foliar canopy, net photosynthesis, light saturation, compensation point.
3. What aspects of temperature affect plant growth and development? Using examples, describe how they affect growth and development and how they can be managed under field conditions.

Nine   Insects, Diseases, and Weeds

Crop producers must wage a continuous, expensive war against three major foes: insects, plant diseases, and weeds. These pests have always attacked crops, but, as a growing world population makes it necessary for agronomists to produce more food per acre of arable land, winning the fight against pests becomes increasingly important. The effort requires cooperative research in a variety of disciplines: entomology (the study of insects) and plant pathology (the study of plant diseases) are two fields that bear directly on the problem. Weed control requires research in several special areas: taxonomy for the accurate identification of weeds; both plant physiology and biochemistry for the actions of herbicides; climatology for the effect of weather on herbicide action; entomology and microbiology for biological weed control; and engineering for the sophisticated equipment currently in use.

Insects, plant diseases, and weeds differ in their effect on crops, but all three result in reduced yields and impaired quality. The magnitude of the yield reduction varies—with the crop, the pest, and other factors—from very slight to total crop failure. Crop quality is impaired through shriveled grain, smaller or discolored fruits and vegetables, changes in taste and smell, or the presence of contaminants from the pest: weed seeds, dead insects, and mold are but a few familiar examples. Regardless of the type of damage, the presence of pests is very costly to the producer and, indirectly, to the consumer.

Insects

Insects are found throughout the world. At any one time there may be as many as a billion individual insects alive. Entomologists esti-

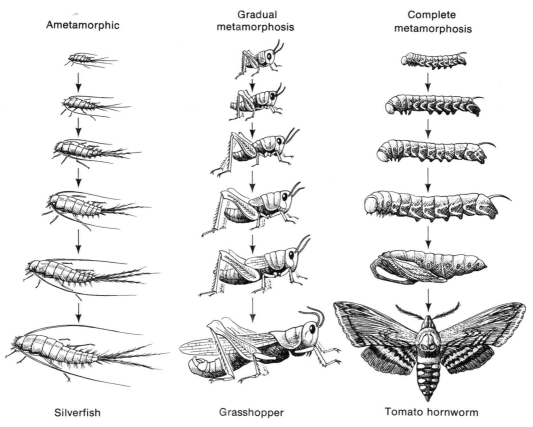

Figure 9-1
The three types of insect life cycles: the simplest is ametamorphic, in which the young closely resemble the adults; the more complex is gradual metamorphic in which the insects mature through successive molts, which transform the nymph (larva) into the adult; the most complex is complete metamorphosis in which there are four distinct stages: the egg, the larval, the pupal, and the adult.

mate that there are about 1.5 million species of insects. More than 80,000 species are found in Mexico and North America; of these, some 10,000 are classified as public enemies because they harm man either directly or indirectly through damage to livestock and crops. The annual cost of controlling the insects that attack food, feed, and fiber crops exceeds $700 million. If the value of the crops and livestock lost because of pests is added to the cost of control, the figure more than doubles!

Insects are classified in the class Insecta, phylum Arthropoda of the animal kingdom. Like most other animals, they reproduce sexually; however, some of the developmental stages of insects do not occur in "higher" animals. The life cycle of an insect follows one of three distinct patterns (Figure 9-1). The more primitive insects follow an ametamorphic pattern of development: the egg hatches into a tiny facsimile of the adult, which grows in size to adulthood but does not pass through different developmental stages. No major agricultural pest follows this pattern, but springtails and silverfish, common household nuisances, are ametamorphic. Many serious agricultural insect pests go through a gradual, incomplete metamorphosis from egg to nymph to adult. The nymphs look much like the adults, but they are not yet sexually mature. The adults cause the most damage to crops. Grasshoppers, leaf hoppers, and true bugs have a gradual metamorphic pattern of development. The most complex pattern of development is the complete metamorphosis that moths, butterflies, beetles, flies, and wasps undergo. These insects pass through four distinct stages as they develop from egg to larva to pupa to adult. Both the larvae and the adults cause damage to crops, although in some instances only the larvae are damaging. For example, both the larvae and the adults of the alfalfa weevil (a beetle) damage alfalfa, but only the larvae of the tomato hornworm and the European corn borer damage crops. The adult moth is harmless except for the fact that it produces and fertilizes eggs.

It is easy to assume that all insects are undesirable, but this is a false assumption. Some insects are in fact essential and highly desirable: they serve as agents of pollination or act as predators of undesirable insects or weeds. An outstanding example is the honeybee, which is essential for the pollination of alfalfa and certain fruit crops. (Without the appropriate insect to carry pollen, pollination does not occur. Without pollination, seed and fruit development cease.) Insects that are desirable predators feed on undesirable insects but do not damage economically important plants. A good example is the ladybug (actually the ladybird beetle), which eats aphids. Aphids are carriers of disease; they also damage crops by feeding on them. The Klamath beetle is a third example of a desirable insect. This bettle feeds exclusively on a serious pest of West Coast rangelands, the Klamath weed. Introducing the beetle to areas infested with the weed has essentially eliminated the weed without harming other, economically important plants.

DAMAGE CAUSED BY INSECTS

The adults of several insect species damage crops directly by chewing, sucking, or boring within or on plants. Chewing damage is most apparent along the edges of leaves or on stems that have been chewed at or below the soil surface. Boring damage may take the form of holes in the leaves, or in the husks of ears of corn, and of grooves or channels cut into leaves, stems, fruit, and the ears of corn. Physical damage from sucking insects is difficult to detect. It is commonly found on the surface of leaves in the form of wounds as tiny as pinpricks, but in some cases tissue around the wounds is damaged, leaving small, necrotic specks.

One shocking example of "chewing" is the damage adult grasshoppers wreak annually on rangelands. Precise losses are virtually impossible to calculate, but the damage to rangelands costs crop producers as much as $10 million per year. Grazing by grasshoppers, in addition to reducing forage yields, causes serious conservation problems. Ranges denuded by grasshoppers may suffer severe erosion from wind and running water. Grasshoppers also cause damage to cultivated crops, mainly cereals, potatoes, and alfalfa; this damage may exceed $18 million annually. The striking financial loss due to grasshopper damage of cultivated crops can be explained in part by the high cost of the extensive pest control that is practiced on these crops. Although more control leads to less damage, the reduced damage is offset to a great extent by the high cost of control. Another example of insect damage to cultivated crops is that caused by the feeding of adult Lygus bugs on alfalfa plants. The bugs may reduce seed yields markedly. In some years they cause more than $10 million worth of damage.

The larvae of many insects cause severe damage to crop plants. Although both the adult alfalfa weevil and the larva feed on plants, the larvae cause much more damage. Both the larvae and the adults of the infamous boll weevil damage cotton crops. The magnitude of the damage varies from year to year, but an annual loss of more than $100 million is not uncommon in the cotton belt states. Losses exceeding $400 million dollars have been reported in epidemic years. The larvae of the European corn borer are also serious pests. The corn borer can cause annual losses to corn crops of more than $100 million, and it attacks other crops as well. The larvae (caterpillars) of many moths, including the coddling moth, the forest gypsy moth, and the armyworm, are serious pests. The larvae of the click beetle, commonly known as wireworms, damage corn, cereals, root crops, vegetables, and flowers. These larvae eat seed before it germinates and also feed on seedlings. The larvae of the Hessian fly (Figure 9-2) are pests in small-seeded cereals, except oats. They feed on young plants and weaken the straw, which results in serious lodging. The larvae of the wheat stem sawfly feed inside the stems of developing plants of wheat and other grasses. The feeding weakens the stems and may cause severe lodging. A 50 percent reduction in wheat yield is not uncommon as a result of invasion by the wheat stem sawfly.

Most of the direct damage done to crop plants by insects is a result of their feeding on or in the plants. But insects also cause damage to crops by laying eggs, transmitting disease, and leaving residue. Though insects do not "care" for their young after birth as "higher" animals do, the female adults of each species lay their eggs in a site that will protect the

Figure 9-2
Adult Hessian fly on a wheat stem. The larvae of this insect cause great crop damage; female adults lay eggs on the upper surface of leaves, along the midrib. [USDA photograph.]

young after birth. In some instances, the eggs themselves or the manner in which they are deposited cause damage to crops. For example, the periodical cicada lays her eggs in the year-old wood of fruit and forest trees. The egg laying itself (not the feeding of the young) causes the young wood to split and die. In a similar way, treehoppers and tree crickets cause damage to the twigs of apple trees, raspberry brambles, and currant shrubs. Other insects deposit their eggs in the flowers or young fruit of a plant and thereby either stop the development of the flowers or lower the quality of the fruit. For example, the strawberry weevil lays her eggs in the unopened buds and then cuts partway through the stem so that the buds will not develop; the plum curculio punctures plums, apples, and peaches, and deposits her eggs in the fruit. Insects act as viral disease carriers or vectors by transferring viruses from diseased to healthy plants during feeding. Leaf hoppers and aphids are prime offenders in this respect. The presence of insect residue known as frass damages crops by lowering the quality of the crop products. Frass often takes the form of "skins" that are shed as an insect develops from egg to larva to pupa to adult. The presence or evidence of living insects in stored crops also lowers the value of the crop product. Problems with frass, living insects, or obvious insect damage to the crop product are frequently encountered in stored grain, wheat, rice, barley, and oats.

CONTROL OF INSECTS

Insects have plagued man since the dawn of history. As man started cultivating plants and storing food, the magnitude of the insect problem increased and the need for insect control became apparent. In the earliest days of agriculture, when sacrifice and the worship of diverse gods proved ineffective in controlling insect pests, great numbers of enslaved workers were sometimes forced to capture and kill insects in a field in order to save a crop. Obviously the effectiveness of this approach was severely limited to the control of large insects in a very small area! However, even as recently as the late nineteenth and early twentieth centuries, large, horse-drawn beaters were employed to stir up insects, which were then captured in huge "scoop nets." Various methods of stirring up heavy dusts to suffocate insect pests were also tried. Neither of these general techniques met with great success.

Regardless of the crop or the insect pest, a crop producer or pest management expert must consider several factors before he can initiate control. First and foremost, he must identify the potential pest. Next, he must determine if there are enough insects to warrant some type of control. If he decides that the insects are damaging the crop or that they represent a significant potential threat, he must correlate the life cycle of the insect pest with the life cycle of the crop to determine at what stages of its development the insect can harm the crop, at what stages it is vulnerable to various types of management or control, and at what stages of development the crop is most vulnerable to damage by the pest. He must also ascertain what damage, if any, various control measures might cause to the crop so that he can predict whether control will result in more serious losses than those caused by the insect itself.

Prevention is less costly and less potentially damaging than cure in pest management. Therefore, control before damage occurs is the goal of forward-looking pest-management programs. Such preventive control may include the elimination of overwintering and egg-laying sites (such as crop residue and weeds) and the timing of seeding and harvesting to minimize overlap of the damaging stages in the life cycle of an insect with the susceptible stages in the life cycle of a crop. It may also require changes in crop rotation patterns as insect populations increase. Any form of control must prevent the insect pest from entering the stage at which it causes damage. Insects in the egg or pupal stage do not harm crops and are not highly susceptible to most insecticides. If an insect causes major damage as an adult, it is desirable to control the "nonadult" forms (nymphs or larvae). If damage is caused by larvae (or nymphs), it is necessary either to kill adults prior to egg laying or to practice control while the insect is damaging the crop. Unfortunately, the latter choice is frequently the only choice available to a crop producer. In the western United States, it has been possible to limit damage from alfalfa weevil larvae by spraying the alfalfa during the winter, thereby killing the adult weevils before they reproduce.

A crop producer who is considering insect control must remember that seeing insects in the field does *not* mean they are damaging a crop, and, vice versa, not seeing them does not guarantee their absence. A grower must be able to identify insects in all stages of their life cycles as well as by the type of damage they do to plants. If damage is done by chewing (as with grasshoppers), control of a sucking insect (like an aphid) is unwarranted.

In recent years, the concept of pest management has been replacing that of complete control. The total eradication of a pest is rarely required to protect a crop from significant losses. Management offers more options to crop producers than does total control, and some of the options it offers are more acceptable from an ecological point of view.

Entomologists recognize four broad categories of pest management and control: genetic, cultural, biological, and chemical. Effective insect management frequently employs methods from two or more of these categories.

Genetic control The development of cultivars of crops that are resistant to major insect pests is one important goal of plant breeders. Success in breeding such cultivars has been quite remarkable with some crops. For example, genes that cause plants to have solid rather than hollow stems have been incorporated into superior cultivars of wheat. The wheat stem sawfly cannot damage solid-stem wheat plants; thus, such plants are resistant to the insect. Similarly, plant breeders have developed cultivars for alfalfa that are resistant to spotted alfalfa aphids. Recently, the genes responsible for resistance to the alfalfa weevil were isolated, and breeding programs are now under way to incorporate this resistance into suitable cultivars of alfalfa. In addition to developing insect-resistant cultivars, plant breeders are seeking to develop alternative crops that are not seriously damaged by insect pests. For example, sainfoin, which has many of the desirable characteristics of alfalfa for hay and pasture, is not damaged by the alfalfa weevil nearly as much as alfalfa is. Although the breeding of insect-resistant cultivars and the development of alternative species may be the best solutions for the prevention of insect pests, they are laborious, time-consuming tasks and cannot provide an immediate cure for an urgent problem.

Cultural control Some insect problems can be alleviated through normal cultural practices, or at least through slight modifications of field operations. Crop rotation can abolish sites in which insects overwinter, nest, and lay eggs. Range renovation practices that destroy annual weeds have effectively eliminated beet leaf hoppers by destroying overwintering host plants. Crop rotation has also been successful in controlling northern corn rootworms and the sod worms that attack corn and soybean crops. Destroying crop residue and pruning have proved to be effective in controlling the corn earworm, the

European corn borer, and the coddling moth in isolated orchards. Deep plowing or burning corn stubble used to be required by law to control the European corn borer in the corn belt.

Adjusting planting and harvest dates can help reduce insect damage. Damage caused by the Hessian fly in winter wheat is reduced by delaying planting until after the "fly-free date." Fly-free dates have been established for most winter-wheat areas based on long-term climatic data. Winter wheat planted after the fly-free date does not emerge until the fall flight of the pest is past. Similarly in New York, the planting of several types of beans is delayed until after the "maggot-free date," which affords protection against the seed-corn maggot. At the opposite extreme, early plantings to avoid insect pests are also of practical importance. In the south, chinch bugs cause less damage to grain sorghums that are planted in midspring (April) than to those that are planted in late spring or early summer. Crops planted in April mature before the bugs have a chance to damage plants seriously. In Texas sorghums are planted in February or March so that plants mature before adult sorghum midges can damage them.

Damage to alfalfa from the alfalfa weevil is reduced by a slightly early initial harvest followed by prompt removal of hay from the field. Although some adult weevils may be removed with the hay, the weevil larvae will be killed by exposure to sunlight or starved due to a scarcity of feed immediately after harvesting.

Biological control The control of harmful organisms through interference with their ecological adjustment is called biological control. In association with genetic and cultural pest management procedures, biological control not only affords adequate protection to crops from insect pests, but also constitutes an ecologically safe method of pest management. To date, researchers have only scratched the surface of the complexities of biological control. Insects, like most organisms, must contend with predators. In some cases, predators can be used to control a specific insect pest. Care must be exercised in releasing predators to prevent the predator from becoming a pest in itself. The common ladybug is a beetle that preys on aphids; in some areas these beetles are released to combat infestations of aphids. Another method of biological control is the introduction of artificially created, genetically sterile mutants into insect populations. The genetic defect (mutation) is spread throughout the population and the reproductive potential of the insect population is greatly reduced. Organisms that cause fatal diseases in specific insects have been released in insect populations and have controlled these populations effectively. The disease must be specific to a particular insect pest. A third form of biological control makes use of the fact that insects are attracted by a number of factors including light, high-frequency sound, and artificial sex attractants. Each of these has been used to attract insects to traps where they are destroyed.

Chemical control In spite of efforts to develop successful, alternative methods of insect management, the most widely employed method of insect control in the field is chemical control through insecticides. Chemical insecticides are applied in the form of sprays, dusts, and gases (which are frequently used as fumigants in stored crops). The agricultural chemical industry has provided crop producers with a wide array of chemicals that have proven extremely effective in controlling most major insect pests; agricultural engineers have developed effi-

Figure 9-3
Special equipment used for dusting tobacco for worm control. [Photograph by J. C. Allen and Son.]

cient methods of applying these chemicals (Figures 9-3 and 9-4). Today, the use of many of these chemical insecticides is being challenged from a safety point of view. Residues of various insecticides remain on produce (fruits and vegetables) or are transmitted from feed through animals to foods such as milk or meat; residues also remain in the soil or are washed into the rivers and lakes that supply water to some segments of the population. Therefore, a rigorous review of the nature, use, and importance of insecticides is mandatory.

However, because serious losses occur in all crops as a result of insects, it has been estimated that, without alternative methods of control, eliminating the use of insecticides would decrease livestock production 25 percent and crop production about 30 percent. These reductions in agricultural production would cause an increase in food prices of from 50 to 75 percent, and the quality of most agricultural produce would be markedly lower. Because the world food supply is already inadequate, such losses cannot be tolerated. Therefore, despite certain detrimental effects, the use of insecticides is necessary in crop production. However, in light of recent findings, more emphasis must be placed on achieving the most efficient use of the safest possible chemicals and on developing alternative measures of control.

There are literally hundreds of thousands of chemical compounds that have insect-

Figure 9-4
Aircraft spraying grain sorghum for insect control. Such aircraft are specially designed for safe, effective, and efficient applications of many types of chemicals, including fertilizers.

killing properties, but only three groups of compounds are used in most insecticides today. The first group consists of chlorinated hydrocarbons, which were used in the first mass-produced, modern, synthetic insecticides. DDT is one of the best-known insecticides in this group. Other common chlorinated hydrocarbon insecticides include chlordane, dieldrin, and lindane. Chlorinated hydrocarbons are extremely effective insecticides: they afford good control of many pests, are reasonably inexpensive, and are quite safe to apply. Unfortunately, chlorinated hydrocarbons have several serious drawbacks. They do not decompose readily into nontoxic substances (are not readily degradable); because they have a long residual life both on plants and in the soil, they can be transmitted from feed through animals to foods such as milk or meat. Moreover, the concentration of chlorinated hydrocarbons builds up in animal tissue. Also, insect populations develop resistance to various chlorinated hydrocarbons so that crop producers must use either increased dosages or new formulations for continued control. For these reasons, the use of many chlorinated hydrocarbons has been restricted or declared illegal.

A second, modern group of insecticides is the organic phosphates. This group includes parathion, malathion, and methyl parathion. Organic phosphate insecticides are quite effective in controlling a wide range of insects; they create less of a residue problem because they decompose more rapidly than chlorinated hydrocarbon insecticides. Unfortunately, some of the organic phosphates are highly toxic to man, and extreme caution must be used in applying them and while working with or around them.

Carbamate compounds constitute a newer group of insecticides that incorporate the desirable traits of both the chlorinated hydrocarbons and the organic phosphates and avoid many of the drawbacks of these other two groups. The carbamate insecticides are effective in the control of many insects, yet they break down to nontoxic forms more rapidly than chlorinated hydrocarbons and are less toxic to man than the organic phosphates. Included in the carbamate group of insecticides are sevin, metacil, and dimetilan.

Chlorinated hydrocarbons, organic phosphates, and carbamates, which are often referred to as second-generation pesticides, have serious drawbacks in addition to those already mentioned. They are broad-spectrum pesticides and as such kill beneficial as well as harmful insects. Perhaps a more serious problem is that insect pests have adapted to prolonged chemical control by developing resistance to even the most toxic, second generation insecticides. At times, some of these resistant pests seem to virtually thrive on various chemicals. The most common example of this phenomenon is the housefly, which has developed resistance to DDT. Superraces of insects might evolve in response to continued chemical control, and such insecticide-resistant pests would cast a frightening shadow over crop production.

Some much older, so-called first-generation insecticides are still in use, including various forms of arsenic (lead arsenate) and fumigants such as methyl bromide; they are effective and are considered relatively safe.

The mode of action of an insecticide describes how the active ingredient affects pests. Many insecticides (including the chlorinated hydrocarbons, the organic phosphates, and the carbamates) act as poisons to the nervous system. Others disrupt metabolic processes. For example, cyanide compounds reduce oxidative respiration by damaging the mitochondria, and fumigants have an adverse effect on respiration in general. Insecticides are either ingested (eaten), absorbed through contact, or "inhaled" by insects. Insecticides that are effective only when eaten are classified as stomach poisons. Contact poisons enter through the "skin" of an insect; most fumigants are inhaled. Knowing how an insecticide is taken in is critical in understanding how and when to apply it. If a stomach poison is to be used (the arsenate compounds are stomach poisons), the plant material on which the pest is feeding must be covered with the insecticide. Contact poisons either must be applied directly to the pest or must leave a residue on plants with which insects will come in contact. Fumigants are useful only in confined areas for the control of insect pests in stored crop products.

The formulation of an insecticide indicates the form in which it is applied. The concentration of the active material may vary from one formulation to another. Common formulations include dusts, wettable powders (material manufactured as a powder that can be mixed with water, but will not dissolve), suspensions (material in a fine powdery form, which is suspended in a liquid for spraying), emulsions (material that blends like oil and water for spraying), and conventional sprays (concentrated material that is diluted with relatively little water). The time required for the breakdown of the active ingredient to a nontoxic substance depends on many factors: the type of material, the concentration, the formulation, and climatic conditions during and after application of the insecticide. Federal, state, and in some cases local rules establish how much insecticide residue is allowable and how soon after application crops can be harvested or fed. These regulations vary from week to week for different crops. Because there is no simple rule, crop producers must check current regulations and follow the manufacturer's instructions for the use of any pesticide. The safe handling, storage, and use of all types of pesticides is of major concern to the consumer as well as to the crop producer. Pesticides can harm man through both direct and indirect contact. Direct contact with a pesticide may result from broken containers, spilling materials, and wind drift or mists during application. These dangers can be avoided by carrying and storing chemicals in appropriate, carefully labeled containers, by wearing protective clothing (gloves, hat, respirator, or even a full rubber suit) when applying chemicals, and by spraying or dusting only in calm weather. Indirect contact results from residues on crops, soil, clothing, and equipment. Indirect contact can be minimized by using pesticides that degrade rapidly to nontoxic materials and by knowing how long it takes various pesticides to degrade under different climatic conditions so that contact with treated materials can be avoided until the pesticides have degraded.

In addition to the obvious health hazard of direct and indirect contacts with pesticides, the effects of pesticides on the environment must be considered. Pollution and other forms of ecological destruction must be minimized. The development for widespread use of nonchemical methods of pest control is essential. If chemical methods are required, then crop producers must use chemicals that degrade rapidly and are specific for the pest in question; that is, they must avoid broad-spectrum pesticides. They must also use insecticides according to directions to avoid contaminating soil and water sources. They must further protect water sources by preventing drift into streams and ponds and by not discarding containers or washing equipment in or near water sources.

The ideal pesticide would have little or no residue, would be virtually nontoxic to man, domestic animals, and game, and would affect only a specified, identified pest. This type of insecticide, a so-called third-generation insecticide, is more than a dream. Insect physiologists have recently found natural hormones that are essential for normal insect development in the early stages of the life cycle. If these same hormones are present during later stages, they disrupt normal maturation and cause the development of sterile adults. The hormones are highly specific for each insect, nontoxic to man, and harmless in the environment. They have already been produced artificially and applied experimentally for insect control. Although more research is required, these hormonal insecticides could provide effective, safe control of several very serious agricultural pests such as the boll weevil, the corn earworm, and the chinch bug.

Insects cost money to both crop producers and consumers by reducing yields, lowering crop quality, spreading diseases (to man, domestic animals, and plants), and necessitating controls. A knowledge of potential insect problems and of their best possible solutions is essential in achieving maximum crop yields of optimum quality.

Diseases

Plants are subject to many diseases that can reduce crop yield and quality. The study of plant diseases is plant pathology. A diseased plant is defined as a plant whose anatomy, morphology, or metabolic processes have been disrupted as a result of the presence or action of a pathogen. Plant pathogens are microorganisms that live on plant hosts. The parasitic relationship is beneficial to the parasite and detrimental to the host. The changes caused in the host by this relationship are the symptoms of disease, and the parasite is termed the causal agent, pathogen, or incitant of disease. Viruses, bacteria and fungi are three large groups of plant pathogens. Nematodes, or eelworms, are sometimes considered to be pathogens too. In addition to diseases caused by microorganisms plant pathologists classify some nutritional deficiency conditions as well as certain physiological abnormalities as diseases.

LOSSES DUE TO CROP DISEASES

Crop losses due to plant diseases in the United States may exceed $3 billion annually, and

Figure 9-5
Foliar symptoms of southern corn leaf blight. From left to right: nearly normal, healthy leaf; moderately infected leaf; and severely infected leaf. [USDA photograph.]

control measures (including those used to destroy nematodes) frequently cost more than $100 million per year. Fungi are the pathogens that cause the most damage to the greatest number of crops. Through sexual reproduction and the genetic recombination associated with meiosis, they can develop many different strains of a single pathogen. Effective control for one strain is often ineffective for another. Wheat stem rust is a leading fungal disease, which can cause losses of $250 million annually in hard red spring and durum wheat alone. This fungus can also cause severe losses in winter wheat. Southern corn leaf blight is another serious fungal disease (Figures 9-5 and 9-6). In 1970, an outbreak of southern corn leaf blight caused an estimated 15-percent reduction in corn yield, which resulted in losses of more than a billion dollars. Other fungal diseases that cause some crop losses each year are ergot (which attacks small grain and forage grasses), root rots, and damping-off diseases. The latter two diseases are caused by varieties of fungi that kill seedlings.

Some of the major diseases caused by bacteria are fire blight of apples and pears, angular leaf spot on cotton, bacterial wilt of alfalfa and other crops, and potato scab. Losses due to these diseases vary from year to year and from one region to the next. Precise figures of economic losses due to bacterial diseases are rare. Losses due to fire blight may take two forms: reduction of yield in any given year (which may be as much as 30 percent) and damage to fruit trees, (which will reduce yields for several years). Similarly, losses due to bacterial wilt take two forms: reduction of yields from infected plants in any given year and ultimate loss of stand over

Figure 9-6
Symptoms of southern corn leaf blight on corn ears: (left and center) ears from severely infected plants; (right) ear from a normal, healthy plant. [USDA photograph.]

a period of several years, which necessitates replanting of the field.

Virologists have studied the viral diseases of plants for less than twenty years so that they are still discovering new diseases as well as reclassifying older disorders. For example, fifteen years ago, what is now known to be the viral, barley stripe mosaic disease was believed to be a physiological disorder that had nothing to do with the presence or effect of a pathogen. Similarly, viruses that cause disease in potatoes are just being identified. Crop producers now realize that potato viruses cause the loss of millions of dollars worth of produce every year.

It is difficult to estimate crop losses due to disease and data are surely not to be taken as absolute values. Losses vary from year to year and from farm to farm. Moreover, the fact that the average reduction in corn yield caused by southern leaf blight in 1970 was in the range of 15 to 20 percent is of little comfort to the corn farmer who lost his entire crop to the disease. However, in spite of the inaccuracies inherent in estimates of average yearly losses due to plant disease, the magnitude of various diseases is well reflected in such values. In the United States and Canada losses in major food and feed crops range from 14 to 15 percent (in barley and corn) to nearly 30 percent (in wheat); losses in cotton hover near 14 percent, and in potatoes near 23 percent. These figures represent not only economic losses to crop producers, but also serious reductions in the potential world supply of food.

TYPES OF DAMAGE

Plant pathogens are parasites. They steal water, minerals, and products of photosynthesis from their plant hosts for their own metabolic use. Frequently, pathogens cause diseased plants to have symptoms that are quite similar to those of plants that are suffering from moisture stress or drought. As a result of the presence and activities of the pathogens infecting crop plants, yields are reduced and quality may be lowered.

As a result of their growth and development, pathogens also damage crop plants directly in other ways. For example, the fungi that cause the rust diseases of cereals destroy cells (including photosynthetic tissue) as they grow throughout a plant. This reduces the photosynthetic capacity of the plant and, thus, reduces the yield. Moreover, in certain of their reproductive stages, the fungi rupture the leaf and stem epidermal tissue, which increases the amount of water lost by the plant. This lowers yields, increases the water requirement, and decreases water-use efficiency. Smuts and ergot reduce grain yield and lower crop quality by replacing the kernels (fruits) of cereals and other grasses with pathogen tissue, often comprising masses of reproductive spores (Figures 9-7 and 9-8). The pathogenic bacteria that cause wilt disease in alfalfa thrive in xylem tissue to the extent that they form a plug, which prevents water from moving upward through the xylem. Thus, in addition to using materials needed by the host plant, these bacteria cause moisture stress in the aboveground parts of the host.

Viruses generally cause physiological damage rather than damage of an anatomical or morphological nature; however, they do cause physical disruption at the cellular or subcellular level by destroying or modifying chloroplasts and mitochondria. The physical disruption results in reduced photosynthetic and respiratory activity. Viral diseases can result in an overall reduction of plant growth (like that caused by the yellow dwarf virus that attacks barley) or a reduction in the size of leaves coupled with a distortion in their shape (such as curly top in sugar beets). Some

Figure 9-7
Spikes of barley showing different levels of damage from Nigra smut: (A) healthy spike; (B) newly emerged, infected spike; (C, D, E, and F) more mature spikes showing replacement of floral parts by fungus growth. [USDA photograph.]

Figure 9-8
The effects of ergot development on barley: (A) healthy spikes and (B) spikes infected with sclerotia (overwintering bodies), which provide the source of primary inoculum. If the sclerotia are harvested with the crop, they contaminate the grain and lower the crop quality. [USDA photograph.]

viruses cause symptoms that are nearly identical with the symptoms that plants exhibit in response to deficiencies of specific minerals or other forms of environmental stress. For example, a deficiency of phosphorus, cool temperatures, and certain viruses produce the same effect in some grasses—a reddening or purpling of leaf tissue.

Identifying the cause of the symptoms requires expertise. If a crop producer cannot determine the cause of a plant abnormality, he must seek advice from either his county agent, a fieldman from a reputable agrichemical firm, or some other authority.

DISEASE CYCLE

All parasitic diseases, regardless of the pathogenic agent (virus, bacterium, or fungus), follow the same basic disease cycle, which interacts with the life cycles of both the pathogen and the host but is most closely associated with the life cycle of the pathogen (Figure 9-9). The disease cycle consists of four phases. First, the pathogen penetrates the host. Examples of penetration are bacteria entering a wound, fungi actually growing into the stomata of a leaf, or viruses being transmitted to a plant by an insect vector. Infection, the second phase, follows penetration and establishes the causal organism in the host. Unless infection occurs, a plant will not become diseased following penetration. Infection does not produce symptoms of disease immediately. The period from infection to the production of symptoms is the incubation period, the third phase. After the incubation

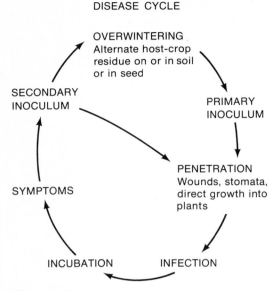

Figure 9-9
General steps in the disease cycle. Disease control consists of breaking this cycle at some stage; if inoculum is not present, plants cannot become infected. Disease-resistant plants prevent either penetration or the actual process of infection.

period, the disease enters the fourth phase of the cycle. This is the disease-development stage in which the pathogen damages the plant so severely that the effects of the pathogen on the host become apparent. A crop producer identifies a plant disease by the symptoms that the pathogen causes, as well as by the presence of the pathogen itself. For example, a grower recognizes bacterial wilt if plants wilt when soil moisture is available and other conditions do not favor wilting. The disease is confirmed by identifying bacteria in the xylem tissue. Ergot can be identified by the presence of sclerotia (ergot bodies) in place of caryopses. The actual damage done to a plant by a pathogen and the symptoms by which the disease is recognized are not necessarily the same. For example, rust diseases are recognized by the presence of pustules of reproductive spores, which spread the fungus but do not harm the plants; rather, the plants are harmed by the parasitic activity of the fungus.

The three groups of pathogenic organisms differ in their methods of penetration. Many fungi can grow directly into host cells; wounds or natural openings are not needed for penetration to occur. This is true of peach leaf curl and powdery mildew. Rusts penetrate the host by growth through the stomata. Most smuts and ergot gain entry through floral parts, frequently the pistil. The bacteria that cause soft rots of fruit, wilt in alfalfa, and galls of fruit trees enter through wounds in the host. The bacteria that cause fire blight of pears and apples enter through nectaries in the blossoms. The bacterial pathogen causing potato scab can penetrate host plants by growing through stomata, entering wounds, or passing directly through the cuticles of young tubers. Viruses enter through wounds. If they are transmitted by means of plant-to-plant contact, slight abrasions (wounds) on the surface of the leaves of both the infected and the healthy plants are the sites of penetration. If the virus is transmitted by an insect vector, penetration is accomplished by means of the wound caused by the feeding insect.

Environmental conditions that favor the germination of overwintering spores, the production of primary inoculum,* the penetration of the host, and the development of a disease vary for different pathogens. The fungus that causes the "damping off" of seedlings is favored by excessive moisture in conjunction with other conditions that slow

---

*The primary inoculum generally comes from the overwintering stage and is capable of infecting a crop plant. Spores produced by the pathogen after it has infected a plant are secondary inoculum and they also can infect crop plants; secondary inoculum is also referred to as repeating inoculum in that it can spread the disease throughout a field.

9 / INSECTS, DISEASES, AND WEEDS

germination and seedling growth of the host. The fungus that causes downy mildew is favored by temperatures ranging from 18° to 25°C (66°–77°F), high relative humidity, fog, and frequent rains. Conditions that favor the development of powdery mildew on cereals and other grasses include cool temperatures (10°C, or 50°F) and moist conditions. Other powdery mildews are favored by dry conditions. The overwintering spores of the fungus that causes ergot (Figure 9-8) must be subjected to temperatures that are below freezing before they will germinate and produce primary inoculum. Then, for infection to occur, cool, cloudy weather is needed during the flowering period of the host plants. The fungi that cause smuts in corn and loose smut in wheat and barley are favored by higher temperatures (26°–32°C, or 79°–88°F). Wheat and barley smut (Figure 9-7) develop only if there is relatively high humidity during flowering. Although urediospores, the secondary inoculum, of stem rust can germinate at temperatures ranging from 5° to 25°C (37°–77°F), the most favorable temperature range for stem rust infection is from 18° to 20°C (68°–72°F). For penetration to occur, a film of water must be present on the host's surface.

In general, bacterial diseases are favored by adequate-to-excessive moisture coupled with conditions that promote rapid growth of the host. For example, the bacteria that cause potato scab thrive when the temperature is between 20° and 22°C (68°–72°F), the soil is well aerated and has a pH ranging from 5 to 8, and moisture conditions favor rapid growth of the host.

Viral diseases depend less on specific climatic conditions for their development than do bacterial and fungal diseases because viruses do not directly require moisture. For example, sugar beet yellows diseases, which are caused by viruses, develop under cool, cloudy conditions; on the other hand, barley stripe mosaic develops best under conditions of bright sunlight and warmth.

LIFE CYCLES OF PATHOGENS

Fungi, bacteria, and viruses have characteristic life cycles that determine the course of the disease cycle to a great extent. Fungi have the most complex life cycles; the life cycles of bacteria and viruses can be considered "simplifications" of the fungal life cycle. It is convenient to describe fungal life cycles by starting with the overwintering stage of the pathogen. This is the stage in which the organism survives when it is not actively growing on a crop plant. Many fungi overwinter as special spores, which remain in the soil, in crop residue, or on seeds. Some fungi overwinter as mycelia or hyphae, the typical fungal "threads." Pathogens differ with respect to the specific climatic conditions they require for their overwintering spores to start growing, but all require temperatures in a fairly narrow range and adequate moisture to initiate such growth.

The full life cycle of the fungus that causes stem rust in wheat (Figure 9-10) illustrates the stages of the life cycles of all fungi. The species name of this pathogenic fungus is *Puccinia graminis*; it is in the class Basidiomycetes. Although fungi that belong to other classes may have different types of spores or fewer stages in their life cycles, they still follow the same pattern of division and reproduction as *P. graminis*.

The fungus that causes stem rust overwinters as a diploid spore called the teliospore. When the spore germinates, meiosis occurs and haploid basidiospores, known as sporidia, are formed. These sporidia do not infect wheat. They infect the common barberry, which serves as an alternate host for the fungus; barberry is essential for the fungus to complete its sexual life cycle. As the fungus grows on the barberry, mating between plus

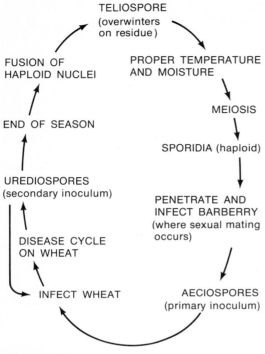

Figure 9-10
Complex life cycle of the fungus *Puccinea graminis*, which causes stem rust on wheat. Note that the fungus must go to the common barberry plant for the sexual stage (mating).

and minus strains occurs. One mating type (either plus or minus) forms pycniospores, which mate with receptive hyphae (fungal threads) of the opposite mating type. This results ultimately in the formation of aeciospores, each of which has two haploid nuclei. These aeciospores are the primary inoculum. They are blown from barberry to wheat plants, and, under the proper conditions, they germinate and penetrate the leaves and stems of the plants. As the fungus grows in the wheat plant, urediospores are produced in pustules on the leaves and stems. One pustule can produce as many as 350,000 urediospores. These spores give the disease its name of rust because they appear as brick red masses.

They are the secondary inoculum. As urediospores are blown from infected wheat plants to healthy plants, they infect the healthy plants, which in turn produce more urediospores to spread the disease. As the grain ripens, the pustules begin to produce another type of spore—the two-celled teliospore. In each cell of the teliospore, two haploid nuclei fuse to form the only diploid nucleus of the life cycle. At the end of the growing season, pustules stop producing urediospores and produce only teliospores. The teliospores overwinter on straw or stubble and germinate in the spring, and the cycle is repeated.

In some areas, climatic conditions allow overwintering of the fungus in the urediospore stage. If the teliospore stage is bypassed, the overwintered urediospores can become the source of primary inoculum. In the absence of barberry, when aeciospores are not formed and the sexual phase of the cycle is eliminated, the urediospores must become the source of primary inoculum if the fungus is to perpetuate itself. Urediospores can be carried great distances by wind and still remain viable.

The fungi (*Tilletia caries* and *T. foetida*) that cause bunt, or stinking smut, of wheat have life cycles that are essentially identical (Figure 9-11); they are less complex than that of *Puccinia graminis*. Neither *T. caries* nor *T. foetida* needs an alternate host to complete its life cycle. The organisms overwinter as teliospores on the surface of grain or in soil. When the seeds (grain) germinate, the teliospores grow and produce primary and secondary sporidia, which fuse to form the primary inoculum—hyphae that can penetrate the seedling directly. The fungus grows in the seedling and invades the developing floral parts where the teliospores are produced.

Bacterial life cycles have fewer stages than fungal life cycles. Descriptions of special spores, alternate hosts, and sexual cycles are not considered essential in describing bacterial life cycles. Instead, emphasis is placed on accounting for overwintering habits, the sources of primary and secondary inoculum, and how the bacteria spread from plant to plant. The life cycle of *Streptomyces scabies*, the bacterium that causes the disease known as potato scab, is typical of the life cycles of many bacteria (Figure 9-12). *S. scabies* overwinter either in the soil or in infected seed potatoes where primary inoculum is produced. If the primary inoculum is in the soil, penetration is through wounds in the seed potato. As an infected potato plant grows, the bacterial pathogen grows and multiplies in its host. Ultimately, a lesion due to bacterial growth (a scab) is formed on the "skin" of the potato. The bacteria then produce spores that are the source of secondary inoculum. Similar life cycles are seen in the bacteria that cause gall diseases, soft rots of fruit, bacterial wilts, and leaf spot diseases such as angular leaf spot of cotton.

Are viruses truly living organisms? Because this question remains unresolved, some biologists prefer to consider the "developmental cycle" rather than the "life cycle" of viruses. Researchers in molecular biology have uncovered many minute details of the developmental cycles of viruses. From an agricultural point of view, the critical questions are how the virus survives and how it is transmitted. Viruses that cause plant diseases survive in living plants. They have no overwintering spores that survive in the soil or on crop residue. They are transferred from diseased plants to healthy plants either through direct contact with infected plants or by transmission from insects which have fed on a diseased plant. In either case, the virus enters the healthy plant through a natural opening or some type of wound, including the wound created by the insect feeding on the healthy host plant. Some viruses are translocated

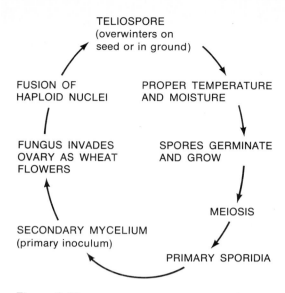

Figure 9-11
Life cycle of the fungus *Tilletia caries*, which causes stinking smut, or bunt, of wheat. This fungus does not require an alternate host; its source of primary inoculum (overwintering spores) may be carried on seeds.

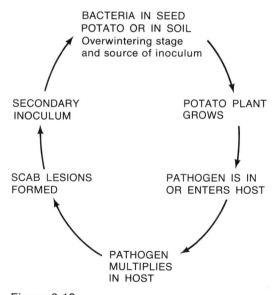

Figure 9-12
Life cycle of the bacterium *Streptomyces scabies*, which causes the potato scab disease. Compared with fungal life cycles (Figures 9-10 and 9-11), this life cycle is quite simple.

from vegetative tissue to seeds and are then carried in seed; plants subsequently grown from such seed are infected with the virus.

Koch's postulates A diseased plant is recognized by a set of symptoms that set it apart from the typical, healthy plant. The symptoms are attributed to the presence or action of some organism, yet this organism is too small to be seen without special equipment. When it can be seen, its direct effect on the host is not always detectable. Given these facts, how does a plant pathologist verify that a specific organism causes a particular disease? In 1881, a German microbiologist, Robert Koch, proposed a procedure to verify whether an organism causes a specific disease. The steps of this procedure are known as Koch's postulates:

1. The organism in question (the pathogen) must always be associated with a specific set of symptoms.
2. The organism in question must be isolated from a diseased plant and grown in pure culture.
3. A healthy plant must be inoculated with the organism grown in pure culture and the healthy plant must develop the characteristic symptoms.
4. The organism in question must then be reisolated from the intentionally inoculated plant and compared with organisms grown in pure culture, and the symptoms of the initial diseased plant and of the intentionally inoculated plant must match before the organism can be claimed to cause the disease.

Koch's postulates are useful in all areas of pathology. They serve as a rigorous guide that helps to prevent a pathologist from jumping to false conclusions about pathogenic cause-and-effect relationships. Although the postulates are simple to state, the precise execution of each one requires a highly skilled scientist.

CONTROL OF PLANT PATHOGENS

The management and control of organisms that cause plant diseases can be achieved through genetic, cultural, or chemical methods, or a combination of the three. Regardless of the method of control, crop producers must understand the disease cycle and the relationship between pathogen and crop life cycles to implement effective control.

The prevention of disease is generally preferable to trying to cure it. Prevention may consist of disrupting the life cycle of the pathogen so that sources of inoculum are not present when crop plants are susceptible. Another form of prevention is to employ cultural practices that preclude plants from being in a vulnerable stage of growth when inoculum is present. Yet another is to use disease-resistant cultivars; plant breeders have developed disease-resistant cultivars of many crop species.

Genetic control The development of disease-resistant crop cultivars seems to be the best solution to plant disease problems. Plant breeders have developed cereal cultivars that are resistant to rusts, smuts, and mildews. However, new strains of many pathogens are generated continually because of genetic recombination in sexual reproduction; so there is a constant need to develop new crop cultivars that are resistant to these new strains.

In recent years, hybrid-corn breeders have been faced with the challenge of developing cultivars that resist a serious, fungus-caused foliar disease—southern corn leaf blight. In 1972, the disease caused major losses in the U.S. corn crop. Although genetic resistance has been isolated and incorporated into high-yielding cultivars, the alleles for resistance have not yet been transferred into male-sterile lines of corn; until corn breeders effect such a transfer, hand detasseling will be required in the production of seed of these new cultivars.

Cultural control In the absence of disease-resistant cultivars, cultural practices that eliminate sources of primary inoculum are effective in the control of many diseases. Because many bacteria overwinter in crop residue, destroying the residue reduces the source of inoculum. The development of ergot in pastures can be controlled by repeated mowing or grazing to prevent grasses from flowering, which prevents sclerotia from developing. Losses due to root rot in winter wheat can be reduced by delaying plantings until the soil temperature falls below 56°F (14°C) at which point the pathogen causing root rot is relatively inactive. Sometimes crop rotation is effective in reducing disease. For example, damage to sugar beets from nematodes can be reduced by growing beets only once every three to four years on a given field. Similarly, the bacteria that cause bean blight can be eliminated from the soil by growing beans only every third year. Unfortunately, considerations other than disease control frequently dictate what crop rotation must be followed.

Using disease-free (certified) seed helps to eliminate diseases that are transmitted in or on the seed. Crop producers can control fungal (e.g., loose smut) and viral (e.g., barley stripe mosaic) diseases in this way. Growers can control many other viral diseases by using virus-free seed, seed potatoes, and grafting stock.

In orchard crops, care in pruning and attention to damaged trees reduce invasions of bacteria. In crops that are cultivated frequently (row crops such as sugar beets), care in cultivation to avoid damaging plants reduces disease potential. Careful management of fertilization and irrigation are also essential in controlling disease: excessive application of either nutrients or water can foster disease almost as readily as a deficiency of either. Management practices that foster vigorous plant growth help minimize losses from diseases.

Chemical control Applying various pesticides is critical in the control of many diseases. Because of the rapid changes in the laws that regulate the use of chemical pesticides, a producer should review the most current regulations before using any pesticide.

Treating seed with mercury-based fungicides is effective in the control of the damping-off diseases of most grasses including cereal crops. It is also effective in the control of covered smut, which is carried on the seed surface. In recent years mercury fungicides have been banned; however, new, less persistent chemicals that are quite effective have been developed.

For economic reasons, chemically controlling field-crop diseases in the field is not practical; however, there are exceptions. Sulfur dust can be used to control mildew. One of the earliest fungicides, Bordeaux mixture (copper sulfate), saved the wine industry in France in the 1880s through the control of downy mildew on grapes. Bacterial leaf spot in peaches can be controlled by applying copper sulfate in the fall to reduce the overwintering bacteria on twigs. A zinc-lime mixture will protect foliage during the active growing season. Some antibiotic sprays have been effective in controlling fire blight of pears.

Fumigants can be used to control some pathogens in the field as well as those that attack stored crops, but the cost of using fumigants is frequently prohibitive. Many diseases of stored crops, such as the soft rots of fruits and vegetables, can be controlled by handling the produce carefully (to avoid bruising) and by regulating the temperature and humidity of the site where the crop is stored.

In summary, disease control is complex; it includes using disease-resistant cultivars and disease-free seed, eliminating sources of inoculum (diseased prunings in orchards and volunteer plants, and alternate hosts for many pathogens), destroying infected crop residue, employing fungicides or other chemicals when and where practical, and exercising care in the handling and storage of crops.

NEMATODES
Three types of specialists study nematodes: the nematologist, who is concerned only with nematodes, the plant pathologist, who holds the opinion that the pest causes disease, and the entomologist, who feels that nematodes are closely related to insects. Nematodes (also called eelworms) are not typical plant pathogens, nor are they insects. They are round worms that are microscopic in size (from about 0.4 to 1.7 mm long) and are classified in the phylum Nematoda; they are not earthworms, which have segmented bodies and belong to the phylum Annelida. Because of their size, nematodes frequently exist unnoticed in the soil. Many are harmless, feeding on dead plant tissue. However, several types of nematodes are serious crop pests. For example, root-knot nematodes burrow into the roots of a number of important crops (sugar beets, cotton, and many legumes) and cause the formation of large galls (or knots); the galls disrupt the development of normal tissue, frequently in vascular areas; this weakens the infested plants and causes them to appear wilted. The presence of the galls is proof of the presence of root-knot nematodes. Plants that have been weakened by root-knot nematodes are more susceptible to infection by pathogens. Other nematodes include the meadow nematode, the sugar-beet nematode, and the nematode that replaces the developing kernels of cereals with a mass of minute pests.

Growers can control nematodes by rotating crops (they must plant crops that are not susceptible to nematodes approximately four out of every five years); by following strict sanitary precautions (to avoid transporting pests on equipment and soil from infested areas to clean fields); and by fumigating which is expensive, but effective). Fumigants include methyl bromide, ethylene dibromide, and similar compounds. If crop rotation is used as a control measure, fields must be cleaned of weeds that serve as hosts for the pest.

Weeds

Weeds are the final group of crop pests. Everyone "knows" what a weed is, yet it is nearly impossible to agree upon a precise definition. A workable definition seems to be: "a weed is a plant growing where it is not wanted." As a rule, weed plants are "tough" and vigorous. They grow rapidly, have efficient root and shoot systems, and are prolific in seed production. Some plants are nearly always considered to be weeds. From an agricultural point of view, pigweed is never wanted; therefore it is always a weed. Other plants are considered to be weeds only because of the time and place in which they are growing. For example, because rye is a valuable cereal crop, rye plants are not weeds to the rye producer. However, rye growing among wheat lowers the market quality of the wheat; so the wheat producer considers rye in a wheat field to be a weed.

DAMAGE CAUSED BY WEEDS
There are four types of damage generally attributed to weeds. First, and most important, weed plants compete with crop plants for water, minerals, and light, thereby causing yields to be reduced. Many common weeds grow rapidly in the early spring before crop plants start active growth; the weeds consume

moisture and minerals so that less of these essential materials is available to the crop plants. Many weeds grow taller than the crop plants and shade them to such an extent that photosynthesis by the crop plants is drastically reduced.

Second, weeds and their seeds contaminate crops and their products. The presence of weeds in hay may have a negative effect on the palatability, flavor, and general quality of the hay. Moreover, if the weeds have a high moisture content when the hay crop is cut, they can cause heating, which could result in a hay fire. Weed seed in cereals lowers the grade of the cereal and, if not screened out before milling, may impart an undesirable flavor or color to the flour.

Third, weeds serve as hosts or alternate hosts for plant pathogens and as homes for insect pests. Common barberry is the alternate host for the fungus that causes stem rust of wheat (as mentioned earlier in this chapter), and the virus that causes the yellow dwarf disease of barley overwinters in wild oats and other weedy grasses. Adult grasshoppers live and lay their eggs in weedy areas.

Fourth, some weeds are poisonous to livestock. Among the offenders are "locoweed," poison hemlock, bracken fern, death camas, larkspur, and lupine. The nature and severity of the poisoning depends on the plant, the time of year (developmental stage of the plant), and the size and species of the animal. As a rule, livestock tend to avoid poisonous plants if adequate, high-quality forage is available; however, when range or pasture becomes overgrazed, the proportion of poisonous plants increases and stock turn to them for feed because high-quality forage is not available.

Besides directly and indirectly causing damage to plants and livestock, weeds are a nuisance. They choke waterways, in addition to consuming water, and generally disrupt efficient farm operations.

MAJOR WEEDS AND LOSSES DUE TO WEEDS

The leading agricultural weed pests in the United States belong to at least eleven botanical families and include both annuals and perennials, and monocots and dicots. The ten most common weeds are (in order of descending importance): pigweeds, crabgrasses, quackgrass, foxtails, thistles, ragweeds, lambsquarters, nutsedges, johnsongrass, and chickweed. These ten species (or, in some cases, groups of species) are serious pests in at least half the United States; in each state in which a particular species is a serious pest, it infests at least 50 percent of the croplands of that state. The relative rankings of the top ten, as well as of the next fifteen most common weeds, vary from region to region (Table 9-1). For example, foxtails, which are the most extensive weeds in the North Central states, are not among the top ten weeds in the South and rank seventh and eighth in the Northeast and the West, respectively. Within the group of weeds that rank fifteenth to twenty-fifth overall, only barnyardgrass and bindweeds are listed among the top ten pests in more than one region. Many of the weeds listed in Table 9-1 include more than a single species (e.g., thistles); this points to the problem of using common names in plant identification (see Chapter 5).

It is impossible to make good estimates of the economic losses sustained by crop producers because of weeds. The extent of losses depends on the type of crop, the type of weed, and environmental factors such as available moisture and nutrients. As a general rule, crop yields are reduced in proportion to the density of weeds in the field (Figure 9-13). The losses, which result from both reduced yields and the lower quality of contaminated crops, vary from year to year and from location to location; however, the summary presented in Table 9-2 gives an idea of the magnitude of the annual economic loss due to weeds.

Table 9-1
Major weeds in the United States (ranked according to the number of acres infested), and the ten most common weeds (by acres infested) in four regions, with the number of states reporting each weed or group of weeds, the percentage of infested acres in which infestation has increased since 1962, and the growth habit of the weed—annual (A) or perennial (P). AP indicates that some species within the group are annual and others are perennial.

Weeds	Nationwide rank by acres infested	No. of infested states	Percentage of infested acres in which infestation has increased	Growth habit	Rank by region (by no. of acres infested)			
					NE	NC	S	W
Pigweeds	1	46	7	A	3	2	2	1
Crabgrasses	2	43	28	A	5	5	1	—
Quackgrass	3	25	8	P	1	3	—	5
Foxtails	4	33	25	AP	7	1	—	8
Thistles	5	37	39	AP	—	4	—	4
Ragweeds	6	34	24	AP	6	7	5	—
Lambsquarters	7	41	<1	A	2	6	—	3
Nutsedges	8	31	89	P	4	—	4	—
Johnsongrass	9	26	52	P	—	—	3	—
Chickweed	10	40	22	A	8	10	9	—
Barnyardgrass	11	34	43	A	—	—	—	2
Bermudagrass	12	18	68	P	—	—	6	—
Dandelion	13	34	74	P	9	9	—	—
Bindweeds	14	23	23	P	—	—	—	6
Cocklebur	15	26	38	A	—	—	7	—
Mustards	16	35	13	A	—	—	—	—
Bromes	17	25	41	AP	—	—	—	7
Bluegrasses	18	28	1	A	—	—	—	10
Purslane	19	21	93	A	—	—	—	—
Morningglories	20	24	41	P	—	—	8	—
Panicums	21	26	56	P	10	—	—	—
Smartweeds	22	25	9	P	—	8	—	—
Kochia	23	12	60	A	—	—	—	9
Henbit	24	18	54	A	—	—	10	—
Docks	25	22	33	AP	—	—	—	—

Source: Data from *U.S. Dep. Agr. ARS* H-1, November, 1972.

Table 9-2
Annual cost of losses due to weeds and of weed control in the United States.

Type of farming enterprise	Losses in yield and crop quality	Cost of control	Total
Field crops	$1,500,000,000	$1,900,000,000	$3,400,000,000
Horticultural crops	255,000,000	300,000,000	555,000,000
Grazing lands	630,000,000	360,000,000	990,000,000
Unfarmed areas	50,000,000	55,500,000	105,500,000
Total	$2,435,000,000	$2,615,500,000	$5,050,500,000

Source: L. O. Baker, Montana State University, 1972.

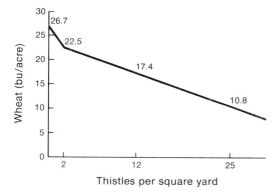

Figure 9-13
Reduction in wheat yield as a result of infestations of Canada thistle. Note that even two plants per square yard reduces yield markedly.

WEED CONTROL

Considering the extent of economic loss due to weeds, it is not surprising that substantial emphasis has been placed on effective weed control. Weed-control practices can be divided into four general categories: (1) hand and mechanical methods; (2) flaming methods; (3) chemical methods; and (4) biological methods.

Mechanical methods In small areas such as gardens, nurseries, or extremely troublesome places in fields, hoeing or pulling weeds by hand is the most effective method of control. Different species of weeds germinate and grow at different times during the growing season of a crop, and, because weed plants are normally prolific, removal is imperative. Annual weeds must be destroyed before they flower; although it is possible to accomplish this with one "treatment," annual weeds are rarely all killed with a single hoeing. Repeated hoeing is necessary for perennial weeds unless the entire plant can be removed or destroyed. The object is to cut off as much top material as possible so that food stored in the roots and crowns is used in developing new top growth. Before this new top growth can replenish the root storage through photosynthesis, it must be cut again. More than 90 percent of Canada thistle can be killed in a single season if the plants are cut off slightly below the soil surface every three weeks during the growing season.

Mechanical methods of weed control simply replace a man and a hoe with a tractor and some form of cultivator (such as a duck foot or a rod weeder). Regardless of how the weeding is done, care must be exercised to avoid "transplanting" weeds by carrying vegetatively reproducing parts, such as rhizomes, from field to field.

Flaming methods The seedlings of crops such as corn, cotton, and, in some cases, sugar beets are not damaged by a "flash" of flame. Therefore, various types of tractor-mounted flame throwers—which use low-pressure gas, kerosene, or other petroleum products—can be used to burn weeds without damaging the young crops. In addition, although the results of this type of burning are unsightly, flaming is an effective method for ridding ditches, roadsides, and fence lines of weeds.

Chemical methods Since 1950, chemists have developed a wide array of selective herbicides that, if properly used, will kill specific weeds or groups of weeds without causing serious harm to crop plants. The selectivity of these chemical herbicides is determined by the fact that different plants absorb and metabolize chemicals differently. The first selective herbicide to be widely used was 2,4-D (2,4-dichlorophenoxyacetic acid). This herbicide is still widely used to kill broad-leaved weeds (like mustard) growing among grasses (like cereal). The selectivity of herbicides like 2,4-D is due in part to the fact that broad-leaved weed plants have more surface area than grasses; therefore, they tend to absorb and translocate more of the chemicals than do grasses. The metabolic responses of broad-leaved weeds to the herbicide are different from those of grasses. The exact nature of these metabolic differences and the reasons for them are not known. The effectiveness of any herbicide depends on the nature and concentration of the active ingredient, the stage of growth of both the weed and the crop, and the temperature and relative humidity at the time of application.

Commercial formulations of herbicides include various types of fillers, wetters, and spreaders in addition to the active ingredient. Most weed killers are distributed in formulations that contain an active ingredient that is dissolved, emulsified, or suspended in a liquid carrier. However, some formulations contain dry granules that can be spread over the surface of the soil, and other formulations are gases that can be injected into the soil or used like a fumigant. The application rates of most herbicides are given in pounds of active ingredients per acre of infested cropland. Because the concentrations of active ingredients vary from herbicide to herbicide, producers must know the concentration of the active ingredient in any given formulation to use the herbicide safely and effectively. For solid materials, the active ingredient is given as either a percentage, an acid equivalent, or a phenol equivalent. For liquids, it is given in pounds of active ingredient per gallon. If, for example, the active ingredient is 2,4-D (both amine and ester forms are common), the acid equivalent simply expresses how much 2,4-D acid there is in an ester or amine formulation. In addition to 2,4-D, herbicides such as 2,3,6-TBA, picloram, and dalapon are expressed in acid equivalents. Phenol equivalents are used in describing the dinitrophenol herbicides.

The use of herbicides has increased dramatically since they were first developed twenty-five years ago. In 1959, about 53 million acres (21.5 million hectares) of crops was treated with herbicides in the United States; by 1968, the number of acres treated was more than 150 million (60 million hectares). The increased use of herbicides is largely attributable to the development of highly selective ones that control specific weeds growing among specific crops. Perhaps the most important of these new weed killers are the preemergence herbicides that are applied before or at the time of planting to control many of the weeds that infest broad-leaved (dicot) crops. The development of herbicides has affected nearly all phases of crop production, including the choice of crop or cultivar to be grown, seedbed preparation, cultivation, fertilization, irrigation practices, and harvesting methods. The most striking example of the way in which new cultural techniques rely on the use of herbicies is the practice of minimum tillage in corn production (see Chapter 14). Herbicides play a major role in increasing yields and reducing labor in crop production (Figure 9-14).

In spite of greater efforts in weed control, major weeds seem to be a continuing problem as indicated in Table 9-1. In part this reflects more accurate estimates of the presence of weeds in recent surveys. In part it reflects the cultivation of new lands, which adds acreage to the areas covered by the surveys. Unfortunately, in large part it also reflects an actual increase in the number of weeds present. In all too many cases, cultural practices that favor a crop also favor weed growth. In addition, the control of one type of weed by herbicides often favors the growth of other weed pests. The use of 2,4-D has greatly reduced infestations of many spring weeds (e.g., pigweed) in cereals and corn, but with competition from these weeds reduced, fall panicum and Kochia have become more serious pests. In Saskatchewan, since 2,4-D has nearly eliminated the growth of wild mustard in wheat, cow cockle has become an increasingly serious problem. As conditions vary, different species increase or decrease in relative importance. Weeds have an alarming reproductive potential and can spread almost explosively. Consider the increase in the number of acres infested with dalmation toadflax in Montana (Table 9-3).

Like other pesticides, herbicides can produce toxic effects on the environment. In recent years, they have been seriously criticized as environmental pollutants. To avoid problems in the use of any chemical, crop producers must follow the manufacturer's directions on the container. In addition, they must be aware of regional restrictions on the use of the pesticide.

Biological methods In light of current concern with the environmental pollution and ecological disruption caused by all pesticides,

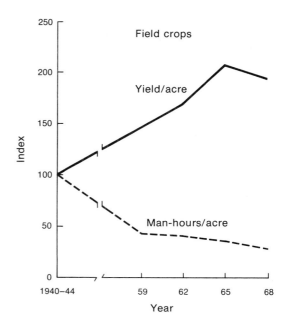

Figure 9-14
The effects of advancing agricultural technology are characterized by an increase in yield per acre and a decrease in man-hours of labor per acre. More crop products are harvested with less total labor. [USDA data.]

Table 9-3
Number of acres infested by leading perennial weeds in Montana (1957–1967).

Weed	No. of acres infested		
	1957	1967	Change
Canada thistle	324,696	489,095	+164,126
Field bindweed	220,830	254,970	+34,140
Leafy spurge	51,718	88,095	+36,377
White top	44,919	29,027	−15,829
Perennial sow thistle	27,043	47,067	+20,024
Russian knapweed	6,302	15,067	+8,765
Dalmation toadflax	109	7,129	+7,020
Total	675,890	930,830	+254,623

Source: L. O. Baker, Montana State University.

agricultural scientists must explore avenues of biological, rather than chemical, control. An outstanding example of an agent of biological weed control is the Klamath beetle, which feeds exclusively on Klamath weed and thus rids croplands of this serious weed pest without damaging other plants.

In addition to control through mechanical methods, flaming, herbicides, and biological agents, crop producers can take other steps to reduce losses due to weeds. They can avoid sowing weed seed when they plant crops. The price of clean (certified) seed is usually more than offset by the decreased cost of weed control. Similarly, growers can avoid bringing weeds into "clean land" in irrigation water. This is difficult to do on an individual basis, but cooperative efforts in keeping banks free of weeds do pay off. Finally, efficient field management overall is an essential component of any weed-prevention program because healthy, well-nourished, well-irrigated crops are better competitors against weeds than sick, nutrient-deficient crops that are suffering from moisture stress.

If crop producers are to meet the needs of the growing world population, continued efforts must be made to reduce losses due to all crop pests. Although the development of any method of pest control is time-consuming and expensive, agricultural researchers contine to seek safer methods for controlling pests. Because of the hazards and shortcomings of other methods, the development of biological controls is potentially one of the most fruitful areas of agricultural research.

Selected References

GENERAL
Protecting our food. 1966. *Yearbook Agr.* USDA.

INSECTS
Insects. 1952. *Yearbook Agr.* USDA.
Metcalf, C. L., and W. P. Flint. 1939. *Destructive and useful insects.* McGraw-Hill.

DISEASES
Plant diseases. 1953. *Yearbook Agr.* USDA.
Roberts, D. A., and C. W. Boothroyd. 1972. *Fundamentals of plant pathology.* W. H. Freeman and Company.
Strobel, G. A., and D. E. Mathre. 1970. *Outlines of plant pathology.* Van Nostrand Reinhold.

WEEDS
Ashton, Floyd M., and Alden S. Crofts. 1973. *Modes of action of herbicides.* Wiley.
Dunham, R. S. 1973. *The weed story.* Institute of Agriculture, University of Minnesota.
Weed control, 1968. *Nat. Acad. Sci. Nat. Res. Council Publ.* 1597.

Study Questions

1. Discuss the types of crop damage caused by insects; cite examples.
2. Discuss the relationship between the crop life cycle and the insect life cycle. How does this relationship affect methods of control?
3. What factors would you consider in determining when and how to control an insect pest?
4. With examples, discuss why it is necessary to understand the life cycles of both the host and the pathogen as well as the disease cycle itself in order to determine appropriate disease control methods.
5. Describe and explain the types of damage caused by a common representative of each of the three major groups of plant pathogens.

6. Describe the types of damage and losses caused by weeds, giving examples of both direct and indirect damage.
7. What are some of the anatomical and/or morphological features that cause typical weeds to be classified as pests?
8. Describe the various methods of weed control and give examples of when and why each method might be used effectively.
9. Discuss the nonchemical, biological, and cultural methods of pest management that can be applied to each category of plant pests (insects, diseases, and weeds).

Ten

Cropping Systems and Resource Conservation

Cropping Systems

INTRODUCTION

A cropping system describes what crop or crops are grown in a particular location as well as how and why they are grown. A particular cropping system may be applied to an extensive geographic area, such as a semiarid region, or to a more restricted area, such as a single farm or ranch. Crop rotation constitutes one part of a cropping system; the precise pattern of rotation is determined only after the overall system has been defined.

Some cropping systems are fairly easy to describe. For example, in the crop-fallow system typical of cereal production in many of the semiarid areas of the United States and Canada, a crop rotates with fallow every other year (see Chapter 7 for details). In some tropical areas, a similarly simple system is followed, regardless of the crop to be grown. The system follows a pattern of clearing land; burning to destory remaining vegetation and to enhance soil fertility; growing crops annually until soil fertility drops below the point at which crop yields justify continued farming, or until native vegetation retakes the land; and then abandoning the "exhausted" land to repeat the cycle in a new location. Other major cropping systems include the systems of tropical or subtropical agriculture typical of Florida and Hawaii, systems of irrigated agriculture of arid lands, and systems that coordinate crop (feed) and livestock production. One of the most rigorous cropping systems is double cropping, a system in which two crops are produced per year on a single field. Double cropping is practiced primarily in regions where there is a relatively long growing season and generally where irrigation is practical. A double cropping system might include fall-sown cereal (barley or wheat), which is harvested in late spring or early summer, followed immediately by a rapidly growing, warm-season crop, such as grain sorghum. Soil fertility management (the proper use of commercial fertilizers) and soil conservation practices are critical factors in managing double cropping systems.

Cropping systems are usually designed to prevent problems, but sometimes a system is devised to solve an existing problem. Saline soils pose peculiar management problems; as salts build up in soils as a result of the flow of salty waters and the accumulation of salts due to evaporation, productive lands are laid waste. In some regions of Montana, North and South Dakota, and adjacent areas in Canada, problems with soil salinity are on the increase. A combination of high water tables, fallowing, sources of salt (mainly natural, associated with the soil's parental material), and patterns of subsurface water movement has caused the spread of saline seep areas (Figure 10-1). Although the problem may be partly alleviated through extensive, costly drainage systems, cropping systems will have to be modified to remedy or prevent the problem. Because under certain conditions fallowing causes water tables to rise, an annual cropping system that does not rely on fallowing would lower water tables and thereby reduce salt residues left by the evaporation of water from the upper layers of the soil. In addition, a cropping system that is developed to combat saline seep must include crop species that both tolerate high levels of salt and root deeply to help lower the water table. Researchers are just starting to identify such species (see Chapter 32).

RESOURCE UTILIZATION

To develop superior cropping systems, existing cropping systems must be evaluated so that viable cropping goals can be established, and the means of attaining these goals can be defined. Both in theory and in practice, a legitimate goal is the optimum, long-term utilization of resources. To achieve optimum resource utilization, cropping systems must be structured so that short-term gains of any kind do not result in major, long-term losses (Figure 10-2). Producers must first analyze what requisite resources are available for the production of essential plant products. They must then develop conservation-oriented cropping systems that foster maximum yields of the plant product without destroying or overconsuming the basic resources.

The dust bowl of the Great Plains represents a tragic example of an unacceptable cropping system. Native rangeland was plowed and seeded to cereals, but the potential for successful production was minimal because of unfavorable moisture conditions. Economic considerations of short-term gains encouraged producers to grow wheat, but drought led to successive crop failures that left the soil bare. The denuded lands eroded badly, and economic as well as ecological disaster resulted. A cropping system that includes growing native perennial grasses was instituted to reclaim parts of the dust bowl; the production of cereals did not effect optimum resource utilization of the Great Plains' cropland because soil conservation required a perennial groundcover.

On the other hand, through such programs as the Central Valley Project in California and such agencies as the Tennessee Valley Au-

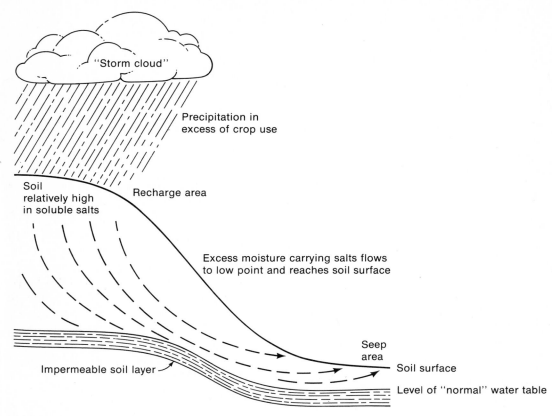

Figure 10-1
Excessive moisture moves through the soil and accumulates salts. When this salty water encounters an impermeable layer, the water moves laterally and may come to rest in a spot where the impermeable layer is near the soil surface. As water evaporates from the soil, the concentration of salt increases.

thority, massive irrigation systems have been developed, making available thousands of acres for high-value cropping systems that optimize resource utilization. In much of California's San Joaquin Valley, resources that create tremendous potential for food production are obviously available (i.e., soil, growing season, and favorable temperature). but a single factor—insufficient moisture—prevented the realization of this potential. As water has been made available through irrigation, the cropping system has changed from dry-land cereals or open rangeland, with comparatively low productivity, to irrigated annual cropping, or even double cropping, with high productivity. Even on a single farm, optimum resource utilization requires that irrigation be developed if factors other than water are not limiting.

CONSERVATION
Agriculturalists have always sought to develop cropping practices that preserve the environment by conserving rather than con-

Figure 10-2
Contour cultivation helps reduce the danger of soil erosion. Part of the field is producing peanuts, the other part corn. Peanuts add nitrogen to the soil and corn requires relative large amounts of nitrogen. Reversing the pattern in the following years allows the corn to utilize the nitrogen fixed by the peanuts. [USDA photograph.]

suming natural resources. Efforts at resource conservation are perhaps most evident in the soil conservation practices designed to prevent erosion and other damage to soil structure. Cereal crops (mainly wheat and barley) are produced in a crop-fallow system in which soil erosion during the fallow period can be a serious problem. A wide variety of cultivation methods have been developed to minimize erosion, which at the same time enhance moisture storage, encourage decomposition of valuable organic matter, and control weeds.

Trashy fallow—produced by cultivation that does not fully turn crop residue into the soil—is effective in conserving soil. Strip cropping—a system in which fairly narrow, long bands of land are planted—is also effective; to reduce losses due to soil erosion even further, the strips are seeded on the contour of the land (Figure 10-3). Plowing and planting around hills rather than over them reduces erosion in all cropping systems, as does the proper timing of cultivations. The prudent use of selective herbicides, which reduces the number of

Figure 10-3
Contour-strip farming is practiced on a large scale. Cultivation on the contour with rows seeded across the slope protects the soil. The alternating cropped and fallow strips supply the normal benefits from fallow. Planted and fallow areas are reversed annually. [USDA photograph.]

cultivations required for weed control, not only helps protect soil structure, but also reduces labor and expenses.

The development of highly selective herbicides has made possible a new cropping system known as minimum-tillage farming. This system, which has been confined mainly to corn production, requires only minimum cultivation for seedbed preparation; seedbeds remain trashy, but the soil is protected (Figure 10-4). All weed control is accomplished by special herbicides instead of through cultivation. Minimum tillage greatly reduces erosion and damage to soil structure (as well as labor) while it increases water infiltration; yields of crops produced under this system are comparable to the yields of crops produced with traditional tillage.

Figure 10-4
Minimum-tillage cropping has become increasingly popular as new herbicides have become available. A chisel planter seeds corn directly into the stubble of a previous crop. There is no seedbed preparation. Soil is conserved and labor and fuel costs are reduced. [USDA, SCS photograph.]

Resource conservation is practiced in a number of other ways. Irrigation systems can be designed to minimize erosion: proper leveling, proper lengths of borders or furrows for specific soils, and safe application rates of water serve to protect the soil; they also help to conserve water. Similarly, the proper selection and application of fertilizers combined with a cropping system that utilizes the applied minerals most effectively conserve the productivity of the soil. In any cropping system, the amount and location of the fertilizer in the soil and the soil's ability to retain fertilizer must be matched with crop requirements. For example, sugar beets need relatively large amounts of nitrogen, but, for highest yields, the nitrogen must be exhausted shortly before harvest. Alfalfa returns nitrogen to the soil but the amount of nitrogen it fixes and thus makes available cannot readily be controlled. Therefore it would not be desirable for sugarbeets to follow alfalfa in a cropping system. However, corn or sorghum, which use large amounts of nitrogen, can well follow alfalfa in a viable system. The nitrogen in the soil is depleted by these crops, and specific amounts of nitrogen fertilizer can then be applied for the sugar beet crop. Keep in mind that, although the alfalfa increases soil nitrogen, it removes large quantities of phosphorus, which must be replaced with commercial fertilizers before another crop can grow.

Sometimes conservation moves beyond the prevention of resource depletion to the actual development of resource potential. For example, the development of watersheds has become increasingly important to agriculturalists and environmentalists alike. A watershed is an area that stores and supplies water from natural sources. Mountainous regions in which the snow that accumulates during the winter provides water for valley rivers and streams in the spring and summer are watersheds. A watershed may be a large geographical area such as the Sierra-Nevada Mountains or a small foothill region on a single farm or ranch.

Because water is a major agricultural resource and because many growers are increasing food production by changing to irrigated cropping systems which require the expansion of existing water supplies, emphasis on watershed development is justifiable. Watershed development encompasses many factors. It may include the construction of large dams and man-made lakes, which store the runoff of major rivers for summer use. It surely includes the control of vegetation in the watershed proper. On much of the semiarid rangeland, brush control not only permits the development of high-quality forage species, but also improves water yield by eliminating plant species that consume large amounts of water and produce low yields. With appropriate brush-control measures (plowing, spraying, using drag lines, bulldozing, and even at times burning), it is not unusual to find that apparently dry springs come "back to life," while the density and yields of desired forage species increase. In any brush removal program, whether it be for watershed improvement or for range renovation, care must be exercised to conserve the soil when brush is removed. After clearing land, it may be necessary to seed annuals that germinate rapidly or short-lived perennials to protect the exposed soil. In some instances, it is advisable to design stripping in which cleared strips alternate with strips left to native vegetation. Watershed development on a local basis includes farming practices that foster water infiltration and storage in the soil. Trashy fallow that holds snow and strip farming in semiarid regions are two such practices, which are also designed to conserve soil.

After a watershed has been successfully developed, farming practices may have to be modified to utilize the water that has been made available: new crop cultivars or species may be required and more fertilizer may be essential; moreover, growers may have to change cropping systems. In the extreme case this could mean a shift from semiarid cropping to irrigated farming.

PEST MANAGEMENT

Pest management and control are essential components of any cropping system. A well-planned system provides for both short- and long-term pest management by incorporating proper cultural practices and suitable patterns of crop rotation. In weed control, for example, continued use of herbicide-sensitive species (e.g., cotton or beans) results in heavier emphasis on cultivation and a possible build-up of weeds. Rotation to less sensitive species (e.g., cereals) allows the effective use of herbicides. On the other hand, row crops such as sugar beets, beans, or corn, which can be cultivated during their growing season, provide an opportunity for nonherbicide weed control, which is essentially impossible with cereals.

To control insects, nematodes, and pathogens, cropping systems must provide for the elimination of the crop residue and volunteer plants that harbor these pests. A cropping program that specifies early hay harvests and prompt removal of hay from the field in conjunction with destruction of unnecessary residues helps control infestations of alfalfa weevils. A cropping system can minimize insect damage to all cereals by adjusting planting dates to prevent plants from being in a vulnerable stage when insect pests are in their most damaging phase. To control infestations of root-knot nematodes, which are serious pests in sugar beets, a cropping system may include a pattern of crop rotation that replaces sugar beets and other susceptible crops with cereals for as many as five years. Following grasses that are susceptible to ergot, deep plowing that buries sclerotia of the fungus that causes the disease should be included in a cropping system if ergot has been a problem. Avoiding grasses in sequence and destroying volunteers also help to control ergot; moreover, the following crop should benefit from the deep plowing. Remember that crop producers must know the life cycles and essential requirements of both the crop and the pest to integrate the most effective form of management into an entire cropping system.

Resource Evaluation

The first step in attaining the goal of optimum resource utilization is to determine which resources limit crop production. The next step is to determine how to augment the limiting resources (if at all possible) to increase production. The final step is to implement modifications. Although the description is short and relatively simple, the job is staggeringly complex.

The major factors that limit world crop production are climate, topography, and soil (fertility, texture, structure, and pH). The limitations imposed on worldwide agricultural potential by climatic conditions can be analyzed by estimating the percentage of total land area that is limited by each factor and combinations of factors. Disregarding pollution (which is a serious problem in some areas), 100 percent of the earth's surface—nearly 34 billion acres (13.8 billion ha) of land—has adequate light to produce some crop; slightly more than 80 percent of this area has a temperature range that can support some form of crop production; and a little more than 30 percent has a suitable pattern of precipitation. When the other major factors are considered along with climate, less than 10 percent of the earth's land surface is ideally suited for cropping.

CLIMATIC FACTORS

In evaluating resources that limit crop production, considerations of climatic factors predominate; many cropping systems are defined, consciously or unconsciously, in terms of climatic limitations. Temperature, precipitation, and light are the main climatic factors that affect crops. The effects of temperature can be measured fairly readily, and, although little can be done to modify natural thermal conditions, various methods (e.g., smudging and irrigation for frost protection) are used to protect crops from slight variations in minimum and maximum temperatures. The components of temperature that have the greatest effect on plant productivity include temperature range (minimum and maximum temperatures), frost-free period, length of growing season (which is generally shorter than the frost-free period), and stability of temperature (diurnal fluctuations and seasonal extremes). Consider an area in which only the frost-free period is limiting; if it has a frost-free period of 90 to 100 days, obviously, many crops are excluded from production (for instance, warm season crops such as rice and cotton). Cropping systems in this area must be confined to cereals and some forages, but which type of crop should be stressed? If forages are stressed, will they be fed to livestock on the farm, or sold? What about transportation for the forage or for the live-stock, or for both? These new, nonclimatic factors, which become limiting, will be discussed later under socioeconomic factors. If the area in which only the frost-free period is limiting has a frost-free period of 200 days or more, a forage cropping system may not represent optimum use of resources. Surely yields would be high, but adequate forage can be produced under more adverse conditions. It might be better to grow cotton, rice, or perhaps soybeans in such an area.

In some cases, temperature has a very favorable, "limiting" effect on cropping systems. Along a very narrow belt between San Francisco and Los Angeles, temperature conditions during the critical flowering period in midsummer are ideal for standard lima bean production. Although other crops can be grown in this region, standard lima beans can be grown here more efficiently than anywhere else. Irrigation systems have been developed to minimize the limiting effects of water so that the production of lima beans has the highest priority for the optimum use of resources. Similarly, the temperature and precipitation conditions in the Napa Valley of California can support many crops, but growers have chosen to cultivate vineyards to support the wine industry, because the crop is valuable economically and uses resources of the area very well.

In evaluating precipitation resources, more than average annual precipitation must be considered. Both the total amount of precipitation and its distribution throughout the growing season must be analyzed to determine to what degree precipitation is a limiting factor. For stability of yield, a reasonably consistent amount and distribution of precipitation is critical. In areas where adequate moisture may occur at irregular intervals, cropping systems must be adjusted to unstable conditions by selecting crop species that tolerate variable amounts of water or by practicing fallow or other methods of moisture conservation. Of course, the total amount of precipitation in a growing season is critical. Precipitation (both rain and snow) is a significant factor in delimiting the tallgrass prairie from the shortgrass prairie. In the absence of irrigation, severe restrictions on the use of other natural resources result from the limited availability of water. For example, the southern end of the Salinas Valley in California (midway between San Francisco and Los Angeles, and from 10 to 40 miles inland from the Pacific coast) is semiarid. Although soils are generally sandy, they are deep, and commercial fertilizers can be used very effectively with proper management. Rarely, are prolonged, subfreezing periods experienced. The growing season is roughly from April through October (about 180 days), depending on the specific location. If this region had more moisture it could support many valuable crops—sugar beets, beans, and some vegetables. Before the development of irrigation, the principle crop was barley grown for livestock feed, but the land use is changing now because of the possibility of irrigated cropping.

Snow is a part of the total annual precipitation and as such must be evaluated as a resource to determine what happens to the moisture it contains: is it stored in the soil for crop use, is it stored in natural or man-made reservoirs, or does it run off and provide no agricultural benefit? In addition to being a water resource, snow is an essential insulator for many fall-sown crops in the north. Winter wheat may survive extremely low air temperatures if covered by several inches of snow. To use a cropping system that depends on this insulation, a grower must be confident that the desired snow will fall at the appropriate time almost every year. This means that long-term weather data must be evaluated to determine if the odds of a certain amount of snow falling during the coldest months are favorable, just as the odds are determined to see if a certain amount of rain will fall during the growing season.

Even fogs and dews must be considered as part of the total precipitation picture. Although fog and dew add relatively little moisture to the soil, they play key roles in modifying temperatures and in reducing transpiration losses. It is in fact a combination of moderate temperatures and fogs that defines the cropping system in the Salinas Valley of California and that favors lima bean production in the area just north of Los Angeles. In some instances fog is not a blessing. To the east of the Salinas Valley, fog in July and August dampens maturing barley and wheat and, daily, delays the start of harvesting operations until midmorning or early afternoon. Also, earlier in the growing season, damp conditions favor the development of some fungal and bacterial diseases. Although conditions of high humidity, or fogs and dew, do not dictate cropping systems, they must be considered both in terms of their impact on other climatic factors (e.g., temperature) as well as in terms of pest management. Sometimes factors such as dew appear to be minor until they are interpreted in terms of their relationships with other factors. Virtually no factor can be considered to be neutral in determining an appropriate cropping system. In addition to adequate total precipitation, the proper distribution of precipitation is of crucial importance. Although all crops utilize the majority of their total water requirement during periods of most active growth, moisture must also be available to mature fruits and seeds. In the absence of irrigation, if the majority of natural precipitation occurs in the winter months, the cropping system must center on either winter annual (spring-maturing) or winter- and/or spring-growing perennial species. If spring and summer rains are the rule, the opposite is true.

In many instances, the limiting effects of water have been circumvented through the development of extensive irrigation systems. Development of high-cost systems requires not only economic analyses, but also further analysis of climatic factors and of soil. If water alone is limiting, then development of massive irrigation units may be justified. The pattern of agriculture in the San Joaquin Valley of California changed dramatically with the development of the Central Valley Project (an irrigation water distribution system) from dryland cereal production to much more extensive cropping systems of row crops (field beans and vegetables). The optimum use of resources (soil and growing season) was more nearly realized when irrigation water became available. Similar stories could be told about Arizona and New Mexico. On the other hand, even though more water would indeed be desirable for cereal production in semiarid areas, the cost of developing extensive irrigation systems may not be justified in an area that has a growing season of less than 100 days.

Although little can be done to alter natural light conditions, both the duration and the quantity of light must be considered in developing a cropping system. The most obvious aspect of light to affect crop production is the photoperiod. Long-day plants, which require lengthening days to flower, cannot be produced successfully near the equator owing to a nearly uniform photoperiod throughout the year. In some regions (the far north for example) relatively short growing seasons are compensated for by extraordinarily long summer days (with up to nearly 24 hours of light). This phenomenon results in the production of giant vegetables; if short-day forage species are grown in these areas, they remain in a vegetative stage almost indefinitely and forage yields are high. Note that few crop species prosper under conditons of continuous light. Thus, even if it were practical, using continuous, artificial light to produce crops is not a solution to food shortages. Photoperiod is also a matter of major concern in seed production.

Species, and in some cases cultivars, must be selected to meet existing photoperiodic conditions. At times, temperature and the length of the growing season must also be matched with photoperiodic requirements for seed production. In northern areas, long days coupled with cool, late-summer temperatures may restrict the seed production of some warm-season crops.

Light quantity, like duration, is difficult to alter under field conditions, but must be considered in selecting crops to achieve optimum yields. Crops such as cotton require high light intensity and cannot be grown efficiently in areas that have many cloudy days during the growing season. As air pollution increases, it, too, may limit cotton production. If other factors are not limiting, high light intensity generally favors high yields, regardless of the crop.

The effects of light are interrelated with the effects of temperature, and both affect the water needs of crop plants. As light drives the photosynthetic process, plants must integrate balanced amounts of all other essential resources to support maximum net photosynthesis, which in turn results in optimum resource utilization and high yields.

SOIL FACTORS

To be sure, the sea has real potential for food production and plants can be grown in water or sand culture, but from an agricultural point of view, land, and thus soil, is the basic resource of crop production. In developing optimum cropping systems, growers must critically evaluate limitations due exclusively to soil while taking into account limitations associated with soil in relation to other factors. The new system of soil taxonomy (see Chapter 6), which integrates soil characteristics and other factors, can simplify the evaluation of some of these limitations. However, the new system of soil classification does not directly reflect how the value of semiarid land rises with irrigation or how the productivity of relatively infertile soil increases with proper fertilizer management.

The soil factors that affect cropping systems include fertility, pH, salinity, texture and moisture-holding capacity, and depth. In the past, natural soil fertility was considered a major factor in determining appropriate cropping systems, but, with the development of complete, artificial (commercial) fertilizers, it has become much less of a limiting factor; however, economic considerations may well become limiting if a soil requires the application of large quantities of commercial fertilizers to produce optimum yields. A cropping system designed for acidic soil may require liming practices to raise soil pH and permit production of a wider array of crop (or hay) species. Otherwise the cropping system may center on a single, acid-tolerant crop such as corn (or rye in cooler regions). On saline soils cropping systems must provide for the reduction of salinity through annual cropping (rather than a crop-fallow system) and may include the development of relatively high cost drainage systems. At the same time, the selection of suitable crops is critical—for example, sugar beets and barley instead of a cereal such as rye.

Soil texture affects the design of a cropping system in several ways: First, it determines the characteristics and method of preparation of the seedbed, which in turn affects the choice of crop. For example, if a heavy soil is worked into the fine seedbed essential for many small-seeded crops (such as forage grasses), crusting, which restricts emergence, may become a serious problem. On the other hand, if a light (sandy) soil is worked into a fine seedbed, crusting will be less of a problem, but, unfortunately, drying may become a problem. Texture also affects the moisture-holding capacity of soil, which is a crucial factor in determining whether irrigation is necessary. For example, safflower can be successfully produced with minimum irrigation on the heavy-textured, clay soils of the rice-growing regions of the Sacramento Valley in California. The high moisture-holding capacity of these clay soils allows safflower to be produced with only a preirrigation. The minimum irrigation serves a dual function: it supplies proper moisture and it helps to control disease in safflower. Safflower is very susceptible to rot diseases caused by fungi. The fungi flourish if the soil surface is excessively moist, which is common during irrigation. Thus, if irrigation is not practiced, conditions that favor the disease are not present. Growing safflower with minimum irrigation is an economically valuable alternative to continuous rice culture on heavy, clay soils.

A soil's moisture-holding capacity must be analyzed in conjunction with natural precipitation (both amount and distribution), the water needs of various crops, and the depth of soil before any cropping system can be established. For example, if the roots of a crop penetrate to a depth of 24 inches (61 cm), as do those of potatoes, and if the soil has a moisture-holding capacity of 2 inches of water per foot of soil depth (5 cm of water per 30 cm of soil), then 6 inches (15 cm) of precipitation would wet the soil to a depth of 3 feet (90 cm), which is deeper than necessary. If on the other hand, the soil holds 3 inches of water per foot of soil depth (7.6 cm/30 cm), then 6 inches (15 cm) of precipitation would wet the soil to the desired depth of 2 feet (60 cm).

The machinery that is necessary to farm efficiently is different for different soils. As a rule the cultivation of heavier soils requires more energy (bigger tractors, or more fuel consumption by small tractors) than the cultivation of light soils. A cropping system must match the equipment available for the soil in question with the cultural requirements of potential crops.

In the final analysis, the problem of how to achieve optimum land utilization can only be solved by determining the best overall use of land for human welfare; this includes making decisions about which land must be used for activities other than crop production (including all types of forage and livestock production). It is relatively easy to identify prime agricultural lands both in the United States and in the rest of the world. Many great river plains (alluvial fans) are recognized as prime agricultural lands as a result of both desirable soil characteristics and favorable climatic factors. At the opposite extreme, some lands are readily identifiable as wild lands that are unsuitable for cropping. Much of the timbered areas of the Rocky Mountains would fall into this category, but even here questions arise when forests are considered to be crops. As a general policy, future urban development should be directed toward those areas in which the sum of environmental and soil factors shows the least potential for cropping. Of course, the land in those areas must be surveyed to discover if it is suitable for building. This means that soil must be examined to verify, for example, that water tables are not too high (i.e., will not cause flooding in basements), or that soil structure and related geologic factors do not make the land so unstable as to preclude the building of secure foundations and roads, or that there is sufficient drainage through the soil to permit proper waste disposal. At present we have at hand the basic knowledge with which to analyze these factors.

Between the two extremes of land that is ideal for the production of many crops and land that has minimum potential for cropping and is therefore more suitable for urban development, there exists a great grey area of land that contains some of the resources necessary for crop production. Such land must be evaluated to determine its best potential use, and all the major limiting factors must be considered before the desirability of altering any one limiting factor can be decided. Through vigorous research, more and more limiting factors are being identified, and improved methods of combining these factors to create optimum cropping systems are under study. The concept that all land must be brought into crop production has been abandoned. The mistakes of the 1920s that caused the disaster of the Great Plains dust bowl will not be repeated.

GEOGRAPHIC AND TOPOGRAPHIC FACTORS

Geographic and topographic factors, which are related to both climate and soil, must be accounted for in any cropping system. Some of these factors are associated with elements of weather. For example, although prevailing winds may not in themselves be a serious limiting factor for crop production, a knowledge of the frequency, magnitude, and direction of winds is necessary to implement optimum resource utilization. Heavy winds can cause severe shattering when grain crops are mature. If the shattering is too severe, and if shatter-resistant cultivars cannot be developed, alternative crops will have to be grown to achieve optimum yields. Moreover, the direction from which the wind generally blows indicates the path of storms and helps to distinguish cool storms, which follow the same path, from warm ones. The direction from which prevailing winds blow also helps to determine where wind barriers should be erected and how sprinkler irrigation systems should be deployed.

Topography, or the "lay of the land," limits cropping systems in many ways. The limitations resulting from major geographic variations have been described in terms of overall temperature, precipitation, and light conditions of an area. On a more local scale, crop producers must consider how the slope of the land affects the growth of crop plants. Variations in temperature between north- and south-facing slopes are directly attributable to exposure or aspect. Growers must also analyze the high and low points of croplands.

Cooler air drains like water into low areas; even in small fields, low cool spots can be identified. The slightly lower temperatures due to cool air drainage have a negative effect on many crops: they delay or cause variable maturity of grain, and they retard the recovery of forages after harvesting. An appraisal of the feasibility of producing the kinds of crops that grow successfully on a terrain must take into account the investments in equipment required to farm the area. The possibilities and desirability of modifying topography (i.e., of land leveling) must be evaluated with respect to costs and returns and with respect to other limiting factors in the overall cropping system (Figure 10-5). For example, topography places serious constraints on the methods of irrigation that can be employed in a given field. In areas where moisture is limiting but other resources are plentiful, thousands of acres have been leveled to allow the development of cropping systems that rely on irrigation. The San Joaquin and Sacramento Valleys in California are two outstanding examples of this type of land use, and there are comparable areas throughout the southwest. Although new types of sprinklers have reduced the necessity of land leveling in some areas, every year, as more water resources are developed, more land is leveled so that total crop production resources can be more fully utilized.

Proximity to large bodies of water is another geographic factor to be considered in establishing a suitable cropping system. Remember that a large body of water acts as a thermal buffer, by cooling slowly in the fall and warming slowly in the spring. Thus, a large lake may afford protection against early fall frosts and provide crops with a longer frost-free season in which to mature; at the same time, it may keep temperatures low enough in the early spring to limit spring growth until the danger of late frosts has passed. Cropping systems around the Great Lakes take advantage of these phenomena, and the sweet cherry industry around Flathead Lake in northwestern Montana is directly dependent on the thermal buffering of the lake.

Localized climatic conditions (such as cool or warm spots, damp or dry spots) reflect the impact of a whole spectrum of topographic and geographic factors that cause variations in the rate of growth and development of crop plants. Each potential variation must be evaluated in developing an appropriate cropping system. For example, it is undesirable for the maturity date of plants to vary in a field of peas grown for canning. If temperature changes caused by topographic factors result in such variation, this specialty crop might be excluded from a particular area; however, this same variation is of minor significance in a pasture and might not be critical in either cereal or soybean crops.

In assaying the potential impact of any factor, a crop producer must consider the magnitude of variation in the factor (e.g., temperature range or distribution of rain), the possibilities and advisability of modifying the factor, and the manner in which crops respond to the factor in question, before he can define an optimum cropping system.

SOCIOECONOMIC FACTORS

It is possible for agronomists to make optimum agricultural use of all natural resources without considering the economic or social consequences of cropping systems, but this is an untenable position. We have made giant strides in understanding and managing the climatic and physical (soil) factors that dictate cropping systems. Certainly, we must consider the human factors, including economics, that have a major bearing on crop production.

Figure 10-5
Equipment used to level fields for surface irrigation. Land levelers are usually quite long so that they span low spots in the field and allow soil to be moved into them.

The crop producer is a businessman. To survive in today's society, he must realize a profit from his efforts. At the same time, he must accept the obligation of efficient utilization and conservation of natural resources. Consider the example of a farmer in a region that can produce reasonable yields of wheat but is also well suited to hay and pasture production. From the point of view of meeting the need for food most efficiently, wheat production may afford the best land use. However, to conserve soil resources, the farmer must return about 1.5 pounds (0.7 kg) of nitrogen to the soil for every bushel of wheat harvested. The cost of this type of fertilizer program may be economically prohibitive if projected yields aren't sufficiently high, but, from the point of view of soil conservation, cereal production without the inclusion of a fertilizer program is undesirable. In view of the foregoing considerations, the farmer may decide that a cropping system of forage production is acceptable, as long as it includes a legume to increase soil nitrogen. The farmer's choice of cropping system changes markedly because of his analysis of economic and conservation factors: instead of producing a cash crop (e.g., wheat) he decides to produce a forage, which can be either marketed as feed or fed directly to livestock.

Cost is a consideration in any cropping system decision. In addition to the cost of land, irrigation, fertilizers, pest control, and labor, the availability and cost of equipment are of major concern. For example, the equipment required for cereal production can be used for nearly all cereals, with the exception of corn, which requires special planting and harvesting equipment. If a grower who has been growing barley wishes to change from a system of annual cropping to double cropping, he will incur less expense if he decides to grow grain sorghum than if he switches to corn. Corn would require a sizeable investment in planting, cultivating, and harvesting equipment, whereas sorghum would require only a new planter because a standard grain combine can be readily modified for harvesting sorghum. Changing from range and pasture to hay also results in requirements for new equipment. Altering a cropping system not only creates (or eliminates) the need for special equipment, it may also change labor, storage, and transportation requirements. Both the cost and the availability of labor may dictate some aspects of acceptable cropping systems. Although advances in agricultural engineering enable more and more jobs to be done by machines, hand labor is still essential for many crops. Until the mid-1960s, tomato production was dependent on the availability of hand labor to harvest the crop. Even where no other factors were limiting, tomato production was precluded in the absence of adequate hand labor. With the advent of mechanical tomato harvesters, and with the development of cultivars that are suitable for mechanical picking, the dependency on "stoop labor" has been eliminated. On the other hand, fruit production (cherries, peaches, etc.) is still dependent on the availability of hand labor for pruning trees and harvesting. Perhaps the most striking conversion from hand to mechanical methods occurred with the development of the mechanical cotton picker.

The farm today supports urban life and, because processing facilities as well as the ultimate marketplace of a crop product are often far from the farm, a grower must evaluate the cost and availability of transportation to determine the economic feasibility of a crop. For example, based on climatic and soil factors, a given location may be well suited for sugar beet production. However, for each 100 pounds (45 kg) of beets produced in the field, less than 20 pounds (8 kg) of sugar is obtained. If the beet processing plant is far from the field, the cost of transportation is prohibitive, and sugar beet production may not be economically feasible in spite of other favorable factors. In general, cropping systems involving crop products that lose much of their high initial volume or weight in processing are developed either around adequate, inexpensive transportation or in close proximity to processing plants. A marked exception to this pattern is found in vegetable production, or "truck farming." The lettuce basket of the United States is the Salinas Valley, about 75 miles (120 km) south of San Francisco; because of the demand for crops that can be produced only in this region, the cost of transportation is offset by its value.

Finally, in determining a cropping system there are completely personal considerations, which might include scheduling periods of peak work loads, wanting more (or less) automation in farming operations, "gambling" on high-cost–high-potential-return crops (e.g., tomatoes), rather than less profitable, but more certain crops (e.g., sorghum), and specializing (growing one crop) or diversifying (growing several types of crops). In the final analysis, a cropping system reflects the environment and man's ability to either adapt to or modify it: it accounts for the physical conditions of the farm (topography) and includes a response to socioeconomic factors as well as an expression of what the individual farmer wants to do within the limits of what can be done.

Summary

A cropping system describes the types of crops to be grown and how to manage them from planting through processing and marketing. The goal of a cropping system is to make maximum use of resources and insure a reasonable profit, while protecting or conserving resources. A crop producer must be a leader in conservation and resource management. To meet this responsibility, he must develop suitable cropping systems for currently cultivated lands as well as for those lands that will be brought under cultivation in the future.

The first step in developing a suitable cropping system is a thorough inventory and analysis of the cropping site. In this inventory a crop producer must evaluate climatic and soil factors to determine what crop species can be grown efficiently. (For example, corn can be grown in Montana but, because of a shorter growing season, yields are lower and less dependable than those from corn crops in Iowa where the growing season is more suitable.) The inventory must identify limiting factors and allow the crop producer to make sound decisions about the practicality of modifying these factors to increase the number or types of crops he can grow. The most obvious examples of the practical modification of limiting factors are the use of fertilizers and the development of extensive irrigation systems; the costs and benefits of such developments are of major concern to crop producers.

After a grower has completed the inventory and its analysis, he must consider the problems of pest management. The overall system and rotation of crops within the system must provide beneficial use of resources and at the same time serve to minimize the buildup of pests. Alternating row and densely seeded crops (e.g., sugar beets and cereals) and where possible warm- and cool-season crops (e.g., corn and barley) provides an opportunity to control weeds through cultivation and chemicals. In addition, by alternating crops a grower guarantees that a field is under cultivation at different times of the year, which also facilitates weed management. Similar consideration must be given to the management of insect pests and plant diseases.

Of course a suitable cropping system depends on available equipment and personal considerations as well. Market or consumer demands play a major role in determining what crops to grow, and certainly the proximity of processing facilities and the availability of transportation for the distribution of the crop product play major roles in determining where various crops are grown.

Looking to the future, existing cropping systems must be evaluated in terms of potential productivity for human welfare; limiting factors must be identified, and then modifications must be made to place all lands under optimum use. The plant sciences in cooperation with many related fields now have the resources to attain this goal. The challenge of increasing crop yields to feed man in the future must be met. We believe that it can be met—and that it will be met.

Selected References

Clawson, Marion D. 1969. "Systems analysis of natural resources and crop production" in *Physiological aspects of crop yield.* American Society of Agronomy.

Janick, J., R. W. Schery, F. W. Woods, and V. W. Ruttan. 1974. *Plant science.* 2d ed. W. H. Freeman and Company.

Saline seep-fallow workshop proceedings. 1971. Sponsored by Highwood Alkali Control Association. Highwood, Montana.

Study Questions

1. What is "an acceptable cropping system"?
2. What factors are critical in determining an acceptable cropping system?
3. What factors that affect cropping systems can be changed most easily?
4. Evaluate local cropping systems in terms of the major factors on which they are based and how these factors can be modified. Use examples.

Part Three

Production Practices

Eleven

Marketing Crop Products

There are three main reasons for a farmer to grow plants: first, he can use them as food for himself and his family, or as feed for his livestock; second, he can appreciate their aesthetic value—many plants, lawns, trees, shrubs, and other ornamentals, are grown more for their beauty than for any other reason; and, third, he can consider them from a purely economic viewpoint—How much is the crop worth in cold, hard cash? In the present era of specialization, the third reason is of significance to every crop producer, because at one time or another each producer sells or markets his plants or their products.

Like other businessmen, farmers strive to make a profit so that they can buy those goods and obtain those services that they cannot produce or provide for themselves. To realize a profit, a farmer must understand the principles of crop growth and development, the economics of crop production, and the marketing of crop products. Frequently, the purely biological requirements of different crop plants help to determine which crop is to be grown (e.g., soybeans or corn) and how it is to be managed (e.g., heavy fertilizer applications and irrigation, or less intensive supplementation and presumably lower yields). The option exercised by a grower is based on economic considerations, a major one of which is the potential marketability of the crop or crop product.

The United States has a "market economy." This means that for most goods and services there is no central authority determining production guidelines: what, how much, or when. The market price mechanism expresses the consumption preferences of individual consumers; it indicates what they will pay, what they want, how much, and when. In other words, prices communicate the relative values of products to the consumer. If people want more of a specific item, the price of that item rises, which encourages producers to divert resources from the manufacture and production of other goods to the production of that particular item. If an excess of an item becomes available, its price drops.

In a competitive marketing system there are many firms so that no one firm can, by itself, affect the market price of an item, and there is free and rapid exchange of information about the market. In the United States the production side of agriculture is relatively competitive; that is, there are so many sellers of raw agricultural products that no one of them has any effect on the market. For example, a decision by the largest wheat producer to sell (or not to sell) his wheat crop at a particular time cannot affect the market significantly. On the other hand, in the U.S. automobile industry there are only three major producers, any one of which can affect the market by changing the amount of automobiles supplied at a given time. This is an example of a much less competitive market.

Marketing Functions

Marketing is the process of moving the product from the producer to the consumer, in the proper amount and form, and at the appropriate time and place. The following activities are marketing functions.

ASSEMBLY

Assembly includes all the steps required to aggregate small lots of products (such as one crop from a single farm) into larger, more homogeneous (e.g., Grade 1 wheat) and more suitable quantities for handling by processors who prepare products for consumers. For example, the assembly of wheat includes collecting the grain at the local elevator, collecting it again at terminals, and finally assembling it for shipment to millers and bakers who process it for consumers.

STANDARDIZING AND GRADING

From the standpoint of making the market work, standardizing and grading are probably two of the most critical functions in the market. They are usually taken for granted in the United States and Canada, but in countries where standardizing and grading are not as highly developed, the market cannot function as efficiently. A uniform grading system makes possible worldwide communication about a product, which is useful in contracting and dealing. For example, USDA Grade Number 2 Yellow Corn describes the product to any buyer whether in Iowa, Montana, Texas, Manitoba, or Japan. A standardized system in which grade accurately indicates quality enables sellers to offer an array of premiums and discounts that can be used to adjust base prices for variation in quality.

The United States and Canada have comparable grade standards (see Chapter 5) for many crops, mainly cereals and feeds. Buyers and sellers in both countries are fully aware of national variations. Through the price system, grading standards encourage the production of the quality of products the consumer wants. Food products that meet U.S. grade standards are graded by a U.S. Department of Agriculture grader, and sellers may indicate the grade designation (the familiar USDA shield) on the product itself or on the package in which the product is marketed at retail. Thus consumers can be guided in their purchases by grade designation.

Although government regulations do not require that the grade of wheat used in bread be indicated to consumers through labeling, the United States Grain Standards Act makes it compulsory for everyone selling wheat by grade and shipping it in interstate commerce to have it inspected for grade (Figure 11-1). The Grain Standards Act also covers corn,

Figure 11-1
Truckload of grain being sampled for grading before shipment abroad. [Photograph courtesy of Port of Seattle.]

oats, rye, barley, grain sorghum, flaxseed, and soybeans. The grading of hay for shipment in interstate commerce is optional. One or more parties interested in a lot of hay may request an official inspector to grade it and issue a grade certificate. Contracts for the purchase of hay of a given grade require official grading. The grading is done by licensed inspectors operating on a fee basis at primary terminal markets; however, marketing hay in this manner is not common.

A number of other inspections, mostly mandatory, are designed to assure the wholesomeness of food. These inspections usually concentrate on products processed for human consumption, although inspection of fresh fruits and vegetables for chemical residues of pesticides is becoming common. In addition, some feeds and feed additives are under scrutiny.

A related type of service offered by the U.S. Department of Agriculture on a voluntary basis is the continuous inspection of fresh and processed fruits, vegetables, dairy and egg products, and other commodities. The inspection is financed through fees paid by those who use the service. The classification of cotton by grade and staple length is an example of the USDA role as a referee in the market place.

STORING

Because agricultural production is highly seasonal, some crop products must be stored for orderly distribution throughout the year. The longer the goods are stored, the greater the risks of spoilage become; moreover, because prices fluctuate with time, the value of the stored crop may change. An efficient market system helps to stabilize prices through speculating and hedging. The costs that arise from the assumption of risk and the accrual of interest on money that is tied up in stored products contribute to storage costs.

The market phenomenon of purchasing commodities when the price is low and holding these commodities until the consumer is willing to pay a higher price is sometimes criticized by both producers and consumers. As a matter of economic fact, this practice provides a service to both because it helps stabilize prices. Formal, theoretical economic proofs would be required to show why this occurs, which is beyond the intended scope of this book.

FUTURES TRADING

Speculation The commodity futures market permits the buying and selling of contracts for goods (wheat, barley, cocoa, eggs, etc.) to be delivered or accepted in the future. Usually buyers and sellers of futures contracts never see the goods, because the contracts are liquidated before it is necessary to make or accept delivery. It is uncommon, but not without precedent, for futures speculators to double their investment overnight. However, it is also very common for them to end up losing money. Studies show that about two-thirds of those who just dabble in futures end up losing. Professional speculators depend on continuous study of the market, systematic trading techniques over a long period, cool judgment, and emotional detachment, rather than luck. The notion that one person's gain is another's loss is reflected quite vividly in the futures market, but in fact it works to stabilize the market system by reducing the fluctuation in prices at both the high and the low ends of cycles and is an important market component.

Corn was the leading commodity traded on the futures market during 1970-71, with 2,748,000 contract transactions (each grain contract is for 5,000 bushels). For only about 5,000 of these contracts was corn actually delivered. The rest of the contracts were traded back and forth but never required delivery. The reason this can happen is that the commodity is not actually bought or sold; rather a promise is made to accept or make delivery at a specific future date at a certain price, which in itself creates a market for the commodity.

Hedging Hedging, sometimes referred to as "insurance hedging," is a way in which producers can use the futures market to reduce the risk of a reduction in the price of a product. A hedger establishes a position in the cash market before harvesting his crop, to insure himself against a future drop in price. Consider the following simplified example. In April a farmer expects to harvest a crop of 50,000 bushels of winter wheat. He determines that it would take $2.45 per bushel (price in Montana) to cover all production costs and give him a reasonable profit for his crop. If he is fearful that the farm market price for wheat will be less than $2.45 per bushel (e.g., $2.35/bu) when he is ready to sell the crop in August, he may sell a futures contract to deliver, for example, in September at $2.75

Table 11-1
Summary of the perfect hedge.

Wheat (in Montana)		
Desired selling value	50,000 bu @ $2.45/bu*	= $122,500
Actual selling value	50,000 bu @ 2.35/bu	= 117,500
Loss		$ 5,000
Futures (in Minneapolis)		
Sold	50,000 bu @ $2.75/bu*	= $137,500
Bought	50,000 bu @ 2.65/bu	= 132,500
Gain		$ 5,000

*Price difference reflects freight cost to central market.

per bushel (Minneapolis price). If the price of wheat on the futures market drops 10¢ per bushel he buys his futures contract back for $2.65 per bushel. By doing this he makes 10¢ per bushel on the futures transaction which balances the 10¢-per-bushel loss he would suffer because of the price drop (see Table 11-1).

The futures price is a central market (e.g., Minneapolis) price, which is higher than the farm price by an amount approximately equal to the freight costs. It should be noted that if, in the preceding example, the price of wheat had gone up between April and September, the hedger would not have gained. By hedging he not only reduced his risk of loss due to a price reduction, but he also reduced his opportunity to gain through a price increase. The above example described a perfect hedge. Gains exactly offset losses. Changes in the price of futures contracts do not always parallel fluctuations in the actual cash price of crop products, as in the above example, but they do tend to move similarly.

FINANCING

Wherever there is any delay in the movement of products from the producer to the consumer, money is invested on which interest accrues; this interest increases the cost of the marketing process. For example, a cotton farmer is paid for his cotton at the gin. Until a gin owner can move the product on to the next stage and be paid for it, he must either use his own money or borrow from credit sources to pay the farmer. If the gin owner borrows money, he pays interest to the lender; if he uses his own money, he foregoes the interest he could receive from investing in something else. The more time it takes to move the product, or the shorter the season of crop production, the more important this investment function is to the holder of the product.

Market communications For the market to work efficiently, market communications must be fast and dependable. There are many institutions participating in market communications: the U.S. Department of Agriculture, the Canadian Department of Agriculture, trade associations, newspapers, and commodity publications. One of the most frustrating problems that agricultural producers have is that it is difficult to obtain reliable information on quantities and prices of goods being bought and sold. Producers must know the market situation with respect to current and predicted demands for their product so that they can make better decisions. The USDA established a marketing service to supply this type of information to everybody concerned so that the people who buy goods directly from farmers and ranchers would not be at an advantage in obtaining market information.

Pricing The price of a commodity expresses its relative value with respect to other goods and services. Pricing is a communicating device in the market because prices rise and fall as the amounts supplied and demanded adjust toward an equilibrium.

Risk-sharing A market tends to exhibit relatively stable prices for goods over the year. Exceptions arise for the more perishable products. Someone must assume the price risks of holding these products as well as the risks associated with the deterioration of the products themselves. This function is performed by people in the market who are willing to take chances on fluctuations in prices, hoping to profit from accepting the risks. Hedging and speculation are two forms of futures trading that risk-sharers engage in.

MERCHANDISING

Merchandising refers to placing goods in convenient locations, displaying them attractively, and making it as convenient as possible for buyers to obtain them.

Transportation Most agricultural products are grown at a considerable distance from the consumers' market. Transportation begins at the farm and continues until the finished products are in the retail markets. Again, there is tremendous variation in the amount of transportation and handling required for different products. For example, one of the factors that contributes to the high cost of marketing wheat (which is passed on to the consumer) is the "man-handling" of finished bakery products. However, more and more bakery goods are being produced in the supermarket itself (Figure 11-2). This cuts down tremendously on the amount of labor required to handle bulky bakery goods. Freight differentials have an important impact on where processing takes place: for example, because of freight rate changes, we have seen a shift in milling sites from wheat-growing communities to major consumption centers.

Freight rates for agricultural products usually remain relatively stable during the first stages of a rise in agricultural prices. Later, rates may rise sharply, possibly after agricultural prices have begun to drop. This relationship may result in a decline in freight rates relative to farm prices during a period when consumer product prices are rising. By the same token when product prices are declining, freight rates may rise relative to farm prices. This freight-rate–price relationship becomes significant for producers who are located long distances from consumers. For example, if there is a sharp drop in the market price of a highly perishable product such as fresh fruit or fresh vegetables, producers may either destroy their crops or leave them unharvested in an attempt to minimize losses, because harvesting and freight costs would result in an actual net dollar loss.

Products that are not highly perishable may be transported long distances on waterways. River and ocean transportation rates are relatively low compared with land transportation rates. Other means of keeping transportation rates as low as possible include: using the transportation facility (truck, railroad car, barge) to the maximum in terms of load hauled and mileage traveled; keeping spoilage, damage, and breakage to a minimum; processing the product, to reduce its bulk; and reducing interstate and international shipment barriers (licensing regulations, weight limits, trade limitations) that increase transportation costs.

Advertising and promotion Advertising alerts buyers to what is available and attempts

Figure 11-2
Baker in a typical large supermarket. In-store preparation of consumer products such as breads and pastries minimizes transportation of finished products and insures freshness.

to motivate them to purchase; it is the art and science of attracting buyers to a product. It is responsible in part for the huge number of different products available to the consumer. People often feel that there is a considerable waste of money in advertising and promoting products. However, consumers must remember that getting information to people about what is available in the market is, at times, essential to the potential marketing and development of new and better goods. Even though it entails a certain amount of waste, advertising fulfills a necessary function in a modern marketing system.

Processing There has been a dramatic increase in recent years in consumer demands for more processing. Agricultural products vary tremendously in the amount of processing they undergo between the farm producer and the consumer. The degree of processing is a major factor in accounting for the difference between the price the farmer receives and what the consumer pays—the marketing spread. For example, considerably more processing is required to move safflower (in the form of oil and meal) from the producer to the consumer than to move hay from the grower to the cattle rancher. Consequently, the marketing spread for safflower is much higher than it is for hay, assuming that other costs (transportation, etc.) are constant.

Packaging Our modern society expects more and more sophistication in the packaging of food products. For example, crackers

Figure 11-3
Rapid unloading of a boxcar of grain is one of the modern advancements in the transportation and handling of crop products. [Photograph courtesy of Port of Seattle.]

are packaged in packages within packages. As more and more fruits and vegetables are sold in frozen form, more packaging is required, and even many fresh vegetables, such as lettuce, celery, spinach, and tomatoes, are now being packaged. Someone must pay for this packaging: the cost becomes a part of the cost to the consumer.

Distribution This function overlaps considerably the transportation, storage, and pricing functions: it refers to the wholesaling and retailing of products to the final consumer, and includes the physical handling and moving of goods from one place to another and the process of exchanging ownership.

PRODUCT DEVELOPMENT AND RESEARCH

Not only do privately owned companies put considerable funds into research and development, but also the U.S. government has made significant investments in this area over a long period. Product development is one of the keys to our enjoyment of so many different kinds of products, as well as to the high quality of products in the United States. Also, research has contributed to the efficiency of our marketing system by devising ways to improve transportability, storability, and the appearance of products.

Marketing Institutions

There are several different ways of classifying marketing institutions. The following is a typical list of classifications.

Local assembly and processing markets Local assembly and processing markets, such as the local grain elevator, are located close to the point of production of plant products. Farmers sell to local buyers who assemble the commodities of a number of producers; sometimes buyers store a commodity until they have accumulated a large enough quantity to make an economical shipment.

District concentration and processing markets The produce from a considerable number of local assembly markets is assembled and processed for shipment to central markets or to consumers' markets.

Primary terminal or central markets Goods from many local assembly or district processing markets are concentrated in central markets for processing (Figure 11-3).

Seaboard markets Just as the name implies, these markets are located on the seaboard where they serve as the point of embarkation for ocean shipment to foreign countries (Figure 11-4).

Secondary processing markets Some plant commodities pass through local and district markets as raw materials. Wheat and cotton are examples of products that are only partly processed in the early stages of marketing. The remainder of the processing is done by processors located in central, secondary processing markets.

Wholesale distribution markets Raw plant products pass through a fan-shaped series of assembly markets to a central market. The finished products pass from central markets and secondary processing markets through a fan-shaped system for dispersion to widely scattered final consumers. The principal wholesale-market middlemen are:

1. Car-lot receivers
2. Jobbers
3. Brokers
4. Commission merchants
5. Auctions

Figure 11-4
Seaboard market provides storage facilities for grain as well as port and loading facilities for large freighters. [Photograph courtesy of Port of Seattle.]

6. Chainstore buying organizations
7. Service wholesalers
8. Speculators

Retail markets Every city, town, and village in which retail stores sell to final consumers is a retail market.

Marketing Agencies

Elevators, creameries, packing sheds, canneries, tobacco warehouses, cotton gins, local buyers, assemblers, auction markets, trucklines, railroads, air cargo companies, commission houses, brokers, organized exchanges, credit institutions, packing plants, flour and textile mills, cigarette factories, wholesalers, jobbers, exporters, converters, factory sales representatives, independent and chain grocers, specialized clothing and general department stores, mail order houses, drugstores, restaurants—these are the kinds of agencies that move plant products to consumers throughout the United States and the rest of the world. Any commercial plant, person, transportation system, or store that assists in moving products to the consumer is a marketing agency. Since the farmer and the consumer have become separated, federal, state, and local governments have had to take on the job of refereeing the marketing system to insure that farmers get fair treatment in the market place and that wholesalers and retailers get what they order and that consumers get the quality of food they demand. For example, consumers and thus stores want a uniform size and grade or quality when buying food. Specifications for long-distance hauling and uniformity have promoted the use of the voluntary grading services provided by the state, provincial, and federal departments of agriculture. A buyer is fairly sure of getting what he orders, thanks in large part to these grades.

Plant Products: Special Marketing Considerations

Many plant products—such as fruits, vegetables, cereal grains, and oil and fiber crops—are marketed directly from the farm and have a small marketing spread. Others—such as seed grains, grasses, and legumes—are marketed indirectly as livestock products (meat, milk, cheese, butter, eggs, etc.) and have a large marketing spread. How directly these goods are marketed affects to a great extent how much a farmer receives for his raw products, and, at the other end of the marketing process, changes in consumer prices depend as much on fluctuations in processing costs as on variations in the prices farmers receive for their raw products. For example, the prices of seed grain, grasses, and legumes are frequently determined by the prices of livestock and livestock products. What the farmer receives for such items depends on their value in the livestock production process because they are merely the raw materials of a product for which the final market will establish the price. This explains why changes in consumer demand for plant products are not reflected as quickly and directly as one might expect in the farm price and, conversely, why sometimes a change in price at the farm is not reflected as quickly and directly as one might expect at the consumer level. The marketing spread for a particular crop product is determined by the number of marketing functions required to get the product from the farmer to the consumer. Each function adds its own cost to the marketing spread. Therefore, if a farmer wishes to reduce the amount of spread in an item that cannot be marketed directly, he has the following alternatives:

1. Perform as many of the functions as possible himself. This means that he must perform them at least as efficiently as is being done at present, or absorb some of the cost himself, which in the end would be of no economic gain to him.
2. Eliminate some of the functions. This implies that some of the functions are not necessary. It also implies that if some of the functions are eliminated, the farmer will reap the benefit of receiving a higher farm price for his product.
3. Exert considerable effort to see that the marketing functions are performed more efficiently, which effort, of course, would come at some cost to the farmer himself. Whether he gains or not depends on the cost of his effort compared with the gain that may accrue to him.

The fact remains that all the functions described are necessary to some degree in the marketing process. Whether or not the marketing spread can be decreased and the benefits accrued to the producer depends entirely on the particular market situation for a particular crop.

FARMER-OWNED MARKETING INSTITUTIONS
Ever since the beginning of the agricultural revolution (about 1860), farmers have at-

tempted to perform their share of the marketing functions and to increase their share of the profits. They have done this through various kinds of marketing associations—cooperatives, and other types of business organizations. Cooperatives are one of the most prevalent types of farm marketing associations and are given special consideration by the U.S. government. They are formed to increase the advantages of both purchasing inputs and selling outputs. The success of these institutions always depends on the amount of mutual effort and support that farmers are willing to give. It also depends on the characteristics of the crop. Cooperative efforts are most successful in the perishable fruits and dairy industries; within these industries, cooperatives seem to be more successful in the handling of specialty crops that require a minimum of processing and can be identified directly by the consumer. For example, cooperatives of walnut growers, Idaho potato growers, and citrus fruit growers have probably shown the greatest success.

Summary

The general ideas of marketing that have been treated in this chapter apply to all kinds of agricultural products, but there are probably as many differences between the marketing of various plant products as there are between the marketing of plant products and the marketing of other kinds of agricultural products. Furthermore, the marketing of many plant products is closely tied to the marketing of animal products. Even cereal grains are closely tied to the marketing of livestock because wheat, for example, can be used as a substitute for corn or barley in feeding. Likewise, some plant products (especially cotton) are closely tied to other industrial products; the marketing of cotton is greatly affected by the marketing of synthetic fibers such as nylon and the polyesters. People involved in crop production must understand the complexities of the agricultural market and the way in which the marketing system operates because what should be grown, how it should be grown, how much should be grown, where it should be grown, and why—all depend on this system.

Selected References

Kohls, R. L., and W. D. Downey. 1972. *Marketing of agricultural products.* 4th ed. Macmillan.
Marketing. 1954. *Yearbook Agr.* USDA.
Sheperd, G. S., and G. A. Futrell. 1969. *Marketing farm products.* 5th ed. Iowa State University Press.
Snodgrass, M. M., and L. T. Wallace. 1964. *Agriculture, economics, and growth.* Appleton-Century-Crofts.

Study Questions

1. How is price affected by a change in demand for a given product?
2. What are the major marketing functions in the process of moving a product from the producer to the consumer?
3. How can hedging be used to reduce risk for a grain producer?
4. What are the important institutions involved in marketing?
5. How can a cooperative assist a producer in the marketing of his produce?

Twelve　　　　　　　　　　Seeds and Seeding

Man's discovery that the seeds of plants can provide sustenance both directly and indirectly (as a source of future plants) gave great impetus to the development of human culture and society by assuring a constant supply of food. As he evolved more effective ways to use seed, man also sought to improve his methods of producing and storing the seeds of select plants so that he could be more confident of growing abundant, high-quality crops.

Today man uses seeds in a variety of ways. He uses them to propagate and multiply many useful plants. He uses them directly as food; the cereal-based diet of millions of human beings consists principally of the edible seeds of rice and wheat. He uses them in the manufacture of many industrial products; the oils and fats produced by such plants as mustard, flax, cotton, soybeans, and peanuts are used in making paint and as lubricants, as well as directly for food. Although seeds have many commercially important uses, the following discussion will confine itself to a consideration of the use of seeds to propagate new plants.

Seeds

All seeds consist of an embryo, a supply of stored food, and a seed coat. In grasses, the food required for seedling growth is stored in the endosperm, which is not part of the embryo; in dicots, food is stored in the cotyledons, which are part of the embryo. The seed coat encloses the stored food together with the embryo. Each part of the seed performs a distinct function: the embryo is a minute plant that will germinate and grow: the endosperm (and the cotyledons) provides energy (food) for germination and sustains the young seedling until it develops green tissue and can manufacture its own food through photosynthesis; and the seed coat protects the embryo and food supply.

GERMINATION OF SEEDS

A seed is essentially a young plant that is in an arrested state of development. It is not the first stage in the life cycle of the plant; rather, each plant begins its life as a single cell, called a zygote, that arises from the union of a male and a female gamete during sexual fertilization. The zygote undergoes a number of mitotic divisions, and, by growth and differentiation, develops into an embryo that is surrounded by a protective coat and has a source of stored food. When the embryo reaches maturity, its growth and development come to a virtual standstill. At this point the young plant (along with some protective tissue from the ovary) enters its "rest period" and is called a seed. There is no further cell division or growth during the rest period. When environmental conditions become favorable for germination, the rest period ends. The resumption of a rapid rate of embryonic development is called germination, which includes biochemical changes that result in embryonic growth. Viable seeds germinate when environmental conditions are favorable and dormancy has been overcome.

Dormancy refers to the period during which a seed does not germinate even though environmental conditions are favorable. Dormancy may occur if the development of the seed is curtailed because of an immature embryo or if an impermeable seed coat prevents the absorption of water and/or oxygen. Chemical changes and hormone balance within the seed affect dormancy. Some seeds have little or no period of dormancy and germinate as soon as they are planted in a favorable environment. Other seeds do not germinate until the specific conditions that remove the physical and biochemical blocks to their germination are established.

A number of conditions must be met before a seed will germinate:

1. The seed (embryo) must be viable or alive.
2. A sufficient supply of water must be available.
3. A favorable temperature must exist.
4. Oxygen must be available.
5. Suitable light conditions must exist for some species.

If any essential condition is lacking, seeds will not germinate.

Germination is entirely a food-utilization process. The stored, insoluble foods in the endosperm or in the cotyledons are changed to soluble foods and are then respired to supply energy for seedling growth. Sufficient food reserves must be available in the seed to meet the energy needs of the new plant through germination; germination is completed when the seedling begins to make and utilize its own food (when the seedling is capable of photosynthesis). The sequence of events that leads to the release of energy for seedling growth includes:

1. Absorption of water
2. Activation of enzymes
3. Hydrolysis of stored foods
4. Translocation of soluble food to growing points
5. Respiration of food and production of ATP
6. Mitosis, development, and cell enlargement
7. Growth and chloroplast development
8. Water and mineral uptake
9. Photosynthesis

Seeds of different species germinate under different temperature conditions. Temperature affects the metabolic rate of the germinating seed and determines the rate of imbibition of water to some extent. Water softens the seed coat, allowing the seed to imbibe water. The embryo swells and the seed coat becomes more permeable to gases. Respiration is stimulated, the protoplasm is activated, and cell division begins.

Soil is not required for germination to occur. However, soil or some other medium containing essential plant nutrients in solution is necessary for a plant to become successfully established and sustain itself.

CHOOSING QUALITY SEED
Producers who rely on growing plants for all or part of their income must keep abreast of the development of new and improved cultivars and new crop species. Plant breeders develop cultivars to thrive in specific sections of the country or on special types of soil, or to produce crop products that meet specific market requirements. A producer must select a cultivar that has been bred to grow in the environmental conditions of his area and to satisfy his production needs. For example, in cold regions alfalfa producers must select cultivars that go dormant as winter approaches; in warm regions this is not desired.

Frequently, farmers find that the lowest-priced seed is of low quality and is therefore not necessarily the most economical, even if it is seed of a suitable cultivar. The main criteria for describing seed quality, assuming that the seed is of an appropriate cultivar, are purity and the percentage of live seed that will germinate. Purity refers to genetic or cultivar purity; it is also a measure of the degree to which seed is free of weed seeds, seeds of other crops, and inert material such as rocks, dirt, or twigs. Purity is expressed on a percentage basis by weight. The germination percentage expresses what proportion of the total number of seeds are alive; it is determined through controlled tests and actually counting the number of seeds that germinate.

In buying seed, a crop producer must be sure that it is seed of an appropriate cultivar and that it is pure and has a high germination percentage. Reliable seed firms always label their seed. Labeling laws vary from state to state, but in all cases the label should include the percentage of purity and germination.

Plumpness or fullness is another desirable seed characteristic; it indicates that the seed will produce vigorous seedlings under favorable conditions. High-quality seed should also be free of disease and insect pests and of various types of mechanical injury that reduce germination or seedling vigor.

SEED CERTIFICATION
The purpose of seed certification is to maintain and make available to the public sources of high-quality, genetically pure seed of superior cultivars. Only those cultivars that are of superior genetic makeup are eligible for certification. Certified seed is high in cultivar (genetic) purity, high in germination, and of good quality (i.e., free of disease and of damaged or immature seeds).

Cultivar purity—which indicates that the seed does not consist of mixtures of various types of cultivars of the same crop—is the first consideration in seed certification, but other factors such as the absence of weeds and diseases, viability, mechanical purity, and grade are also important. One of the most effective ways to limit the distribution of weeds is to plant weed-free seed. Losses from many plant diseases can be reduced by planting disease-free seed. Finally, properly cleaned and graded seed is easier to plant and gives a more uniform stand. Seed certification is designed, therefore, to maintain not only the genetic purity of superior cultivars, but also reasonable standards of seed condition and quality.

Most states have seed growers associations that are recognized as the legal certifying agency for seed grown in that state. These associations, in cooperation with scientists who develop cultivars and seed producers who meet rigorous standards, make available reliable sources of seed. The whole process of developing and increasing a supply of pure, high-quality seed, from the selection of superior plants in the field by the plant breeder (see Chapter 32) to the production of certified seed for planting on the farm, is rigidly controlled in the field, at harvest, and in storage through inspections by skilled representatives of certifying agencies. After seed has been certified, it is up to each grower to keep it pure and free from mixtures through planting, so that he can reap the benefits of planting certified seed in the form of higher yields, higher crop quality, and an overall greater return per dollar invested in producing the crop.

CLEANING SEEDS

In spite of measures taken to insure purity, seed as it comes from the field is almost always contaminated to some extent with a variety of materials. These contaminants include weed seed, seed of other crops (or of different cultivars of the same crop), chaff, dirt and dust, and other inert material. Producers must separate these contaminants from the crop seed to insure a supply of pure seed that has a high germination percentage. There are many external physical or morphological differences in the seeds of different crops that enable growers to separate desirable seed from contaminants. These differences include size, shape, weight, surface area, specific gravity, color, and seed-coat texture. A number of mechanical devices have been devised to take advantage of these properties in cleaning and sorting seeds.

The cleaning process may begin with scalping, which removes material (such as awns) that is coarse enough to be easily separated by screens, or it may begin with hulling (removing adhering floral parts), which may be part of the threshing operation. The next step is to put the seed through what is commonly referred to as a fanning mill. This is an air-screen cleaner that separates the bulk of the foreign material from the seed by screens and moving air (Figure 12-1).

More sophisticated separation can be achieved by taking advantage of physical differences among seeds. Specific-gravity separators utilize differences in seed density to separate weed seeds and foreign material from seeds of desired species (Figure 12-2). Cylinder separators with indented discs separate short seeds from long ones; separation by seed length can be used to sort cultivars of a crop as well as to separate different crop or weed seed from desired seed. The resistance of seeds to air flow varies with their density, shape, and texture; an aspirator sorts seeds according to differences in their resistance to air. Smooth seeds can be separated from rough seeds by using a velvet roll separator. Seeds of different colors can be separated by an instrument that is sensitive to light and dark; this method can be used not only to separate cultivars but also to identify diseased seed (e.g., the presence of ergot sclerotia).

Cleaning seed may appear to be an expensive operation. However, in the long run, the best control of weeds is to prevent them in the first place by planting clean seed. The high cost of controlling weeds after they have become established (see Chapter 9) exceeds the cost of clean seed.

Scarification is not actually part of the cleaning process, but if needed it is done in conjunction with cleaning. Scarification consists of scratching or nicking the seed coat of hard-seeded cultivars or species to hasten water uptake and germination in the planted seed. The seed of several forage legumes (e.g., vetches) is scarified.

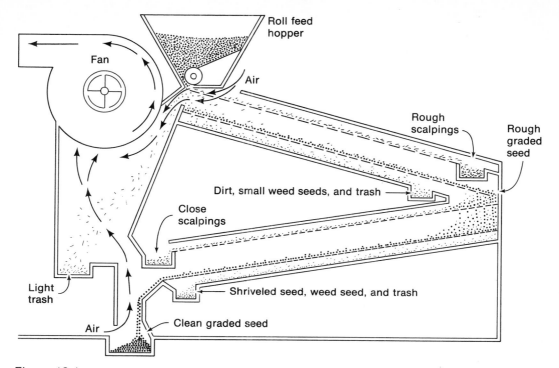

Figure 12-1
A fanning mill, which is a cleaner that is used to separate seeds from foreign material and to separate seeds of different sizes and weights. [From *U.S. Dep. Agr. ARS Handbook* 179.]

STORAGE OF SEED

Seeds are alive and their viability depends in large measure on the storage environment. Temperature and humidity are the two key factors that must be controlled to insure safe seed storage. As a general rule, seeds should be stored in a dry place and extremes in temperature should be avoided. The lower the temperature at which the seeds are stored, the higher the relative humidity permissible for safe storage. However, low temperature and low relative humidity combine to create ideal storage conditions for maintaining seed viability. Controlling the level of oxygen in the storage environment is another means of prolonging seed viability; a low level of oxygen forces the seed to reduce its rate of respiration. Regulating and monitoring the level of oxygen in a storage environment is practical only for

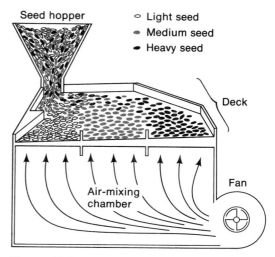

Figure 12-2
A specific-gravity separator used to separate seeds of different densities. [From *U.S. Dept. Agr. ARS Handbook* 179.]

the preservation of seeds of especially scarce and valuable genetic stocks.

Because stored seed is a potential target for both insects and diseases, pest control during storage is essential. Under warm, moist conditions, insects reproduce at a rapid rate in stored seed and can soon reduce the germination percentage in a lot of seed through feeding and egg laying. Fumigants are available to control these insects, but caution must always be exercised in using chemicals. Because seeds are alive they must be protected as insect pests are destroyed. Sealed storage, or storage under inert gases, helps to prevent mold growth and insect activity in stored seeds. However, oxygen must be available for respiration to continue; thus, caution must be observed in storing seeds under these conditions.

In warehouses where seeds are stored, rodents are frequently a serious problem. The best control is to construct storage buildings in such a way as to exclude rodents, and to exercise extreme care to prevent rodents from slipping into the storage area through open doors and cracks in floors or walls. If rodents get into the storage area in spite of efforts to exclude them, they must be exterminated either by traps or poison.

SEED TREATMENT

Plant pathogens are often carried on or in the seed or in the soil. The same temperature and moisture conditions in the soil that favor germination may also allow the pathogens to become active. Many crops are more susceptible to infection in the seedling stage than at any other stage in their life cycle. Therefore, if a plant can be protected from a pathogen during the seedling stage, it has a better chance of remaining healthy throughout its life.

Coating seeds of various species with specific chemicals that prevent or control the organisms that cause smuts and other seed- and soil-borne diseases has long been recommended. Although plant pathogens are not found in every lot of seed or in all soils, most producers view seed treatment (which is relatively inexpensive) as an economical method for reducing losses due to disease. Blanket seed treatment has been effective for many crops. For example, treatment of cereal grains with organic mercury fungicides was so extensive and effective from 1950 to 1970 that covered or stinking smut of wheat was virtually eliminated as a problem. However, in 1971 the Environmental Protection Agency (EPA) rescinded the permit for interstate shipment of most mercury fungicides because such mercury compounds had been found to be highly toxic to humans. This caused a shift in the types of chemicals available for treating seed. Hexachlorobenzene (HCB), pentachloronitrobenzene (PCNB), manet, captan, thiram, and other nonmercury fungicides were developed to replace the organic mercury compounds in the control of smuts and many seedling rots. Each new compound must be approved by EPA.

The chemicals used in seed treatment can be applied in dust, liquid, or slurry form. The liquid or slurry form can be used most easily by commercial seed treaters such as those found at many elevators and seed processing plants. Some growers prefer to apply dust formulations to seed in the drill or seeder box, but most growers find this somewhat inconvenient and inefficient; thus, most seed is treated at the elevator.

The compounds used to treat seed are generally toxic to both man and livestock as well as to the organisms they control. It is therefore necessary to use extreme caution in treating seed and in handling the treated seed. Some of the precautions that must be observed during and after the application of chemicals are:

1. Read and follow carefully the directions regarding application rates.
2. Treat only seed that you intend to plant.

(Although growers have not found that nonmercuric seed treatments cause a deterioration in the viability of stored seed, they have confirmed that organic mercury treatments reduce the germination percentage of stored seed.)

3. Treat seed that has a low moisture content.
4. Do not treat physically damaged seed.
5. Do not expose your skin to or breathe fungicide dust during seed treatment; aspirators are available and should be used.
6. Do not mix treated seed with untreated seed that is to be marketed. (If kernels of treated seed are found in a lot of grain being sold, the whole load of seed may be condemned.)
7. Do not feed treated seed to livestock.

As a general rule it is not necessary to treat forage legume seeds unless their germination percentage is very low. On the other hand, forage grasses and cereal grains should be treated in most cases to assure a good stand of disease-free plants.

INOCULATION

Plants in the legume family produce foodstuffs for both man and animals. In addition, they have the potential to utilize atmospheric nitrogen indirectly through a symbiotic relationship with certain soil bacteria. The bacteria reduce atmospheric nitrogen to ammonia (NH_3) by using sugars and other compounds produced by their legume hosts as a source of energy; they can convert atmospheric nitrogen into a form usable by plants only in the protected environment of the legume's root nodule. Legumes use the nitrogen fixed by the bacteria for their own vital processes. To assure legume crops of the nitrogen-rich environment in which they thrive, producers add pure cultures of specific bacteria to crop seed before planting; this process is known as inoculation.

The symbiotic relationship between bacteria of the genus *Rhizobium* and legume plants was discovered by German scientists in 1888. Two years later, experiments showed that the transfer of soil from a field in which legume plants had previously flourished benefits legume growth. This procedure of soil transfer was the first crude method of inoculation. In 1890, *Rhizobium* were cultured and added to legume seed just before planting. This process of introducing pure cultures of *Rhizobium* to the legume seed is the basis of the present inoculation industry.

Nodulation The *Rhizobium* bacteria used for legume inoculation are tiny (0.5–0.9 μm), single-celled, motile microorganisms that invade the root hairs. Under appropriate conditions, legume plants react to these bacteria by forming nodules (Figure 12-3). The pH of the soil affects the vitality of *Rhizobium*: the lower the pH of the soil, the less desirable is the environment for these beneficial organisms. If the pH falls below 5.5, the bacteria can no longer survive. It may therefore be necessary to raise the pH of acidic soils by liming before a crop can benefit from inoculation.

Rhizobium are specific to the crop species (and possibly to the cultivar) in which they effectively cause formation of the nodules that provide the site for their symbiotic nitrogen fixation. Certain *Rhizobium* cause legumes to produce numerous small, white nodules, but the bacteria fail to fix very much atmospheric nitrogen in these nodules. Effective strains of bacteria cause legumes to produce large, pinkish nodules on their roots; the bacteria fix large amounts of free nitrogen in this type of nodule. If the proper strain of bacteria is present and nodules form, the legume plants use little soil nitrogen. As new roots develop, the nodules formed on older roots slough off, and the nitrogen fixed in these nodules (nitrogen previously in a form unavailable for plant

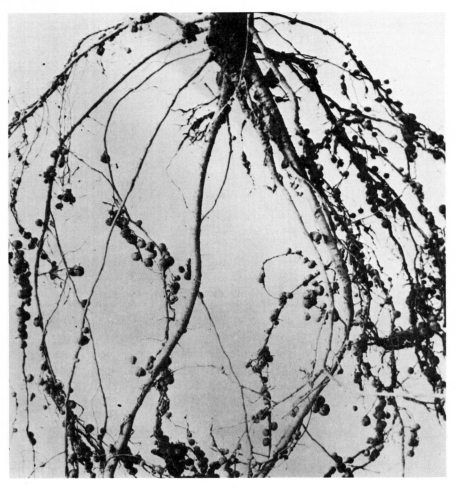

Figure 12-3
Nodulation of roots of a typical legume plant caused by the presence and activity of a specific strain of nitrogen-fixing bacteria of the genus *Rhizobium*. [USDA photograph.]

growth) becomes available to plants as part of the soil's organic matter.

To help crop producers select the proper strain of *Rhizobium* with which to inoculate a stock of legume seeds, inoculation specialists have classified legumes according to cross-inoculation groups (Table 12-1). Specific strains of *Rhizobium* are effective within each cross-inoculation group. To insure rapid establishment of a legume stand in the northern United States and Canada, growers must not only choose the proper strain of *Rhizobium* for a particular legume species, but also select bacteria that are capable of infecting young seedlings at low soil temperatures (12°–15°C).

Seed for large-scale planting is usually inoculated by the seller or the seed processor. However, sometimes a crop producer inocu-

Table 12-1
Cross-inoculation groups for major legumes.

Group	Legumes included in group
Alfalfa	Alfalfa, sweetclover, burclovers, spotted clover, black medic, and sourclover
Clover	Red, crimson, alsike, white, strawberry, subterranean, zigzag, and hop clovers
Pea	Peas, vetch, lentils
Bean	Garden, navy, Great Northern, pinto, and other beans
Soybean	All cultivars of soybeans
Cowpea	Cowpea, lespedeza, peanuts, and some beans
Lupine	Lupine
Birdsfoot trefoil	Birdsfoot trefoil
Big trefoil	Big trefoil
Sainfoin	Sainfoin
Vetch	Milk and crown vetches

lates seed just before planting. He must be sure to use the proper strain for the crop he is seeding and to follow the directions on the label. Also, inoculum must be fresh; the expiration date on the container must be checked.

Methods of inoculation The purpose of legume inoculation is to add fresh cultures of effective strains of *Rhizobium* to the seed. It is not considered practical from an economic standpoint to add bacteria to the soil. Methods of applying the inoculum to the seed are:

1. Slurry, in which a mixture of water and the proper inoculum is mixed with the seed. (Seed should be planted within 48 hours of treatment.)
2. Dry, in which a mixture of the inoculum and peat moss or clay are mixed with the seed until the seed is covered by the inoculum. (Seed should be planted within 48 hours of treatment.)
3. Wet, in which the legume seed is wetted before being mixed thoroughly with the inoculum or a mixture of clay and inoculum. (Seed should be planted within 48 hours of treatment.)
4. Commercially coated, in which a sticking agent and the inoculum are applied to the seed before it is sold. The sticking agent makes the inoculum adhere to the seed and is reported to extend the life of the bacteria so that the treated seed need not be planted immediately. The length of time these bacteria remain viable has not been well established; it is therefore recommended that commercially coated seed be planted as soon after inoculation as possible for best results.
5. Vacuum, in which the seed is coated with a slurry of bacteria and then placed in a vacuum. When the vacuum is released, the bacteria are carried into the seed itself. This process is reported to extend the length of time the bacteria remain viable. However, it is recommended that the seed be planted as soon after inoculation as possible because there is little evidence that the bacteria remain active for an extended period.

Reasons for inoculation Seeds are inoculated to prevent nitrogen deficiency, to increase yields, to improve the quality of a crop, and to enrich the soil.

Prevention of nitrogen deficiency. Each seed contains the nitrogen compounds necessary for its initial growth, but, after the embryo has depleted this supply, the seedling must draw its nitrogen from the soil. If the nitrogen level of the soil is low, the plant may suffer nitrogen deficiency, which reduces its vigor and productivity. If the proper bacterial inoculum has been applied, the seedling will use its reserve of nitrogen first and then it will use the nitrogen resulting from symbiosis. In addition to assuring its own supply of available nitrogen, a well-nodulated, infected plant increases the supply of usable nitrogen in the soil. Alfalfa that has been inoculated may fix as much as 194 pounds of nitrogen per acre (217 kg/ha) annually.

Crop yield. The yield of a legume crop grown on nitrogen-deficient soil increases proportionately more with inoculation than the yield of a crop grown on soil that has a high nitrogen content. On soils of average fertility, increases in yield ranging from 15 to 20 percent have been reported due to inoculation.

Crop quality. Legumes are a rich source of plant protein. Foods such as peas, beans, peanuts, and lentils supply a relatively inexpensive form of protein for the human diet. Forage legumes are one of the most economical sources of protein for livestock. The protein content of legumes is affected by the amount of available nitrogen: the greater the supply of nitrogen, the higher the protein content. A poorly nodulated legume crop not only depletes the soil's supply of nitrogen, but also produces less protein than does a properly inoculated plant.

Soil enrichment. Nodulated legumes make excellent green manure. Their organic matter not only improves the physical properties of the soil, but also returns more nitrogen to it than was taken from it for plant growth. If the soil is to receive the greatest benefit from leguminous green manure, all top growth should be turned under because about 80 percent of the nitrogen is in the leaves and stems; turning under only the legume roots after harvesting a crop does not add nearly as much nitrogen to the soil.

Stand Establishment

To produce maximum yields, a grower must establish good stands. He can obtain good stands by seeding high-quality seed into a well-prepared seedbed at the recommended rates, by seeding at proper depths and at the optimum time, by covering seed properly, by using appropriate weed-control techniques, and by following other agronomic recommendations. Before discussing seeding techniques and equipment, it is necessary to consider some seedbed-preparation practices that result in better stands.

SEEDBED PREPARATION

Good stands start with good seed. However, even the best seed will not produce optimum yields if other factors are not properly managed. A suitable seedbed that supplies the appropriate environment for germination is of the utmost importance in establishing an optimum stand. Although precise requirements vary from crop to crop, the soil of the seedbed must be worked to insure adequate moisture for germination at the appropriate seeding depth. The soil particles surrounding the seed must contact as high a percentage of the outer surface of the seed coat as possible to allow a maximum rate of water absorption, which promotes rapid germination. The structure, texture, and tilth of the soil affect seed-soil contact, as do such characteristics of the seed coat as hairiness and other irregularities. (Tilth refers to the physical condition of the soil in relation to plant growth: granulation, stability of aggregation, moisture content, aeration, rate of water infiltration, drainage, and capillary capacity.)

Before planting, a grower must perform cultivations that insure the proper environment for germination and at the same time preserve soil structure and minimize water loss due to evaporation from the soil surface. The soil should not be too finely crushed or pulverized or crusting may occur. The soil should not be cultivated while wet (i.e., immediately following a rain or irrigation) as this can also destroy soil structure. In addition to creating a suitable environment for germi-

Figure 12-4
Disc plow used in initial steps of seedbed preparation for many crops. [Courtesy of Deere and Company.]

nation, seedbed preparation should be a major part of a continuing weed control effort.

In a cropping sequence, the first operation following harvest is the destruction of crop residue and the incorporation of this residue into the soil. This operation may be accomplished with a variety of implements. The moldboard plow is commonly used. However, disc plows (Figure 12-4) and stubble discs (which serve to incorporate only part of the stubble) are also widely used. In some regions, producers destroy and incorporate crop residue in the fall; in regions that have short growing seasons, the operation is delayed until early spring. Early cultivation is desirable because it allows more time for organic matter to decompose; however, sound soil-conservation practices may require that some stubble be left on the soil to protect it from wind, rain, and the running water produced by melting snow and to maximize the absorption of water into the soil.

Final seedbed preparation is completed in the spring (or late summer for fall- or winter-sown crops). The first step is to cultivate with discs (Figure 12-5) or harrows (or both). The surface soil must be worked finely enough to minimize excessive evaporation and to insure adequate contact of the seed with moist soil. In addition, it must be cultivated deeply enough to uproot or cut off and kill early growing weeds. Weed control at this time is crucial. Weeds that are not killed before seeding have a head start that places the crop seedlings at a disadvantage in competing for moisture, nutrients, and light. After cultivation, fields are sometimes rolled to insure a firm seedbed; this is quite common in prepar-

Figure 12-5
Tandem discs used in final steps of seedbed preparation and for weed control. [Courtesy of Deere and Company.]

ing the seedbed for small-seeded grasses (most forage grasses) and for many forage legumes. For crops that require irrigation or planting on beds, growers must dig furrows or form beds before rolling. To conserve moisture, planting should immediately follow final seedbed preparation.

Putting the soil in proper physical condition and supplying adequate moisture do not by themselves insure a good stand. Seed is alive and requires oxygen. Therefore, flooded or saturated soils inhibit germination. Soil temperature is also critical. The optimum temperature required for germination varies widely for different crops. There are basic differences between cool-season crops (some cereals and many forage grasses) and warm-season crops (corn and rice, for example) in this respect. Finally, some seed requires light for germination so that producers may have to adjust seedbed preparation and seeding methods to insure penetration of light to the seed.

SEEDING
Two critical factors must be controlled during seeding: depth and rate. Planting at a proper and uniform depth into a well-prepared seedbed that has an adequate supply of moisture favors vigorous seedlings and the establishment of a good stand. The proper seeding depth for any crop depends on the type of soil and the size of the seed. As a rule, large-seeded crops are seeded more deeply than small-seeded crops (e.g., corn more deeply than wheat, beans more deeply than clover) because larger seeds have relatively more stored food for the seedling to utilize while growing from a greater depth.

Accurate control of the seeding rate also favors the establishment of a high-quality stand. Both overseeding and underseeding waste seed and reduce potential yield. Underseeding does not allow full utilization of the available moisture, nutrients, and light; overseeding creates an excessive demand on these factors and reduces the yields of all plants. The seeding rate of a crop should be based on the pure, live seed ratio; that is, on the ratio of the weight of viable seeds of the desired cultivar to the total weight of the seed stock, which may include nonviable seeds, weed seeds, and inert matter along with the viable seeds. If 80 percent of the seed is viable and it is 95 percent pure, the pure, live seed ratio is 76 percent ($0.80 \times 0.95 = 0.76$); thus, in 100 pounds (45.4 kg) of this seed, there are 76 pounds (34.5 kg) of pure, live seed of the desired cultivar. If the recommended seeding rate based on 100 percent pure, live seed is 6 pounds per acre (6.7 kg/ha), which is a common rate for some small-seeded range grasses, then the appropriate seeding rate for this crop is slightly less than 8 pounds per acre (9 kg/ha):

Actual seeding rate

$$= \frac{\text{rate based on 100\% pure, live seed}}{\text{pure, live seed}}$$

$$= \frac{6}{0.76}$$

$$= 7.9 \text{ pounds per acre (8.9 kg/ha)}$$

To minimize seeding costs and to obtain good stands, it is important to maintain good seeding machinery in top working condition. A good seeder places seed at a uniform, controlled depth, has a positive control over seeding rate (e.g., pounds per acre, kilograms per hectare, or seeds per foot of row), and does not damage seed. There are three main types of seeding machinery available commercially: grain drills with attachments for seeding legumes and grasses, seeders that are designed exclusively for seeding small-seeded grasses and legumes, and broadcast seeders such as endgate seeders or centrifugal throwers. Frequently, seeders are joined with other implements (e.g., fertilizer-application equipment or implements to cover the seed after planting) so that the machinery can accomplish several jobs in a single operation. Let us examine each of the three types of seeders in detail.

Grain and fertilizer-grain drills are generally designed to seed rows at definite intervals; they can be adjusted to space rows 6, 7, 8, 10, or 14 inches (15, 18, 20, 25, or 36 cm) apart. Grain drills come equipped with a positive seed-metering device (either a fluted wheel or a double run) to control the seeding rate and distinctive seed-furrow openers (such as the single-disc, double-disc, shovel, or hoe type); a selection of devices that cover the seed is usually available as are press wheels that firm the soil around the seed for better seed-soil contact. Grain drills can be equipped with attachments for seeding legumes and other small, smooth-seeded crops; these attachments come with positive-metering devices such as small, fluted wheels. Several manufacturers also supply grass-seeding attachments that are equipped with a free-flow type of metering device under which seed flow continues as long as there is agitation in the box.

Crop producers can obtain grass seeders as separate machines that are equipped with one of three seed-metering devices: (1) the fluted-wheel, (2) wheels with cells and brush cut-outs, or (3) agitated free-flow through an opening in the bottom of the hopper. These grass seeders are also equipped with either double-disc openers, rollers or press wheels for covering and firming the soil, or a combination of both disc openers and rollers.

Broadcast seeders such as endgate or knapsack seeders scatter seed over prepared soil. This machine is one of the earliest developed seeders. Because it only distributes the seed, a second operation must be performed to cover the seed by rolling, harrowing with a spring-tooth harrow, or disc harrowing; the method used depends on the depth of covering desired. Growers should observe one precaution if they use broadcast seeders of the centrifugal-throw type to plant mixed seeds such as legumes and grasses, or oats, legumes, and grasses; oat and legume seeds are heavier than grass seeds and are thus thrown a greater distance, which results in strips bare of grass. To overcome this problem, it is necessary to overlap the seeding by about one half of the lateral seeding span. To avoid overseeding the legumes and oats with such an overlap, the rate of seeding for the oats must be cut in half and the amount of legume seed in the grass-legume mixture must be reduced by one-half. If the normal seeding rate is 50 pounds of oats, 10 pounds of clover, and 10 pounds of grass per acre (56 kg of oats, 11 kg of clover, and 11 kg of grass/ha), then a grower should use 25 pounds of oats, 5 pounds of clover, and 10 pounds of grass per acre (28 kg of oats, 6 kg of clover, and 11 kg of grass/ha) if he is using the overlap method of seeding.

In seeding cereal grains, oats, barley, or wheat, it is not difficult to maintain the desirable depth of covering soil (1.5-2.0 in., or 4-5 cm) with any of the seed furrow openers available for grain drills. It is more difficult to maintain the 0.50- to 0.75-inch (1-2 cm) planting depth desired in seeding forage legumes and grasses. Depth bands can be installed on disc openers to help maintain shallow planting depths. Most grain drills use drag chains to cover the seed once it has been placed at the proper depth; they also frequently use press wheels to firm the covering soil (Figure 12-6). Pressing is particularly beneficial on clay soils; it helps to establish good contact between moist soil and seed. Drills with press wheels have greater control over the depth of seeding than do other drills. Depending on the type of soil, rolling after seeding may be desirable. Depth of seeding with broadcast seeders or endgate seeders is extremely variable. Harrowing after broadcasting covers some seeds at the proper depth, but places others at depths that are either too shallow or too deep. Rolling after harrowing does not appreciably change the depth of the seed. However some types of grass seeders come equipped with special rollers that manipulate the soil to cover the seed. With these seeders, the depth of seeding can be varied according to the manufacturer's instructions.

Calibration of seeders Grain drills with the force-feed types of metering devices can be calibrated for seeding rate according to the following procedure, which is valid for small grains, legumes, and grasses in any of the force-feed metering mechanisms. (The examples below are computed in pounds of seed per acre; they can be converted to kilograms per hectare by using the appropriate conversion factors.)

1. Calculate the number of revolutions of the drill wheel required to plant one acre. On a data sheet record the number of revolutions required to cover 0.25 acre. In practice, calibration is frequently based on only one-half of the drill; in this case, only one-half of the total drill width is used in calculating wheel revolutions.
2. Calculate approximate wheel rpm of the seeder corresponding to tractor speed at which drill is to be used, and record on data sheet.
3. Put seed or diluted seed mixture in seed box.
4. Jack up drill and place mark on wheel rim.
5. Set indicator to give desired rate of seeding (according to manufacturer's calibration).
6. Place small bag under each seed tube, numbering bags from left to right for later identification.
7. Turn wheel the number of revolutions required to cover 0.25 acre, maintain-

Figure 12-6
Three modern tractor-drawn grain drill units, each with double hoppers—one for seed and the other for fertilizer. Grain drills are equipped with press wheels to insure optimum seed-soil contact for uniform germination. Note the weed-free seedbed. [Courtesy of Deere and Company.]

ing wheel rpm at about the value calculated above (within ± 25 percent is satisfactory).

8. Weigh seed from each bag separately and record weights on a data sheet. If the weight of any bag differs greatly from the average, investigate the seeding mechanism and seed tube from which that bag was taken.

9. Obtain sum of weights of bags, which is pounds collected per 0.25 acre. Multiply by 4 to get rate in pounds per acre. (Convert to bushels if computing on a volumetric basis.)

10. Calculate correction factor as follows:

Correction factor =

$$\frac{\text{rate indicated by drill setting}}{\text{actual rate obtained}}$$

11. Then, in order to drill seed at some desired rate, set the drill feed as follows:

Desired rate × correction factor

12. For accuracy, repeat calibration test for the corrected setting.

Constants for calibrating equipment:

1 bushel = approximately 1.25 cubic feet

or 32 quarts dry measure

or 37.5 quarts liquid measure

Drill width = space between furrow openers (ft)

× number of furrow openers

Distance traveled per acre =

$$\frac{43{,}560}{\text{drill width (ft)}}$$

Number of revolutions of wheel per acre =

$$\frac{\text{distance traveled per acre}}{\text{wheel circumference}}$$

Wheel rpm =

$$\frac{\text{mph} \times 5{,}280}{60 \times \text{wheel circumference (ft)}}$$

Free-flow mechanisms can be calibrated only by determining the ratio of field speed to rate of flow of material through the opening in the bottom of the hopper. To calculate this ratio, it is necessary to measure the free-flow for a period during which there is agitation in the hopper. Then, to obtain the seeding rate in pounds per acre for this type of seeding mechanism, multiply the rate of seed flow in pounds per minute by 43,560 (which is the number of square feet per acre). and divide the product by the product of the distance traveled in one minute multiplied by the width of the machine in feet:

Pounds per acre =

$$\frac{\text{pounds per minute} \times 43{,}560}{\text{distance traveled in 1 min} \times \text{machine width (ft)}}$$

This formula can be reduced to seed flow in pounds per minute multiplied by 495, divided by expected field speed in miles per hour times machine width in feet:

Pounds per acre =

$$\frac{\text{pounds per minute} \times 495}{\text{mph} \times \text{machine width (ft)}}$$

Row-crop planters In addition to the three general types of seeders, there are a number of special types of row-crop seeders·that vary in design from simple single-row, hand-operated planters to mechanical planters that seed eight or more rows at a time. They function like grain and grass seeders. Each row-crop seeder places seed at a specific depth in the soil, and has a seed-feeding device, a furrow opener (usually some type of hoe or shovel), and a covering device. Frequently, row planters are also equipped with press wheels that firmly pack the soil over the seed. To accommodate the wider row spacings required by row crops, compared with those required by small grains (e.g., 24–42 in. or 61–107 cm versus 6–14 in. or 15–36 cm), row planters generally have a separate seed hopper for each row. In addition, because row crops require more exact plant spacing (both between and within rows) than do wheat, barley, or forage crops, the seed-metering devices of row-crop seeders are usually more precise than those found in grain drills. There is a metering device for each row; it generally consists of a plate that is held either horizontally or vertically in the hopper. Notches to accommodate seed of a specific size (e.g., corn, beans, cotton, or sugar beets) are cut into the plate at specific intervals. The plate revolves as the seeder is drawn forward. The distance between the notches allows single seeds to be dropped at precise intervals. Many row planters can be used for a variety of crops by changing row spacings and plates.

A relatively new development in row-crop planters is the plateless planter, which is replacing the conventional planter in some areas of the corn belt. One of the big advantages of this type of planter is that it can handle a large variety of seed sizes and shapes and therefore eliminates much of the extremely accurate seed grading and sizing required to operate a plate type of planter. The plateless planter enables farmers to grow their own hybrid corn seed because there is no need for them to be concerned about the size and uniformity of the seed to be planted.

Like other seeders, row seeders can be adapted to carry out several jobs in a single operation; they can even be adjusted to apply both pesticides and fertilizers during the seeding operation.

Fertilization It is usually a good practice to fertilize while seeding. With grass seedings, nitrogen applied at seeding time is especially effective in improving stand establishment and seedling vigor. This starter fertilizer, sometimes referred to as "pop up" fertilizer, gives most effective results in early spring when soil temperatures are low and the natural release of nitrogen in the soil is inhibited. A complete fertilizer, properly placed, is generally recommended. Broadcasting the fertilizer on new seedings is not as effective as banding it about one inch below and to one side of the seed when it is drilled. With banding, the new seedlings make better use of the fertilizer so that less of it goes to produce vigorous weeds. Drills that band fertilizer are available. Seed and fertilizer should not be in direct contact with each other when drilled into the soil because high concentrations of fertilizer materials (especially urea and ammonium nitrate) immediately surrounding the new seedling may be toxic or lethal. The same precaution should be observed when incorporating insecticides or herbicides into the soil at seeding time. The germinating seed and new seedling are especially sensitive to toxic substances in the soil solution.

Starter fertilizer is usually not sufficient to meet the needs of the forage crop for the full season. The main purpose of the starter fertilizer is to get the seedlings off to a quick start to make them more competitive with weeds and to quickly establish a stand of vigorous, rapidly growing plants. The starter fertilizer should be followed with a top dressing of fertilizer that will supply the nutrient needs of the stand throughout the season. Fertilization rates should be determined on the basis of soil-test results.

The development of special planting and fertilizing equipment is another example of the way in which diverse fields of research (e.g., engineering) contribute to increasing crop production. Comparable contributions have been made in harvesting and cleaning equipment.

Harvesting Seed

Seeds of different species vary greatly in their pattern of maturity. Most grain and oil seed crops reach physiological maturity (a developmental stage after which no further increase in dry matter occurs) at about the hard dough stage. At this stage, the seeds are firm but an indentation can be made in them with the thumbnail; the moisture content is usually about 40 percent, depending on the species and on weather conditions. After this point, further ripening consists mainly of desiccation with little transfer of nutrients into the seeds or fruit.

In some cases the seeds on a plant do not ripen uniformly. Late tillers in cereal crops ripen later than the early tillers. Other crops with indeterminate flowering habits, such as birdsfoot trefoil, produce seeds that are in various stages of maturity. Frequently, by the time the late seeds are mature, the seeds that ripened earlier have dropped to the ground.

To determine the optimum dates on which to harvest a seed crop, a producer must evaluate the potential viability of the harvested seed, the shattering characteristics of the crop, and alternative harvesting methods. Harvesting seed prematurely not only reduces the yield, but also has a potentially adverse effect on the quality of the seed: it can diminish the viability of the seed and the vigor of the seedling. Delaying harvest until the seed is mature and dry may result in yield losses due to lodging, shattering, or dehiscence. In determining the harvest date, a producer must also evaluate factors that affect the ease or difficulty with which the seeds can be harvested: the height and size of the plant; the weight, size, and shape of the seed; the uniformity of maturity of the seed head; shattering resistance; and weather conditions. Unfavorable weather at harvest time has, on many occasions, severely reduced the yield and quality of the seed produced.

Engineers and agronomists continue to develop improved harvesting equipment and methods. At present, the principal harvesting machine is the combine, which can be adjusted to harvest cereal grain, oil seed crops, forage grasses, and legumes. Adjustments that make it possible to harvest such a broad variation of seeds include: varying the speed of the air fan; replacing rasp bars with rubber-covered angle bars or other special bars; changing the sieves according to the size, shape, and weight of the seed being harvested; varying the cylinder speed; and inserting a sheet-metal pan under the tailing extension. An experienced operator can change the concave clearance, cylinder speed, amount of air, and forward speed to adjust to moisture conditions and the kind of seed being harvested. Improper adjustment and operation of the combine can result in excessive loss of seed, contamination, physical damage, and diminished viability.

Other lesser-used harvesting machines include the binder, the separator, and the stripper. The binder may be used to cut and bundle a seed crop. The bundles are shocked and allowed to cure, after which they are put through a threshing machine or separator that separates the grain from the straw and chaff. The stripper removes seed from the plant and at the same time allows the plant to continue growing in the field. Pneumatic strippers use air to harvest seeds that are very light in weight and that shatter easily. Stripped seed is not dry enough for storage and must be dried before processing. Bluegrass is the principal crop harvested by stripping.

Seeds of other crops are harvested with specialized equipment, such as corn picker-shellers, or by hand.

When harvesting the seed of any crop, it is important not to damage the seed physically. Damaged seed usually has a lower germination percentage, which impairs stand establishment. Properly harvested seed must be processed and stored with care to maintain the quality of the seed that is to be planted.

Drying Seed

It is necessary to dry moist seed before processing and storing to preserve its viability and diminish its susceptibility to physical damage. The germination of seed that is stored with a high moisture content can be severely reduced. If the moisture content exceeds 20 percent, the respiration of the seeds and of the microorganisms present in and on the seed may produce enough heat to raise the temperature sufficiently to kill the embryo and, in severe cases, to cause fire. Seeds with a high moisture content are also more susceptible to physical damage during processing; such damage reduces viability and encourages the formation of molds. Excess moisture also favors infestations of insect pests. Weevils and other insects that damage seed multiply only in seed that has a moisture content of more than 8 percent. Furthermore, too much moisture causes seeds to stick together so that they do not feed through processing equipment easily.

Because of these difficulties, seeds are dried. They can be dried naturally by allowing air to move around damp seed that has been spread out to dry, or they can be dried artificially by mechanically forcing heated or unheated air through the seed mass to remove excess moisture. Because high temperatures can injure the seed, it is recommended that the air temperature for drying not exceed a range of 90° to 110°F (32°–44°C). If artificial drying is carried out too rapidly, it results in the shrinking and splitting of some seed. Rapid drying can also cause physical and chemical changes that make the seed coat impermeable to water. Conversely, if drying takes place too slowly, molds that can reduce the vigor of seed may become active.

Commercial Production of Seed

The seed industry is one of the most profitable enterprises in agriculture. In the United States and Canada, the commercial production of seed is a multibillion-dollar business.

Vegetable seeds are marketed by several firms, most of which are located in the western United States where they contract with independent farmers to produce most of their seed stocks. The contracts benefit both the seed firm and the grower: they allow the firm to select seed grown in areas best adapted for the production of various species, and they assure the farmer of a market for his produce. Contracting seed firms employ fieldmen who advise farmers on various production practices such as pest control, fertilization, cultural practices, irrigation, time and method of harvest, and other practices that assure good yields of high-quality seed.

The production of flower seed is a little more specialized than the production of vegetable seed. Some firms specialize in supplying growers with the seed of bedding plants; others supply florists and producers of cut flowers. Producers and distributors of flower seeds must build a reputation on consistent production of high-quality seed.

Producing, processing, and distributing field-crop seeds is less well-defined. Some seeds are produced and seeded on the same farm, with any leftover seed being sold to one of the neighbors. Small grain seed is frequently produced and distributed in this way. In contrast, the production of hybrid-corn and sorghum seed is a highly structured and specialized business. The method of producing hybrid seed precludes planting back seed produced on the farm because of genetic considerations (see Chapter 32). Crop producers must either purchase commercial hybrid seed or produce their own hybrid seed each year. As hybrid wheat and other hybrid cereal cultivars are developed, they will probably be produced and marketed similarly to hybrid corn.

There are, in general, two types of producers of small-seeded forage grass and legume seed: casual producers and specialists. For small crop producers who may either harvest a crop for seed or cut it for hay, depending on seed prices, weather, and outside demands on their time during the growing season, seed production is just one of several alternative uses for the crop product. This type of casual seed producer is found, for the most part, in the western half of the United States and Canada. The second type of forage-seed producer specializes in producing seed of one species. For example, there are farmers in California and other western states who specialize in the production of alfalfa seed. They usually produce seed under carefully controlled conditions to assure its eligibility for certification. A specialized producer may be under contract to a wholesaler or to a member of a marketing cooperative, or he may be independent and sell on the open market. Although casual producers have tended to dominate the production of forage grass and legume seed, they are being replaced more and more by specialists.

The seeds of cotton and tobacco are grown and distributed through a highly developed seed production and marketing system. Most of the cotton and tobacco cultivars that are grown in the South have been developed by plant breeders employed by private firms. In contrast, most of the recommended cultivars of cotton that are grown in the Southwest have been developed by publicly supported, agricultural experiment stations.

The production, distribution, and marketing of seed are vitally important to the agricultural industry; they play a crucial role in maintaining a viable economy in the predominantly agricultural regions of the United States and Canada. Large quantities of seed are traded back and forth between these two countries, and Canada has recently joined the United States in buying and selling forage grass seed in the international market. North American seed imports from and exports to Europe and several Asian and South American countries constitute a significant part of the international seed trade.

Selected References

Harmand, J.E., L. M. Klein, and N. R. Brandenburg. 1961. Seed cleaning and handling. *U.S. Dep. Agr. ARS Handbook* 179.

Strobel, G. A., and D. E. Mathre. 1970. *Outlines of plant pathology.* Van Nostrand Reinhold.

Robins, W. W., T. E. Weier, and C. R. Stocking. 1966 *Botany,* 3d ed. Wiley.

Seeds. 1961. *Yearbook Agr.* USDA.

Study Questions

1. Diagram a typical monocot seed and label the following parts: plumule, radicle, scutellum.
2. What conditions must exist before a seed can germinate?
3. How is germination affected by differences in temperature?
4. What factors determine seed quality?
5. How does certified seed differ from non-certified seed?
6. How can seed and seedlings be protected against seed- and soil-borne pathogens?
7. Define and describe the symbiotic relationship between plants in the legume family and bacteria of the *Rhizobium* species?
8. Why are the depth and rate of seeding important to stand establishment?
9. How would you go about setting your seeder to make sure you were seeding at the proper rate?

Thirteen　　　　　　　　　　Cereals and Man

Cereals and Social Evolution

For centuries, cereals have supplied the means for sustaining human life. A cereal is generally defined as a grass grown for its small, edible seed. Wheat, rice, barley, rye, and sorghum are the major cereals. Corn may also be considered a cereal, even though it has comparatively large seed. There is conclusive evidence from archaeological and botanical research that the transition from hunting and gathering food to producing and storing it—and the subsequent emergence of primitive societies—was closely connected to the biological (organic) evolution of cereals and to their cultivation.

Archeologists have found a wealth of evidence in the ruins of ancient settlements that wheat was grown in the earliest sites of civilization in the Fertile Crescent of Mesopotamia. This region, which comprises the valleys and surrounding foothills of the Tigris and Euphrates rivers, now includes parts of Turkey, Iran, Iraq, and Syria (see Figure 1-4). Diggings at Jarmo (in Iraq) indicate that prehistoric man settled there from 10,000 to 16,000 years ago and began to grow his own food. The evidence uncovered at this site virtually proves that the major crop was wheat, or at least a known ancestor of wheat. Barley was also apparently important. Carbonized remains of both crops have been clearly identified. In a span of approximately 3,000 to 4,000 years (from about 8,000 to 4,000 years ago), man was dependent on cereals for sustenance. During this period he developed what are believed to be the first true settlements and primitive societies. Man changed more in the 3,000 to 4,000 years in which the cultivation of cereals evolved than in the preceding 250,000 years of his existence.

Other societies developed independently in other regions of the world. Although the evidence is perhaps not as conclusive, it strongly suggests that wherever societies developed, cereal of one kind or another played a major role. Indications are that rice was the major cereal for early societies in China and sorghum for early societies in Africa. It is possible that the societies in Asia and the Orient developed later than those in the Near East and that they adopted methods

Table 13-1
World production in 1971 of the major cereal crops used for food.

Continent and country	Crop production (thousands of metric tons)				
	Barley	Corn	Oats	Rough rice	Wheat
Africa	4,560	22,177	178	7,592	8,981
Algeria					1,235
Egypt		2,342		2,534	1,729
Ethiopia	1,565				
Kenya		1,400			1,620
Madagascar				1,873	
Malawi		1,100			
Morocco	2,350				2,300
Nigeria		1,220			
South Africa		8,600	97		
South Rhodesia		1,179			
America, North	23,474	155,814	18,341	5,357	60,917
Canada	13,099	2,946	5,606		14,412
Mexico		9,886		469	1,853
United States	10,070	140,728	12,712	3,824	44,620
America, South	1,192	27,685	690	9,760	9,618
Argentina	553	9,930	475		5,440
Brazil		14,307		7,111	2,132
Columbia		915		841	1,368
Asia	31,220	48,848	3,106	280,678	84,809
Afghanistan				2,166	
Burma				8,178	
China				104,000	32,000
India	2,865	6,500		64,000	23,247
Indonesia		2,632		18,663	
Iran	850			1,100	3,700
Japan				14,139	
Khmer Republic				2,732	
Korea, North				2,800	
Korea, South	1,857	1,800		5,556	
Malaysia, West				1,545	
Nepal				2,353	

for cereal production that originated in the Fertile Crescent.

Civilization in the New World had more recent beginnings—5,000 to 9,000 years ago. Archeological evidence indicates that some of the earliest societies developed in Mexico; their major cereal was corn, or a closely related ancestor (Figure 1-5). Elsewhere in Central and South America, societies evolved around the domestication of other species, including beans (both common and limas) and tuber crops. Given the known centers of origin of various cereal species, it is not surprising to find that cultural evolution followed the domestication of different types of plants from one region to another.

The Importance of Cereals Today

The unquestionable significance of cereals to modern man is clearly reflected in the importance of cereals in human diets throughout the world: in the United States, cereal products comprise between 20 and 25 percent of the average diet; in central and western Europe, they comprise as much as 50 percent; and, in much of Asia (where rice is the major cereal) and Africa, they comprise 80 percent or more.

Early man no doubt tried to use many plants for food; cereals best suited his needs merely by chance. The characteristics that made cereals valuable to early man make

Table 13-1 (*Continued*)

Continent and country	Crop production (thousands of metric tons)				
	Barley	Corn	Oats	Rough rice	Wheat
Pakistan				18,000	
Philippines		2,013		5,100	
Sri Lanka				1,398	
Thailand		2,300		13,510	
Turkey	4,170	1,135	455		13,594
Vietnam, North				6,000	
Vietnam, South				6,324	
Europe	52,322	41,189	19,107	1,820	81,280
Austria	1,016				
Bulgaria	1,253	2,518			3,095
Czechoslavakia	1,851				3,878
Denmark	5,458				
France	8,950	8,782	2,539		15,360
Finland	1,054		1,424		
Germany, East	2,286				2,490
Germany, West	5,774		3,037		7,142
Greece					1,905
Hungary		4,732			3,922
Ireland	943				
Italy		4,469		892	10,070
Poland	2,449		3,195		5,456
Romania		7,762			5,585
Spain	4,783	2,058			5,456
Sweden	2,049		1,867		
United Kingdom	8,558		1,360		4,815
Yugoslavia		7,442			5,605
Oceania	3,332	278	1,326	318	8,998
Australia	3,105	212	1,277	300	8,674
USSR	34,500	8,600	14,600	1,420	98,700
Ukraine	5,982	5,921	1,299	1,169	21,977

Source: *Statistical Yearbook, 1973.* (United Nations, 1974).

them of great value today: cereals are a high-energy food that is easily stored and easily transported.

Cereals as a group are the most widely adapted crop species. If only a single food crop can be grown in a given environment, it is usually a cereal. Cereals are both warm-season (e.g., rice and corn) and cool-season crops (e.g., oats and rye). The major one—wheat—is the world's most widely cultivated plant. The breadth of its adaptation and the importance of its various forms can be dramatically illustrated by the fact that somewhere in the world wheat is being harvested every month of the year. A comparison of the amount of wheat harvested with the amounts of other cereals harvested in the world is given in Table 13-1.

Although yields may be low, cereals can be grown under adverse conditions; as conditions improve, yields increase. Thus, with modern technology, cereals respond favorably to intense field management, including an increased use of commercial fertilizers and irrigation.

Because cereals require relatively little labor, large areas of land can be cropped, and the return in terms of food per unit of labor is high. In addition, many of the steps in cereal production have become highly mechanized.

The ease with which food crops can be

stored and still retain their value is as important today as it was for ancient man. Because the moisture content of cereals is relatively low when they are harvested (less then 15 percent if the grain is to be used as food), handling and storing are comparatively easy and the grain stores well. Like any crop, however, cereals must be protected from storage pests such as insects and rodents.

The broad range of adaptation, the efficiency of production, and the ease with which cereals can be stored make cereal crops a dependable source of nutrition. A cereal grain can supply carbohydrates, proteins, fats, minerals, and vitamins if these nutrients are not destroyed by overprocessing. The cereal grain (caryopsis) is a source of food for the embryo that it contains. The food stored for the developing embryo can also be used by man. In spite of their many virtues, cereals are not a perfect food. Although they supply most of the calories in many diets throughout the world, they do not supply the dietary balance required for proper nutrition. Consider the dietary needs of a working man: He should consume between 2,500 and 3,500 calories per day, depending on his size and the type of work he does. However, even though he takes in 3,500 calories, he may be undernourished owing to an improperly balanced diet. Many diets that consist primarily of cereals are too high in carbohydrates and deficient in both vitamins and proteins. The symptoms of vitamin deficiency in man are quite well known, easily identified, and readily remedied by supplementation. The symptoms of protein deficiency are harder to identify. For an adult man to obtain the 65 to 80 grams of protein that he needs per day from cereals would require an excessively high caloric intake; other foods supply protein far more efficiently (Figure 13-1). In addition, most plant protein is nutritionally incomplete; it does not contain all the essential amino acids. Recent research and advances in plant breeding have led to the development of cereal cultivars that have more protein of higher quality. The development of corn that is high in lysine is an excellent example of the improvement of protein quality because most cereal proteins are deficient in lysine, one of the essential amino acids. Edible legumes such as field beans and soybeans are the most complete sources of plant protein (see Chapters 19, 20, and 22).

Many types of plant products have the potential for contributing large quantities of energy (calories) to human nutrition. No single crop can supply all the protein required. However, mixtures of grains and legumes can supply the correct combination of amino acids in a diet. Generally, grains (cereals, including corn) are deficient in isoleucine and lysine, two essential amino acids. Edible legumes are deficient in the essential amino acid tryptophan and in the sulfur-containing amino acids (e.g., cystine). Diets that include appropriate mixtures of cereals and legumes can supply all the essential amino acids: beans and corn or rice (typical of parts of South America) or soybeans and rice (typical of parts of Asia). Essential amino acids can be supplied through animal sources—meat, poultry, fish, milk, and so forth—but animals ultimately depend on plants for essential amino acids.

Common Characteristics of Cereal Plants

CLASSIFICATION

All cereals are members of the grass family (Gramineae) and as such have many important features in common. They are in the class Angiospermeae and subclass Monocotyledonae. Each cereal species and indeed each cultivar of a species has specific traits that will be discussed separately in subsequent chapters.

It is essential for an agriculturalist to un-

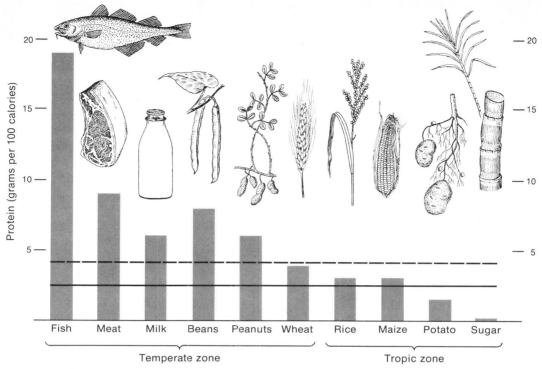

Figure 13-1
Relationship between calories and proteins for several foods. Adults require about 2.5 grams of protein per 100 calories (solid horizontal line) and infants about 4 grams of protein per 100 calories (dashed horizontal line). In both temperate and tropic zones, plant products can meet the needs of adults; however, protein quality must still be improved. [From Kwashiorkor by Hugh C. Trowell. Copyright © 1954 by Scientific American, Inc. All rights reserved.]

derstand the concept of a crop cultivar and the characteristics that distinguish cultivars within each crop species. The crop cultivar (or variety) is not considered to be a formal subdivision in the scientific system of plant classification (review Chapter 5 for this concept), but it is often included in both technical and popular agricultural literature. Cultivars of specific crops are not interrelated on an evolutionary scale in the manner in which orders, families, genera, and species are supposedly related; they are generally the products of the efforts of plant breeders and to this extent may be considered the products of man-directed evolution. Crop cultivars differ genetically by one or more agronomically important traits. For example, wheat cultivars may differ in their resistance to specific diseases or to races of specific diseases. The differences may be simple and clear-cut, such as differences in disease resistance, or they may be complex and not as clear-cut, such as differences in yield or crop quality. Differences among cultivars may affect such factors as the baking quality of wheat flour or the malting quality of barley. Physiological traits that result in greater photosynthetic potential or more efficient water use also separate cultivars. In addition, cultivars frequently differ by traits of apparently minor agronomic importance, such as the presence or absence of awns; these traits, which are useful in identifying cultivars in the field, usually involve inflorescence morphology. Except for differences in response to some foliar diseases and vague differences in plant color, cultivars are difficult to identify during the vegetative stage. Similarly, identifying many cultivars by fruit or seed traits is difficult and if required should be done by an experienced seed-testing technician.

STRUCTURE

The edible seed of a cereal plant is, in fact, a fruit. It is a simple, dry, indehiscent structure called a caryopsis. In popular terminology, it is referred to as a grain, a kernel, or a seed. It is composed of several outer layers and of at least two components, the endosperm and the embryo, of major interest from a nutritional-processing point of view. The entire pericarp, the testa, and usually the aleurone layer—collectively termed bran—are removed in milling. The embryo is also removed and may be included in the bran. Bran that includes the embryo has a higher protein content than the carbohydrate-rich endosperm because the embryo, or germ, contains more protein than any other part of the seed. This should not be surprising because of the concentration of enzymes (proteins) in the embryo, which are essential for germination, growth, and development. Graham flour, made from the entire wheat kernel, is of greater nutritional value than white flour, from which both the bran and the germ are removed during processing (Figure 16-4). Rice is lower in protein than wheat, and polished rice is lower in essential vitamins (e.g., thiamine) than brown rice because milling removes the bran layers that contain the vitamins. The characteristics of the kernel that play an important role in determining the potential uses and market value of all cereals are considered for specific crops in Chapters 14 through 17.

All of the cereal crop species are annuals; they complete their life cycle in a single growing season. The cool-season species (wheat, barley, oats, and rye) are planted in the fall or spring and harvested in mid-to-late summer. The warm-season species (corn, sorghum, and rice) are planted in late spring or early summer and harvested into the fall.

STAGES OF GROWTH

The pattern of growth and development of all cereals is similar (Figure 13-2). Germination starts with the imbibition of water and activation of enzymes, which ultimately allow the energy stored in the endosperm to be used in respiration. The specific conditions required to initiate germination vary from species to species and are discussed separately for each crop. Germination continues with the penetration of the testa and the pericarp by the radicle, and shortly thereafter with the establishment of the primary root system, which develops from the seminal roots. The primary root system is not particularly extensive and does not support the plant throughout most of its life. A secondary root system composed of adventitious roots develops early in the life cycle; this fibrous root system becomes the permanent root system. Unfavorable environmental conditions (particularly drought) during root development cause the development of shallow, restricted roots that inhibit growth and ultimately result in reduced yields. Quality seed, proper seedbed preparation, and environmental conditions that favor rapid, uniform germination and vigorous seedling growth are essential for optimum stand establishment.

Germination is completed when the coleoptile breaks through the surface of the soil and the primary culm emerges and elongates in the typical pattern of intercalary growth. Shortly after the emergence of the primary culm, buds in the axils of the lower leaves are activated and tillering is initiated. The basal internodes do not usually elongate during tillering so that all tillers appear to arise from the base or the crown of the plant. If competition from other crop plants and weeds is held to a minimum at this stage and other conditions are favorable, plants can offset an insuf-

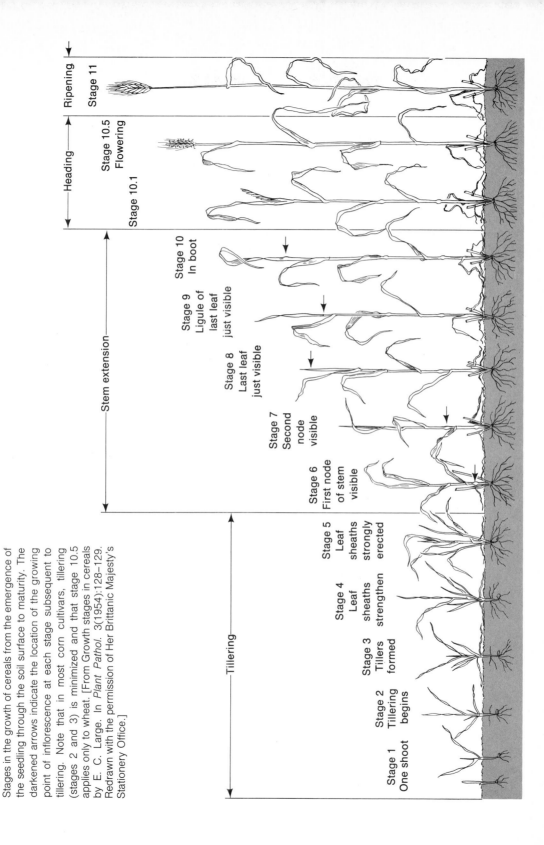

Figure 13-2
Stages in the growth of cereals from the emergence of the seedling through the soil surface to maturity. The darkened arrows indicate the location of the growing point of inflorescence at each stage subsequent to tillering. Note that in most corn cultivars, tillering (stages 2 and 3) is minimized and that stage 10.5 applies only to wheat. [From Growth stages in cereals by E. C. Large. In *Plant Pathol.* 3(1954):128–129. Redrawn with the permission of Her Brittanic Majesty's Stationery Office.]

ficient or uneven seeding density by producing many tillers. For example, given adequate space and favorable environmental conditions, the smaller-seeded cereals (e.g., wheat and barley) tiller prolifically and compensate for thin or uneven stands; conversely, moisture stress, disease, or other unfavorable conditions at the tillering stage reduce the ability of the plants to compensate for sparse stands, and yields may be reduced as a result of poor tillering. Corn is an exception; many hybrid cultivars currently seeded have been developed to produce only a single culm, but some recently developed, high-yielding hybrid cultivars have plants that produce several tillers.

Tillering is essentially completed before the culms elongate markedly, but the leaves do grow in length to produce a basal cluster or rosette of leaves during tillering. Because the growing points that differentiate into floral structures are protected below the surface of the soil, exposed leaf tissue can be moderately damaged (e.g., frozen) or grazed without causing irreparable damage to the crop; however, yields may be reduced. Grazing fall-sown cereals increases tillering and is a common practice in some areas. Not only does this supply fall forage, it may also increase cereal yields as a result of prolific tillering. However, if growing points are injured or destroyed, late tillers may develop, which is undesirable in cereal crops because the grain that the late tillers bear matures late and disrupts harvesting. The high moisture content of later-maturing, wet grain that is harvested and stored with earlier-maturing, drier grain causes storage losses due to mold formation or damage from heating.

A crop makes its greatest demand on soil moisture and minerals during the jointing stage of development. At this time there is a burst of growth and the culms elongate rapidly. Inflorescences are differentiated early in this stage of development. Stress of any kind—moisture deficiency, mineral deficiencies, or disease—can result in stunted plants and less-than-maximum inflorescence development; such developmental impediments cause the crop to produce barren (or poorly filled) spikes or panicles or unevenly filled ears of corn at maturity.

As jointing is completed, the inflorescence is pushed through the sheath ("boot") of the uppermost leaf, called the flag leaf. The boot stage of development immediately precedes the emergence of the inflorescence from the sheath of the flag leaf. In most cereals, flowering occurs between the time the inflorescence is in the boot and the time it emerges from the boot, or shortly thereafter. Elongation of the upper internode (peduncle) and flowering are temperature sensitive; even moderate changes in temperature can disrupt the flowering-pollination-fertilization process. Cool weather at the boot stage delays flowering, and microspores or pollen may be rendered sterile by both low and high temperatures. The specific temperatures that limit flowering and fertilization vary from species to species and even among cultivars of a particular species.

Corn differs from the other cereals in that each flower is imperfect—that is, either male or female, but not both. The plants are monoecious; that is, a single plant bears both male flowers (tassles) and female flowers (ears, including silks). The tassles emerge much as the inflorescences of other grasses do. The silks are the stigmas, one from each female flower (on the ear or cob). The cob is in fact the woody central axis of a spike; thus, the ear is a spike with only female flowers. Each female flower can become a corn kernel. In the floral development of a single plant, pollen tends to mature and be shed slightly before the silks of the same plant are receptive. However, there is a sufficient overlap of maturity between pistils and stamens to insure pollination. Corn and rye are the only cereals that are cross-pollinated; the other species are largely, but not exclusively, self-pollinated.

Following pollination and fertilization, the fruit develops as described in Chapter 4. Although the plant's life cycle is nearing completion, the growth and development of the caryopsis depend on the plant. Leaf tissue, starting at the lower leaves, dies rapidly as the grain matures, and the need for minerals diminishes. The products of photosynthesis are channelled predominately to the developing embryo and to the enlarging endosperm. For plump, heavy-test-weight, high-quality grain, adequate moisture is essential after pollination. If natural precipitation and moisture stored in the soil are inadequate, irrigation may be required late in the season.

Cereals are generally harvested when the endosperm hardens sufficiently and the moisture content of the grain drops to 18 percent or below. As the kernel develops following fertilization, the endosperm goes through a series of stages during which the relative viscosity and moisture content of the endosperm tissue decrease as its stiffness increases. The earliest identifiable stage is the milk stage. If a kernel is crushed between the fingernails at the milk stage, the endosperm flows easily and is the consistency of a watery paste or a heavy, lumpy cream. At the soft-dough stage, the endosperm can be squeezed from a kernel, but it does not flow easily even though it is soft. At the hard-dough stage, a kernel can be dented with a fingernail, but it is difficult to squeeze the endosperm from the kernel. Sometimes grain is cut and windrowed before the endosperm has reached the hard-dough stage and the moisture content has dropped below 18 percent; it is then allowed to dry, before it is threshed. Grain harvested at the soft-dough stage, or even at the milk stage, may be viable if allowed to dry slowly at moderate temperatures; however, the germination percentage of grain harvested with a high moisture content is usually low, as is its milling quality. In addition, overly moist grain may heat and spoil unless artificially dried.

Common Cultural Practices and Problems

The general cultural requirements for barley, oats, rye, and wheat are quite similar from seedbed preparation through harvesting. All require moderately fine, but not pulverized seedbeds that have well-aggregated soil particles so that crusting is minimized. Weed control is an important function of seedbed preparation for these crops. Seedbed preparation should preserve moisture near the soil surface so that seed is placed into moist soil. The seeding depth for cereals is from 1 to 3 inches (2.5–7.6 cm), rarely deeper. Shallower seeding subjects the seeds and seedlings to the danger of inadequate moisture as a result of evaporation from the soil surface.

Although exact seeding rates and row spacings vary from crop to crop and from region to region, the smaller-seeded cereals are all seeded in closely spaced rows that are planted from as little as 6 inches to as much as 12 or 14 inches apart (15.2 cm to 30.5–35.6 cm). Seeding rates may result in from 1 to 10 seedlings per foot (30.5 cm) of row. In the absence of environmental stress, the tillering capacity of individual plants yields a surprisingly uniform number of tillers in a given area in spite of wide differences in seeding rates.

Seeding, except for corn, is nearly always done with a variation of the standard type of grain drill (Figure 12-6). In some areas, however, seed is broadcast either by surface equipment or by aircraft, and then covered by harrowing. Seeding rates must be increased from 10 to 20 percent with this method, and even with increased rates, stands are often irregular and yields are reduced.

Because of the narrow row spacings, cereals are rarely cultivated during the growing season. Weed control is accomplished through proper seedbed preparation and the use of herbicides. Several weeds—for example, pigweed and wild oats—are common pests in all cereals. However, the growth habit of cereals planted in close rows affords them significant competitive ability to crowd out weeds, after the seedling stage. Like many grasses, cereals tolerate a wide array of herbicides. Both pre-emergence and postemergence weed killers are used effectively; more than half of the total acres planted to barley, oats, rye, and wheat are treated with some type of herbicide.

All cereals require adequate moisture and balanced mineral nutrition. However, the precise needs and responses to deficiencies differ for each species (and are described in subsequent chapters). A problem common to all cereals is lodging—a result of excessively high rates of nitrogen fertilizer and heavy seeding rates, which may lead to tall, weak-stemmed plants. The development and use of dwarf-type, stiff-strawed cultivars permits the use of higher rates of nitrogen fertilizer, but lodging and shattering are less likely. The general response of all cereal crops to drought is reduced yields and lighter test weights.

Harvesting methods and equipment are quite similar for all cereals except corn. An increasing percentage of the total acreage of all cereals is harvested with some type of self-propelled combine (or picker). These modern machines reduce labor to a minimum, and at the same time preserve crop quality by minimizing damage to kernels and removing a great deal of foreign matter. (See Chapter 5 for a brief discussion of grain grading.) Because of a tendency to shatter at maturity, cereals may be windrowed with a swather and then threshed with a combine. The use of binders and stationary threshers is virtually a thing of the past.

Because similar cultural methods and equipment can be used in producing barley, oats, and wheat, a single producer may grow all three crops. In determining which cereals to grow and the sequence in which to grow them, a producer must seek to minimize potential contamination of fields and equipment with mixtures of various species (e.g., rye in wheat). Care in cleaning and in storage is also essential to minimize contamination.

Producers of grain sorghum may employ cultural practices quite similar to those described for barley, oats, rye, and wheat, or they may follow the cultural practices used in corn production. In any case, grain sorghum is harvested with a standard grain combine. Corn is cultivated more as a row crop than as a closely drilled cereal. Special equipment is required for all phases of its production, from planting through harvest (see Chapter 14). Rice is treated as an aquatic species and its production requires many special cultural practices (see Chapter 18).

The storage requirements for all cereals are generally uniform. The grain must be dry: 15 percent moisture or less is highly desirable; more than 18 percent moisture is not acceptable for safe storage. If the grain is not sufficiently dry after harvest, it must be dried artificially before storage. Artificial drying is common for both rice and corn. Once dry, the grain must be stored in an area that protects it and keeps it dry. Properly dried grain is not subject to significant damage from moderate temperature variation, but extreme heating should be avoided. Subfreezing temperatures generally are not harmful. Stored grain must also be protected from diseases and from insect and rodent infestations. Prevention is the best form of protection and cleanliness at the bin or elevator is the best form of prevention. To minimize infestations of rodents, storage areas should be constructed to prevent entry through ceilings, walls, floors, and doors. Cleanliness also helps to limit the spread of seed-borne pathogens. Infected and uninfected seed should be stored separately. Storage problems can start in the field (in a dirty combine, a grain truck, or an on-farm storage bin or shed) and continue through transportation to the mill. Constant control of storage conditions is essential to insure the protection and maintenance of crop quality.

Common Quality Factors

The quality of all cereals is affected by the same factors that determine the quality of wheat as graded by the U.S. Department of Agriculture (see Chapter 5). Chief among these factors are test weight, defects, and amount of mixtures. Production practices can be managed so that the highest possible grade is attained. Although without irrigation little can be done to minimize the impact of an unexpected moisture stress on test weight, planting rates can be adjusted for expected amounts of precipitation, to control competition among crop plants, and to foster higher test weight. With all cereals, damage to the grain can be held to a minimum through the proper care and adjustment of equipment and by harvesting only properly matured grain (or drying grain before storage if it is too damp). Foreign material can be eliminated as a serious problem by starting with a clean field and by harvesting properly. Choice of a suitable cultivar and cleanliness in processing and storage can be helpful in reducing losses in the field (and in storage) due to pests and other hazards.

Crop quality involves more than USDA grade designation. A grower may seek to produce a crop of a specific nature or quality (e.g., high-protein wheat for bread flour, or low-protein barley for malting) to meet a particular market demand. Understanding the needs of the crop enhances his chances of attaining the desired quality.

Conclusion

Cereal crops are grown extensively worldwide. Although several cereals can produce low-to-moderate yields under what would be considered poor management or stress conditions for many crops, growers can only achieve maximum yields by producing each crop under specific environmental conditions and rigorous field management. In spite of the fact that cereals do not supply all the protein and vitamins necessary for a balanced diet, cereal crops will continue to be a major source of food for an expanding world population in the foreseeable future. To help prevent malnutrition and starvation, plant scientists must continue to extend their knowledge of how best to grow these crops.

Selected References

Hoff, J. E., and J. Janick, eds. 1973. *Food*. W. H. Freeman and Company.

Janick, J., R. W. Schery, F. W. Woods, and R. W. Ruttan. 1974. *Plant science*, 2d ed. W. H. Freeman and Company.

Wilsie, C. P. 1962. *Crop adaptation and distribution*. W. H. Freeman and Company.

Study Questions

1. Discuss the association between the development of human society and the organic evolution of cereals.
2. What major anatomical and morphological features do cereal crops share?
3. What features make cereal crops of major importance to mankind?
4. Why are cereals not perfect foods?
5. Describe briefly the process of milling grain.

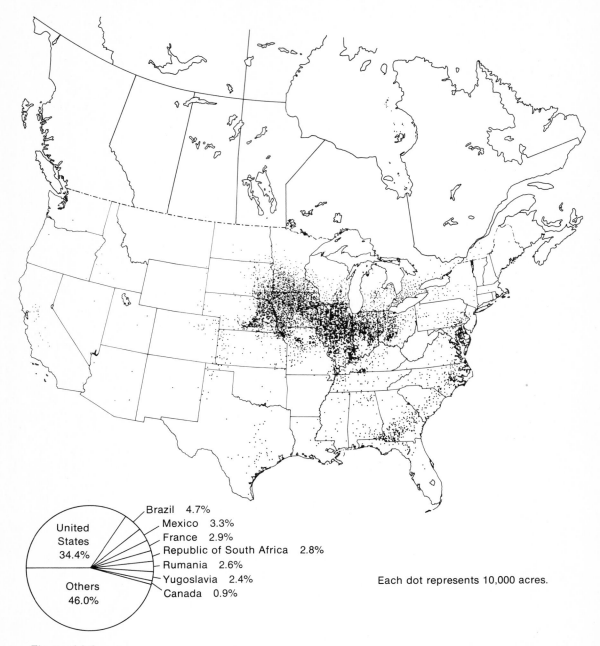

Figure 14-1
Major corn-producing areas in North America and contributions to total world production. [Data from U.S. Departments of Agriculture and Commerce and the United Nations.]

Fourteen Corn and Sorghum

Origin and History of Culture

Corn Corn (*Zea mays*) is the only important cereal thought to have evolved in the New World, reportedly in Mexico. It is also the most highly domesticated of all field crops. Corn does not (and apparently cannot) survive in nature; it survives only under cultivation. Wild ancestors of corn are not found today; yet the evolution of corn has been traced through archeological and paleobotanical studies. These studies indicate that corn was a significant crop in Mexico 5,000 years ago and perhaps earlier. American Indians grew, and selectively improved, corn from 3400 B.C. to 1500 A.D. These early agriculturalists were the first plant breeders working with corn. Columbus reported that corn was grown in Cuba in 1492. Explorations of North and South America in the fifteenth and sixteenth centuries indicated that corn was grown extensively in Chile and as far north as the Great Lakes region. Today, corn is grown in most parts of the world, a result of the efforts of plant breeders to modify its adaptation. Corn is a major food and feed crop worldwide; its annual production closely follows that of rice.

Corn is the leading feed crop in the United States. The history of the development of hybrid corn and of changes in production practices, which have led to nearly unbelievable increases in yield and unparalleled growth in livestock production, is the brightest type of agricultural success story. Since its evolution in the unrecorded past, corn has contributed in a large way to the success of civilization in many areas of the New World. Its importance to human welfare is not apt to diminish in the foreseeable future.

Sorghum Sorghum (*Sorghum bicolor*) is less important than corn in the United States. However, in Africa, where it evolved, various types of sorghum are major food crops. Wild ancestors of sorghum, which resemble the commercial cultivars grown today, are found in Africa. Unlike corn, sorghum can survive in nature. It was introduced into the United States from France in the mid-1850s.

Botanical Characteristics

Corn Corn is the only cereal that does not have a typical grass flower. A corn plant is monoecious: its flowers are both incomplete and imperfect. The staminate (male) flowers (florets) are borne on a panicle at the apex of the vegetative axis of the plant. Spikelets, with a single floret, are borne in pairs, one sessile, the other pedicellate. This floral structure is commonly referred to as the tassel. The "ear" of corn is the pistillate (female) inflorescence and is a spike with a woody rachis, the cob proper. Spikelets with one fertile and one sterile floret are born in pairs along the "cob." This paired arrangement explains why there is generally an even number of rows of kernels on a corn cob. The "silks" of the ear are long styles, one from each fertile floret. Each floret has a single pistil, the ovary of which develops into the kernel or "seed." Culms range in height from 2 to more than 20 feet (0.6–6.0 m) and are grooved on alternating sides of each internode. Most corn cultivars have a single culm with a single ear. Leaf blades are long and wide, and the margins are generally smooth. A pronounced ligule is found at the junction of the leaf blade and the sheath, and the culm is grasped by large auricles.

Many corn plants develop a pronounced brace root system—adventitious roots that arise from nodes that are above the surface of the soil. There are seven types of corn, which can be recognized by their kernel characteristics: (1) dent corn, the prevalent type in the corn belt of the United States, identified by a pronounced dent in the top of each kernel; (2) popcorn, the kernels of which are flowery in the middle and surrounded by a hard (flinty) outer layer, (3) flour corn, which has soft kernels; (4) flint corn, which consists mainly of flinty or hard endosperm; (5) sweet corn, which has translucent kernels that wrinkle at maturity; (6) waxy corn, and (7) pod corn (Figure 14-2). Sweet corn is familiar to the home gardener and of major importance to the frozen and canned food industries. Waxy corns and pod corns are of minimum interest commercially.

Sorghum The inflorescence of a sorghum plant is a panicle. It may be either compact or loosely branched. In either case, two spikelets, each having a single floret, are borne on each branch. The sessile spikelet bears a fertile floret, whereas the floret in the pedicellate spikelet is either sterile or staminate. Sorghum has wide leaves, quite similar to those of corn, but it is readily distinguished from corn by the toothed margins of the leaves. The culms of sorghum, like those of corn, are oval in shape and have a pronounced groove or indentation. Sorghum tillers much more than corn, and culm height, depending on the type and cultivar, ranges from 2 feet to more than

Figure 14-2
Ears of various types of corn, from left to right: pop corn, sweet corn, flour corn, flint corn, dent corn, and pod corn. [USDA photograph.]

12 feet (0.6–3.6 m). A dwarf sorghum plant is shown in Figure 14-3.

The major grain-sorghum groups are kafir, hegari, milo, and feterita. Durra and shallu are also grain-sorghum groups. These groups differ in vegetative characteristics, inflorescence morphology, and kernel qualities. They also differ with respect to their major agronomic use or value. For example, cultivars of the milo group are used nearly exclusively for grain; the stubble (stover) is rarely grazed. The typical inflorescence of a plant in the milo group is awned, quite compact or dense (short and thick), and oval in outline. The kernels are relatively large and either white or yellow. The kafir group is used for both grain and forage. The stems are sweet and juicy, the leaves are large, the inflorescences are awnless, and the kernels vary in color from white to red. In addition to the grain sorghums, there are forage and sugar sorghums classified as *S. bicolor*. Sugar sorghums are also called sorgos or "cane."

In the milo group, many popular cultivars are dwarf-type hybrids. The dwarf characteristic has allowed the development of plant types that are more readily harvested with standard grain combines. The genus *Sorghum*

Figure 14-3
Typical dwarf sorghum plant. Note the compact panicle. [USDA photograph.]

includes several species of agronomic importance other than *S. bicolor*: sudangrass (*S. sudanense*) and johnsongrass (*S. halepense*) are well known. The plants in these species are similar to those described above, but their leaf blades are narrower.

Environmental Requirements

Corn Corn is a warm-season, annual crop. Although cultivars have been developed that can be grown in most of the world, corn is best suited to regions where for three or four consecutive months the average temperature is from 70° to 90°F (21°–32°C). (Refer to Figure 14-1 for U.S. and Canadian distribution.) Planting is generally delayed until the soil temperature is 55°F (13°C) or higher. Temperatures in the mid-80s (about 30°C) are optimum for plant growth during flowering and grain ripening. Little growth occurs if the temperature is below 65°F (18°C), and prolonged exposure to temperatures below 45°F (7°C) may be lethal.

In the United States, corn is produced in regions that have from 20 to 40 inches (51–102 cm) of annual precipitation, or under irrigation to supplement natural rainfall. Moisture in the summer is critical: from 6 to 8 inches (15–20 cm) of rain (or irrigation) is essential in the preflowering period. Corn uses water relatively efficiently. Large quantities are needed for high yields. Drought conditions lead to poor seed set and light-test-weight grain, thus to generally lower yields and lower quality grain.

Because of plant breeding, corn can be grown in all fifty states in the United States and in southern Canada; however, the major area of production is the U.S. corn belt. Iowa, Nebraska, Indiana, and Illinois lie in the center of this region, which extends south to eastern Texas and east to the Atlantic coastal states, excluding New England. Precipitation in the corn belt is adequate (yet irrigation is good insurance), temperatures are high, and the frost-free period is long enough for excellent growth. The crop accumulates enough degree days so that it is mature enough to withstand frost several weeks before the first frost usually occurs.

Corn is best adapted to well-drained, loamy soils. It can be grown successfully in soils whose pH ranges from 5.5 (rather acidic) to 8.0 (moderately basic). Corn requires large amounts of essential minerals; thus, soil fertility must be high to obtain high yields. Generally, fertilizers must be used to insure that soil fertility is high enough. Under ideal climatic conditions in which moisture and temperature are optimum, applications of 100 to 150 pounds of nitrogen (actual N) and of more than 100 pounds of phosphorus (P, not P_2O_5) per acre (112–168 kg of N and 112 kg of P/ha) are not uncommon to obtain yields of more than 200 bushels of grain per acre (12,000 kg/ha). Large amounts of fertilizer are used with high seeding rates. Fertilizer balance is essential; the addition of a single mineral may not greatly change the yield, but combinations have a dramatic effect, as indicated in Table 14-1.

Corn requires the greatest amount of nitrogen from about two weeks before tasseling (before pollen is formed and shed) until three weeks after tasseling (in the early stages of caryopsis development and growth). Of course, adequate nitrogen is essential throughout the life cycle, but to insure proper anther development, pollen formation, and pollination (which are necessary for high yields), mineral balance, including adequate nitrogen, in the late vegetative stage of growth is critical. Corn can use as much as 4 pounds of nitrogen

Table 14-1
Yield of corn in response to different combinations of fertilizers averaged over eight years in Illinois.

Fertilizer (lb/acre)			Yield (8-yr avg in bu/acre)
N	P₂O₅	K₂O	
0	0	0	80
80	0	0	84
80	80	0	83
80	80	80	100
0	80	0	80
0	0	80	93

Source: From B. Koehler. Corn stalk rots in Illinois. *Illinois, Univ., Agr. Exp. Sta., Bull.* 658 (1960):59.

per week per acre in the five-week period before and after flowering. Although most fertilizer is applied before or at the time of seeding, if nitrogen-related deficiencies are noticed, a side dressing of nitrogen fertilizer applied before the initiation of floral development (or before the jointing stage of growth when culms elongate rapidly) will increase yields. Overfertilization causes a number of problems: late maturity; tall, weak-stemmed plants and lodging; and excessive, inefficient water use (to support the excessive vegetative growth).

Phosphorus deficiency delays maturity—specifically, silking. Without adequate phosphorus, pollen may be shed before the silks are mature and receptive. This results in poor seed set on the ear, unfilled florets at the tip of the ear, and misshapen ears. Potassium deficiency is expressed in a similar manner.

Informed crop management is critical when high seeding rates and heavy fertilization are combined to achieve high yields. Crop management includes all phases of production: seedbed preparation, fertilization, irrigation, cultivar selection, weed and pest control, harvesting, and storage.

Sorghum The environmental conditions required for optimum yields of grain sorghum are quite similar to those described for corn. Sorghum cannot tolerate low temperatures. If it is exposed to temperatures below 49°F, prussic acid accumulates in its vegetation and may reach levels that are toxic to livestock. This is a serious problem in forage sorghums and in sudangrass. It may also be a problem in grazing-sorghum stubble. (This problem is discussed in greater detail in Chapter 29.) Like corn, grain sorghum requires large amounts of essential minerals for satisfactory yields. However, sorghums survive and produce satisfactorily under a wider range of conditions than does corn. Water requirements for sorghum, which are less than those for corn, are close to those for wheat (see Chapter 16). In addition, sorghum plants are quite drought-resistant; drought conditions cause their leaves to curl up and form a closed tube that shields the stomata and reduces transpiration. Of course, because grain sorghum is a warm-season crop, either summer precipitation or irrigation is essential. Unlike corn, grain sorghum can be produced on soils that have a pH ranging from 5.0 to 8.5. Sorghums can tolerate high temperatures throughout their life cycle better than any other cereal; they can also tolerate flood conditions better than any other cereal, except rice.

Although grain sorghum can be produced under a wide range of conditions, the actual areas of production are fairly well defined (Figure 14-4). The major region of sorghum production is a north-south belt in the southern Great Plains that extends from eastern New Mexico east into Arkansas and Missouri and from southern Texas north into South Dakota. This geographic concentration of

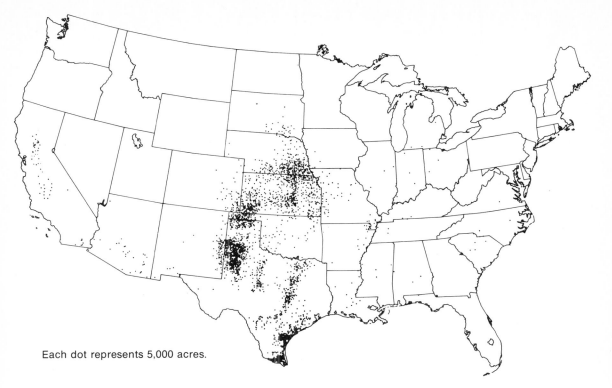

Each dot represents 5,000 acres.

Figure 14-4
Major sorghum-producing areas in the United States [Data from U.S. Department of Commerce.]

sorghum production is a direct reflection of the crop's drought resistance. There are also islands of production in central Arizona, in the corner where Arizona meets California, and in the Central Valley of California. In these latter regions, where irrigation is essential, sorghum is frequently grown as a second crop in a double cropping system.

Production Practices

SEEDBED PREPARATION AND SEEDING

Corn Seedbed preparation for seeding corn usually starts in the fall. Fall plowing is especially important when corn follows corn in a rotation sequence because the heavy stubble from the previous crop must be worked into the soil before seeding. Fall plowing is also important in controlling late-growing annual weeds and many perennial weeds. However, care must be exercised to protect the soil from wind and water erosion during the winter and spring. In the spring, the field is cultivated and tilled into a fine seedbed. Cultivation equipment usually includes discs and spike-tooth harrows. A roller may be used to compact the seedbed and eliminate air pockets. This procedure helps to insure contact between the seed and moist soil, which is necessary for germination.

With the development of new, highly effective herbicides, minimum-tillage methods of corn production have become increasingly popular. The minimum-tillage method applies more to cultivation for weed control than to preparation of the seedbed, but, under this method, even cultivation for seedbed preparation is held to a minimum. Such methods have been practiced in the South and the Southeast where corn follows a sod crop, such as orchardgrass, or a grass-legume mixture. To prepare the seedbed in these areas, a narrow slit is cut through the sod and the soil underneath is cultivated for seeding. Any remaining surface vegetation is killed by preplanting herbicides, such as atrozine or paraquat; these herbicides are also used to control weeds throughout the growing season. Although the covering vegetation is killed, a dense mat of plant debris, which helps to conserve moisture and provide nutrients, remains to protect the soil from wind and water erosion. Regardless of the method of cultivation, adequate moisture and essential minerals must be available for normal plant growth. With minimum-tillage techniques, losses due to erosion are greatly reduced and crop yields increase because of the increased availability of water and minerals (phosphorus and, in some cases, calcium). Minimum tillage may also increase soil pH. However, the "chemical cultivation" associated with minimum tillage can cause moisture to evaporate from the surface soil; in heavy, clay soils that crack when dry, moisture is lost through the cracks. The water losses caused by minimum tillage are not a serious enough reason to abandon the method. In addition to the economic advantages of increased crop yields, minimum tillage results in economic savings by eliminating tractor operations.

Crops that are rotated with corn benefit from minimum-tillage farming methods: stand establishment and seedling growth of forage crops—orchardgrass and red clover, for example—improve when these crops follow corn produced under minimum-tillage management.

Seeding depth for corn is deeper than that for the smaller-seeded cereals; average depths range from 2 to 3 inches (5.0-7.5 cm), but depths of slightly more than 4 inches (10 cm) are acceptable. In the United States, most corn is planted with a planter designed to place the seed with precision. Fertilizer attachments often accompany the corn planter or modified, row-crop seeder (Figure 14-5). Sometimes, seed is planted in wheel tracks to insure a firm seedbed. More commonly, planters are equipped with press wheels that assure seed contact with the soil. Rows are planted from 20 to 44 inches (51-112 cm) apart, with a range of 30 to 36 inches (76-91 cm) being the most common. Narrower row spacings are used for extremely high seeding rates, and very wide spacings are used for light seeding rates in areas where yield potential is limited. In addition to the commonly used planters, both hill planters and check-row planters are used. Check-row planting allows cultivation in two directions (along and across rows) for weed control. Since the development of effective herbicides and the advent of heavier seeding rates, this planting method has been rarely used.

Corn seeding rates, unlike those of most other crops, are expressed in terms of plant density, or number of plants per acre (Table 14-2). Today optimum rates are approximately 20,000 plants per acre (50,000 plants/ha), but, in areas of extremely high yield potential, with intense management seeding rates may be as high as 30,000 to 35,000 plants per acre (75,000-87,500 plants/ha); the seeding rate of sweet corn can reach 60,000 to 70,000 plants per acre (150,000-175,000 plants/ha). These higher seeding rates are obtained either by leaving less space between rows of plants or

Figure 14-5
Modern eight-row corn planter with fertilizer attachments for precise placement of seed and fertilizer. [Courtesy of Deere and Company.]

by planting more plants in a row; yields increase more with narrower spacing between rows than with more plants per foot of row (Table 14-3). However, as plant density increases, competition among plants for moisture and essential minerals increases, and more fertilizer and water are required. In addition, shading becomes a serious problem because it may result in tall, weak-stemmed plants, lodging, and reduced yields. These undesirable effects can be minimized by planting cultivars that tolerate heavier seeding rates; that is, cultivars that do not become etiolated with shade. Such plants may have thinner leaves, partly because of a light epidermal layer, and a thinner cutin layer than plants that are not shade tolerant. The leaves of many shade-tolerant plants are arranged so that self-shading is minimized. For example, in cereals that tolerate shade (including corn), leaf blades are more erect or parallel to the culm, rather than horizontal. The plants of shade-tolerant corn cultivars are shorter and have narrower leaves than those of cultivars that do not tolerate shade. Leaves of shade-tolerant plants may have a lower light saturation point (i.e., require less light energy for maximum photosynthesis) than those of nontolerant plants, and the higher planting rates permitted with such plants may allow a

Table 14-2
Number of corn seeds planted per acre in relation to the distance between rows and the distance between seeds in a row.

Inches between seeds in a row	Number of seeds per acre by inches between rows							
	28	30	32	34	36	38	40	42
6	37,340	34,850	32,670	30,750	29,040	27,540	26,130	24,890
7	32,000	29,870	28,000	26,340	24,890	23,630	22,410	21,330
8	28,000	26,140	24,500	23,060	21,780	20,640	19,600	18,670
9	24,890	23,230	21,780	20,500	19,360	18,340	17,424	16,590
10	22,400	20,910	19,600	18,450	17,420	16,510	15,680	14,930
12	18,670	17,420	16,330	15,360	14,520	13,750	13,070	12,450
14	16,000	14,930	14,000	13,180	12,450	11,790	11,200	10,670
16	14,000	13,010	12,250	11,530	10,890	10,317	9,800	9,330
18	12,450	11,620	10,890	10,250	9,680	9,170	8,710	8,300
20	11,200	10,450	9,800	9,220	8,710	8,250	7,840	7,470

Source: E. C. Rossman and R. L. Cook. Soil preparation and planting. In *Advances in corn production,* ed. W. H. Pierre, S. A. Aldrich, and W. P. Martin. © 1966 by Iowa State University Press, Ames, Iowa.

Table 14-3
Variation in bushels-per-acre yield of corn for crops planted with narrow or wide spacing between rows, at four different planting rates, in East Lansing, Michigan.

Planting rate (plants/acre)	Spacing between rows	
	28 inches	36–40 inches
9,200	68	66
13,800	81	76
18,400	88	83
23,000	83	80
Average	80	76

Source: E. C. Rossman and R. L. Cook. Soil preparation and planting. In *Advances in corn production,* ed. W. H. Pierre, S. A. Aldrich, and W. P. Martin. © 1966 by Iowa State University Press, Ames, Iowa.

higher LAI (more leaf surface area per unit of ground area). Shade-tolerant plants may produce less total vegetative growth than their nontolerant counterparts; therefore, they may require less water and essential minerals for comparable yields, and their water-use efficiency may be relatively higher. If a suitable cultivar is selected and intense field management is practiced, yields of up to 400 bushels per acre (25,000 kg/ha) are theoretically possible.

Planting dates vary with location and the cultivar to be planted. Soil temperature is critical in determining planting dates. Yields may be reduced if crops are planted too early or too late (Table 14-4). In Texas, planting starts in early February. For most major corn-producing regions in North America, the optimum planting dates range from late April to mid-May. As a rule of thumb, planting dates are delayed from south to north, following Hopkin's Bioclimatic Law. It takes corn from 90 to 150 days to mature, depending on region, temperatures, and the number of degree days required by specific cultivars. In regions that have relatively short growing seasons (or if corn is planted late), shorter-season cultivars are grown. However, with chemical weed control and improved seed quality, earlier seeding may be practical even in areas that have a short, cool growing season.

Sorghum Seedbed preparation for grain sorghum is frequently started in early spring,

Table 14-4
Yield of corn (bu/acre) for crops planted at five different dates in the fourteen-year span from 1949 to 1963, at East Lansing, Michigan.

Year	April 16–29	May 1–9	May 12–20	May 22–31	June 1–11	Mean
1963	108	112	103	91	83	99.4
1962	94	87	58	51	48	67.6
1961	141	138	128	124	111	128.4
1960	90	97	86	78	79	86.0
1959	129	143	135	127	125	131.8
1958	108	112	116	102	87	105.0
1957	109	106	93	81	65	90.8
1956	—	96	91	77	55	79.7
1955	102	112	105	100	82	100.2
1954	86	92	81	80	70	81.8
1953	—	108	103	100	68	94.7
1952	69	89	80	83	49	74.0
1950	—	78	—	70	44	64.0
1949	—	85	—	75	89	83.0
10-yr mean	104	109	99	92	80	96.8
12-yr mean	—	108	98	91	77	
14-yr mean	—	104	—	89	75	

Source: E. C. Rossman and R. L. Cook. Soil preparation and planting. In *Advances in corn production,* ed. W. H. Pierre, S. A. Aldrich, and W. P. Martin. © 1966 by Iowa State University Press, Ames, Iowa.

Note: The best mean yields in a fourteen-year period were obtained from crops planted in early May. In 1962, 1956, and 1952, yields were poor, whereas, in 1961 and 1959, they were good.

rather than in the fall. If the sorghum is grown as a second crop following barley or other early-maturing crops (usually cereals), the residue of the previous crop is plowed into the soil as the first step in seedbed preparation. Fields may be disc plowed or disc harrowed. If the sorghum crop is to be planted on beds or ridges for irrigation, these beds or ridges are frequently shaped with a lister as the crop is planted. Because of differences in the size and weight of seed among cultivars of grain sorghum, planting rates are measured in plants per acre, as they are for corn, with the average rate being approximately 75,000 plants per acre (187,500 plants/ha). This seeding rate is between 8 and 15 pounds per acre (9–17 kg/ha). Depending on moisture availability (and to some extent on equipment), rows of sorghum are planted from 20 to 40 inches (51–102 cm) apart; this spacing directly affects yield (Figure 14-6). Optimum seeding depths for sorghum, which are comparable to those for wheat and barley, range from 1 to 2 inches (2.5–5.0 cm). Planting dates for grain sorghum are generally seven to fourteen days later than for corn grown in the same area. Thus, in Texas, grain sorghum is planted in mid- to late-February; it may be planted well into June in northern regions or in regions where it is grown as a second crop.

HARVESTING, STORAGE, AND ROTATIONS

Corn Corn is harvested for grain, silage, and stover; the grain is used for both food and feed, whereas the stover and silage are used

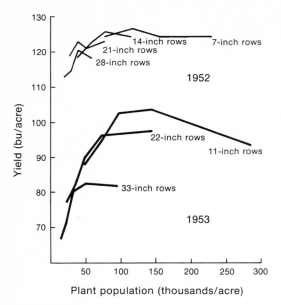

Figure 14-6
Response of irrigated grain sorghum to different row spacings and planting rates in Kansas (solid line represents 1952; dashed line, 1953). [From D. W. Grimes, and J. T. Musick. Effect of plant spacing, fertility, and irrigation managements on grain sorghum production. *Agron. J.* 52 (1960):647–650.]

method of harvesting is not common in the United States or Canada, or in other countries where agricultural production has been mechanized. Because lodging of the crop seriously reduces the efficiency of mechanical pickers, production practices that limit lodging—such as proper seeding rates, fertilization, and cultivar selection—are essential for the efficient use of modern harvesting equipment. Corn that is harvested for silage is harvested when the grain is in the glaze stage (when the endosperm is hardening); it is then chopped and stored in a silo. The process of silage production is described in detail in Chapter 31. For stover, the stalks and leaves of the corn plants are harvested, and then cut and fed through a shredder.

In the U.S. corn belt, a common pattern of crop rotation is corn followed by corn. Other rotations include corn-corn-oats (or another cereal), and corn-corn-oats followed by one to five years of meadow. Soybeans are commonly included in corn rotations. One reason to rotate from corn is disease and insect control. Certain pests persist in corn stubble and can best be controlled by not planting corn for several years.

only for feed. More than 80 percent of the corn grown in the United States is harvested as grain. The majority of this is harvested with a mechanical picker-sheller (Figure 14-8) that removes the ear from the plant, removes the husk from each ear, and shells the grain (i.e., removes the kernels from the cob). The grain is shelled at 27 percent moisture, or less. It is dried to 13 or 14 percent moisture before storage. The artificial drying of corn is becoming common practice. Sometimes picking and husking are done in one operation and shelling in a second operation. In countries that do not use mechanical harvesting equipment, corn is cut, bound, and shocked before the ears are picked, husked, and shelled; this

Sorghum The vast majority of grain sorghum is harvested with a standard grain combine (Figure 14-9). The development of shorter (dwarf) hybrids has increased the efficiency of combine harvesting. Although sorghum kernels will thresh free from the inflorescence when their moisture content is as high as 30 percent, harvesting in southern areas is usually delayed until the grain moisture is less than 18 percent, unless the grain is to be artificially dried. In areas that have a short growing season (such as the north and central Great Plains), harvesting is started after crop growth has been stopped by a frost.

In the Great Plains area, grain moisture content must be less than 13 percent for safe

Figure 14-7
Corn field about six weeks before harvest.

storage; in the south, because of high relative humidities, grain moisture content must be 11 percent or less, and drying is sometimes required before a crop can be safely stored.

Although crop rotation systems for grain sorghum vary from region to region, they are patterned after those for corn. A major consideration in determining which pattern to follow is the temporary nitrogen-depleting effect of sorghum stubble. (This is only a minor consideration for corn rotation.) The stalks of sorghum contain large amounts of sugar and, when the stubble is plowed into the soil, the bacteria involved in decaying the residue use soil nitrogen for their metabolism. Although the bacteria tie up the nitrogen only temporarily, the soil may be deficient in nitrogen for a brief period following a sorghum harvest. This deficiency can be overcome by applying nitrogen fertilizers or by allowing complete decay of the stubble so that the nitrogen tied up by the bacteria becomes available in the soil for the succeeding crop. Under a system of double cropping (with irrigation), grain sorghum follows a fall-sown, early-summer-harvested cereal such as wheat, oats, or barley. After the sorghum crop is harvested in September or October, the stubble is worked into the soil, and the field is frequently left fallow until the following spring when it is planted with a high-cash-value row crop, such as beans, sugar beets, tomatoes, or safflower. In nonirrigated, single-crop regions, grain sorghum is rotated either with itself or with other cereals (e.g., barley). In areas that have adequate moisture, a typical corn rotation pattern may be followed with grain sorghum replacing corn. In drier (semiarid) areas, a crop-fallow rotation system is often practiced.

Figure 14-8
Modern corn picker-sheller unloading shelled grain for transport to storage area. [Courtesy of International Harvester, Inc.]

Diseases and Pests

DISEASES

Corn Although corn is susceptible to a number of diseases, damage to this leading crop by disease is limited. Research resulting in an extensive knowledge of corn genetics has led to the development and use of disease-resistant corn cultivars. Until 1970, corn smut was the worst corn disease. Corn smut is a fungal disease in which the ear is replaced with a mass of spores called a gall. The fungus, which overwinters in the soil, infects seedlings. Yields were reduced as much as 30 percent in epidemic corn smut years. The disease is controlled by rotating crops and by planting cultivars that are resistant to the fungus. Southern corn leaf blight (Figure 14-10), also a fungal disease, caused near catastrophic losses of corn in 1970. The disease, whose development went unnoticed for several years, suddenly reached epidemic proportions as it moved from south to north. The epidemic was due largely to the fact that many of the

Figure 14-9
Modern grain combine harvesting grain sorghum near Ralls, Texas. The stubble remaining after harvesting protects the soil from wind erosion. [USDA photograph.]

cultivars of corn were developed using one source of cytoplasmic male sterility, the Texas source. Disease susceptibility is associated with the Texas cytoplasmic sterility-fertility restorer system (see Chapter 32); thus, cultivars developed using this system are uniformly susceptible to it. The genetic resistance to the disease that does exist is not associated with any cytoplasmic male sterility-fertility restorer system; genotypes that are resistant to the disease but do not have the desired male sterility-restorer system must be detasseled by hand (an expensive procedure) to produce commercial hybrid seed. In response to the 1970 southern corn leaf blight epidemic, plant breeders and plant pathologists developed specific disease-resistant hybrid cultivars that had to be detassled by hand. Today, with the immediate threat of the disease eliminated, plant breeders are developing cytoplasmic male-sterile lines, which include fertility-restorer systems that are resistant to corn blight. These new cultivars will once again eliminate hand detasseling in the production of commercial hybrid seed.

Corn is also susceptible to several rot dis-

Figure 14-10
Symptoms of southern corn leaf blight on husks and ears of corn. [USDA photograph.]

eases that infect either the roots or the stalks and ears. Seed treatment with new, approved fungicides affords reasonable control of these rots.

Sorghum Kernel smut is one of the most serious diseases affecting grain sorghum, but it is a minor concern with forage and sugar sorghums. Although treating the seeds with fungicides affords moderate control of smut, the best control is the use of disease-resistant cultivars. Such resistance has been developed in the milo, feterita, and hagari groups of sorghum. Other diseases are various fungal root and stalk rots and bacterial leaf-spot diseases. These diseases can be controlled by using disease-resistant cultivars. Head smut, which affects inflorescences, can become a problem, but serious epidemics are not common.

WEEDS

Corn Because corn is planted late in the spring and because corn crops have wider row spacing than most other cereals, cultivation is one method of weed control. Early annual and perennial weeds are controlled during seedbed preparation. Weeds that persist during the growing season are controlled by cultivating down the rows. Appropriate implements, such as duck-foot-type shovels, are commonly used for row-crop cultivation. In addition to weed control by cultivation, chemical control is also important. Preemergence herbicides, such as Atrazine, are used to eliminate weeds that grow earlier than, or along with, corn seedlings; 2,4-D and related compounds are employed to control later, broadleaf weeds. Herbicides play a major role in weed control in corn production. More than 75 percent of the acreage of corn (in 1968, nearly 49 million acres, or 19.8 million hectares) is treated with some type of herbicide.

About half of the treated acreage receives only preemergence herbicides. Considerably more than a third, but less than half, of the treated acreage receives a postemergence herbicide, and slightly less than one-fifth receives both pre- and postemergence treatment. The major weed pests include pigweeds, crabgrasses, lambsquarters, Canada thistle, morningglories, panicums, kochias, and switchgrass. Weed control practices have reduced the acres infested with shattercane and sandburs.

Sorghum Weed pests that compete with and weaken grain sorghum plants are generally the same as those discussed for corn. One addition is field bindweed, which has become a serious sorghum pest since 1965. Herbicides are being used more and more to control weeds that grow in fields of grain sorghum. More than two-fifths of the acreage harvested is treated with either a preemergence herbicide, a postemergence herbicide, or a combination of both preemergence and postemergence sprays. Combinations prove to be most effective.

INSECTS

Corn Of all of the cereal crops, corn is the most severely damaged by a wide array of insect pests. Because corn is consumed directly as both feed and food, chemical control of insects must be approached with great caution to avoid contaminating crops with insecticides or their toxic residues.

The European corn borer, a major pest, destroys as much as $35 million worth of corn crop per year. The larvae of the corn borer feed in the plants and destroy the stalks, flowers, and ears (Figure 14-11). The exact life cycle of the insect varies from location to location, but the full-grown larvae of all corn

Figure 14-11
Corn borers feeding on ears of flint corn. [USDA photograph.]

borers overwinter in stubble. In areas that have a long growing season, such as Texas, as many as four broods, or generations, may be produced per year. Further north (e.g., in Iowa), only two broods are produced per year. Control of the corn borer consists of three main components: first, proper cultivation is important. Because the larvae overwinter in stubble and weeds, cultivation that buries the larvae before they pupate (anytime between April and July, depending on location) can prevent the insects from maturing and reproducing. Second, cultivar selection is important. Nearly all hybrid cultivars are resistant to the first brood of larvae so that, if succeeding broods of insects are eliminated by cultivation, plants will not be damaged. Third, in cases of severe infestation, chemical control is sometimes used, particularly with sweet corn. Because of residue problems, chlorinated hydrocarbon insecticides are not reccommended, but some organic phosphates and the newer carbamate insecticides are allowable. Before applying any insecticide, it is essential to check with local authorities to determine what types and amounts of chemicals can be used. All practices that favor vigorous plants, including proper fertilization and irrigation, also help to control pests. Finally, weather may also be a controlling factor: cool summer weather may reduce the number of larvae that pupate and become reproducing adults.

In the southwestern United States, the corn earworm (also known as the cotton bollworm) is a serious pest. Damage caused by insect larvae feeding on developing ears may reduce annual yields by 1 to 2 percent, producing a potential loss of nearly $1 million. Plants infested by the corn earworm are more susceptible to infection by several parasitic fungi. Cultivar selection is a major component in controlling the corn earworm; corn cultivars with heavy husks tend to resist invasions of this pest. The high cash value of corn used for canning and of sweet corn sometimes justifies chemical control of the corn earworm.

Various types of cutworms and moth larvae also cause significant damage to corn crops by cutting seedlings at the soil surface. Effective control is accomplished by banding insecticide with the seed, in the same way that fertilizer is applied with the seed. Armyworms also cause damage to all types of corn; they are serious pests in the late-planted corn grown in northern regions of the United States. Armyworms can be controlled effectively by various insecticides, but they must be used with caution to avoid toxic residues.

The chinch bug is another pest that causes serious damage to corn crops, particularly in the southeastern United States (Ohio and Missouri). The adult insects migrate to corn fields either by crawling or flying from adjacent barley fields and feed on the aboveground parts of the corn plants, particularly the leaves. Chinch bugs are not killed by cold weather; they overwinter as adults. Chinch bug infestation is favored by above-average temperatures and light rainfall from early March through October.

Crawling insects can be controlled by encircling uninfested fields with a furrow filled with coal tar creosote, which serves to trap the pests. Other barriers that are effective include a band of dinitro compounds and chlorinated hydrocarbon insecticides. Sprays of these compounds will partly control the flying, adult insects. Early planting is an effective means of preventing damage from chinch bugs, as is cultivation that destroys or buries overwintering adults.

Other insects that infest corn include grasshoppers and rootworms. Corn, like other cereals, is also subject to losses from insect infestation during storage. The control of insects, such as granary weevils, that damage corn in storage, is discussed in Chapter 12.

Sorghum Only two insects cause major losses in grain sorghum: the chinch bug and the southern midge. Chinch bug damage (as well as its control) is similar to that described for corn. The larvae of the southern midge damage sorghum crops by feeding in florets and preventing grain formation. This pest can be controlled through early plantings so that flowering is completed before the larvae are ready to feed; if infestations become severe, they must be controlled with insecticide sprays. The quality of grain sorghum can be impaired and yields reduced by severe aphid infestations. Field control of the pests by aerial spraying may be practical.

Various birds—crows and blackbirds among others—also cause serious losses by feeding on the crop in the field. The use of noisemakers, as well as other techniques, has proved to be only partly effective in bird control.

Crop Utilitzation

Corn Corn is the leading cash crop in the United States; its value to producers approaches $7 billion annually. About 85 percent of the approximately 60 million acres (23 million ha) planted each year is harvested for grain. The remaining acres are harvested for silage, or grazed, or used in other forms for forage. As much as 80 percent of the entire U.S. crop is used for livestock feed: hogs are the primary consumers (40 percent), followed by other livestock (beef cattle and sheep), and then poultry. In the United States, sweet corn is produced for the fresh vegetable market or processed (frozen and canned) for food, but relatively little land is used for this crop (less than a million acres). Although in North America corn is considered mainly a feed crop, on a worldwide scale it is a major food crop. Only about 10 percent of the world crop is processed; nevertheless, many important products are produced from processed corn. For example, wet milling of the grain yields starches, syrups, media for drugs and feed production (from steep waters), protein (gluten), oil for feed supplements, and corn oil, which is a valuable cooking oil used in the preparation of many foods.

Sorghum As a livestock feed, the grain from grain sorghum is equal to corn in total nutritional value. Because the sorghum is slightly higher in protein (2 percent) than corn and a little lower in fat (1 percent), livestock gain more efficiently on corn than on grain sorghum; hogs and cattle require more sorghum than corn per pound of gain. Sorghum is also an important forage. The roles of various types of sorghums in forage production are discussed in Chapter 30. Outside the United States, grain sorghum is an important world food crop, most notably in Africa. Recently, plant breeders have identified genetic factors for high lysine content in grain sorghum. Because lysine is one of the eight amino acids that cannot be synthesized by humans and because it is necessary for human protein utilization, plant breeders are working to develop high-lysine cultivars of grain sorghum that can be grown as relatively high protein food crops. The amount of grain sorghum processed for industrial purposes has increased in the past several years. The processed grain yields starches that are used in adhesives, cloth and paper sizing, and wallboard. Sorghum grain is also fermented to produce alcohol. In this respect, grain sorghum is as important as corn and wheat.

Sorghums other than the grain and forage varieties are utilized in diverse ways: the sorgos or sweet sorghums yield sugar and syrups; broomcorn is a true sorghum with an inflorescence that can be made into the common kitchen broom.

Conclusion

Corn and sorghum are grown in North America mainly as feed crops. The fact that large areas of prime agricultural lands can be used to grow feed rather than food crops reflects the high level of meat-protein consumption and the well-being of agriculture in the United States and Canada. Only countries that have a wealth of natural resources and efficient, mechanized crop and livestock production can afford the biological, energy-conversion inefficiencies of growing high-quality plants as feed for animals that are consumed as food.

Selected References

Leonard, W. H., and J. H. Martin. 1963. *Cereal crops.* Macmillan.

Pierre, W. H., S. A. Aldrich, and W. P. Martin, eds. 1966. *Advances in corn production.* Iowa State University Press.

Sprague, G. F., ed. 1955. *Corn and corn improvement.* Agronomy 5. Academic Press.

Study Questions

1. Describe the major morphological features of a corn plant and then contrast the morphology of this annual grass plant with that of grasses such as wheat, oats, grain sorghum, and rice.
2. What environmental factors determine where most corn is grown in the United States?
3. If heavier seeding rates are used, is it best to reduce spacings between rows or to increase the number of seeds per foot of row in order to achieve maximum yields? Explain and justify your answer.
4. What are the major benefits of minimum-tillage methods of crop production?
5. What factors would you consider in planning a crop rotation sequence with corn as the major crop? Give practical examples.
6. Under what environmental conditions might grain sorghum be grown in preference to any other crop? Give examples.

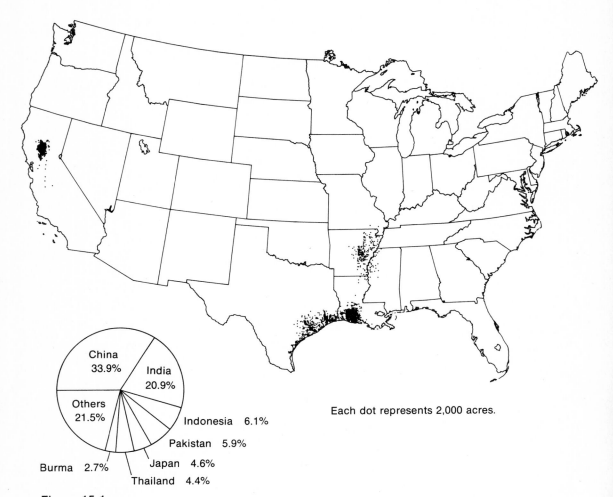

Figure 15-1
Major rice-producing areas in the United States and contributions to world production. [Data from U.S. Department of Commerce and the United Nations.]

Fifteen   Rice

Origin and History of Culture

Rice is the most important food crop and the major dietary component for more than a third of the world's population. Rice production is concentrated in Asia where more than 90 percent of the world's supply is produced. Rice is also grown in certain areas of the United States (California, Texas, Louisiana, Arkansas, and Mississippi), where it is a major crop, and in the Mediterranean countries, in Latin America, and in Africa.

The rice cultivated today, *Oryza sativa*, reportedly originated in Asia from crosses between wild ancestors. It became an important crop in China about 3000 B.C. and spread to Europe by 1000 A.D. Although the species evolved at about the same time as cultivated wheat (*T. aestivum*), there is not as much evidence linking the development of human society to the cultivation of rice as there is showing the significance of wheat in this respect. There are three plausible explanations for this: either societies did not in fact develop around the cultivation of rice as they apparently did around the cultivation of wheat; or the evidence of such cultivation has been less completely preserved because of the climate in areas where rice is grown; or research has yet to unearth evidence revealing ties between the early cultivation of rice and the evolution of society.

Rice was first grown in the United States in South Carolina about 1685. Shortly thereafter its cultivation spread to the delta lands of North Carolina and Georgia. Rice production was confined to these areas until after the Civil War. Then in the 1880s farmers began to grow rice along the Mississippi River in Louisiana and on the Gulf coast in both Louisiana and Texas. Although California is currently a leading rice-producing state, rice was not produced in the Sacramento Valley until 1912.

Botanical Characteristics

Rice is the only major crop species that is aquatic; rice plants can grow successfully in standing water. However, a rice crop does not have to live in water. The water requirement of rice, as measured by the transpiration ratio, is similar to that of other major crops. Rice plants can transport oxygen or oxidized compounds from the leaves to the roots and into the rhizosphere. (Normally oxygen is taken in through the roots and transported to the leaves.) The oxygen in the leaves of rice plants comes from two sources: atmospheric oxygen absorbed by the leaves and oxygen released in photosynthesis through the photolysis of water. Rice has a shallow root system and is therefore very susceptible to water stress; it usually yields better on flooded soils.

Like oats, rice has a loose, freely branched, panicle-type inflorescence. Each branch of the panicle bears several spikelets containing a single floret. Unlike other cereals, each rice floret has six (rather than three) stamens and two, long styles. Glumes are relatively small. When rice is threshed, the lemma and palea generally remain fused to the grain as hulls. Different types of rice are recognized by the length of the rice grain: short-grain rice has kernels about 5.5 mm long; medium-grain rice has kernels about 6.6 mm long; long-grain rice has kernels about 7 or 8 mm long. In the past, short-grain rice was referred to as the japonica type and long-grain rice as the indica type. As a result of extensive intercrossing between these types, the terms are nearly obsolete. Short-grain rice is grown primarily in California and other subtropical northern areas. It has shorter straw and is less prone to lodging under conditions of high soil fertility than other types of rice. Long-grain rice is grown in tropical regions and is better adapted to less fertile soil conditions and less moisture. Breeders have developed new cultivars that incorporate the desirable traits of both short- and long-grain types.

Vegetatively, rice is similar to oats in that it has a long ligule and no auricles. Stems are generally hollow and the number of nodes varies: late-maturing cultivars have more nodes and are therefore usually taller than early-maturing cultivars. Depending on the cultivar and the growing conditions, rice plants tiller prolifically and vary in height from 2 to 6 feet (60 to 180 cm).

Environmental Requirements

Rice is the leading cereal crop in subtropical and tropical regions. It is a warm-season, short-day crop; however, cultivars differ in their sensitivity to photoperiod. Successful rice production requires a growing season of 110 to 180 days with an average temperature of 70°F (21°C) and warmer temperatures during the middle of the growing season. In the southern United States, rice cultivars are divided into four groups according to the minimum length of their growing seasons: very early types require from 110 to 115 days to mature; early types from 116 to 130 days; midseason types from 131 to 155 days; and late types more than 156 days to mature. Rice germinates best at 70°F (21°C); minimum temperatures for germination and seedling growth are from 52° to 54°F (11°–12°C). Root development may be retarded if soil temperatures rise above the mid-80s during seedling growth. Fertility (reproduction) is reduced if the temperature drops below 75° to 77°F (24°–25°C) during the flowering period. Temperatures above 100°F do not apparently adversely affect the growth and development of rice plants; however, the optimum temperature range for maximum yields is in the 90s. Minimum temperatures for vegetative growth, flowering, and kernel development are in the low 70s. Because rice can be grown in standing water during nearly its entire life cycle, it is produced in warm regions that have either 40 to 60 inches (100–150 cm) of annual rainfall or plentiful and relatively inexpensive irrigation water. Irrigation requirements, depending on soil and climatic conditions, can be as high as 90 or more acre-inches per acre per year. In Southeast Asia, the average rainfall required for rice production is close to 150 centimeters per crop. In the United States the major factors limiting rice production are temperature, growing season, and the availability of water for irrigation.

Because rice requires a constant and plentiful supply of water, it is frequently grown on heavy clay soils that have an impervious, subsoil layer (hard pan) that limits drainage. Rice tolerates acidic conditions and is productive on soils with a pH ranging from 4.5 to 7.5. Soils on which rice is grown are often flooded, which causes a series of physical and chemical changes that result in a completely new set of soil-plant relationships. In most soils, these changes are beneficial to rice. Flooding increases soil pH (from slightly acidic to neutral or slightly basic) by causing reactions that reduce iron oxides and increase the concentration of carbon dioxide in the soil. In addition to increasing soil pH, flooding leaches salts from the soil; thus, rice production can be used to reclaim saline soils. Of course, irrigation water must be free of excessive salts to realize the benefits of leaching; low-quality water (e.g., water containing excessive amounts of sodium or boron) can cause a toxic buildup of materials and a reduction of yields. The results of leaching, even if high-quality water is used, are not all beneficial: minerals, including essential micronutrients, may be leached from the soil. Thus, the types of fertilizers and rates of application must compensate for leaching losses, which may vary with the soil type.

Production Practices

SEEDBED PREPARATION AND SEEDING

In the United States, rice is grown in standing water. Fields are leveled into paddies or basins that follow the contours of the land (Figure 15-2). Semipermanent levees from 4 to 6 feet wide (1.2–1.8 m) and 3 feet (1 m) high are made with a modified lister or border disc harrow. The seedbed itself is prepared in a manner similar to that for any other small-seeded cereal, Rarely are fields chiseled because chiseling can break up the hard pan below the rooting zone, which would allow excessive water to be lost from the paddies.

The paddies are flooded in April and early May and are planted from early April in the South to mid- or late May in California when the water temperature in the paddy is above 60°F (15°C). If rice is seeded into water, the water temperature should be above 55°F (10°–15°C). In parts of the Sacramento Valley in California, irrigation water comes from deep storage lakes or reservoirs and is unusually cold; therefore at times it is necessary to flood the paddies and then allow the shallow water to warm before seeding. Seeding into flooded paddies is done by air (Figure 15-3). Because rice seeds cannot germinate in flooded soil because of a lack of oxygen, rice that is planted by aircraft is soaked in water before seeding. Soaking causes the initiation of germination. The soaked seed is ready for planting when the tip of the radicle pushes through the seed (caryopsis) hull. It is removed from the water and air dried immediately before planting so that it spreads uniformly as it drops from the aircraft. Seed that has been

Figure 15-2
Typical rice paddies. Water is maintained at a uniform level by allowing it to flow between paddies. Note that levees are becoming weed infested; rice is seeded by air and so it also grows on the levees. [USDA SCS photograph.]

soaked sinks in the paddy and is rapidly anchored in place as roots grow and develop. Because rice is an aquatic species, the seedlings survive with less free oxygen than the seedlings of other crops; the oxygen necessary to support seedling respiration is released through enzymatic activity within the plant. Rice seedlings can therefore survive total submersion longer than other seedlings; however, prolonged submersion (of more than two weeks) can weaken or kill rice seedlings. (Seedlings of other cereals tolerate submersion for less than a week.)

In parts of the rice-producing areas along the Gulf coast and in California, rice is seeded into dry soil with a standard grain drill. Seeding depths vary from 1 to 3 inches (2.5–7.6 cm).

Paddies are flooded immediately after seeding.

Seeding rates for rice vary from about 120 to 150 pounds per acre (134–168 kg/ha). The heavier rates are used with aerial broadcast seeding. With lighter seeding rates (or wider row spacings), yields decrease but the protein content of the grain increases. With heavier seeding rates and the use of more nitrogen fertilizer, the total amount of leaf area per given unit of ground area—known as the leaf area index, or LAI—increases; yields generally increase as the LAI increases. As is the case with other cereals, heavy seeding rates and large quantities of nitrogen fertilizer can cause serious lodging that results in reduced yields. Growers can resolve this problem by planting newly developed short-strawed and

Figure 15-3
Aerial seeding of rice on flooded paddy. Note pattern of seeds hitting water beneath the aircraft. Flagman, lower left, indicates the width of each strip planted so that a uniform stand is obtained. [Courtesy of University of California Agr. Ext. Serv.]

achieve maximum yields with the most economic use of fertilizers. The type and amount of fertilizer to be applied should be determined by a soil test. However, flooding changes soil chemistry so that soil tests alone may be inadequate to develop specific fertilizer recommendations; field experiments and experience may also be necessary. Because of leaching, most rice crops require from 90 to 150 pounds of nitrogen per acre (100–170 kg/ha). In some areas, adequate phosphorus may not be available, and, in some, micronutrients may be inadequate; for example, in Texas and California, rice is grown on soils that are deficient in zinc.

The time at which nitrogen should be applied is determined by the developmental stage of the plant. Growers should apply about 50 percent of the required nitrogen either at seeding or when tillering begins, and the other 50 percent when the plants initiate the development of panicles after the internodes of the primary tiller or culm have elongated sufficiently. The precise internode length depends on the cultivar (Table 15-1). Nitrogen applied at the panicle-initiation stage of development increases grain yield but does not contribute to excessive vegetative growth; thus, it fosters the development of shorter straw, which reduces the potential for lodging. The second fertilizer application (and sometimes the first) is applied on standing water either by airplane, as is done in the United States, or by manual broadcasting.

high-yielding cultivars that do not lodge with heavy applications of nitrogen fertilizer.

Although paddies usually remain flooded throughout most of the growing season, tests are being made to determine the advisability and practicality of draining paddies periodically during the growing season.

Applying the proper type and amount of fertilizer at the appropriate time is critical to

HARVESTING, STORAGE, AND ROTATIONS

Toward the end of the growing season (usually in September) as rice grain is maturing, paddies are drained and the soil dries. The rice is harvested with a modified grain combine on which the wheels have either been modified or replaced with short tracks (Figure 15-4). At harvest, the moisture content of the grain must be 20 percent or less. After harvest,

Table 15-1
Correlation between time of midseason nitrogen application and internode length for Vergold rice grown near Stuttgart, Arkansas.

Year	Time of nitrogen application (days after emergence)				
	35	40	45	50	55
	Grain yield (kg/ha)				
1963	3,485	4,467	4,518	4,993	3,651
1964	4,960	4,675	4,684	5,015	4,753
1965	6,039	6,145	6,828	6,699	6,242
Average	4,828	5,096	5,343	5,569	4,882
	Median internode length (mm)				
1963	1.8	3.6	7.3	9.0	31.5
1964	2.0	2.9	2.6	4.4	20.5
1965	1.3	4.1	25.0	43.5	68.0
Average	1.7	3.5	11.6	19.0	40.1

Source: V. L. Hall, J. L. Sims, and T. H. Johnston. Timing of nitrogen fertilization on rice: II. Culm elongation as a guide to optimum timing of applications near midseason. *Agron. J.* 60(1968):450–453.
Note: Although yields differed from year to year, applying nitrogen from 45 to 50 days after the emergence of rice plants generally resulted in the highest yields; at this time, internodes are moderately elongated and panicles are initiated.

before the grain can be effectively processed or safely stored, it must be artificially dried until its moisture content drops to less than 15 percent. Rice as it comes from the field, or rice that has had less than 50 percent of the hulls (palea and lemma) removed, is classified as paddy or rough rice. It is milled and processed further for human consumption.

The aquatic nature of rice and the rather specific soil conditions required for rice production affect all aspects of the crop's culture, including rotation. Where climatic conditions are suitable, two rice crops may be grown on the same field in a year, but this is rare in the United States. In many areas, rice follows rice for two or more years in succession. This practice is justified by the high cost of surveying and leveling fields and of building levees on an annual basis; the paddies essential for rice production are not needed for alternative crops and may interfere with them. After two or more years (particularly in the United States), levees are destroyed and alternative crops are grown for one or more years. One reason for destroying the levees is weed control. Although herbicides are effective, satisfactory control requires periodic cultivation. Soil conservation also demands crop rotation.

By and large, crops rotated with rice are warm-season crops that are adapted to heavier, poorly drained soils. The choice of which crop to rotate with rice depends as much on available equipment as on other factors. Rice is rotated with soybeans in Arkansas. After the soybeans are harvested, fall-sown oats are interseeded with lespedeza. The legume serves as a green-manure crop. Along the Gulf coast, land on which rice is grown is rotated to pasture and then back to rice. In California, safflower, which thrives on heavier soils (see Chapter 27), is commonly rotated with rice. Vetch is also included in rice rotations as a green-manure crop. When rice follows vetch, requirements for nitrogen fertilizer

Figure 15-4
Combine adapted for harvesting rice. Note that the crop is badly lodged. The special vehicle in the background is used to haul rice from the combine to trucks, which cannot drive directly to the combine because the fields are wet at harvest. [Courtesy of University of California Agr. Ext. Serv.]

may be reduced from 130 pounds to 90 pounds of nitrogen per acre (from 145 to 100 kg/ha). Furthermore, the organic matter improves the structure and texture of the soil.

One pattern of rice rotation that has proved economically profitable is rice-fish-fallow (or another crop) and back to rice! Rice paddies are highly suitable habitats for some sport fish, and rice growers in some areas have become commercial fish-pond operators; in Arkansas, for example, they raise catfish and they "harvest" them by draining the paddy. Rice stubble also provides exceptional cover and a suitable habitat for water fowl and certain upland game birds.

Diseases and Pests

DISEASES

Rice is susceptible to several fungal diseases. Worldwide, the major fungus that attacks rice is *Pyricularia oryzae,* causing the disease known as blast. Breeding for resistance is the only way to control this disease. This is a constant task because new, virulent races of the fungus evolve continually.

Seedling blights, caused by several different fungi, and root rots are also serious rice diseases, particularly under cool conditions. These diseases can be controlled by treating seed with the appropriate fungicides and following cultural practices that encourage seedling vigor; such practices include planting when soil is at the optimum germination temperature, maintaining adequate fertility, seeding at a shallow depth when temperatures are low, and flooding paddies immediately after seeding if the crop has not been sown on flooded paddies.

In the southern United States, brown leaf spot is a serious fungal disease. Both seedlings and mature plants are susceptible to this disease. Fungicides can control brown leaf spot in seedlings; resistant cultivars are the only solution to the problem in mature plants. In the United States, plant breeders have not yet developed any commercial cultivars that are completely resistant to brown leaf spot.

Other fungal diseases of rice include narrow brown leaf spot, stem rot, and gibberella blight. These diseases can all be fairly well controlled through the treatment of seed and, in the Orient, through the employment of cultural practices to control water levels in paddies and allow the soil to dry before epidemics are apparent.

In the United States, viral diseases of rice are of minor importance. However, the virus that causes "noja blanca" nearly destroyed the rice crop in the southern United States in the 1950s. The rice industry in this region was saved by the development of new, disease-resistant cultivars. In the Orient, yellow dwarf and stripe diseases cause significant losses.

WEEDS

Many weeds that compete successfully with other cereals pose only a minor threat to rice crops because they cannot survive submersion. However, these same weeds frequently become pests along levees and irrigation ditch banks. In addition, several weeds tolerate flooding and are therefore major pests in rice production. The most frequently reported troublesome weed in the United States is barnyardgrass which tolerates flooded conditions reasonably well. Other weed pests include bulrushes, ducksalad, red rice, hemp sesbania, signalgrasses, and spangletops. Worldwide, sedges are serious weedy pests. None of the major weed pests of rice has diminished in importance in the past ten years; in fact, bulrushes, signalgrasses, and spangletops have spread.

Weed control starts with seedbed preparation. Where rice follows rice in a rotation sequence, weed control on levees and ditches is critical because the levees are not plowed out annually. More than 80 percent of the area in the United States that is used for growing rice is treated with some type of herbicide. Pre-emergence treatment alone is relatively uncommon. The vast majority of the treated fields receive a postemergence herbicide. Only in California are pre- and postemergence sprays used extensively. Rice crops in Texas have recently suffered from severe weed problems, and available herbicides have not provided adequate weed control.

Because rice is grown in standing water, the dangers of environmental pollution from any chemical applied to the crop (or field) are magnified. When the water is drained from the paddy, chemicals move with it. Thus, extreme caution must be exercised in the use of any pesticide.

INSECTS

In California, insects are of minor importance in rice production, although the "tadpole shrimp," which is not an insect, can cause serious losses. In the southern United States, the rice stinkbug causes appreciable damage to crops every year. Other insects that attack rice include stem borers, sugar cane beetles, and leafhoppers. These insects can all be controlled with appropriate chemicals; however, producers must exercise extreme caution to avoid draining insecticide-contaminated water from paddies into rivers and lakes.

Crop Utilization

Rice is primarily a high-energy or high-calorie food. It contains less protein than wheat. Brown rice from which all of the hulls have been removed, is relatively high in proteins, in vitamins B_1 and B_2, and in calcium and iron; unfortunately, it is not readily accepted by the North American or Asian consumer. Milled rice loses valuable proteins and vitamins in the milling process during which the embryo and the darker, outer layers of the caryopsis (the aleurone layer) are removed. Much of the loss of nutrients can be prevented through parboiling, a common process. Of course, any additional step in processing increases the cost of the milled rice. In both the United States and Canada, most rice is consumed as boiled rice. Some is converted into breakfast cereals and some is processed, precooked, and packaged as specialty dishes. A little is used in both the brewing and the distilling industries. Little rice is converted into flour in North America.

The milling properties of rice depend in part on the type of kernel. Short-grain cultivars, which are produced predominately in California, have a low amylose content and high milling quality. When cooked, however, the kernels tend to become sticky and lose their shape. Processing techniques and special cultivars are being developed to minimize these problems. Customs and preferences seem to be the major factors determining the use and acceptance of long- or short-grain rice.

The by-products of rice milling are used for a variety of purposes. Rice bran is used as cattle feed. Rice germs (embryos) are a source of oil used in the manufacture of soap. Rice hulls can be used as wood substitutes in the manufacture of insulation materials and cardboard; they are also used in manufacturing linoleum and rayon.

Rice is a significant component, though not a staple, in the North American diet. Although North America produces less than 10 percent of the world supply of rice, it does export some rice. In addition, researchers in the United States are working to develop improved cultivars and production practices that may help rice-dependent nations become self-sustaining. One of the major rice research programs in the world is the American-supported International Rice Research Institute (IRRI), which coordinates world research on rice from its main offices in Manila in the Philippine Islands.

Selected References

Fagade, S. O., and S. K. Datta. 1971. Leaf area index, tillering capacity, and grain yield of tropical rice as affected by plant density and nitrogen level. *Agron. J.* 63:503–506.

Hall, V. L., and R. M. Railey. 1964. Timing nitrogen fertilization of rice with the morphological development of the plant. *Rice J.* 67(12):6–9.

Hall, V. L., J. L. Sims, and T. H. Johnston. 1968. Timing of nitrogen fertilization of rice: II. Culm elongation as a guide to optimum timing of applications near midseason. *Agron. J.* 60:450–453.

Leonard, W. H., and J. H. Martin. 1963. *Cereal crops.* Macmillan.

Study Questions

1. What cultural practices are peculiar to rice compared with other cereals?
2. How and why does rice tolerate flooding?
3. Discuss the environmental factors that limit the distribution of rice.
4. How is the time of fertilizer application determined for rice?

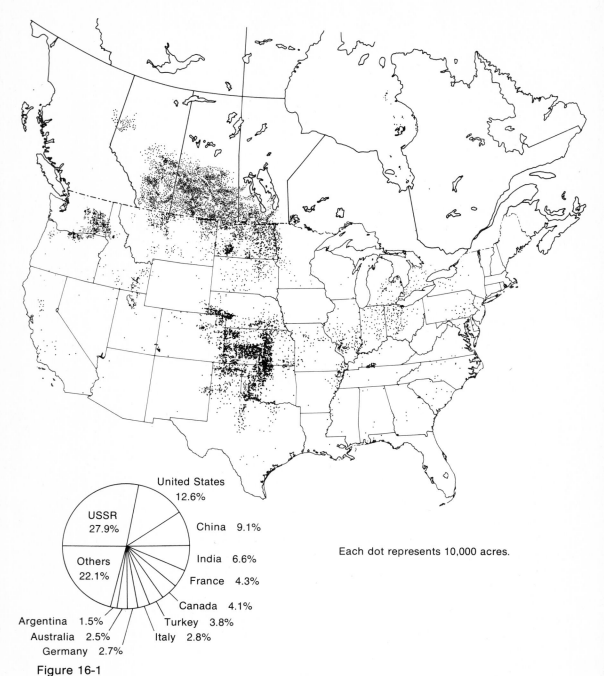

Figure 16-1
Major wheat-producing areas in North America and contributions to world production. [Data from U.S. Departments of Commerce and Agriculture and the United Nations.]

Sixteen   Wheat

Origin and History of Culture

Wheat (*Triticum aestivum*) as it is known today evolved through the intercrossing of at least three related wild ancestors. Its center of origin is reported to be in southwestern Asia, and evidence shows that it has been a major agricultural commodity since prehistoric times. Early explorers and merchants introduced wheat into Europe, and colonists brought it to North America early in the seventeenth century. Wheat production in the United States started in Massachusetts and moved westward with settlers during the next two hundred years. In terms of acreage planted, wheat is the second most important cereal crop in the United States, being exceeded only by corn. In Canada, it is the foremost cash crop. In both Canada and the United States, wheat is the leading cereal crop (Figure 16-1).

Botanical Characteristics

Wheat has a spike type of inflorescence (Figure 16-2). Typically it has a single spikelet per node of the rachis, and three or more florets per spikelet, with at least one being sterile in most cases (Figure 16-3). The wheat spike is indeterminate in growth; it does not terminate with a spikelet. Glumes in wheat are fairly large and may have an awn; the grain (caryopsis) threshes free from the lemma and palea. In the vegetative stage, wheat is narrow-leaved like other cereals, except for corn and sorghum. Wheat plants have neither ligules nor conspicuous auricles. Wheat seedlings differ from other cereal seedlings in that the first internode (the coleoptilar internode) in wheat seedlings elongates very little; thus, the secondary roots form essentially where the seed is planted. In other cereals, the first internode elongates, and secondary roots form above the

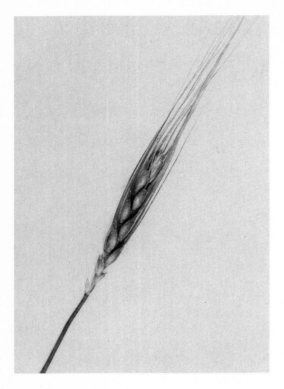

Figure 16-2
Typical wheat spike. Note that not all cultivars have awns.

Figure 16-3
Single wheat spikelet. There are four obvious florets; the smallest one is sterile.

seed proper. The extent of elongation of the coleoptilar internode is genetically determined and varies among cultivars of winter wheat. Recent studies show that there is a correlation between very slight elongation and winter survival or cold hardiness of seedlings. Apparently, when the coleoptilar internode does not elongate, the delicate, primordial floral tissue is held deeper in the soil where it is more protected from temperature extremes. Plant breeders may use this trait to develop cultivars that are more winter hardy than existing ones. Planting depth and soil temperature as well as light affect the elongation of the coleoptilar internode.

Wheat, like all grasses, has a caryopsis type of fruit, but the characteristics of the wheat kernel are commonly described in terminology that pertains only to wheat. This is a result, in part, of the separation of the components of the kernel in the process of milling wheat (making flour).

The typical wheat kernel is from 3 to 10 mm in length and from 3 to 5 mm in diameter. The approximate composition of a fully developed kernel (Figure 16-4) is:

1. The germ (the embryo, consisting of plumule, scutellum, radicle, and hypocotyl) comprises about 2.5 percent. It is high in proteins and fats and is removed before milling.
2. The bran (pericarp, testa, nucellus, and aleurone layer) comprises as much as 14 percent. It is a by-product of milling and is used in dairy and poultry feeds. Small amounts are used in breakfast cereals.
3. The starchy endosperm (the storage part of the caryopsis that develops from the union of polar nuclei with the endosperm nucleus) comprises from 83 to 87 percent. The cells of the endosperm are large, loosely packed, and filled with starch granules; they are held together by gluten, a protein cementing substance. The amount of gluten contained in the endosperm is determined by the cultivar and the environment in which the crop grows. The percentage of gluten greatly affects the baking properties of flour; bread wheats (HRS and HRW—see Chapter 5) are awarded a premium for having a high protein content (i.e., a high percentage of gluten); the durum wheats used in pastas are low in protein. Gluten gives wheat flour the strength to stretch and retain gases as fermenting dough expands. Only rye and a new, man-made cereal known as triticale also contain gluten, or like compounds.

Differences in inflorescence morphology and kernel characteristics are used to distinguish classes and cultivars of wheat. These differences are also significant in grading grain. Classes of wheat are recognized in part by the color (red or white) and the texture (soft or hard) of the kernel: Hard Red Wheat, Soft Red Wheat, Hard White Wheat, and so forth. In addition, three features distinguish winter from spring wheat. Spring types have a relatively large germ; that of winter types is smaller. (Durum wheats have pointed germs.) All wheat kernels have a pronounced groove (crease) running the length of the kernel. Winter wheat kernels have a deep crease; spring wheat kernels have a relatively shallow one. The tissue surrounding the crease is called the cheeks of the kernel. The cheeks of winter wheat kernels are rounded, whereas those of spring wheat kernels are somewhat angular.

Some kernel characteristics are also useful in separating cultivars, although the major differences between cultivars are more fre-

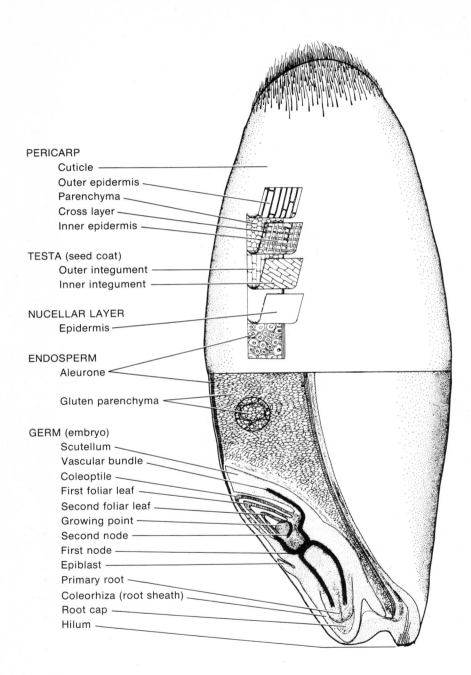

Figure 16-4
Detail of a single wheat kernel. [USDA.]

Figure 16-5
Winter wheat nearing maturity at Bozeman, Montana. Plants tiller and utilize available resources.

quently associated with inflorescence morphology; the presence or absence of awns on glumes and/or lemmas, the shape of glumes, lemmas, and paleas, the color of lemmas or glumes, the presence and/or patterns of pubescence on glumes, lemmas, paleas, and sometimes on leaves and stems are but a few of the features that serve to identify cultivars. The distinguishing characteristics that are important agronomically are yield, quality, seed size (i.e., test weight), and resistance to specific diseases.

The morphology of wheat plants is typical of most grasses. Wheat plants range in height from 1 to 4 feet (30.5–111.9 cm), depending on the cultivar and the environmental conditions under which a crop grows. Plants have anywhere from two to fifty tillers, depending on available space. Under field conditions, from two to five tillers are common (Figure 16-5). Plant breeders have been striving to develop short-strawed, high-yielding wheat cultivars. In 1970, Norman Borlaug was awarded the Nobel Prize for his accomplishments in this field of plant breeding (see Chapter 32).

Environmental Requirements

There are two general types of wheat: winter and spring. Besides having distinctive kernel characteristics, they require different environmental conditions for normal growth and development and are thus planted at different

Table 16-1
Winter survival and yield of winter wheat planted at different times at three locations in three different years.

Month planted	1959–60 Hansil Valley		1960–61 Clarkston		1965–66 Blue Creek	
	Survival (%)	Yield (kg/ha)	Survival (%)	Yield (kg/ha)	Survival (%)	Yield (kg/ha)
June	90	403	50	34	15	155
July	—	—	50	376	60	961
August	—	—	65	524	90	2,937
September	85	1,035	80	974	95	3,037
October	95	981	95	1,297	90	1,942

Source: W. G. Dewey and R. F. Nielson. Effect of early summer seeding of winter wheat on yield, soil moisture, and soil nitrate. *Agron. J.* 61(1969):51–55.

Note: Regardless of the year or the location, early planting (June or July) reduced yield, and, in two instances, late planting also reduced yield. The impact of planting date on winter survival seems to depend on either the year or the location, but in general early planting reduces the potential for winter survival.

---

times of year. Both types are classified as *Triticum aestivum*.

In the Northern Hemisphere, winter wheat is planted in the early fall when the soil temperature drops below 55°F (13°C); after germination, it overwinters as a seedling and then resumes growth with the onset of spring. It tillers, elongates, and then flowers in late spring or early summer, and ripens in June, July, or August, depending on the latitude (Table 16-1). Winter wheat requires vernalization—exposure to a prolonged, cold period during the seedling stage—to induce flowering. Cold temperatures along with short days also favor the initiation of tillering. If winter wheat is planted in the spring (i.e., if it is not vernalized) seeds germinate, but inflorescences do not form and culms do not elongate; the plants remain in a short, vegetative cluster, called a rosette, and do not flower. This is in part an auxin-regulated phenomenon. Although winter cultivars differ in their cold hardiness, all cultivars benefit from the insulation afforded by a winter snow cover, which acts much as an igloo over the seedlings. In the absence of a snow cover, serious losses due to freezing and desiccation may occur; the exposed seedlings cannot obtain water from frozen soil. Losses are intensified by low temperatures (from 10° to 20°F, or −12° to −7°C) and wind. During spring growth, winter wheat is damaged if the temperature falls below 32°F; freezing temperatures cause stamen and pistil malformation and pollen sterility. Cold weather during flowering, although rare, also causes pollen malfunction or sterility.

Spring wheat, as the name implies, is sown in the early spring; it germinates, grows, and ripens during the spring and summer, and is harvested in late summer or early fall. Spring wheat production in North America is limited to Montana, North and South Dakota, Minnesota, and the adjacent Canadian provinces of Alberta, Saskatchewan, Manitoba, and Ontario. It is produced in regions in which winters are too severe for winter wheat to survive. Because the yields of winter wheat may be more than 10 percent greater than those of spring wheat, winter wheat is preferred to spring wheat in areas where it is adapted.

Spring wheat is generally planted as soon as a seedbed can be prepared after the soil temperature rises above 34°F (1°C). It can grow significantly at temperatures near freezing; thus, it is one major crop that can be grown in areas with a frost-free period of only 90 to 100 days. Early planting of spring wheat is desirable to insure a maximum growing season. Yields are reduced by one or more bushels per acre for each 10-day delay in planting during May. If spring wheat is planted after mid-May, yield reductions are even greater because of a shortened growing season, higher temperatures, higher water requirements for seedlings, and related stresses. Like winter wheat, spring wheat can be damaged by severe freezing during growth and flowering.

All wheat grows best on loamy soils that are medium in texture, fairly high in organic matter, and fertile. Wheat is not well adapted to acidic soils; the most favorable pH range for wheat production is from 7.0 (neutral) to 8.5 (slightly alkaline).

Because most wheat is produced on dry land, the availability of moisture is a major factor in wheat production. Both the amount and distribution of natural precipitation are of prime concern to wheat producers. In areas, where wheat is the major crop, irrigation is still relatively uncommon because of the large acreages planted by each producer; however, the development of large, self-propelled sprinkler systems and of other types of sprinklers (e.g., center-pivot types) that can irrigate large areas has increased the acreage of irrigated wheat. In areas where wheat is not the major crop, but is a member of a crop rotation sequence that includes higher-cash-value crops (row crops such as sugar beets or tomatoes), wheat may be irrigated so that the distribution of natural precipitation is not critical. Wheat requires from 10 to 25 inches (25.4 to 63.5 cm) of annual precipitation distributed throughout its growing season. Moisture must be available in the soil at planting. In the crop-fallow rotation system typical of most semiarid, wheat producing areas, profitable crops are produced with as little as 8 inches of annual precipitation as a result of moisture stored in the soil during the fallow period; however, distribution is still critical. Rain during the late vegetative stage of growth can cause lodging; excessive moisture favors some foliar disease; and, late rains interrupt or delay harvesting. In parts of the Great Plains (e.g., Oklahoma), excessive moisture reduces wheat yields. Moisture requirements, including moisture stored through fallow, vary from north to south. Because of lower maximum and average temperatures and more favorable (higher) relative humidity, less water is required to obtain comparable wheat yields in the north than in the south. Remember that high temperatures, wind, and lower relative humidity increase a plant's water needs.

Like any other plant, wheat cannot germinate in dry soil. With spring wheat, moisture at planting is rarely a problem because of the availability of moisture accumulated throughout the winter. Moisture for germination may be a problem with winter wheat. In the absence of adequate early fall rains before planting, winter wheat may be sown into dry soil. If rains arrive too late, the seedlings may not be vigorous enough to survive the winter. If there is inadequate moisture for germination and seedling development, the seed may rot in the ground, causing a crop failure. In northern areas where spring wheat is adapted, it is sometimes seeded following the failure of winter wheat.

After seeds start to germinate, favorable conditions enable them to utilize their reserve energy (endosperm) to produce the most vigorous seedlings possible. Roots must develop adequately to provide water and minerals for the top (leaf) growth that will provide energy for respiration during the winter and at the onset of spring growth. In addition, root growth must be extensive enough to anchor the seedling firmly and to support spring growth. Seedlings must have adequate moisture before the onset of winter dormancy to minimize losses due to desiccation; one cause of desiccation is warm air that creates a moisture demand on the seedling while soil is still frozen so that water is not available. Any condition that reduces root growth (e.g., moisture stress, improper balance of nutrients, or pests) reduces the survival potential of a crop by causing the growth of weaker seedlings. Moisture stress during tillering reduces the number of tillers per plant and thus reduces yield. Similarly, moisture stress during the development of floral parts reduces the number of spikelets per spike or the number of florets per spikelet, depending on when the stress occurs. Stress during flowering can cause pollen sterility. Moisture stress during any of the above periods reduces yield by reducing the number of kernels produced. Finally, moisture stress after pollination, during caryopsis development, causes plants to develop pinched or shriveled kernels, which results in a low-quality, low-test-weight crop.

In summary, wheat does not produce well in warm, humid areas. It requires moderately cool conditions for germination, followed by a cooler period for tiller formation (or for vernalization of winter wheat); grain matures best under warm, dry conditions that are also ideal for harvesting the crop.

Production Practices

SEEDBED PREPARATION AND SEEDING

Seedbed preparation for wheat is similar to that of other cereals, except perhaps corn. Fields are first cultivated, with a moldboard plow or disc, to a depth of 4 to 12 inches (10–30 cm); the exact depth depends on the location and the cropping history of each field. Remember that it is desirable to change the plowing depth (or deep plow) periodically to prevent the development of a compact plowsole or hard pan immediately below the plowing depth. In the Great Plains region, chisel plows and sweep implements have largely replaced moldboard and one-way plows. Plowing operations must provide for the incorporation of crop residue and the conservation of soil and moisture. The initial plowing spreads stubble over the surface of the soil to insure protection from wind and water erosion and to trap snow and thus foster water accumulation and storage. This type of cultivation is referred to as stubble mulching or trashy fallow. Plowing is followed by harrowing immediately before planting. Fields may be chiseled periodically to loosen the subsoil. In dry areas where wind erosion is a serious threat, a Noble-blade is used to control weeds after harvest and throughout the fallow period. The use of this implement also maintains a trashy fallow that fosters snow accumulation.

For spring wheat, an initial cultivation may be made in the fall, after harvest. Then, in the spring, the final seedbed preparation is completed as soon as fields can be worked efficiently and the crop is planted as early as possible. For winter wheat, it is desirable to prepare the seedbed as early as possible (in July and August) to allow for earlier planting. Seeding early in September permits seeds and seedlings to take advantage of fall rains, which encourage vigorous growth. In addition, early seedbed preparation with a cultivation immediately before planting is effective in controlling fall-germinating annual weeds, such as cheatgrass; it also allows time for organic matter to decompose in the soil, and it may enhance the infiltration of late summer rains into the planting (and seedling-rooting) zone of the soil. Early preparation of the seedbed for winter wheat can increase yields from 15 to 25 percent.

Because seedlings must be as vigorous as possible to survive the winter, the seedbed must provide optimum conditions for seedling growth. It must contain the proper balance of all necessary minerals. Seedbed preparation should insure adequately moist soil within 2 inches (5 cm) of the soil surface; the seedbed should be fine enough to minimize the loss of water from the soil and to allow seed-soil contact so that the seed can absorb water for germination. At the same time, the seedbed must not be overcultivated—the soil should not be pulverized so that crusting occurs and limits emergence. A moderately cloddy seedbed helps to prevent crusting of the surface soil caused by rain after planting, which may restrict seedling emergence.

The seeding depth for wheat varies from 1 to 3 inches (2.5-7.6 cm), depending on the type of soil (seeds are planted deeper in lighter, sandy soils than in heavy, clay soils), the seed size (the larger the seed, the greater the allowable seeding depth), and the level at which adequate moisture is available for germination. It is desirable to plant into moist soil, but not if the planting depth exceeds about 3 inches (7.6 cm).

Seeding methods vary with soil types, local geography, and customs. Broadcast seeding, either on the ground or from the air, is practiced in areas where wet ground limits access to fields, or where seed is planted into heavy stubble, or where extremely large areas must be planted rapidly. Even when broadcasting is followed immediately by harrowing, seeding rates must be increased from 25 to 50 percent and yields are still frequently lower than those normally achieved with more refined seeding methods. Broadcasting, even when followed by harrowing, fails to bring the seed into close contact with soil moisture; thus, germination is low, and poor stands result in reduced yields.

In addition, broadcasting leaves too much seed on the surface of the soil, subject to pests (birds) and losses due to blowing or washing. Producers most often seed wheat with some type of grain drill (Figure 16-6). Grain drills can plant at a precise, uniform depth, with accurate row spacing, and controlled seeding rates. For example, the furrow drill opens a furrow into which the seed is dropped through a boot typical of standard grain drills, thereby planting seed into the moist soil of the furrow; after planting, snow accumulates in the furrow, affording extra insulation for winter wheat seedlings. In northern areas, yields increase markedly if crops are seeded with furrow drills. The development of the furrow drill, coupled with efficient fallowing, has extended the range of production of winter wheat (in preference to spring wheat) north into Montana, North Dakota, and Alberta.

Planting rates for wheat depend principally on predicted available moisture. In semiarid regions that have only 10 to 12 inches (25.4-30.5 cm) of rain per year, seeding rates may be as low as 20 pounds of seed per acre (22.4 kg/ha). All seeding rates should be based on a pure, live seed rating (see Chapter 12). In areas that have a high annual precipitation, planting rates can be as high as 100 pounds per acre (112 kg/ha). With higher planting rates, row spacing may be as narrow as 4 inches (10.2 cm); with lower planting rates, rows are planted from 12 to 18 inches (30.5-45.7 cm) apart. Commonly, row spacings on commercial grain drills range from 7 to 9 inches (17.8-22.9 cm).

Wheat, like many grasses, has a remarkable capacity to adjust or respond to planting rates. If the planting rate is light, each seedling produces an abundance of tillers so that the number of stems in a given area is comparable to the number produced by a crop seeded

Figure 16-6
One of several types of drills used to seed wheat. [Courtesy of Deere and Company.]

under a heavier planting rate. For example, crops planted at 40 pounds of seed per acre (44.8 kg/ha) produce approximately the same number of tillers per square yard as crops seeded at 60 pounds per acre (66.2 kg/ha), as long as adequate moisture and nutrients are available. With lower planting rates, more tillers develop because more light reaches critical sites of auxin accumulation, which reduces apical dominance of existing stems and allows axillary buds to develop and grow. One problem with late tillering is that the later maturity of tillers interrupts harvesting, and grain from later tillers may be small and excessively damp. Cultivars differ in the maximum number of tillers produced per plant and in the extent to which they adjust their tillering in response to variations in planting rates. The choice of cultivar therefore affects the calculation of appropriate seeding rates.

Excessively high planting rates can cause a number of problems. Essential materials such as water and minerals may be adequate for seedling growth, but, as plants grow and compete more for light, moisture, and nutrients, they may deplete the supply of one or more of these essential factors. If adequate supplies are not available throughout the entire life of the plants, the yields of all plants are reduced. The specific cause of the yield reduction depends on when a factor is depleted. Plants that suffer from moisture stress early in their life cycle produce fewer tillers; if moisture stress occurs later, the number of florets per spikelet or spikelets per spike is reduced. Moisture stress after flowering results in smaller or lighter kernels (reduced test weight). In addition to these adverse effects, excessively high planting rates cause plants to grow tall, with weak stems. Because of shading between plants, auxin-regulated stem elongation is abnormal and stems become etiolated, much as they would be if plants were grown in the dark. Tall, weak-stemmed plants tend to lodge, which causes problems in harvesting and results in reduced yields. Tall plants also

Table 16-2
Variations in yields of spring wheat in Montana in response to different amounts and combinations of fertilizers.

Treatment	Element (lb/acre)			Yield (bu/acre)
	Nitrogen	Phosphorus	Potassium	
1	0	0	0	29.0
2	0	20	25	30.0
3	20	20	25	37.4
4	40	20	25	44.5
5	60	20	25	48.2
6	80	20	25	57.0
7	100	20	25	56.0
8	40	20	0	49.1
9	40	0	25	44.7

Source: Unpublished data courtesy of P. L. Brown.
Note: In these experiments, phosphorus and potassium had little effect on yield; however, depending on the composition of the soil, in other locations any essential element might limit yields.

shatter more, reducing yields even further.

In the final analysis, the proper planting rate is one that allows optimum and balanced use of environmental resources, mainly available moisture and essential minerals. It should also minimize competition between wheat plants and allow wheat plants to compete successfully with weeds.

FERTILIZERS

Wheat generally responds favorably to applications of both nitrogen and phosphorus fertilizers and may respond to applications of other essential minerals, depending on the soil. Specific fertilizer recommendations should be based on the results of soil tests. Table 16-2 shows that the application of nitrogen fertilizer increased grain yields in a linear pattern from 29 bushels per acre with no fertilizer to 56 bushels per acre with from 80 to 100 pounds of nitrogen per acre (compare treatment 2 with treatments 3, 4, 5, 6, and 7). Note that there was a uniform application of phosphorus (in the form of P_2O_5) and of potassium (in the form of K_2O) and that, because adequate amounts of these minerals were available in the soil, the addition of these minerals did not greatly affect yield (compare treatments 1 and 2 and treatments 8 and 9).

There is no set of rules establishing the best time to apply fertilizers and the types and amounts to be applied to achieve maximum yields. However, several general principles afford reasonable guidelines. One principle is that all plants require a balanced supply of minerals. The benefits of nitrogen fertilizer may be lost if other essential minerals are not available in adequate quantities. In general, applying from 40 to 80 pounds of nitrogen (N) per acre (45–90 kg/ha) usually increases wheat yields, and heavier application rates are common where environmental conditions favor high yields. Responses to more nitrogen and other minerals vary with the soil and with the environmental conditions under which the crop is produced. Moisture is a critical factor; generally, heavier applications of fertilizer are used with heavier seeding rates when there is abundant moisture.

With winter wheat, part of the required nitrogen and all other required minerals are commonly applied in the fall when the crop is sown. The remainder of the nitrogen is applied in the spring when probable available moisture for the crop can be estimated more accurately; hence, fertilizers and available moisture can be better balanced. In the fall, wheat requires nitrogen for vegetative (stem and leaf) growth and for tillering; in the spring it is essential in reproductive growth. With winter wheat, nitrogen deficiency can occur after the onset of spring growth, even though adequate nitrogen had been available for fall (seedling) growth. Although yields may be reduced as a result of the deficiency, applying side-dressings of nitrogen fertilizer in the spring until plants are in the boot stage of development can alleviate deficiency symptoms and reduce yield losses. High levels of available nitrogen in the soil are essential during and after flowering for the production of grain protein. Phosphorus, which is applied in the fall, is essential for vigorous root growth of seedlings, but if phosphorus is deficient, spring applications are of little benefit. For spring wheat, all fertilizers are applied at or immediately before planting because spring wheat plants develop and grow more rapidly than winter wheat (Figure 16-7).

Excessive amounts of nitrogen can have an adverse effect on the growth of wheat plants. If adequate supplies of other essential minerals are available, too much nitrogen may result in excessive vegetative growth; tall, weak-stemmed plants that tend to lodge; and delayed maturity. The excessive growth causes plants to have abnormally high water requirements and can lead to the premature depletion of soil moisture, which produces drought conditions as the crop matures. Delayed maturity may also force plants to flower late, when the weather is warm; high temperatures cause poor pollination and seed set, resulting in reduced yields. Excessive nitrogen does not contribute to an increase in yield; it may, in fact, contribute to a reduction in yield.

Phosphorus deficiency in the fall for winter wheat results in poor root development, weak seedlings, and, ultimately, reduced stands and poor yields. Spring applications of phosphorus on winter wheat are of little value.

In general, the yield of a plant is reduced if the plant is deficient in any element at any time during its life cycle, even though the deficiency is alleviated by the application of fertilizer. The earlier the deficiency is corrected, the less yields will be reduced.

HARVESTING, STORAGE, AND ROTATIONS

Wheat is harvested some place in the world every month of the year. In North America, harvesting starts as early as the end of May or the beginning of June in the Deep South and ends in late September or early October in the northern spring wheat areas of the United States and Canada. Most wheat is harvested with a self-propelled combine (Figure 16-8). For combining, grain should be from about 12 to 14 percent moisture. In areas such as North and South Dakota, where shattering due to wind and hail is a serious threat, some wheat is windrowed at higher moisture levels, and then allowed to dry before being threshed with a combine. To minimize losses due to shattering in areas of Canada where shattering is a problem, crops are swathed when kernels are about 18 percent moisture and then threshed with a combine equipped with

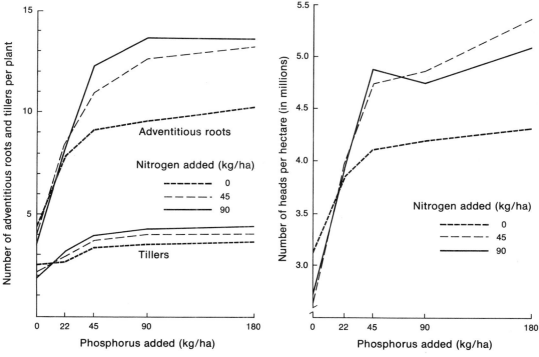

Figure 16-7
The effects of different rates of nitrogen and phosphorus fertilization on spring wheat: (A) vegetative traits; and (B) reproductive trait that determines ultimate yield. [From A. L. Black. Adventitious roots, tillers, and grain yield as influenced by N-P fertilization. *Agron. J.* 62 (1970): 32–36.]

a pickup attachment. On rare occasions, standing threshers are still used. Binders are rarely used today. Modern technology has automated most of the harvesting and handling of wheat. However, producers must adjust and use equipment properly to insure a high-quality crop: cleanliness of equipment, speed of operation, cylinder and concave adjustment for speed and clearance, proper screen size, and air-flow adjustment for cleaning the grain during harvest are all important. When in doubt, County Extension agents and implement service- or salesmen can supply information for specific situations. In the absence of disease or other crop disasters, wheat yields vary from as low as 12 bushels per acre (806 kg/ha) to as high as 100 bushels per acre (6,720 kg/ha) or more. Yields depend on seeding rates, soil fertility, moisture availability and distribution, and the effect of various pests.

Proper crop rotation is an important management consideration in any farming enterprise; it is essential to insure continuous high yields, to protect the soils, and to control many crop pests. In areas where wheat is the princi-

Figure 16-8
Cutaway view of self-propelled combine showing mechanism (1) that feeds unthreshed grain into the cylinder (2 and 3) where it is threshed, and cleaning devices (4–7) that separate the grain from straw and chaff. Grain is discharged by means of an auger system into a storage bin, and the straw and chaff are blown from the rear of the combine. [Courtesy of Deere and Company.]

pal crop and where average annual rainfall is more than 20 inches (50 cm), wheat is grown annually in a so-called wheat-wheat rotation. In semiarid regions, wheat and fallow or wheat and two fallow crop years are common rotations. In the wheat belt of the United States, other rotations include wheat-wheat-peas, or a year or two of wheat followed by some legume other than peas.

Diseases and Pests

DISEASES

All three major groups of pathogens—fungi, bacteria, and viruses—attack wheat. (See Chapter 9 for a general review of plant pathogens and other crop pests.)

The several species of fungi that cause rust diseases in wheat are in the genus *Puccinia*. Rust diseases are favored by cool, moist conditions. They are recognized by pustule formation on stems, leaves, sheaths, and inflorescences. These pustules are the sites in which reproductive spores form; the reproductive spores can infect healthy plants. The *Puccinia* fungi cause plants to use water inefficiently; they also cause the destruction of plant tissue. Plants infected with the fungi produce fewer and lighter kernels, and the kernels look as if they were formed on plants suffering from moisture stress. Various fungicides are effective in controlling the rust-causing fungi; however, the most efficient way to control rust diseases is to grow genetically resistant cultivars. Annual losses due to rust diseases vary from nearly 0 to 100 percent. On the average, more than 10 percent of the wheat grown in the United States is lost to rust diseases annually.

Other serious fungal diseases of wheat include loose and covered smut (the latter is also known as stinking smut or bunt). The fungi that cause both types of smut infect floral parts and reduce yields by replacing kernels with spore masses. The fungus that causes loose smut is difficult to control because it is carried in the caryopsis. The most desirable control is genetic resistance in the host wheat plant. Loose smut can also be controlled by treating seed with hot water (41°C), which kills the fungus without markedly damaging the seed. Hot water treatment is relatively expensive and therefore not commonly used; researchers have developed systemic fungicides that make the hot water treatment for loose smut virtually obsolete. Covered smut is spread by spores carried on the surface of the seed; the spores are the primary inoculum. Covered smut is well controlled by genetic resistance in different cultivars. However, a number of seed treatments are also highly effective in controlling the disease. In the past, a number of mercury compounds were used to control covered smut, but the use of these compounds is now forbidden by federal regulation. New compounds, such as hexachlorobenzene (HCB), are proving to be almost as effective as the mercury compounds in controlling covered smut.

The damping-off and root-rot groups of fungal diseases cause losses of as much as 30 percent of the U.S. wheat crop each year. Treating seed with mercury afforded reasonable control in the past, but such treatment is no longer permitted. Suitable alternatives have not been developed, although hexachlorobenzene affords some control. Also, genetic resistance for some races has been identified by plant scientists in Canada. Delaying planting until soil temperatures fall below 55°F (13°C) achieves some degree of control in winter wheat because the fungi are less active at lower temperatures.

Bacterial diseases of wheat are of relatively minor significance. Viral diseases, such as wheat streak mosaic disease, affect crops sporadically. This viral disease is transmitted by the spider mite.

Diseased plants suffer from damage inflicted in one (or more) of several ways, depending on the pathogen. All pathogens are parasites; they consume the products of photosynthesis that would normally contribute to crop yield. In addition, fungi, and to a lesser extent bacteria, steal needed moisture from the plant; thus, many diseased plants appear to suffer from drought conditions. Pathogens reduce the water-use efficiency of plants by increasing their total water requirement. Finally, pathogens destroy cellular organelles, tissues, and organs. Viruses are notorious for the destruction of chloroplasts, as are some fungi. This reduces the photosynthetic capacity of the plant, thereby lowering yield. As both fungi and bacteria grow, they may break the epidermis of leaves and stems, which further disrupts normal water balance in the plant and reduces yields. Of course, evidence of pathogens in grain also lowers its market value.

WEEDS

Weeds that grow in wheat are a continuing problem. Specific weeds are regional in distribution, yet some general principles apply to all weeds. Weeds damage crops through competition for water, essential minerals, and light; they may also contaminate crops.

Major weed pests of wheat include both grass and broadleaf (dicot) species. Some of the most common weeds are thistles, chickweed, bindweeds, mustards, bromegrasses, and wild oats. Relative newcomers to the list of weed pests that compete seriously with wheat include thistles (several species), kochia, cockles, field pennycress, pepperweeds, and wild radish. Foxtail, lambsquarters, quackgrass, and ragweeds have become less serious pests nationally in the United States.

The control of weeds that grow in wheat is effected through cultivation and the use of herbicides. Both preemergence and postemergence herbicides are used; nearly 40 percent of the total U.S. wheat acreage is treated with some herbicide. Postemergence herbicides are far more widely used than preemergence compounds because they are generally more effective in weed control. In 1968, for example, when more than 55 million acres (22.6 million ha) of wheat were harvested in the United States, more than 20 million acres (8.2 million ha) received a postemergence herbicide treatment. In the same year, only about 600,000 acres (246,000 ha) received preemergence treatment and only 300,000 acres (123,000 ha) were treated with both pre- and postemergence herbicides.

Weed control by cultivation is limited to seedbed preparation and fallow practices. Cultivations should be timed to destroy weeds before they reproduce and when they are in their most vulnerable or weakest stage of development. For annuals, this often means destroying weeds at the seedling stage when plants are expending seed energy (stored in cotyledons or endosperm) and seedlings are not yet well established and self-supporting through photosynthesis. Perennials, such as Canada thistle, are most vulnerable during flowering and seed development when crown and root energy reserves, essential for regrowth, are at the lowest level; therefore,

control measures at these stages are most effective. Special attention must be given to weeds that can spread vegetatively. Care must be taken to destroy rhizomes, stolons, and other plant parts from which new plants may arise and to avoid transporting these parts to uninfested fields on equipment. Appropriate herbicides (e.g., 2,4-D for mustard) must be applied at recommended rates when weed plants are in a vulnerable condition and crop plants will not be harmed. Herbicides are effective when weeds are growing actively because the chemicals are more readily incorporated into the plants.

Annual, broadleaf weeds such as field bindweed, pigweed, and kochia can be destroyed through a combination of cultivation before planting (which destroys weed seedlings) and herbicides such as 2,4-D, MCPA, and DNPB. Time, frequency, and rate of herbicide application depend on the specific weed or weeds to be controlled and on climatic conditions. The control measures for perennial, broadleaf weeds, such as Canada thistle, are similar to those for many annuals. For thorough control, however, cultivation and herbicide application must be practiced for 3 to 5 years.

One aspect of weed control that cannot be overemphasized is the use of clean, weed-free seed. The savings realized by reducing losses due to weeds more than offset the slightly higher prices of certified seed. Finally, all field practices that promote vigorous, healthy plants help to minimize losses from weeds.

INSECTS

The wheat stem sawfly causes significant losses by seriously damaging the stems of wheat plants. The insect lays its eggs inside stems causing them to split. After the eggs hatch, the larvae further damage stems through feeding. The damage caused by the sawfly ultimately results in lodging. Control of this pest is best achieved by growing cultivars with solid stems in which larvae cannot feed and by rotating crops, which denies the pest a suitable host.

The Hessian fly is a periodic pest of wheat. Larvae feed in individual tillers causing lodging or death. Control of the Hessian fly is achieved principally through regulating planting dates. If planting is delayed until after the adults have died, seedlings cannot be damaged by egg laying and the feeding of larvae. Appropriate planting dates to avoid Hessian fly damage vary from south to north, and from year to year, depending on the life cycles of the insect pest. Aphids can be a pest to wheat either by feeding on and damaging the crop directly or by transmitting viral diseases. At present there is no practical way to control aphids in wheat crops. Grasshoppers also cause serious losses in wheat through feeding on plants. Chemical control is uncommon; plowing to a depth of 5 inches (12.7 cm) affords some protection by burying eggs.

Crop Utilization

Unlike many cereal crops, wheat is mainly a food, rather than a feed, crop. After processing, the crop product is consumed directly by man usually as a flour product. Mixed wheat is the only market class of wheat generally used for feed. However, plant breeders in Canada have recently developed cultivars to be grown specifically for feed. These feed cultivars are finding a market in both Canada and the United States; this new market outlet could lead to the formal recognition of a new market class of wheat (see Chapter 5). Each market class of wheat has peculiar traits that make it desirable for specific purposes. For example, because of its high protein (gluten) content, Hard Red Spring wheat yields a flour that is superior for baking bread. This wheat (or the flour made from it) is mixed with Hard Red Winter wheat to obtain flour with a protein content suitable for bread flour. The flour made from Soft Red Winter wheat has a lower protein content and is therefore used more for cakes, cake mixes, and similar products. Durum wheats yield flour that is not suitable for bread but is ideal for pastas. Remember that wheat quality can be greatly affected by environmental factors and production practices such as fertilization and irrigation (Table 16-3).

Wheat is milled to produce flour. In producing white flour, the germ is removed and as much as 30 percent of the weight of the grain goes to by-products: bran, middling, and shorts. These particles contain different proportions of the outermost layers of the kernel and bits of endosperm starch. They are used in various feeds.

The milling process to produce white flour follows specific steps. First the grain is washed and dried to remove foreign material; it is also scoured to remove fuzz and other material adhering to kernels. Next it is tempered by soaking to toughen the bran and condition the endosperm starch for milling. After being tempered for up to six hours, the grain is passed through a series of corrugated rollers. This flattens and crushes kernels without grinding them. The bran layers and embryo are separated from the starchy endosperm by a series of screening and blowing devices. The outer layers are separated into bran, middling, and shorts, and the embryo is separated for processing into wheat germ compounds (e.g., wheat germ oil). The starchy endosperm material is passed through more rollers to remove as much bran material as possible; materials are recycled through rollers to remove even more bran particles.

Table 16-3
Average values for grain yield and traits determining baking quality from eleven fertilizer tests involving a total of twenty-four Hard Red Winter and Hard Red Spring wheat cultivars in Montana.

Fertilization	Grain yield (bu/acre)	Test weight (lb/bu)	Grain protein (%)	Loaf volume (cm^3)
Check: no nitrogen applied	27.0	62.4	11.4	682
Largest amount of nitrogen applied	44.2	60.5	15.3	936

Source: C. F. McGuire, J. R. Sims, and P. L. Brown, Fertilizing Montana wheat to improve grain yields and milling and baking quality. *Montana State Univ. Agr. Exp. Sta. Bull.* 674(1974).
Note: There is a correlation between high test weight and low grain protein and between loaf volume and percentage of protein.

Finally, the starchy material, after grinding, is passed through a fine cloth sieve. This last step yields flour. About 75 percent of the initial grain becomes flour. The flour may be further sifted or refined; it is generally bleached to improve flour quality. Bleaching replaces aging, which requires several months. Many flours are enriched by the addition of vitamins and minerals lost in the milling process.

Variations in the milling process yield other types of flour. The inclusion of bran materials and germs in the final product yields whole wheat flours that are usually brown. These flours do not satisfy the tastes of many members of an affluent society, but they have a higher nutritional value.

Conclusion

In North America, wheat is the staff of life, in spite of the fact that the North American diet is relatively high in animal protein. As a source of protein, wheat is superior to rice, the dietary staple of many Asian nations. With increased international trade, wheat has become the major world food crop, and the contributions of wheat to human nutrition are not likely to diminish in the foreseeable future.

Selected References

Dewey, W. G., and R. F. Nielson. 1969. Effect of early summer seeding of winter wheat on yield, soil moisture, and soil nitrate. *Agron. J.* 61:51–55.

Leonard, W. H., and J. H. Martin. 1963. *Cereal crops.* Macmillan.

Peterson, R. F. 1965. *Wheat.* World Crops Books, ed. L. N. Polunin. Interscience.

Study Questions

1. Describe briefly the morphological features of the wheat kernel, the vegetative part of the plant, and the inflorescence.
2. Discuss how the date and rate of planting, irrigation, and fertilization affect the growth of wheat plants as well as the yield and quality of the grain.
3. What are the major pests of wheat and how do they affect or harm the plant? Use examples.

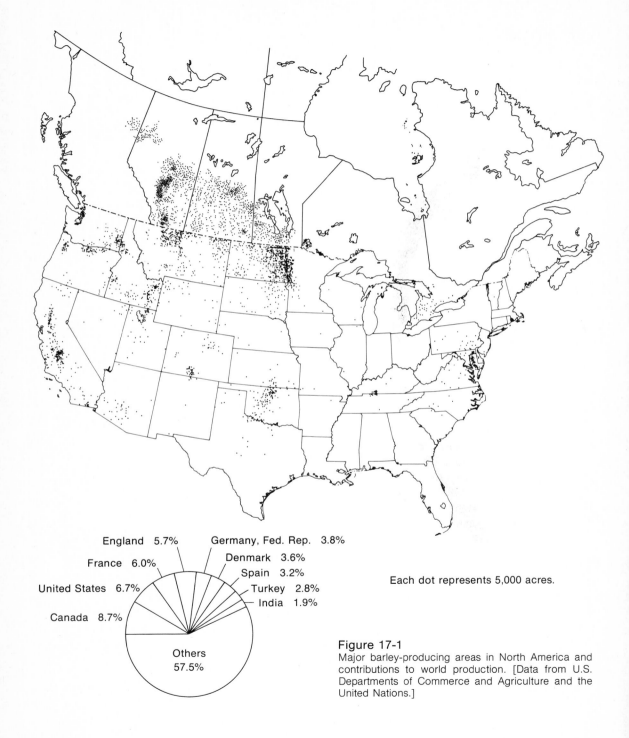

Figure 17-1
Major barley-producing areas in North America and contributions to world production. [Data from U.S. Departments of Commerce and Agriculture and the United Nations.]

Seventeen Barley

Origin and History of Culture

Cultivated barley, *Hordeum vulgare,* probably evolved from a wild ancestor in the Near East, although suggestions have been made that it originated in Tibet. The immediate ancestor would depend on the precise center of origin. It is possible that different types of barley originated in different places.

Until the fifteenth century, barley was ground for flour used in baking bread. English settlers brought two-row types of barley to New England in the mid-seventeenth century. Six-row types were introduced into the New World by the Dutch. Early Spanish settlers brought barley to the southwestern United States and to the West Coast in the eighteenth century.

Barley is a major source of food today for large numbers of people living in the cooler, semiarid areas of the world where wheat and other cereals are less well adapted. In many countries of the Middle East and northern Africa, barley constitutes the major part of total cereal production. Other large producers of barley include South Korea, Iran, India, Turkey, and Ethiopia. In Canada, barley is currently the second most important cereal crop. It is of lesser importance in the United States, but must still be considered a major field crop (Figure 17-1). The majority of the barley produced in North America is used as livestock feed, although a premium is paid by malting houses for the special, malting-quality barley. Plant breeders have developed special cultivars for the malting industry; these cultivars may yield less than feed cultivars grown

Figure 17-2
Spikes of six-row barley (left) and two-row barley (right). Note that the two-row type has sterile, lateral florets.

under comparable conditions. Malting cultivars of barley are grown in Canada, but little of the grain produced is of a quality that is satisfactory for malting. Recently, higher-yielding feed cultivars have been developed and are commonly sown because of their yield advantage. A small amount of barley is processed for prepared food in North America.

Botanical Characteristics

Barley, like wheat, has a spike type of inflorescence. The barley spike has three spikelets per rachis node, and each spikelet has a single floret. The barley spike terminates in a spikelet; thus, it is determinate in growth. Two distinct types of spikes are recognized in barley: six-row types and two-row types (Figure 17-2). Six-row types appear to have a whorl of six florets at each rachis node, but there are in fact only three spikelets at each node, each having only a single floret. The floret in each of the three spikelets at each rachis node is fertile. Thus, three kernels are formed at each node. In two-row types, the lateral spikelets bear a sterile floret so that the mature spike appears to have only two rows of kernels. In contrast to wheat, the glumes of both types of barley spikelets are greatly reduced and may be nearly hairlike. Normally, awns are not formed on the glumes; however, unlike wheat, nearly all barley cultivars have awns on the lemmas of each floret. The morphological characteristics of awns, such as smoothness or roughness, are used to distinguish cultivars. The hull (lemma and palea) adheres to the caryopsis throughout harvesting. There are a few cultivars of naked barley in which the caryopsis threshes free of the hull, as it does in wheat.

Barley leaves are distinguished from the leaves of other cereal plants by the presence of large auricles and the absence of a prominent ligule. In terms of vegetative growth, root development, tillering, and plant height, barley is quite similar to other small-seeded, annual grasses (e.g., wheat). Note that the coleoptilar node does elongate in barley, whereas in wheat it generally does not. There are both spring and winter types of barley.

Environmental Requirements

Barley is the most widely distributed cereal crop because of its tolerance to adverse climatic conditions. It is grown in all crop areas from the North to the South Pole. It is well adapted to regions where other cool-season crops are produced (Figure 17-3). If moisture is a limiting growth factor (under semiarid conditions), barley is the most productive of any cool-season cereal; yields under dryland conditions surpass those of wheat, oats, and rye. Barley tolerates high temperatures (above 90°F or 32°C) under dry conditions but is less tolerant of heat under conditions of high humidity. Thus, barley is poorly adapted to the hot, humid, nearly subtropical region of the southeastern United States.

Barley grows well on a fairly wide range of soils. It is best suited to heavier soils that have a high moisture-holding capacity and a neutral to slightly basic pH (between 7.0 and 8.0). Barley is the most salt-tolerant of the cereal crops. Compared with other cereals, it is also moderately drought- and frost-tolerant.

In spite of its wide range of adaptation, barley, like any other organism, responds negatively to less-than-optimum conditions. The general response of barley to moisture and temperature stress (either heat or cold) is quite similar to that described for wheat. Because barley has only a single floret per spikelet, environmental conditions that cause a reduction in the number of florets per spikelet in wheat may cause a direct reduction in the number of spikelets per spike in barley.

The distinction between winter and spring types of barley is not as clear-cut as it is for wheat. Barley grows in temperatures ranging from 38° to 100°F (3°–38°C); during the vegetative stage of growth the optimum temperature is in the 70s (24°C), and during flowering the optimum temperature is in the high 80s (30°C). For germination, temperatures must be between 40°F (4°C) and 85°F (30°C). A cool period with short days during vegetative growth enhances tillering as it does for wheat; light (the photoperiod) is more of a factor than temperature in this regard. Seed maturation is favored by warm, dry weather. Planting dates are dictated by weather conditions, much as they are for wheat. In general, optimum planting dates for winter and spring types of wheat and barley coincide quite closely. In regions where there is little danger of severe winter freezing, growers may plant spring-type barley in the fall, well into December. If winters are not severe, fall-planted, spring-type barley grows more vigorously and produces more abundantly because the crop has a longer period of active growth if it germinates in the fall. Remember that winter barley, unlike spring barley, is dormant for several months during the cold season. Spring types of barley are planted in the spring in areas that have more severe winters because winter barley may be killed by severe winter conditions even though winter wheat may survive in the same area. In general, winter barleys are not as winter hardy as the hardier winter wheats.

Production Practices

SEEDBED PREPARATION AND SEEDING

Barley has the same general seedbed requirements as wheat. The methods, equipment, and timing of operations are also quite similar. Because of its greater drought tolerance, barley may be produced under slightly dryer conditions than wheat; thus, cultivations must be designed and timed to minimize water loss or maximize water infiltration and retention. (These factors must of course be carefully considered for any crop produced without irrigation in semiarid to subhumid regions.)

Planting rates vary from area to area, depending on potential moisture. With irrigation, or with an assurance of 18 to 20 inches (45.7–50.8 cm) of rainfall during the growing season, planting rates can be above 90 pounds per acre (100 kg/ha). Under conditions of restricted moisture, rates drop to below 30 pounds per acre (34 kg/ha) to conserve available moisture and help insure that existing

Figure 17-3
Nearly mature, high-yielding barley grown in Dubois County, Indiana. [USDA photograph.]

plants have adequate water throughout the growing season. Of course, yields are reduced as planting rates are reduced; yet, to achieve maximum yields under existing conditions, heavier planting rates that produce stands for which adequate moisture is not available must be avoided (Table 17-1). Even with abundant, available moisture, excessively high planting rates may result in the growth of tall, weak-stemmed plants that lodge easily. Row spacings for wheat and barley are similar, as are seeding methods.

Barley requires balanced mineral nutrition, just as any other crop. Inadequate supplies of nitrogen and phosphorus, and to a lesser extent potassium, can limit yields. Fertilization must be balanced with expected available moisture to insure highest yields and efficient use of fertilizers. With barley grown for feed, it is desirable to apply fertilizers at the rates recommended for highest yields, because the amount of protein contained in the grain is not as critical in feed barley as it is in malting barley (which must be low in protein) or in grain that is to be milled. An application of 50 pounds of nitrogen per acre (56 kg/ha) usually increases yields. Increased yields are also realized with the application of other essential minerals. Specific recommendations must be based on the type of barley and the results of representative soil tests. Barley responds to deficiencies and excesses of minerals similarly to wheat.

HARVESTING, STORAGE, AND ROTATIONS

Like wheat, the vast majority of barley grown in the United States is harvested with a self-propelled grain combine. The grain usually has a moisture content of 15 percent or less at harvest (Figure 17-4). However, in regions where shattering is a serious problem, barley can be windrowed when the grain is in the

Table 17-1
Effect of different times and rates of planting on yield of Arivat barley grown in Mesa, Arizona.

Planting date	Planting rate (kg/ha)	Yield (kg/ha)
November	11	5,186
	22	5,085
	45	4,983
	67	4,881
	90	4,931
	123	4,474
December	11	3,508
	22	3,864
	45	4,118
	67	4,271
	90	4,373
	123	4,627
January	11	2,745
	22	3,152
	45	3,660
	67	3,814
	90	4,068
	123	3,864

Source: A. D. Day and R. K. Thompson. Dates and rates of seeding fall-planted barley (*Hordeum vulgare* L.) in irrigated areas. *Agron. J.* 62(1970):729–730.
Note: Yields decrease with late planting; the highest yields come from crops seeded at lighter rates in November.

hard-dough stage of development and then threshed after the kernels harden and dry.

In recent years, increasing emphasis has been given to harvesting high-moisture barley—harvesting when grain moisture is from 30 to 40 percent. This practice results in yield increases on a dry-weight basis of as much as 30 percent, with an average increase of more than 15 percent. High-moisture barley generally has a higher percentage of plump (well-filled) kernels than does barley harvested at maturity. Early harvesting is another advantage of high-moisture barley; on the average, high-moisture crops can be harvested more than a week earlier than normally mature crops. Losses due to shattering are also reduced. Standard equipment is used in har-

Figure 17-4
Modern, self-propelled combine harvesting barley near Walnut Grove, California. For efficient harvesting, grain must be dry. [Courtesy of Deere and Company.]

vesting high-moisture barley, but extra caution is required in storing the crop. Because the grain has a high moisture level, heating and subsequent spoiling of the crops, or even fire from spontaneous combustion, are real threats. Thus, high-moisture grain must be treated much the same as silage. It must be stored in airtight silos that exclude oxygen and thus prevent aerobic respiration. The respiration of high-moisture grain and microbial activity produce the heat that damages stored crops. High-moisture grain can be stored safely in grain bins or wooden silos if organic acids are used as preservatives. A mixture of propionic and acetic acids prevents the growth of bacteria and molds.

Rotation sequences for barley are similar to those for wheat. Under dryland, semiarid conditions, barley-barley or barley-fallow is a common rotation. Where moisture is more abundant, or where barley is grown as a relatively low-value crop on irrigated land, it may follow two or more years of row crops, such as sugar beets or tomatoes. In such cases, barley is not the major crop.

MALTING BARLEY
Producing malting barley requires special management efforts, but a superior manager can realize extra profit by producing this specialty crop. The malting industry has established rigid specifications for barley destined to become malt. To realize the added profit of a crop acceptable for malting, all specifications must be met. The production of malt is unlike the milling of wheat for flour in

Table 17-2
Preferred and generally acceptable protein levels in malting barley for common malting cultivars of both six- and two-row types.

Malting-barley class	Upper protein limit preferred	Percentage of dry weight acceptable
Midwestern six-row types: Conquest, Dickson, Larker, Traill, Beacon	11.5–13.0	13.5
Western two-row types: Betzes, Firlbecks III, Klages, Hannchen, Miravian, Pirolene	10.0–12.0	12.5

Source: Malting Barley Improvement Association.

that mixtures of high- and low-quality grain (common in milling) are not acceptable for malting. Grain must be of uniform, high quality for malting. It must be plump, free from broken and skinned kernels, and bright in color (indicating that healthy, mature, dry grain was harvested and properly stored). For safe storage, grain should contain no more than 13 percent moisture. In addition, a low percentage of grain protein is absolutely necessary. Excess protein in the grain causes excessive enzymatic activity during the malting process; this ultimately lowers the quality of the end product. Preferred protein levels for grain used in malting range from 11.5–13.0 percent for six-row cultivars to 10.0–12.0 percent for two-row cultivars; upper limits are 13.5 and 12.5 percent, respectively (Table 17-2). The germination precentage of the grain must be high and the rate of germination uniform because controlled germination is an important step in the malting process. Any factor that reduces germination or makes it nonuniform can cause a lot of barley to be disqualified for malting purposes and to be relegated to use as feed.

The requirements for growing malting barley are similar to those previously described for barley in general; however, more rigorous management is required at each step of production. First, a producer must select a cultivar that has been deemed acceptable by the malting industry and that is adapted to the specific environmental conditions in his area. Both six-row and two-row types are accepted by the malting industry. As a rule, six-row types (cultivars such as Larker, Dickson, and Conquest) are grown more in the plains. Two-row cultivars (e.g., Piroline, Firlbecks III, Hannchen, and Betzes) are produced in California, the Northwest, and the intermountain regions (Figure 17-5).

Seedbed preparation for malting barley is quite similar to that for feed barley or other small-seeded cereals. Research in both Canada and the United States shows that early seeding on nonfallow land helps to reduce the protein content of grain. The early seeding allows the crop to grow during cool periods when water needs are lower and danger from moisture stress is reduced. Remember that moisture stress in barley (as in other cereals) has an adverse effect on the development of the endosperm and causes plants to produce shrunken kernels with an increased percentage of protein as a result of relatively more embryo, which is high in protein compared with the endosperm. If moisture conditions are favorable, nitrogen fertilizers can be used to increase yields without causing grain protein to reach excessively high levels. Thus early seeding, which enables plants to germin-

Figure 17-5
Distribution of malting-barley cultivars in the major malting-barley areas of the United States. [Courtesy of the Malting Barley Improvement Association.]

ate and grow under more suitable moisture conditions, allows the use of nitrogen fertilizer, resulting in higher yields of grain acceptable for malting (Table 17-3).

One of the benefits of fallow is the decay of organic matter which releases minerals, including nitrogen, for plant use. Because of the decay of organic matter, fallow can result in high levels of soil nitrogen which in turn can cause grain to have unacceptably high levels of protein. Planting on nonfallow land circumvents this problem. Of course, crops must have adequate moisture to insure the maturation of plump, high-quality grain.

Conditions at every stage of a crop's development affect the amount of protein contained in the grain. If conditions during the vegetative stage favor lush plant growth, unfavorable conditions in the postflowering period can result in excessively high grain protein for malting. Conditions that must be avoided throughout the life cycle of the crop include moisture stress, diseases, insect infestations, competition from weeds, deficiencies of essential minerals or imbalance of available minerals; weed control in malting barley is essential throughout the growing season to minimize competition by weeds for soil moisture.

Producing a high-quality crop in the field

Table 17-3
The effects of varying rates of nitrogen and potassium fertilization and different planting dates (early or late) on yield and percentage of grain protein in barley grown in North Dakota.

Rate of application of nitrogen or potassium (kg/ha)	Yield (kg/ha)		Average percentage of protein	
	Early-seeded	Late-seeded	Early-seeded	Late-seeded
0.0 N	2,613	2,381	12.8	13.4
22.4 N	2,866	2,645	13.1	13.6
44.8 N	3,023	2,684	13.2	13.9
67.2 N	3,147	2,770	13.5	14.2
0.0 K	2,882	2,569	13.2	13.8
14.0 K	2,924	2,640	13.1	13.7
28.0 K	2,930	2,650	13.1	13.8

Source: J. C. Zubriski, E. H. Vasey, and E. B. Norum. Influence of nitrogen and potassium fertilizers and dates of seeding on yield and quality of malting barley. *Agron. J.* 62(1970):216–219.

Note: In general, potassium has little effect. With late plantings, even the lowest rates of nitrogen fertilization result in unacceptably high levels of grain protein for malting; with early plantings, yields increase with applications of nitrogen fertilizer, and even at the highest rate of nitrogen fertilization, the percentage of grain protein is at least acceptable for malting.

is the first step toward realizing cash premiums, but all the effort of producing high-quality malting barley can be lost through careless harvesting. Careful adjustment and frequent inspection of combines are essential (Figure 17-6). As conditions change in the course of a day (the grain dries more later in the day), it may be necessary to readjust harvesting equipment. The specific adjustment depends on the particular make and model of the combine, but several generalizations can be made.

First, it is essential to watch the speed of the combine cylinder. High cylinder speeds cause excessive damage to kernels. As a rule, cylinder speeds that are suitable and safe for harvesting wheat are too high for malting barley. In addition, cylinder speeds for two-row types should be lower than for six-row types. The two-row types thresh easily and have more fragile hulls; therefore, they are more vulnerable to damage. Also, because they are produced in drier regions, two-row types of barley are drier and more brittle at maturity than six-row types, which are grown in the more humid midwestern areas. Proper concave adjustment is also crucial to minimize damage to kernels. The concave adjustment must be correlated with the cylinder speed: close concave settings require slower speeds. Settings have to be changed as the grain dries during the day. Although the persistence of a short segment of awn with the grain reduces test weight, it does not affect plumpness, which is measured by grain that passes through (or does not pass through) sieves of specific sizes and shapes. Harvest procedures that leave a piece of awn generally help hold kernel damage to a minimum. Finally, to preserve the quality of the grain as it comes from the field, both sieves and wind must be properly adjusted. Tailing returns should be held to a minimum by proper sieve adjustment. Wind speed should be great enough to separate grain and lift the chaff over the shoe. As conditions change throughout the day, these adjustments must be monitored. Even after harvest, care must be taken to

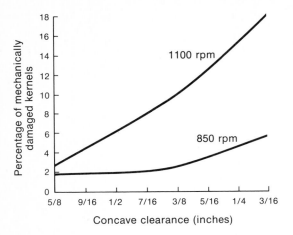

Figure 17-6
Relation between concave clearance, cylinder speed, and damage to barley kernels. The malting industry demands minimum damage. [Courtesy of the Malting Barley Improvement Association.]

minimize damage to grain as it is moved to and from storage areas and to protect it from pests and other damage in storage.

Diseases and Pests

DISEASES

Barley is susceptible to a wide range of fungal diseases; smut is very common. In plants suffering from covered smut, kernels are replaced by a hard mass of black spores encased in a membranelike sheath. The disease becomes apparent when plants reach the heading stage of development and fungal spore masses become evident. Spores are carried on the surface of the kernel where they germinate and infect young seedlings. Covered smut is more prevalent in crops grown on acidic soils than in those grown on neutral or alkaline soils. Some genetic resistance to the disease has been reported, but the most common control is through seed treatment. Organic mercury compounds have been very effective; however, they are forbidden under current federal regulations. Treating seed with other compounds, such as hexachlorobenzene (HCB), effectively controls covered smut, and new, systemic fungicides also provide control of the disease.

Nuda loose smut is another serious fungal disease of barley. In plants suffering from loose smut, entire flowers are replaced by masses of spores. When these masses rupture, the spores, carried by wind or rain, infect other plants by penetrating the stigma and surviving in the developing kernel. The pathogen is carried inside the seed. The most effective control of loose smut is the use of genetically resistant barley cultivars. Because the pathogen is carried in the seed, chemical seed treatment is ineffective. Carefully controlled hot water seed treatments can also control the disease. Seed is soaked for five hours at 70°F (21°C), then drained and soaked for one minute at 120°F (49°C), and then transferred and soaked at 126°F (53°C) for 11 minutes. After soaking at 126°F (53°C) for 11 minutes, the seed is plunged into a cold bath. Because temperature control during hot water treatments is critical, they are used only for small lots of seed. Anaerobic soaking treatments at lower temperatures are also used to control loose smut.

Barley is also susceptible to a number of other diseases caused by fungi. Stem rust, leaf rust, and stripe rust cause periodic losses, mostly in winter barley grown in the more humid areas typical of the southern United States, Europe, and Asia. The only economical control of these rust diseases is the use of resistant cultivars. Of all the cereal crops, barley is the most susceptible to and suffers the greatest losses from powdery mildew. The disease is favored by conditions, such as excessive nitrogen, that result in lush vegetative growth. Powdery mildew can be controlled with sulfur dust, but like many diseases, the best control is the use of resistant cultivars.

Root rots and damping-off diseases also cause losses in barley. New compounds, such as HCB and carbamates, are moderately effective in disease control.

Ergot can be a problem in barley. In plants suffering from the disease, the grain is replaced with a hard, horny spore mass—a sclerotia—that is potentially toxic to humans and domestic animals. Resistant cultivars have not been developed to control ergot and there are no chemicals that control the disease. Cultivation to bury the sclerotia and eliminate the source of primary inoculum combined with conditions favoring pollination minimize losses. Cultivars that have open florets (in which the lemma and the palea separate as a result of the swelling of the lodicules) are more susceptible to infection because the fungal spores can readily contact the pistil.

Other diseases of barley include scab, scald, net blotch, and viral diseases such as yellow dwarf and barley stripe mosaic.

WEEDS

Like all other field crops, barley is subject to losses from the presence of weeds. Weed control starts with seedbed preparation (or with fallow operations in many semiarid regions) and includes planting high-quality, weed-free (certified) seed. During the growing season, barley crops, like all closely planted crops, are not cultivated because of narrow row spacing. Herbicides are also important in controlling weeds in barley. Both preemergence and postemergence types are used. More than half the acreage harvested in the United States each year is treated with some type of herbicide. Postemergence herbicides are commonly (and effectively) used. Preemergence chemicals or combinations of pre- and postemergence sprays are not widely used.

Weeds that commonly compete with wheat (Chapter 16) are also serious pests in barley. These include pigweeds, lambsquarters, thistles, bindweeds, dock, wild buckwheat, and wild oats. Several weeds have become significant in the United States in the past ten years: for example, pigweeds, kochia, thistles, and wild radish. However, as a result of the development of effective weed management programs, a few weeds have decreased in significance. These weeds include quackgrass, ragweeds, cheatgrass (downy brome), and knawel.

INSECTS

In the central United States, chinch bugs damage barley as well as other cereals. There is no effective, economic control for this pest. In this same region and further south, the green bug may be a serious pest in winter barley. Genetic resistance to this pest has been identified, but it has not yet been widely incorporated into cultivars adapted to the area. Grasshoppers also cause significant damage to barley periodically. If infestations are heavy, chemical control is practical. Chlorinated hydrocarbon insecticides are effective in controlling grasshoppers, but their use is currently restricted. Organic phosphates and carbamate compounds provide reasonably effective control as do various types of poison baits. Aphids also damage barley both by feeding and by transmitting viral diseases. Control for aphids under normal conditions is not practical.

Crop Utilization

There are three major uses for barley in North America. More than 50 percent of both the American and the Canadian crop is used as feed. Barley is fed as whole grain to sheep; before being fed to other types of livestock, it is usually ground or rolled. The grain may also be steamed to soften it so that it can be more readily rolled into flakes. The feed value of barley is about five percent less than that of corn. The nutritional value of the grain is slightly reduced by the presence of the relatively nonnutritious hulls (lemma and palea), which comprise about 10 percent of the grain. Recent advances in plant breeding hold high promise for the development of barley cultivars with higher protein content and increased amounts of specific, essential amino acids (e.g., lysine).

Some barley is pearled and used in processed foods or exported as a rice substitute. Pearling is a type of milling that removes the hulls, bran layers (including aleurone), and germ from the caryopsis, leaving a pellet of starchy, high-energy endosperm. To be processed in this manner, barley must be plump and have a high test weight. The grain of some barley cultivars has a characteristic color; this coloring is frequently associated with the aleurone layer. If this coloring cannot be easily and completely removed in processing, the grain may be considered unacceptable for pearling.

In the United States, as much as 25 percent of the total barley crop is used in making malt. Less of the Canadian crop is converted into malt, although cultivars suitable for malting barley are widely grown. Nearly all of the malt produced in both countries is used in making beer and whiskey. The production of malt takes advantage of enzyme-regulated changes that occur during germination, modifying starch stored in the endosperm to readily fermentable materials (a fermentable material is one that can be converted into ethyl alcohol). The malting process requires that grain germinate under closely regulated conditions so that enzymatic activity can be controlled to produce a uniform product. The key enzyme is diastase, which regulates the conversion of starch into dextrin and maltose (two readily fermentable sugars). Plump kernels are desirable because they have a proportionately large quantity of starch (endosperm) that can be enzymatically converted into fermentable sugars.

The first step in producing malt is to soak (steep) clean grain for two or three days until the kernels have a uniform moisture content of 44 to 46 percent. During steeping, water may be drained periodically. Uniformity in rate of water uptake is essential; thus, damaged or skinned kernels that absorb water more rapidly are undesirable. After steeping, the grain is drained and transferred to either a rotating drum or a special room where it is spread in a uniform layer about a foot deep and stirred periodically. The temperature in the room is maintained at 68°F (20°C), and the grain is allowed to germinate for about six days. Germination activates the diastase enzyme system, catalyzing the conversion of stored starches into sugars normally used by the developing seedling as a source of energy for growth. Uniformity in the rate of germination is essential. Damaged kernels that germinate too rapidly (or not at all) lower the quality of the malt. When the sprouts (referred to as acrospires in the malting industry) are between 75 and 100 percent as long as the kernel, the germination process is terminated by artificially drying the germinated seed in a kiln until the moisture content drops to between 4 and 5 percent. The dried kernels minus the malt sprouts (rootlets), which are knocked off in handling the dried grain, are known as dried malt. A bushel of dried malt weighs about 34 pounds. The quality of the malt, affected by conditions during steeping and germination, depends largely on cultivar characteristics.

In brewing, malt is ground, mixed with water and other grain products that provide additional fermentable materials, and cooked. The cooked material is added to pure malt that has been held in water maintained at 48°C; this temperature favors the breakdown of proteins. The temperature of the new liquid mixture is elevated twice (first to 65°C, then to 75°C) to complete the conversion of starches into sugars and to denature the proteins. The liquid (first wort) ultimately becomes beer. During the heating stages, the protein level of the malt is critical. Excess protein not only causes excessive enzymatic activity, but also reduces the quality of the wort, hence of the beer produced directly from it. Thus the quality of the grain in the field is ultimately reflected in the end product on the supermarket shelf.

Small amounts of barley are used in the baking industry in North America; in Europe, barley flour is used in small quantities in producing black breads. As the nutritional quality of barley is improved, it may again become a major food, as well as a major feed crop. Because barley is the most widely adapted cereal species, it will continue to be an important crop worldwide.

Selected References

Leonard, W. H., and J. H. Martin. 1963. *Cereal crops.* Macmillan.

Malting Barley Improvement Association. *Grains of truth about malting barley: malting barley protein content.*

Malting Barley Improvement Association. *Know your 1973 malting barley varieties.*

Malting Barley Improvement Association. *Threshing malting barley.*

Manitoba Department of Agriculture. 1974. *Field crop recommendations for Manitoba.*

Zubriski, J. C., E. H. Vasey, and E. B. Norum. 1970. Influence of nitrogen and potassium fertilizers and dates of seeding on yield and quality of malting barley. *Agron. J.* 62:216–219.

Study Questions

1. Why and where might barley be grown in preference to wheat?
2. What factors affect the malting quality of barley and how can they be managed in the field?
3. Describe the major differences between a spike of six-row barley and one of two-row barley; between barley and wheat.
4. What are the major diseases and pests of barley? How can they be managed or controlled?

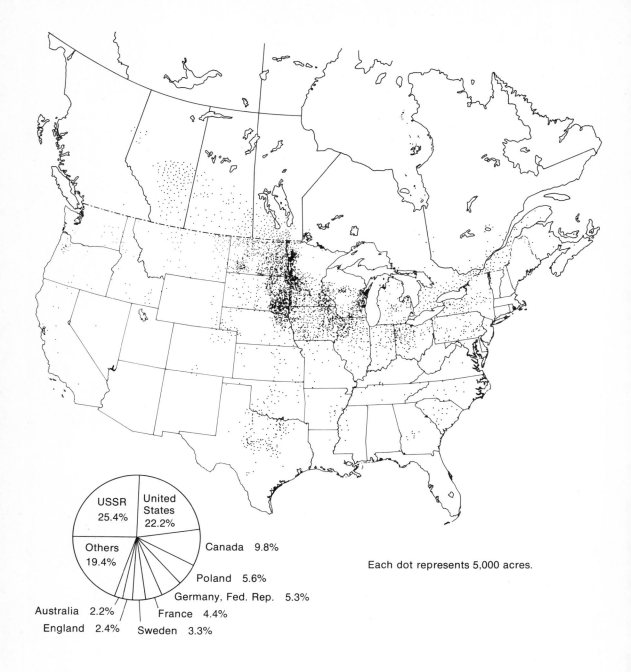

Figure 18-1
Major oat-producing regions in North America and contributions to world production. [Data from U.S. Departments of Commerce and Agriculture and the United Nations.]

Eighteen

Oats and Rye

Origin and History of Culture

Oats Two closely related species of oats are cultivated today—*Avena sativa* (common oats) and *Avena byzantina* (Mediterranean oats). At one time it was believed that these species evolved in Europe, but N.I. Vavilov established that they evolved from several species of wild oats about five thousand years ago in central Asia (northwestern India and Afghanistan), where common wheat and two-row barley had evolved two or three thousand years earlier.

Compared with wheat, oats came under cultivation relatively recently. In Europe, archeological findings indicate that oats emerged as a cultivated crop at about the same time as barley, nearly four thousand years ago. Evidence also shows that cave dwellers in what is now Switzerland cultivated wild oats before 1000 B.C. However, little is known about domesticated oats before the Christian Era. Ancient Egyptian and Hebrew writings contain no record of the cultivation of oats. Some agricultural historians maintain that the earliest written record of oats is found in the works of Greek philosophers and naturalists who wrote between 400 and 250 B.C. Available records indicate that some type of oats was a weed pest in barley and wheat fields at the beginning of the Christian Era; early writers even hypothesized that both wheat and barley somehow degenerate into oats! (Wild oats are still a major weedy pest of other small grains.) Early colonists brought oats to the New World along with other grains. They first grew them on an island off the coast of Massachusetts at the turn of the seventeenth century (about 1602). By 1611, the Jamestown colonists were growing oats on the mainland.

Today, the acreage planted to oats (Figure 18-1) is variable and on the decline. Although oats are suitable for milling, they do not equal

Figure 18-2
Major rye-producing regions in North America. [Data from the U.S. Departments of Commerce and Agriculture.]

wheat in nutritional value for humans. They are nevertheless a superior feed for horses, dairy cows, poultry, and breeding stock in general. Oat grain may be higher in crude protein and fat than the grain of barley. (Remember that the percentage of grain protein is markedly affected by environmental factors. Nitrogen fertilizer tends to increase grain protein, and grain produced under drought conditions may be relatively high in protein because endosperm, or starch development is restricted.) Also, cultivars differ in percentage of grain protein. In regions in which annual hay is produced, oat hay is popular and oat straw is a valuable bedding material. In spite of its nutritional value and its broad adaptation, oats rank a distant fifth in world cereal production, behind rice, wheat, corn, and barley.

Rye Cultivated rye (*Secale cereale*) originated in the same general area as wheat, barley, and oats. Like oats, its evolution followed that of wheat and barley. Archeological evidence indicates that rye was the last of the cereals to be cultivated. Thus, cultivated rye evolved long after wheat and barley, but at about the same time or shortly after oats. The failure of rye to become a major food crop may in part be explained by the fact that both wheat and barley were relatively intensely cultivated by the time rye reportedly evolved from a wild ancestor, *S. montanum*.

Rye has never been a major world crop, but until the early twentieth century, the worldwide acreage used for rye production was almost half that of wheat; more than 95 percent of the rye was produced and consumed in Europe. Leading producers were Russia (more than 50 percent of the total produced) and Germany (about 25 percent). As the production of wheat spread with the development of better-adapted cultivars and modern technology, the acreage planted to rye continuously decreased. Today, rye is of minor importance in Canada and the United States (Figure 18-2). However, in eastern Europe and the Scandinavian countries, rye is still a leading source of flour, second only to wheat.

In North America, rye is grown for grain; it is also used as an annual hay or pasture crop (as are oats). The giant types of rye are used as green-manure crops or as cover crops.

Botanical Characteristics

Oats Cultivated oats have a panicle type of inflorescence (Figure 18-3) that distinguishes them from the other cool-season cereals. Generally, the panicle is open and highly branched. The spikelet is formed at the end of a pedicel and usually has two fertile, lateral florets and a sterile, central floret. The central floret may be staminate (have functional anthers). Many cultivars have fairly conspicuous awns. In the vegetative stage, oats are recognized by the presence of a very conspicuous ligule and the complete absence of auricles.

Depending on cultivar characteristics and environmental conditions (available moisture and minerals, diseases, and planting rate), oat plants range from slightly less than 2 feet to 5 feet (60–150 cm) in height; they occasionally grow taller than 5 feet. Under typical field conditions (Figure 18-4), each plant has from three to five tillers. Oat stems are hollow. Leaves are up to 12 inches (30.5 cm) long and more than 0.5 inch (1.3 cm) wide. The root system is typical of most annual grasses.

At maturity a single oat plant can produce more than 100 viable kernels. Kernels are light yellow to white, but hulls may be yellow, red, or, in rare instances, black. The hulls (palea

Figure 18-3
Typical oat panicles. Glumes are relatively large. Pairs of florets and single florets are shown at the bottom. [USDA photograph.]

Figure 18-4
Field of oats nearly ripe for grain harvest.

and lemma) do not thresh free of the caryopsis at maturity, with one exception: the kernels of the species *Avena nuda* do thresh free. Although this species is of no significant economic importance, the so-called naked trait is of interest to some plant breeders and geneticists.

Rye Rye, like wheat and barley, has a spike type of inflorescence (Figure 18-5). Like wheat, it has a single spikelet per node of the rachis. Unlike wheat, it has three florets per spikelet; the two outer spikelets are fertile and the central one is sterile. In the vegetative stage, wheat and rye are difficult to separate. Rye is typically more bluish green and much taller than wheat.

Rye plants grow taller than any other cereal except corn. Plants are typically more than 5 feet (1.5 m) tall. Some giant (tetraploid) cultivars can grow taller than 6 feet (2 m). As with other cereals, tillering depends on environmental conditions. The prolific vegetative growth typical of rye makes it well suited for use as a green-manure crop.

Kernels of rye are quite similar to those of wheat. The size, shape, and texture are so similar that rye kernels are extremely difficult to separate from wheat. Therefore, rye is generally not grown on fields on which wheat is produced because volunteer rye plants contaminate the wheat and lower its grade.

Environmental Requirements

Oats Oats are adapted to cool, temperate regions with 25 or more inches (63.5 cm) of precipitation per year, or with irrigation. Temperatures above 33°C at flowering cause "blasting," or the dropping of florets. Thus, oats cannot be considered to be as heat toler-

Figure 18-5
Spikes of typical rye plants. [USDA photograph.]

ant as wheat or barley. Mediterranean (red) oats are more heat tolerant than white or common oats. If oats grow under cool conditions in nitrogen-rich soil, the plants accumulate enough nitrates to be toxic to livestock fed hay or silage made from such material.

Soil pH is relatively noncritical in oat production; compared with other cereals, the crop tolerates a wide pH range. However, soil texture is more critical. Oats tend to lodge badly if grown on heavy soils; thus, light- to medium-textured soils are recommended for oat production. Oats require moderately high soil fertility for high yields. Care must be exercised to balance all essential minerals and to avoid excess nitrogen. Applying more than 60 pounds of nitrogen per acre (about 70 kg/ha) can result in nitrate accumulations and/or lodging because of lush vegetative growth. In more humid areas, a lack of phosphorus may limit yields; favorable economic returns are common if phosphorus fertilizers are used. As is the case for all crops, proper fertilizer applications depend on the results of soil tests.

Although oats are broadly adapted and can be grown in a wide range of environments, more than 75 percent of the total acreage planted to oats in the United States is in the northern central states, with more than 50 percent of the acreage located in Minnesota, Wisconsin, Iowa, and North and South Dakota. Oats are produced in adjacent regions of Canada, where adequate moisture is available.

Rye Rye is the most cold hardy of all cereal crops. It germinates at temperatures at or close to freezing. On cold, acidic, unfertile soils, rye is the best-adapted cereal. However, for optimum growth and development, rye requires the same balance of essential minerals as wheat. Specific recommendations for fertilization depend on soil analysis and on local climatic conditions.

Production Practices

SEEDBED PREPARATION AND SEEDING

Oats Seedbed preparation for oats is the same as that described for other cool-season cereals. Fall-sown oats should be seeded from mid-September through mid-October. Yields are markedly reduced if crops are planted earlier or later, with the exception of California, where oats are seeded in November. In areas with mild winters, oats can be sown through December. Seeding in early-to-mid-fall fosters vigorous seedling development and winter hardiness. Dates of spring seeding vary from south to north; it may begin as early as mid-February in southeastern Texas and as late as mid-May in North Dakota and Minnesota (and adjacent regions in Canada). Spring-sown oats should be seeded as early as possible to avoid drought and heat damage (flower blasting) during heading. For optimum growth and highest yields, they should be seeded before the soil temperature reaches 50°F (10°C).

Depending on location and predicted availability of moisture during the growing season, planting rates range from as high as 60 or 70 pounds of seed per acre to as low as 30 pounds or less per acre (78.4–33.6 kg/ha); yields depend both on row spacings and on seeding rates (Table 18-1). Oats are commonly seeded with a standard grain drill in rows from 7 to 12 inches apart (17.8–30.5 cm). Where oats follow corn in a rotation system, broadcasting is often practiced to avoid fouling the drill with corn stubble.

Rye The procedures followed in the seedbed preparation and seeding of rye are essentially

Table 18-1
Effect of variations in planting rate and row spacing on yield (in bushels per acre) of oats.

Seeding rate (lb/acre)	Row spacings (inches)			
	8	16	24	32
64	90.1	—	—	—
32	89.5	75.1	—	—
21.3	—	—	64.3	—
16	—	70.5	—	54.5
10.7	—	—	58.9	—
8	—	—	—	47.6

Source: H. L. Shands and W. H. Chapman. Culture and production of oats in North America. In *Oats and oat improvement*, ed. F. A. Coffman (American Society of Agronomy, 1961).
Note: Highest yields were obtained with 8-inch row spacing, but at this spacing seeding rate had little effect.

the same as those described for winter wheat. Winter rye is seeded from late summer to late fall. Fall seeding should be done early enough to insure good stand establishment and seedling vigor before winter freezing. Spring rye should be seeded as early in the spring as soil conditions and temperatures allow. Seeding rates for rye range from 20 to slightly more than 30 pounds per acre (22.4–33.6 kg/ha).

HARVESTING, STORAGE, AND ROTATIONS

Oats Oats, like wheat and barley, are frequently harvested with a self-propelled combine. Obviously, equipment must be cleaned and adjusted for the specific crop to avoid excessive losses in the field. Although the kernels of all grasses are botanically the same, variations in size, weight, and texture must be considered in adjusting equipment. Oats are susceptible to excessive losses from shattering at maturity. In parts of the United States and Canada where shattering can be a severe problem, oats are windrowed and allowed to dry before being combined. If a crop is produced for hay, highest yields are obtained by harvesting when the kernels are in the milk or very soft-dough stage of development. The management and specific problems of oats and other cereals grown for pasture and hay are discussed in Chapter 31.

Rotation sequences for oats are similar to those previously described for wheat and barley. Because oats are produced in more humid regions, fallow is of much less importance. In the corn belt, where oats are preferred to wheat and barley in crop rotations, a common rotation is corn–corn–oats. Oats are often planted with an annual or biennial legume (e.g., sweetclover); the oats are harvested for grain, and the companion legume grows and serves as a green-manure crop, which is turned into the soil in the fall. Recall that legumes fix nitrogen; thus, they increase soil fertility. The crop residue improves the physical properties of the soil.

Rye Like other small-seeded cereals, rye is harvested with a combine. Rotations for rye are similar to rotations for other cereals grown in semihumid areas, except that rye is rarely included in wheat rotations because volunteer rye lowers the quality of the wheat and is extremely difficult to separate from it.

Diseases and Pests

DISEASES

Oats Oats are susceptible to a number of fungal and viral diseases. Diseases caused by bacteria are of less importance. In humid and subhumid areas, crown rust is a significant fungal disease of oats. This disease is found in all the oat-producing areas of the world. Infected plants ripen prematurely, produce grain having a low test weight, yield less overall, and lodge severely. The fungus causes a reduction in test weight and yield and increases the water requirements of the plants. The only effective control of the disease is the use of resistant cultivars. Eradicating buckthorn, the alternate host for the fungus, is also important in the control of the pathogen. Stem rust is another fungal disease that affects oats. Again, growing resistant cultivars is the only effective control, but no oat cultivars are resistant to all races of the fungus that cause the disease.

Oat seedlings, and to a lesser extent the tillers of more mature plants, are susceptible to Victoria blight, which is caused by a fungus carried on the grain; the fungus infects and kills seedlings. Seed treatment with approved fungicides affords some protection against the diseases; however, disease-resistant cultivars offer the greatest protection. Unfortunately, genetic resistance to Victoria blight is associated with susceptibility to crown rust, and breeders have not yet developed oat cultivars resistant to both diseases.

Oats are also susceptible to both loose and covered smut. In both diseases, fungal spore masses coat or replace floral parts. Partial control of smut diseases can be achieved by treating seeds with new compounds such as HCB. As is the case with most diseases of cereal crops, the use of resistant cultivars is the most effective method of controlling loose and covered smut of oats.

In the humid, oat-producing areas along the Gulf of Mexico, *Helminthosporium* leaf blotch, a fungal disease, is a serious problem. The fungus causes the destruction of leaf tissue and subsequent yield losses. Control methods include crop rotation to destroy spores on crop residue, seed treatment, and the use of resistant cultivars.

The main viral pathogen that infects oats is the yellow dwarf virus that attacks wheat and barley. Aphids transmit the virus while feeding on healthy plants. The virus causes destruction of chloroplasts and stunting of plants, thereby reducing both grain and hay yields. The only effective control is the use of resistant cultivars currently being developed by plant breeders. Field control of the vector aphid is not effective because the insects feed on healthy plants and transmit the virus before insecticides can destroy the insects.

Rye Ergot, a fungal disease, is the most infamous disease of rye. Fungal spores produced by overwintering bodies (sclerotia) enter rye florets during the normal pollination period, germinate, and ultimately invade the ovaries. If conditions for pollination are favorable, losses are reduced. Remember, rye is cross pollinated. Ergot is recognized by the presence of sclerotia, which replace kernels. These fungal reproductive bodies are toxic to livestock; if milled, they cause the flour produced to be unsuitable for human consumption. The source of toxicity in sclerotia is lysergic acid, a major component of the illegal, hallucenogenic drug LSD. In northern Europe, periodic outbreaks of ergot in rye have caused extensive epidemics of ergot poisoning and death among humans. Ergot is not all bad: properly refined, it can be used for a variety of medical purposes. There is no satisfactory chemical control for ergot nor have resistant cultivars been developed. The best control of the disease is deep plowing to bury sclerotia and eliminate sources of primary inoculum, and cultural practices that insure good pollination. Such cultural practices include proper seedbed preparation and seeding to insure vigorous, healthy plants and balanced mineral nutrition. Temperature extremes during flowering can cause pollen sterility and result indirectly in ergot infection. Although temperature cannot be modified on a large scale, this problem can be managed by selecting cultivars that flower when extreme temperatures are not expected and by regulating planting dates. Drought or moisture stress should be avoided if possible.

Other diseases that affect rye include stem smut, anthracnose, and rusts. If the spores of fungi causing stem smut are carried on the seed, the disease is controlled by seed treatment. If the spores are soil-borne, chemical control is less effective, but smut-resistant cultivars afford good control of the disease, regardless of the source of inoculum. Anthracnose is controlled by maintaining adequate soil fertility and through crop rotation to destroy sources of inoculum. No effective controls of rye rusts are available.

WEEDS

Oats Oats are plagued by the same weeds that are pests in other cool-season cereals. General control methods are essentially the same as those described for wheat and barley. Because oats are produced in areas of relatively high rainfall, fallow is not a significant weed-control practice with oats, as it is with wheat or barley. Wild oats are a particular problem in cultivated oats because the kernels of wild and cultivated oats are very similar, which makes removing the seeds of wild oats from cultivated oat seed difficult. Wild oats are also alternate hosts for many diseases of cultivated oats (e.g., yellow dwarf virus).

Rye Weed pests of rye are generally the same as those for other cool-season cereals, and control measures are the same as those previously described. Rye can be a weed in other cereals; for example, in wheat.

INSECTS

Oats Only three insects are serious field pests of oats. The green bug, also a pest of barley, may cause significant losses in the southern central states, where chemical control of this insect may be justified. In addition, aphids cause damage by transmitting viral diseases and, to a lesser extent, through feeding. From time to time, grasshoppers cause serious losses, too. Insects cause as much as a 5 percent annual loss of stored oats; they can be controlled in the granary through cleanliness and fumigation. Cleanliness is essential; fumigation of the granary and of the stored grain may also be necessary. As is the case with any pesticide, fumigants must be carefully and properly used for the safety of the fumigator and the protection of the stored crop.

Rye Many of the insect pests that attack other cereals also damage rye. Among the leading insect pests are Hessian flies, sawflies, chinch bugs, and grasshoppers. In general, insects do more damage to other small-seeded cereals than to rye. Little field control of insect pests of rye is practiced.

Crop Utilization

Oats About 90 percent of the oats produced in Canada and the United States is used for livestock feed. The majority of this is fed on the farm. Thus oats play a comparatively minor role in grain marketing. Oats contain a higher percentage of protein and oil than barley, and are therefore slightly more nutritious. Oats are either fed whole or rolled before feeding.

A small fraction of the oats produced in North America is milled for oat flour or processed for dried, packaged breakfast cereals. The oat husks or hulls, a by-product of the milling process, have a number of commercial uses because of the furfural (an oily liquid aldehyde) they contain. Furfural, or its chemical derivatives, is used extensively as a raw material in the manufacture of nylon, as a solvent essential in the manufacture of synthetic rubber, in refining petroleum oils and resins, and in pesticides, disinfectants, and preservatives. Oat husks are also used as filters in the brewing industry, and as fillers and sources of fiber in manufacturing paper and paper products.

Oats are frequently grown as a companion crop in the establishment of perennial grasses and legumes; they are also grown as a cover crop to protect the soil. In addition, oats are a suitable hay crop. Oat straw is a valuable bedding material and a satisfactory roughage for livestock.

Rye Most rye is milled for flour, coarse rye meals, and rye flakes; only a small percentage of the world crop is consumed as livestock feed or used by the North American malting industry. Rye flour is used in the baking of dark breads (e.g., pumpernickel). Bread made from rye flour is heavy and quite dense. Rye flour is also used as a filler in prepared soups, sauces, and custards, and starch from rye flour is a major constituent of commercial adhesives. Coarse rye meal is used in the production of hard breads, and rye flakes are used in producing breakfast cereals.

Rye straw has a low nutritional value and is not suitable for livestock feed; it is used mainly as bedding material. At one time, it was in demand for the manufacture of hats, harness padding, baskets, and similar items. It was also used in the production of coarse, brown paper, but has been replaced by wood pulp. Today, rye straw is used as packing material for a variety of items; however, synthetic materials have reduced its role in this area.

Because rye grows at very low temperatures and tillers more prolifically in the early spring than other cereals, it produces relatively abundant, early vegetative growth; this trait makes rye the most suitable (and least harmed) cereal crop for early spring grazing

in crops to be harvested for grain. Remember that plants must retain enough leaf tissue for adequate photosynthesis. Overgrazing must be avoided or serious yield losses will occur.

Triticale, a man-made species derived from rye-wheat hybrids, is a potentially valuable crop. Plant breeders have attempted to combine the tolerance of rye to severe environmental conditions with the nutritional, milling, and baking qualities of wheat. This new crop is discussed in Chapter 32.

Selected References

Bland, Brian F. 1971. *Crop production.* Academic Press.

Coffman, Franklin A., ed. 1961. *Oats and oat improvement.* American Society of Agronomy.

Leonard, W. H., and J. H. Martin. 1963. *Cereal crops.* Macmillan.

Study Questions

1. Compare the environmental requirements of rye with those of other cereals, and discuss the conditions under which rye might be an important crop.
2. Characterize the inflorescences of wheat, barley, and rye, and distinguish those of rice and oats.
3. What are the major pests of oats and of rye? How can they be managed or controlled?
4. Describe and explain the major similarities and differences between the production practices applied to oats and rye and those used in growing other cereal crops.

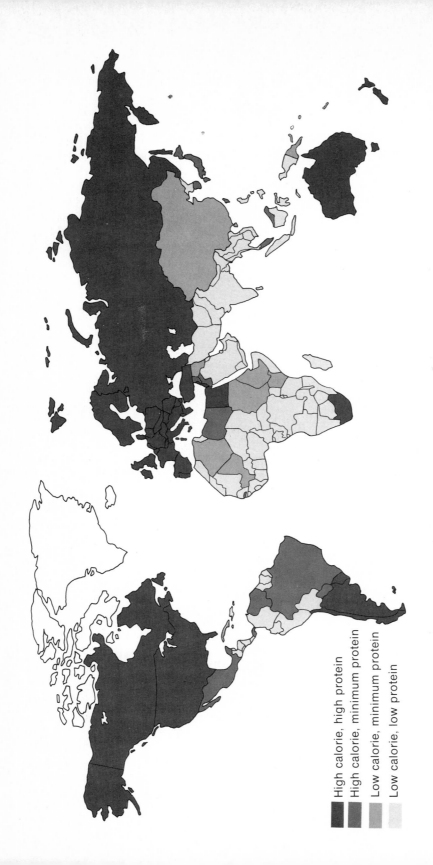

Figure 19-1
Availability of kilocalories and of protein for the human population. Note the dietary wealth of North America versus the dietary poverty of much of Africa. Even countries such as Mexico and Brazil have poor, protein-deficient diets. [From *Population, Resources, Environment*, 2d ed., by Paul Ehrlich and Anne H. Ehrlich. W. H. Freeman and Company. Copyright © 1972.]

- ■ High calorie, high protein
- ■ High calorie, minimum protein
- ■ Low calorie, minimum protein
- ■ Low calorie, low protein

Nineteen   Edible Legumes and Man

Introduction

The frequently quoted axiom "man does not live by bread alone" is of more than philosophical interest; it is a biological fact! To prevent malnutrition, a diet must provide both sufficient calories and nutritional balance. In many areas where the average daily diet includes 2200 kilocalories, malnutrition is still prevalent because the diet is poorly balanced (Figure 19-1). Most commonly, the average diet is high in starch and low in protein. In Asia, the ratio of starch to animal protein (including dairy and poultry products) to nonstarch plant products (including fruits, vegetables and nuts) is about 75:10:15 by weight. In North America, the ratio is about 40:35:25.

The problem of protein deficiency is complex. Although the prevention and relief of serious deficiency fall mainly in the realm of the nutritional and medical sciences, the amount and the quality of protein produced by plants are two basic concerns of plant scientists. Plant species differ with respect to the amount and the quality of protein they provide. The amount of protein is quite simple to calculate; it is expressed as a percentage of the total weight of a food, or as the number of milligrams of protein per gram of food (or as some other weight-to-weight ratio). Quality is more complex to compute. The availability of specific amino acids, the building blocks of proteins, determines protein quality. The human body uses more than twenty different amino acids to build the various proteins it needs for growth, development, and maintenance. Of these, ten are considered essential to the human diet because they cannot be synthesized by the body. The ten essential amino acids are, in alphabetical order: arginine, histidine, isoleucine, leucine, lysine, methionine, phenylalanine, threonine, tryptophan, and valine. Protein from any source (plant or animal) is classified as complete (provides all essential amino acids in adequate concentration), partially complete (provides all but one essential amino acid, or provides all essential amino acids but in low concentration for normal growth), or as totally incomplete (does not provide an adequate supply of essential amino acids).

In terms of human nutrition, cereals are excellent sources of energy (starch) but comparatively poor sources of protein, whereas several genera in the legume (Leguminosae) family provide large amounts of high-quality protein (see Figure 13-1). Protein from both field beans and soybeans is nutritionally complete when supplemented with methionine. Soybean meal (see Chapter 20) is as much as 40 percent protein (comparable to some sources of animal protein and far less expensive, but more efficient, to produce). Peanuts are also an exceptionally good source of high-quality protein.

Animal protein, particularly meat, must be considered a luxury. Only those countries rich in plant resources can afford to feed plant products to livestock and then indirectly consume the plant product in the form of meat. Amino acid requirements (protein requirements), which many believed could be satisfied only with meat or animal products, can be fully satisfied with a balanced cereal-legume diet, or with a legume diet supplemented with specific amino acids. Combinations of soybeans and rice, wheat and peas, or corn and beans provide adequate, well-balanced nutrition. In the face of increasing populations and decreasing space in which to produce food, feed, and fiber, plant protein for direct human consumption will become increasingly important. The best sources of plant protein are found in the legume family; thus, a knowledge of this important plant family and of some of the major genera in it is essential to plant scientists and crop producers.

Origin and History of Culture

The legume family (Leguminosae) consists of more than 10,000 species; the most important edible legumes belong to 14 species in 10 genera. Common names for edible legumes confuse their generic relationships. Field and garden peas are classified as *Pisium sativum*. Chickpeas, also known as garbanzo beans, are neither true peas, nor beans; they are classified as *Cicer arietinum*. Common, or field, beans and lima beans are two species in the genus *Phaseolus*. The term "pulses" refers to the seeds of legume plants consumed directly as food. In Europe, pulses include a wide array of species; in the United States, the term is used less commonly and generally refers to peas and all types of beans. The castor bean (*Ricinus communis*), grown mainly for its oil (which is used for industrial purposes), is not a legume, let alone a bean; it is a member of the Euphorbiaceae family!

One or more important legume species evolved in each of the eight major centers of origin described by Vavilov (Table 19-1). Thus edible legumes originated in the same regions in which the major cereals evolved. The combination of cereals and legumes made available the plants necessary for a balanced diet. Early society probably evolved around cereal culture, but the growth of civilized centers must have also depended upon the cultivation of various legumes. Although early man knew nothing of protein balance and amino acid requirements, surely more than blind luck or fate brought cereals and legumes into his diet. By trial and error, he must have realized that some combinations of plants were better for him than others. (We now know that legumes complement cereals in supplying adequate amounts of essential amino acids.)

Archeological evidence shows that legumes were cultivated before 5000 B.C. The earliest

Table 19-1
Centers of origin of important edible legume species.

Center of origin	Common name	Species
China	Soybean	*Glycine max*
	Chickpea	*Cicer arietinum*
India	Urd bean	*Phaseolus mungo*
	Mung bean	*Phaseolus aureus*
	Cowpea	*Vigna sinensis*
	Pea	*Pisum sativum*
	Lentil	*Lens esculenta*
Central Asia	Horse bean	*Vicia faba*
	Chickpea	*Cicer arietinum*
	Mung bean	*Phaseolus aureus*
Near East	Lentil	*Lens esculenta*
Mediterranean	Pea	*Pisum sativum* (large-seeded)
Abyssinia	Cowpea	*Vigna sinensis*
South Mexico and Central America	Common beans	*Phaseolus vulgaris*
	Lima bean	*Phaseolus lunatus*
South America	Common bean	*Phaseolus vulgaris* (secondary center)
	Lima bean	*Phaseolus lunatus* (secondary center)

Source: Adapted from *The origin, variation, immunity and breeding of cultivated plants* by N. I. Vavilov. Translated by K. Starr Chester (Ronald Press, 1951).

written record of the culture of soybeans—perhaps the oldest cultivated legume—dates back to 2838 B.C. in China. Evidence also indicates that some legumes (beans in the genus *Phaseolus*) were cultivated in Central and South America from four to six thousand years ago, before the cultivation of corn (the only cereal to originate in the New World). As early as 300 B.C., the Greek forerunners of modern plant scientists recorded that several legumes were used in crop rotations and for feed.

Because the various genera of legumes originated in widely diverse regions of the world, no single genus has dominated world legume production. However, soybeans, which are becoming an increasingly important worldwide source of high-quality protein, are the most important edible legume produced today.

The United States and China account for more than 90 percent of world soybean production. Soybeans are also produced in Japan and, in limited quantities, in Brazil.

Peanuts are another major edible legume worldwide. India accounts for more than 30 percent of world peanut production; Africa (Nigeria and Senegal), China, and South America (Brazil and Argentina) also produce significant amounts. The southern United States produces some peanuts. Lentils (*Lens esculenta*), which originated in Southeast Asia and have been an important food throughout history, continue to be of significance in Asia and, to a lesser extent, in parts of Europe; they are more of a specialty in North America. Cowpeas (*Vigna sinensis*) are grown in the southern United States.

Various species of the genera *Phaseolus*, *Vicia*, and *Cicer* are important sources of protein throughout Asia, Africa, and Europe. Both the black gram, or urd bean, and mung beans (*Phaseolus mungo* and *Phaseolus aureus*) are grown extensively in India where they supply essential proteins to many vegetarians. The broad bean, also known as the horse or Windsor bean (*Vicia faba*), is well adapted to northern Africa; it was introduced into Europe before the introduction of common and lima beans (*Phaseolus vulgaris* and

Phaseolus lunatus), which originated in the New World. Common beans, along with soybeans, are the most important edible legumes produced in North America. In the United States, the chickpea, commonly called the garbanzo or gram (*Cicer arietinum*), is grown nearly exclusively in California. Mexico, Spain, and Turkey produce some chickpeas, but India is the major producer.

Botanical Characteristics

Because they are members of the same family, all species of legumes have certain features in common. Legumes can be annuals, biennials, or perennials. The major edible legumes are predominately annuals; under appropriate environmental conditions, some species become slightly biennial or perennial, but among edible legumes this is of essentially no practical importance. The truly biennial and perennial species of legumes are important forage species throughout the world (see Chapters 28 and 29).

The vegetative growth of edible legume plants is either bunch-type (determinate), vine-type (indeterminate), or intermediate. (Flowering is generally indeterminate.) Leaves are compound with petioles and obvious stipules. The leaves of most important edible legumes are palmately compound with three net-veined (palmately veined) leaflets; the leaves of some species are pinnately compound with pinnate venation.

All edible legume plants develop a taproot system. Under favorable environmental conditions, plants are usually deeply rooted. Some secondary roots branch from the primary roots and develop freely, but a massive mat of roots does not form. If inoculated with an appropriate strain of *Rhizobium* bacteria (see Chapter 12), the roots of legume plants form nodules in which the symbiotic bacteria live and fix atmospheric nitrogen. Thus, edible legumes contribute to soil fertility.

Although all legumes have an indeterminate pattern of flowering, the various genera differ with respect to floral groupings (type of inflorescence). Flowers of edible legumes are borne singly, or in groups, in leaf axils, in racemes, or in spikelike clusters. In spite of differences in inflorescences, all members of the legume family have a very typical, distinctive flower (Figure 19-2). The legume flower is complete and perfect. Although in most species self-pollination is possible, cross-pollination, usually with an insect pollen vector, is common. The typical legume flower has a calyx composed of five fused sepals. The corolla is unusual: it is composed of five petals of unequal size and shape. As a result the flower is papilionaceous (shaped like a butterfly). There is a single, predominate petal called the banner or standard. On both sides of the banner there are single wing petals. Opposite the banner, the two smallest petals fuse to form the keel. All the petals may be fused to the calyx as well as jointly fused at their bases to form a corolla tube.

The pistil of a legume flower is composed of a superior, simple ovary with a fairly long style and a somewhat blunt, knob-shaped stigma that is frequently sticky. Three or more seeds generally form in the ovary (in peanuts, normally only one to three seeds form). The ovary matures into a simple, dry fruit, the familiar bean or pea pod, which is also called a legume. Normally, this fruit is classified as dehiscent, but the pods of many edible legumes are indehiscent, a trait that is agronomically desirable because it reduces shattering.

The legume flower has ten stamens. The filaments of nine of the stamens are fused throughout most of their length and form a sheathlike structure around the ovary. The tenth stamen is free. (For details of floral structure, see Chapter 4.)

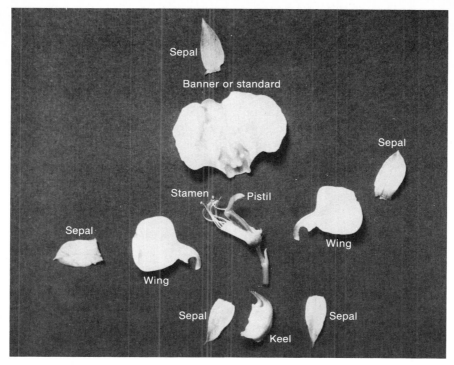

Figure 19-2
Typical legume flower. Note prominent standard or banner petal, two wing petals, and fused petals forming the keel in the lower center of the flower. There are five sepals, but only three are visible at the base of the flower.

Environmental Requirements

Because various genera of edible legumes have evolved in widely diverse areas of the world, it is difficult to generalize about environmental requirements. The environmental requirements of some important species have been modified and their ranges of adaptation extended through plant breeding (see Chapter 32). The major edible legumes are warm-season species; they are planted in the spring or early summer and harvested from late summer to well into the fall. In the United States, planting time and appropriate soil temperatures closely approximate those for corn. However, unlike corn, some species (notably lima beans) do not tolerate excessively high temperatures during flowering. Because they are planted in the spring, it should not be surprising to note that most species of edible legumes are either short-day or day-neutral in their photoperiodic response.

Edible legumes require between 18 inches (46 cm) and 40 inches (100 cm) of precipitation annually. Availability of moisture throughout the growing season is important; adequate moisture immediately prior and subsequent to flowering is critical. Edible legume crops can be produced successfully with irrigation.

Production Practices

The seeds of most edible legumes are larger than the kernels of most cereals (except corn); thus, the seeding depth for legumes is generally deeper than for cereals. Initial seedbed preparation for most edible legumes is comparable to that for corn or cotton and quite similar to that for the smaller-seeded cereals. Unlike wheat and barley, the edible legumes are seeded in fairly widely spaced rows; wider row spacings are necessary for normal plant development. Growers use some type of row-crop seeder to achieve suitable spacing. Currently, there is a trend toward narrower row spacings because research suggests that higher yields are achieved with more narrowly spaced rows; nevertheless, most row spacings for edible legumes are wider than those commonly used with cereals.

Soil fertility must be managed for each crop on the basis of soil tests. Most legumes, with appropriate inoculation, fix nitrogen; therefore, nitrogen fertilizer is generally not required. However, crops respond favorably to applications of phosphorus, potassium, and calcium. Sometimes calcium is essential both as a plant nutrient and to control pH through liming (see Chapter 6). The requirements for other minerals, including micronutrients, vary with the crop species and the particular field, but the quality of the crop may be greatly affected by the application of fertilizers.

Many of the harvesters used for edible legumes are similar in function to the grain combine, but modified in terms of how the plant is cut and lifted to the cylinder and in the design of the cylinder and the concaves.

Diseases and Pests

Like other crop species, edible legumes are subject to a wide array of diseases, which vary from crop to crop (see Chapters 8, 20, 21, and 22). Similarly, various legumes are subject to attacks by a wide array of insect pests. The use of pesticides to control diseases and insects in the field is more common with edible legumes than with cereals, but because legume crops are consumed directly as food, with little or no processing, extreme caution is required in applying pesticides to avoid the accumulation of dangerous residues.

Weeds are a serious problem in fields of all edible legumes, just as they are with all crops. Young legume seedlings planted in widely spaced rows do not severely shade or otherwise suppress weeds through competition. However, the wide spaces between rows make cultivation, particularly early in the growing season, a feasible, though expensive, method of weed control. Extra care must be exercised during cultivation to avoid damaging roots, and reproductive (fruiting) organs in peanuts (see Chapter 21). Herbicides are also used to control weeds in legumes, but, because edible legumes are dicots, they are potentially more susceptible to herbicide damage than many other crops. Therefore, specific herbicides have been developed to control particular weeds in various legume crops at different stages of growth.

The field management of edible legume crops is more rigorous than for cereal crops. This is the result of several factors. The potential cash value per acre of the legume crop is frequently greater than for cereals. Thus, greater production expenses may be well justified. The wider rows necessary for optimum growth make cultivation for weed control during at least part of the growing season both possible and essential. In addition, there are more insect pests subject to field control in legumes; thus, insecticide management becomes more critical than in cereals. Finally, the crop product is consumed directly, without the processing through which dietary supplements are added to cereal products such as flour; thus, the nutritional quality of the crop when it is harvested equals the maximum nutritional quality of the produce consumed by the ultimate consumer. Therefore, growers must optimize field management to produce the best legume crop possible.

Crop Utilization

Like the kernels of cereal crops, the seeds of edible legumes can be stored quite safely for long periods if they are adequately dried at harvest or immediately thereafter. However, because legume seeds have a high oil content, light, humidity, and temperature must be more carefully controlled in storage than with cereals to prevent the seeds (oil) from turning rancid. Of course, any crop product must also be protected from insects and other pests.

Most cereals are processed into flour or meal before they are consumed. Although the seeds of edible legumes can be processed into a variety of products (from flours and meals to peanut butter), processing is not required. Dried seeds of edible legumes enter the diet directly as boiled, baked, or roasted peas and beans. In more than a few instances they are consumed raw, as salads. Nonetheless, the natural nutritive value of legumes is more fully retained with minimum processing.

Many of the edible legumes are considered field crops if the dry seed is harvested at maturity, but they are considered vegetables if the crop is harvested before it matures. Seed harvested before maturity, and sometimes the entire fruit, appears on the grocery shelf as fresh vegetables. Of course, these same crop products can be processed to become the canned or frozen vegetables (beans and peas) familiar to many homemakers throughout North America. Current processing methods help retain much of the nutritive value of the food, but the cost of processing is relatively high.

In a world where more than one-third of the population suffers from hunger and dietary deficiencies, the major role of edible legumes must be as a source of high-quality protein. In addition, many edible legumes are superior sources of edible oil (e.g., peanuts), and, after processing, the by-products of the seeds of many species (e.g., soybean meal) provide a valuable, protein-rich feed supplement for livestock. The vegetative parts (vines) of many legumes are suitable forage. Finally, legumes improve soil fertility through nitrogen fixation and are therefore valuable components of many crop rotations and cropping systems.

Although cereals may continue to constitute the major part of the human diet, improving the quality of that diet in the future may well depend on the production of edible legumes. Since the mid-1960s, meat substitutes derived from the seeds of edible legumes have been remarkably improved and widely accepted even in well-nourished nations, such as the United States and Canada.

Selected References

Heiser, Charles B., Jr. 1973. *Seed to civilization*. W. H. Freeman and Company.

Lappe, Frances Moore. 1971. *Diet for a small planet*. Friends of the Earth/Ballantine.

Wilsie, Carroll P. 1962. *Crop adaptation and distribution*. W. H. Freeman and Company.

Study Questions

1. Why might edible legumes become increasingly important in the human diet? Describe the characteristics of the legume family that make different legumes important nutritionally.
2. Describe the features of legume plants that are common to nearly all edible legumes.
3. Compared with animal protein sources, are legumes an efficient source of protein?

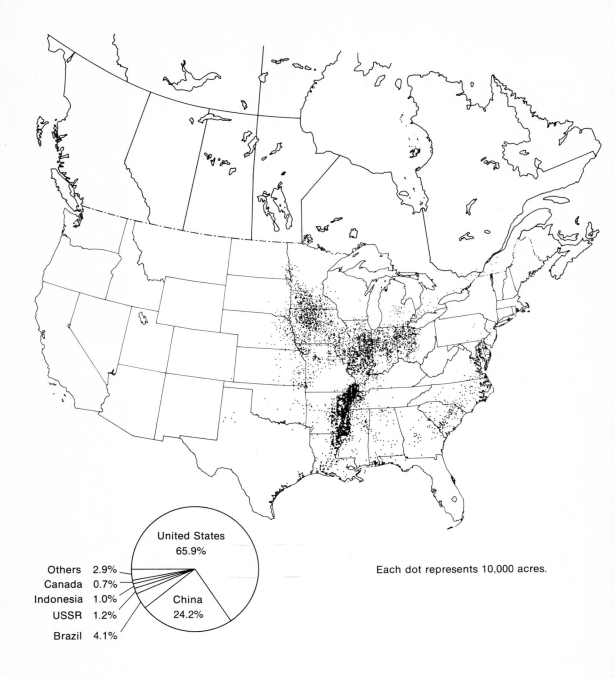

Figure 20-1
Major soybean-producing regions in the United States and contributions to world production. [Data from U.S. Department of Commerce and the United Nations.]

Twenty Soybeans

Origin and History of Culture

Of all of the crops produced in the United States since the mid-1960s, none has increased in importance and in acreage as much as soybeans. In areas where as recently as 1955 corn predominated, soybeans are rapidly becoming equally important (Figure 20-1).

Soybeans, *Glycine max*, evolved in Southeast Asia or China. Ancestors are found today in China and Korea. The earliest record of the culture of soybeans comes from China and dates back nearly three thousand years. The crop was first grown in the southern United States shortly after 1800, where early producers grew soybeans mainly as an annual hay crop or, in some instances, as green manure. With the development of oil extraction methods, soybeans were also produced for commercial purposes. The United States first exported soybean oil to Manchuria in 1910. Oil was initially extracted from soybeans on a local basis with equipment (expellers) designed for processing cottonseed.

Between 1939 and 1969, the percentage of the world soybean crop harvested for seed increased from about 40 percent to nearly 100 percent. In the same general period (1940 to 1970), the contribution of the United States to world production increased from less than 40 percent to more than 60 percent. A major cause for this change was the demand for soybean oil during World War II. Soybean oil serves a dual purpose: it is used as an inedible, drying oil in paints and ink, competing with linseed oil (obtained from flaxseed, see Chapter 27); and it is also a valuable, edible, nondrying oil, competing with other vegetable oils, such as cottonseed oil.

In recent years, more emphasis has been placed on producing and processing soybeans for high-protein meal. Soybean meal is a major component of the meat substitutes that are becoming increasingly popular in the United

States. Some people believe these high-protein meat substitutes may be a partial answer to improving the quality of the diets of people in undernourished nations (Figure 19-1). Through the financial support of federal, state, and industrial organizations in the United States, researchers are working to improve the quality of soybean protein and to increase both the quantity of seed protein and the overall yield of soybean crops.

Botanical Characteristics

The soybean is a warm-season, herbaceous annual (Figure 20-2). Plants are more or less erect, with branches coming from a main stem. The extent of branching depends on environmental conditions (light and available moisture) and on planting rates. Mature plants vary in height from 1 to 6 feet (30 to 180 cm), depending on the cultivar, planting date, and planting rate; average plants are from 3 to 4 feet (90 to 120 cm) tall. Heavy planting rates result in taller plants and frequently in reduced branching and weaker stems, which leads to lodging. Soybeans have palmately compound leaves with three leaflets. The leaflets are usually oval, but they can be elongated or lance-shaped in some cultivars. The stems, leaves, and pods of the plants of most cultivars are covered with a fine pubescence.

Depending on the cultivar, soybean plants are either determinate or indeterminate in growth habit: earlier-maturing cultivars (Group IV or earlier) are indeterminate; later-maturing ones (Groups V–VIII) are determinate. Indeterminate types sometimes appear to be determinate because of the short internodes and small axillary racemes on some actually indeterminate stems.

Soybean plants, which have typical, small legume flowers (see Chapter 19), are predominately self-pollinating. However, cross-pollination by insects does occur and may be a problem in maintaining cultivar purity in seed fields (see Chapter 12 for a discussion of certified seed). Flowers are purple or white; groups of from eight to sixteen flowers are borne in terminal or axillary racemes. Typical of the legume family, the pistil is simple and the ovary matures into the familiar legume pod. At maturity pods usually contain two or three seeds, but they can contain as many as five. Seeds vary in shape from nearly spherical to somewhat flattened discs and in color from pale green and yellow to dark brown.

Environmental Requirements

Soybeans thrive under conditions favorable for corn production, and, in fact, have become an important crop in the North Central states and the corn-producing areas from the mid-Atlantic coast to the Mississippi Delta. Plant breeders have also developed cultivars that are adapted to regions in Canada where corn is produced.

Soybeans are a warm-season crop. Soil temperatures of 60°F or above favor rapid germination and vigorous seedling growth, which is essential for successful competition with weeds and therefore for weed control. However, very warm midsummer temperatures (above 90°F, or 32°C) reduce yields and lower oil quality. Sustained temperatures below 75°F (24°C) during this same period delay flowering. The minimum temperature for effective growth is about 50°F (10°C). Like corn, soybeans are produced in areas with a minimum frost-free period of 120 days and a mean summer temperature above 70°F (21°C). Cultivars differ with respect to the minimum growing season required for maturation. In the seedling stage, or after flowering when the

Figure 20-2
A nearly mature soybean plant; seeds in lower right are not fully mature. Note compound leaves and pubescence on stems, leaves, and pods. [USDA photograph.]

crop is ripening, soybeans can tolerate frost better than corn.

Soybean plants are the most sensitive of all crop plants to light duration (photoperiod) and are sensitive to light quantity. They are short-day plants, but cultivars differ markedly with respect to the minimum dark period required to induce flowering. In addition to controlling the initiation of flowering, the photoperiod (which varies with changes in latitude) affects the development of soybean plants. For example, plants that develop normally near Washington, D.C., tend to be shorter and to mature more rapidly if they are grown further south in the Gulf Coast region where days are relatively shorter. All cultivars do not respond in the same way to variations in photoperiod in terms of floral initation and pattern of vegetative growth. Changes in planting date, which expose plants to different photoperiods, can have the same effect on plant growth as changes in latitude (Figure 20-3).

In North America, plant scientists classify soybean cultivars into ten maturity groups based on the region and the day length to which each cultivar is adapted. Classification is from adaptation to long days in the north (Groups 00, 0, and 1) to adaptation to shorter days in the south (Groups V–VIII; cultivars in Group VIII are the latest in flowering and are grown in the southernmost regions). Within each maturity group, cultivars may differ in maturity by as many as fifteen days. Plants that flower later than cultivars in Group VIII have been identified; these types, although not currently available commercially, are classified as Groups IX and X. Because of photoperiodically related responses, a change in latitude of less than 100 miles (161 km) may necessitate growing a different cultivar.

Soybean plants require high light intensity for vigorous growth. They suffer from shading and competition for light from taller-growing weeds; thus, for highest yields, strict weed control is essential. Depending on local conditions and the cultivar, extra care must be exercised to use appropriate planting rates and row spacing to prevent self-shading of plants. Soybeans, like cotton, suffer from excessive cloudy weather and are thus best adapted to areas that have few cloudy days during the summer months.

To produce optimum yields, soybeans require between 20 and 30 inches (51–76 cm) of water—either natural precipitation, irrigation, or a combination of the two. Soybean crops are generally produced in regions where annual rainfall is adequate, but the availability of irrigation is comforting insurance for producers. Soybeans are sensitive to both the amount and distribution of moisture. Germination is reduced by either excessive or insufficient soil moisture. Adequate available moisture is also critical during flowering and immediately thereafter, during pod filling, and during seed development.

Soybeans are adapted to heavy clay soils and even to muck (highly organic soils), and they produce more on these soils than do either cotton or corn. The optimum soil pH ranges from 6.0 to 6.5. A pH of 6.5 is recommended. On some soils liming (see Chapter 6) may be required before soybeans can be grown.

Production Practices

SEEDBED PREPARATION AND SEEDING

Seedbed preparation for soybeans generally follows the pattern discussed earlier for corn: fall or early-spring plowing to incorporate crop residue followed by discing and harrowing to prepare a firm seedbed for planting. As is the case for all crops, a primary objective in seedbed preparation is to destroy weeds before planting.

Row spacings vary from as close as 18 inches to as far apart as 40 inches (45–100 cm). Spacings of 24 to 36 inches (60–90 cm) are common. Row spacing is often determined by the availability of equipment used for corn production so that corn and soybean row spacings are frequently the same. Currently, there is a trend toward narrower row spacings. Regardless of row spacing, from six to twelve seeds per foot (30 cm) of row are planted; however, because of branching, planting from four to eight plants per foot of row results in essentially the same yield (Table 20-1). With fewer plants per foot of row, plants tend

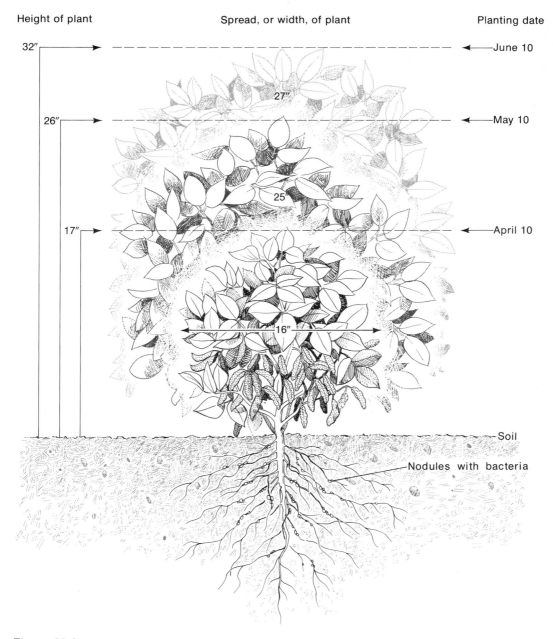

Figure 20-3
Changes in the pattern and amount of growth of a single soybean plant in response to date of planting (a photoperiodic response) at one location. Similar changes occur in plants seeded on the same date at different latitudes. [Redrawn from J. L. Carter and E. E. Hartwig. The management of soybeans. *Advan. Agron.* 14 (1962):359–412.]

Table 20-1
Average yields of four types of soybean plants seeded at different rates and row spacings.

Row width (inches)	Seeding rate (lb/acre)	Yield (lb/acre)
30	75	2,515
10	75	2,604
10	150	2,604
10	225	2,587

Source: D. R. Hicks, J. W. Pendleton, R. L. Bernard, and T. J. Johnson. Response of soybean plant types to planting pattern. *Agron. J.* 61(1961):292.

to branch more prolifically and utilize the extra space.

Seeding rate should be based on desired plant populations per acre or hectare, as for corn, and not on pounds of seed per acre, or kilograms per hectare. Of course, seeding rates vary with seed weight and row spacing: with rows planted 18 inches (45 cm) apart, the seeding rate ranges from 85 to 100 pounds per acre (95–112 kg/ha); with 40-inch (101 cm) rows, the rate ranges from 40 to 50 pounds per acre (45–56 kg/ha). Heavier seeding rates per row are favored in the north where cultivars are shorter and may have smaller leaves (thus more plants are required to achieve an optimum use of available light), or if row spacings are wide or plantings are late. Heavy seeding rates can result in taller, sparsely branched plants that tend to lodge easily and in plants that set few pods at lower internodes. On the other hand, lighter seeding rates favor weeds (because of reduced competition), decrease plant height, and encourage branching. Wider row spacings and lighter seeding rates are used in dryer regions. In the final analysis, the best seeding rate depends on cultivar characteristics (e.g., determinate versus indeterminate growth habit), environmental factors (e.g., available moisture, competition with weeds), and available equipment. The seeding depth for soybeans varies from about 1 to 2 inches (2.5–5.0 cm). Emergence decreases markedly if seeds are sown deeper than 2 inches (5 cm).

The planting date varies from region to region and depends on the cultivar, the soil temperature, and the cropping system. The best time for planting soybeans in most areas is early May. However, in Maryland yields decrease if crops are seeded after early April. The magnitude of the decrease depends on the cultivar. Soybean crops can be successfully harvested from plantings as late as mid-June, but yields are drastically reduced. In the Gulf Coast states and along the Mississippi Delta, plantings can be made as late as early July, but planting this late is done only under a double cropping system in which soybeans follow a winter cereal in the rotation.

FERTILIZERS

As with nearly all members of the legume family, soybeans fix atmospheric nitrogen if the proper strain of *Rhizobium* bacteria is present in the soil or if the seed is properly inoculated. Nitrogen fertilizer is rarely required for successful production. However, even though the bacteria can persist in the soil under favorable conditions for ten years or more, planting inoculated seed is strongly urged. On fields that are low in nitrogen, inoculation has increased total yield by more than 30 percent and its protein content by more than 10 percent; it may increase the amount of protein produced per acre by nearly 50 percent compared with yields from plantings of uninoculated seed. The strain of *Rhizobium* effective on soybeans is specific for soybeans; this legume does not cross-inoculate with any other crop (see Chapter 12 for a discussion of legume inoculation).

Soybeans are better adapted to poor soils than corn and cotton. Soybean plants, like peanut plants (see Chapter 21), apparently have the ability to utilize fertilizer residues that are normally not available to other crops. Both types of plant may excrete special enzymes that break down soil-fertilizer complexes that hold essential minerals in a form unavailable to most plants. Regardless of their ability to grow on soil that has low fertility, soybeans do remove essential minerals from the soil, and these must be replaced if soil productivity is to be maintained.

As a rule of thumb, soybeans remove about 0.9 pounds (408 gm) of phosphorus per bushel of beans harvested. Applications of 30 to 40 pounds of phosphorus pentoxide (P_2O_5) per acre (34–45 kg/ha) are recommended to insure good yields. Soybeans use large amounts of potassium, removing about 1.2 pounds (545 gms) of potassium from the soil for each bushel of beans harvested. Many soils on which soybeans are produced may be deficient in potassium. Although precise fertilizer recommendations must be based on the results of valid soil tests, applications of 300 to 400 pounds per acre (336–448 kg/ha) of 0-20-20 commercial fertilizer are common in many soybean-producing areas. In some areas, both calcium and magnesium may limit crop growth and must, therefore, be provided in fertilizers. Soybean seedlings are extremely sensitive to fertilizer salts. Banding fertilizers with or near the seed is not recommended. The use of "pop-up" fertilizers, advantageous in some crops, can reduce soybean stands by as much as 50 percent. In the midwestern United States, fall applications of fertilizers result in slightly higher yields than spring applications; conversely, in some southern areas, spring broadcast applications are preferable to fall applications. Dry fertilizers should be worked into the soil by cultivating to a depth of 6 to 8 inches (15–20 cm).

Because of their symbiotic relationship with *Rhizobium* bacteria, soybeans add nitrogen to the soil and are thus a suitable green-manure crop. Even if harvested for seed, soybeans benefit the subsequent crop in a rotation. As a green-manure crop, soybeans return as much as 90 pounds of nitrogen per acre (107 kg/ha) to the soil; if harvested for seed, they return about 15 pounds of nitrogen per acre (18 kg/ha).

HARVESTING, STORAGE, AND ROTATIONS

Fall harvesting of soybeans grown for seed begins when pods are yellow and dry, and beans or seeds are in the hard-dough stage (Figure 20-4); the beans should have a moisture content of less than 15 percent. Losses during harvest are a serious problem with soybeans. Under extremely unfavorable conditions, losses can run as high as 20 percent of the crop. Losses are due to four major causes: (1) shattering in the field, (2) shattering at the cutterbar, (3) threshing, and (4) separating and cleaning. If grown for hay, soybeans are harvested anytime after pods form. Most soybean crops are harvested with grain combines (Figure 20-5). Proper cylinder speed and cylinder adjustment are major factors in minimizing seed damage and harvest losses. To prevent seed damage, cylinder speeds must be reduced, compared with speeds used for harvesting cereals. Seed yields from 35 to 40 bushels per acre (2,350–2,690 kg/ha) are common; however, yields are increasing as superior new cultivars are developed. Soybeans are harvested for both their oil and their protein. In recent years, a major plant-breeding effort has been directed toward increasing the percentage of protein in the seed, in addition to increasing overall yield.

Because of their high oil content, soybeans must be stored with special care. They must be less than 14 percent moisture, the level generally deemed safe for wheat or corn. If the beans are 14 percent moisture, they can be safely stored for only about a month. If the

Figure 20-4
Soybean plants that are ripe and ready for harvest. Note that nearly all of the leaves have dropped. [Photograph by J. C. Allen and Son.]

Figure 20-5
Modern combine harvesting soybeans. [Courtesy of Deere and Company.]

moisture content is lowered to 12 percent, they can be stored for as many as three years. If the moisture level is reduced to 10 percent, safe storage is extended to as many as ten years. Of course, temperatures must be relatively low, and, as is the case for all crop products, the beans must be protected from pests.

In areas in which annual cropping is the rule, such as throughout most of the corn belt, soybeans are frequently included in systems of corn rotation. (As the value of soybeans increases, it may become difficult to say whether corn or soybeans is the primary cash crop.) Rotations follow the pattern of two years of corn separated by a year of soybeans, a year of cereal, another (winter) legume, and back to corn. In a corn-oats-legume rotation, soybeans can replace oats. In the Deep South, soybeans are rotated with both cotton and rice, either on an alternate year scheme, or more commonly once every three or four years.

A major consideration in developing a crop rotation that includes soybeans is the effect of the legume on the fertilizer needs of the subsequent crop. Yields of both corn and small grains (wheat or barley) increase when the crops follow soybeans in rotation. The magnitude of the yield increase depends on the soil, environmental conditions, and the crop, but increases in yield that range from 5 to 10 percent are common.

Continuous planting to soybeans is not recommended. As they do in any monocropping sequence, crop pests proliferate even with chemical control. Specific rotations depend on what other major cash crops are adapted to a given area and what equipment is available.

Soybeans can contribute to soil erosion because they leave soil in a loose, friable condition. However, recent evidence shows that, by employing sensible cultural practices, and by rotating soybeans with corn and pasture crops, soybeans do not cause more significant problems of erosion than do similar crops.

In areas of intensive farming where double cropping is practiced, soybeans are grown as a summer crop following winter cereals or peas produced for canning or freezing. With double cropping, yields of both crops are reduced as much as 10 percent. It is essential to harvest each crop as soon as it is ripe, so that the second crop can be sown as close as possible to an optimum (or at least an acceptable) planting date. Double cropping makes extreme demands on the soil; fertilizer management and soil conservation practices therefore become increasingly important. In some areas of the corn belt as well as in other soybean-producing regions, the cultural practices used in soybean production cause the formation of a compacted plow pan, which restricts root growth and thereby reduces yields. The problem has become so severe that annual chiseling to a depth of 14 inches (36 cm) is necessary. A partial remedy in double cropping with cereals is the use of minimum-tillage practices that allow the soybeans to be seeded directly into the cereal stubble.

Diseases and Pests

DISEASES

Nearly fifty pathogens cause periodic, regional losses in soybean crops. However, there are no major diseases of soybeans comparable, for example, to the rust diseases of wheat.

As might be expected, fungal diseases are most prevalent. Brown stem rot causes weakening of the stem, wilting and lodging, and of course, reductions in yield. If the disease becomes severe, because the pathogen persists in the soil, long-term rotations away from soybeans may be required for control. In stem canker, another fungal disease, the pathogen is carried in the seed and in crop residue; seedling stems become girdled and the plants die from the disease. Deep plowing to bury crop residue and rotations away from soybeans provide partial control; also, some cultivars are partly resistant to stem canker disease. Various pod and stem blights, caused by different fungi, can result in serious losses in the corn belt. They either kill plants outright or reduce seed yields without killing the plant. Pathogens persist in crop residue and can be carried on the seed. Crop rotation and the use of certified, disease-free seed are effective methods of controlling these blights. Resistant cultivars are available for the control of other fungal diseases, such as frogeye leaf spot, brown rot, and downy mildew.

The bacterial disease, wildfire, can be a serious foliar disease of soybeans grown in the South. Bacteria are carried in seed and infected tissue. Control includes disease-resistant cultivars, using disease-free seed, and rotating crops. Other, less important bacterial diseases include bacterial blight and bacterial pustule diseases.

Viral mosaic may reduce yields in infected plants as much as 75 percent. The virus is carried in the seed. To date, this disease has been serious only in Illinois.

Seed treatment for disease control is generally not recommended with soybeans. However, if plantings are made when soil temperatures are low, or if other factors (e.g., low quality, old seed), favor slow germination, seed treatment with an approved fungicide to control seed- and soil-borne pathogens can be of some benefit. If seed is treated, it should be done before inoculum (*Rhizobium*) is applied.

WEEDS

Weed control is an important consideration in soybean production. To many soybean producers, the biggest management challenge is to have a weedfree field. Losses in yield and crop quality due to weeds approach 20 percent annually. As it does for all crops, weed control starts with thorough seedbed preparation and includes the use of high-quality, weed-free (i.e., certified) seed. Because soybeans are planted in relatively widely spaced rows, cultivation during the growing season is a significant part of the weed-control program. Weekly, or more frequent, cultivations with a rotary hoe between the time seedlings emerge and the time plants reach a height of about 6 inches (15 cm) provide effective weed control in many areas. After plants are well established (more than 6 inches, or 15 cm tall) they can shade and effectively control newly emerging weed seedlings by competing successfully for light.

Herbicides are also important in the control of weeds in soybeans. About 60 percent of the soybean acreage in the United States is treated with some type of herbicide. Preemergence herbicides, postemergence herbicides, and combinations of both are used. About 70 percent of the acreage treated receives only preemergence sprays; combinations are used on about 20 percent; and less than 10 percent receives only a postemergence spray. In the past ten years, chemists have developed new preemergence sprays that are markedly more effective than those available prior to 1965.

The leading weed pests of soybeans vary from region to region, but they include the following: pigweeds, crabgrasses, lambsquarters, foxtails, nutsedges, ragweeds, johnsongrass, cockleburs, morningglories, smartweeds, jimsonweed, and velvet leaf. These weeds are persistent pests in soybeans; in recent years no additions to or deletions from the list of important weeds have been made; although no major weed pest has been eliminated or significantly reduced, at least no new pests have been reported.

INSECTS

There are no major insect pests of soybeans. Many pests cause minor damage, but extensive field control is unnecessary. Insects that damage soybeans include grasshoppers, leafhoppers, stinkbugs, velvet bean caterpillars, the Mexican bean beetle, and the bean leaf weevil. Some mites also attack the crop, and in some regions, rabbits and groundhogs feeding on the young plants can cause serious losses.

Keep in mind that, as the acreage of soybeans continues to increase, additional problems may arise. This is typical; if one crop is replaced by another (e.g., corn by soybeans), or if the acreage planted to one crop increases rapidly, pests (e.g., insects) change their habits. Looking to the future, researchers and producers must be prepared to identify and remedy potential problems before they become critical.

Crop Utilization

Soybeans are a versatile crop. Until World War II they were used mainly as a forage crop and for green manure. During and immediately after the war, the crop was harvested for the seeds (beans), which were processed for oil. Today most soybean crops are processed to produce a high-protein meal used in meat substitutes and as a supplement in livestock feed; soybean oil is extracted during the processing.

About 90 percent of all soybean oil is used as an edible oil in the production of mayonnaise, margarine, and similar products. The drying fraction of the oil is used in paints, varnishes, and ink. Three processes are used in extracting oil from the seeds. The expeller (screw) process and the solvent extraction process are used by mills that specialize in soybean processing. Oil can also be extracted by means of the hydraulic press method, which is commonly used in processing oilseed crops (see Chapter 27). The processing method affects both the quantity and the quality of the oil and the meal recovered. Mechanical methods yield less oil and more meal, but the meal is lower in protein compared with meal produced by solvent extraction methods. If processors use a mechnical method of extraction, they recover about 15 pounds (6.8 kg) of oil and 80 pounds (36 kg) of meal that is 41 percent protein from 100 pounds (45.4 kg) of seed; there is about a 5-pound (2.3 kg) loss. Using a solvent method of extraction, they obtain 18 pounds (8.2 kg) of oil and 76 pounds (34.5 kg) of meal that is about 44 percent protein. The solvent extraction method thus yields 33.4 pounds (15.2 kg) of protein per 100 pounds (45.4 kg) of seed, whereas the mechanical method yields 32.8 pounds (14.9 kg) of protein for the same amount of seed. Therefore, the solvent extraction method is becoming more extensively used.

In the past, as much as 98 percent of the soybean meal produced was used in livestock feed. Today, a much greater proportion is used in producing meat substitutes. Many problems that lowered consumer acceptance of soymeal meat substitutes have been overcome, and more and more soymeal "meats" are appearing in supermarkets. Soybeans are also used in specialty foods. Soybean products, from "milk" to bread to "ice cream," replace cereals and common dariy products in the diets of many people who have severe food allergies. The soy products are hypoallergenic and highly nutritious. In looking to the future, soybeans must be considered a crop with great potential for improving both the quantity and the quality of the human diet.

Selected References

Carlson, John B. 1973. Morphology. In *Soybeans*, ed. E. B. Caldwell. Agronomy Monograph No. 16, pp. 17-66. American Society of Agronomy.

Carter, J. L., and E. E. Hartwig. 1962. The management of soybeans. *Advan. Agron.* 14:360-412.

Pendleton, J. W., and E. E. Hartwig. 1973. Management. In *Soybeans*, ed. E. B. Caldwell. Agronomy Monograph No. 16, pp. 211-231. American Society of Agronomy.

Probst, A. H., and R. W. Judd. 1973. Origin, U.S. history and development, and world distribution. In *Soybeans*, ed. E. B. Caldwell. Agronomy Monograph No. 16, pp. 1-12. American Society of Agronomy.

Study Questions

1. Why have soybeans become increasingly important in the past twenty years?
2. What environmental factors limit the distribution of soybean production?
3. How and why are soybean cultivars classified into "maturity groups"?
4. What factors would you consider in fertilizer management for soybean production?
5. Briefly discuss the factors that determine planting rate and pattern in soybeans.
6. Briefly discuss the major crop pests and diseases of soybeans.

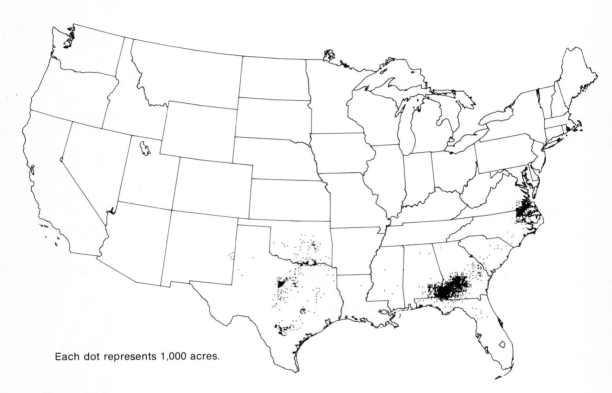

Figure 21-1
Major peanut-producing regions in the United States. [Data from U.S. Department of Commerce.]

Twenty-One   Peanuts

Origin and History of Culture

Eating salted peanuts seems to be an essential part of spectator sports in the United States, and eating peanut-butter sandwiches, an essential part of growing up. But despite the fact that so many peanuts are being consumed plain, roasted, or processed, peanuts are not a major commercial crop in this country, and the United States produces only about 5 percent of the total world crop.

Peanuts appear to have originated or evolved in Brazil and Paraguay, where closely related ancestors have been identified. Like tobacco and corn, the peanut has become a major crop far from its center of origin. Apparently, peanuts were carried by early travelers from South America to Africa and the East, and they have become an important part of diets in South Africa, India, China, and Southeast Asia. During the American colonial period, slave traders brought them from Africa to North America.

Peanuts did not become a crop of commercial importance in the United States until about 1876. By 1900, peanuts had become a major crop in parts of the cotton belt, where, because of crop losses from the boll weevil, they were produced instead of cotton. During both world wars, peanut acreages increased markedly to meet demands for vegetable oils, plant proteins, and related plant products. In the past ten years, peanut yields have more than doubled as a result of disease control, development of new cultivars, proper fertilizer management, and mechanization. Currently, peanuts are grown nearly exclusively as a food crop. Because the peanut seed is a good source of high-quality protein and is high in superior-quality oil, peanuts are being viewed with renewed interest because of their potential as a source of protein for a malnourished world.

Although peanuts are an important crop in some parts of the South and Southeast, as a cash crop they rank generally behind cotton, soybeans, corn, and tobacco in areas where all of these crops are grown. The major producers are North Carolina, Alabama, Georgia, Florida, Virginia, Oklahoma, and Texas. Certain types of peanuts are grown under irrigation in parts of New Mexico (Figure 21-1).

Botanical Characteristics

The binomial designation for the peanut, *Arachis hypogaea,* describes the most peculiar trait of the species—subterranian fruit formation (*hypo,* meaning under, and *gaea,* meaning ground). Common names include goober, pinder, and groundnut. Peanuts are an annual, warm-season crop. A peanut plant has a central, upright stem and many lateral branches. When these branches are upright, the plant is designated a bunch type; when horizontal, the plant is a runner type. The Spanish and Valencia groups are bunch types and tend to mature earlier than the Virginia group (which consists of both bunch and runner types). The distinction between "bunch" and "runner" is not absolute; many superior cultivars have intermediate, or semierect, growth habits. Although runner types generally have smaller seeds, no real distinction can be made on that basis.

The peanut plant has a relatively deep taproot system with a well-developed lateral (secondary) root system. If properly inoculated, nodules form on the roots. The leaves are pinnately compound and are usually composed of two pairs of leaflets; at times there is a single, fifth leaflet. The leaf is borne on a slender petiole.

The flower morphology of the peanut plant is fairly typical of legumes. The calyx forms a long, slender tube with the corolla at the tip. The flowers are sessile and are borne in leaf axils either singly or in groups of up to three. Generally, flowers are self-pollinated.

Figure 21-2
Roots and pods of a peanut plant. [USDA photograph.]

After pollination, the perianth withers, and several peculiar events occur. At the base of the ovary, a meristematic region grows and becomes a stalklike structure (the gynophore) that bends downward (a geotropic reaction) and forces the ovary into the soil. The gynophore is commonly referred to as the peg, and the stage of plant development at which the gynophore is activated and elongates is referred to as the pegging stage. The ovary matures underground into a pod—the common unshelled peanut (Figure 21-2). As noted earlier, the specific epithet *hypogaea* (below ground) reflects this peculiar behavior. (Recall that hypogeal germination is a pattern of germination in dicots in which the cotyledons remain below the soil surface.) The mature pod contains from three to six seeds, or "peanuts." Spanish types usually have relatively small pods with two seeds, or nuts; Virginia types have larger, two-seeded pods. The seed coat (testa) encloses the individual peanut in the familiar papery layer, which varies in color from purple and red to tan. The seed coat is removed in processing.

Environmental Requirements

Although the general environmental requirements for peanuts are quire similar to those for both cotton and tobacco, the crop is not produced uniformly throughout the cotton belt (see Chapters 23 and 24). Even in the major producing states, there are only certain areas that have the most suitable conditions. There are three, fairly distinct, major production regions: (1) Virginia and North and South Carolina, where Virginia cultivars (both runner and bunch) are grown; (2) the Deep South—Georgia, Alabama, and Florida—where southeastern runner type (Virginia runners) and Spanish cultivars are produced; and (3) a southwestern region that includes Texas and Oklahoma, where Spanish (bunch type) cultivars are grown, and New Mexico, where Valencia (bunch type) cultivars are produced under irrigation.

TEMPERATURE
Peanuts are definitely a warm-season crop and generally require a relatively long frost-free period. A frost-free period of about 200 days is desirable (although cultivars that mature in 110 to 140 days have been developed). Also, production is usually limited to areas with a mean annual temperature above 45°F (7°C) and a mean July temperature above 75°F (24°C).

MOISTURE
Peanuts are grown in regions in which the annual precipitation is more than 40 inches (102 cm), although with adequate irrigation they can be successfully produced in drier regions. Distribution of moisture is important. The plant requires adequate moisture throughout its life, and adequate available moisture is critical at pegging and fruiting time. Generally, a minimum of 19 inches (48 cm) of moisture distributed from late April through the summer is required for satisfactory yields. Excessive moisture (saturated or flooded soils) and high temperatures reduce yields, and under such conditions, bunch-type (Spanish and Valencia) cultivars may be superior to the runner and semierect types.

SOIL
Peanuts are best adapted to light, well-drained loam soils. They tolerate acidic soils better than cotton and are not adversely affected by aluminum toxicity, which is frequently associated with acidic soils in the South. Root growth and development are not seriously restricted by acidic conditions, as they are in cotton. The optimum pH range is from 6.0 to 6.5. Seedlings are more tolerant of salts than are mature plants.

The crop is not well suited for soils that are high in organic matter. Organic matter harbors many pests that are harmful to peanuts and may cause seed discoloration and consequent lower grade designation.

Production Practices

SEEDBED PREPARATION AND SEEDING
Careful and thorough seedbed preparation is critical for maximum yields. A mellow, moderately firm seedbed is essential, but packing, which might ultimately inhibit pegging, should be avoided. Weed control is an essential part of seedbed preparation. Deep plowing in the fall or early spring is recommended to bury crop residue—a practice of major importance in controlling pathogens that live on

crop residue (corn, cotton, and other crops) and seriously damage peanuts. The initial step in seedbed preparation is discing, followed in a month or more by deep plowing with a mold-board (or bottom) plow. Final seedbed preparation consists of cultivation with a spike-tooth harrow. This is normally completed immediately before planting for maximum weed control. Peanuts are usually planted flat, although low beds are used under extremely wet conditions. Where the crop is irrigated, peanuts are planted on beds, except when sprinklers are used.

Peanuts grown in southern areas are planted earlier in the year than those grown in northern areas. The planting dates, like those of most warm-season crops, reflect the times at which soils are warm enough to support germination. However, planting dates can be altered because of rain, which prevents seedbed preparation or seeding operations.

In the Deep South, planting starts in mid-March. Spanish cultivars will mature in this region when seeding is as late as the end of June, but late planting reduces yields. In other areas, plantings are made from April through early May, except in Texas, where plantings closely follow the planting times for cotton; that is, from March through April (Table 21-1). During optimum planting times, soil temperatures range from 60° to 70°F (16°–21°C).

Planting in most commercial fields is done with cotton or corn planters, using special plates for peanuts. Seeding rates depend on two factors: the growth habit of the cultivar (bunch or runner) and the field spacing (both the distance between rows and the distance between plants in a row). Bunch-type cultivars are usually seeded at heavier rates than runner types. Row spacings for bunch types vary from 24 to 36 inches (60–90 cm); seed spacings vary from 4 to 8 inches (10–20 cm). Row spacings for runner types are 30 to 36 inches (77–91 cm); seed spacings are 6 to 14 inches (15–35 cm). Seeding depths range from 1.0 to 1.5 inches (2.5–3.8 cm) in heavier soils and to 3 inches (7.6 cm) in lighter soils when large seeds are used.

As seeding rates increase, yields per acre increase but yield per plant decreases. However, the percentage of large seeds increases as the seeding rate increases. When the seeding rate is increased, yields can be increased more by reducing the distance between rows from 36 to 30 inches (91–77 cm) than by reducing the space between seeds in a row. Heavy seeding rates tend to delay maturity, which can be a serious consideration in areas with a minimum frost-free period. Planting rates for shelled seed range from 70 to 100 pounds per acre (78–112 kg/ha) for runner types and from 90 to 110 pounds per acre (100–123 kg/ha) for bunch types. In many areas, both bunch and runner types are seeded at the same rate (note that rates for the two types overlap).

Seed quality and size are important factors in stand establishment and seedling vigor. Some producers plant back seed they have grown—an acceptable practice as long as high-quality seed is saved as planting stock. Occasionally, small, immature seeds, sometimes called "pegs," are saved. Germination and emergence relate to seed size; generally, the smaller the seed, the lower the germination and emergence. The field emergence of these small, immature seeds is up to 10 percent lower than the emergence of certified, grade-1 seeds. There are about 830 seeds per pound (1720 seeds/kg) of grade 1, and 1700 seeds or more per pound (3740 seeds/kg) of pegs. The number of seeds per unit of weight depends on seed size; Virginia groups have larger seeds and thus relatively fewer seeds per unit of weight. In addition to a higher rate of emergence, seedlings from larger (grade 1) seeds are more vigorous; they develop more rapidly and tend to resist adverse conditions better. Seed

Table 21-1
Effects of different planting and harvesting dates on peanut yield in Virginia.

Planting date	Harvesting date	Yield (lb/acre)				
		1952	1953	1954	1955	Average
4/22	9/26	3660	2684	4233	2928	3376
	10/5	4444	2905	4668	3196	3803
	10/15	4189	3254	4429	3450	3831
5/2	9/30	4599	3219	4919	3721	4115
	10/10	4451	3664	4631	3449	4048
	10/21	3849	4354	4735	3071	4002
5/12	10/5	3912	3265	4271	3057	3626
	10/15	4239	3240	4695	3234	3852
	10/25	4251	4073	4417	3032	3943
5/22	10/10	4011	2916	4483	2948	3589
	10/21	3887	2976	4759	3080	3675
	10/30	3687	3023	4817	3378	3726

Source: G. M. Shear and L. I. Miller. Influence of time of planting and digging on the jumbo runner peanut. *Agron. J.* 51(1959):30–32.
Note: Late plantings delay harvesting; the best time to plant in this location is in early May.

treatment with an approved fungicide is highly recommended to help minimize losses from various seed and seedling diseases. Dormancy in fresh seeds can be a problem and is more pronounced in most runner types than in Spanish and Valencia groups. In the runner types, it commonly ranges from three to six months, although it may persist for nearly two years; there are runners, however, in which dormancy is not a problem. Dormancy is not a problem in the Spanish (bunch) group. Under suitable storage conditions (see Chapter 12), peanuts retain their viability for as many as six years, but the percentage of germination decreases with age after two or three years.

FERTILIZERS

Fertilizer management of peanuts is rather peculiar. Because peanut seedlings are quite sensitive to fertilizer salts, fertilizers are either broadcast or placed deep in the soil so that the seedlings are not exposed to high salt concentrations. Peanuts use relatively large amounts of essential minerals. Unlike other crops such as corn and cotton, peanuts can use the residues of fertilizers applied to crops that precede them in rotation; they can take advantage of minerals that are not available to other crops (Table 21-2). This capacity may be due in part to the extensive root development of the peanut plant; it may also be the result of some enzymelike secretion from its roots, which makes minerals bound to soil particles available to the plant. Thus, growing a crop that requires large amounts of fertilizers (such as corn) immediately before peanuts is a widely recommended practice. Of course, the preceding crop must be fertilized adequately, according to soil-test results.

Peanuts, like other legumes, can fix nitrogen; thus, nitrogen fertilizer is rarely necessary. A proper balance of nitrogen and phosphorus is essential for early maturity—a particularly critical consideration in parts of Virginia and North Carolina where the growing season is short. Calcium fertilization, in addition to liming to increase soil pH, is important; adequate available calcium is essential for high yields and disease resistance. Excessive amounts of potassium may limit calcium uptake by the plant; thus, the

Table 21-2
Effects of fertilizers and fertilizer residues on peanut yield and seed size. Peanuts were grown in two types of soil and followed corn in a rotation.

	Galestown fine sand		Woodstown loamy fine sand	
	Fertilizer applied		Fertilizer applied	
	only to corn	to both crops	only to corn	to both crops
Yield (lb/acre)	2845	2980	3690	3585
Extra large seeds (%)	37.6	34.4	38.3	35.7
Mature seeds (%)	71.2	70.2	72.6	71.9
Shriveled seeds (%)	1.7	2.1	1.4	1.7

Source: D. L. Hallock, Effect of time and rate of fertilizer application on yield and seed size of jumbo runner peanuts. *Agron. J.* 54(1962):428–430.
Note: Results indicate that, at best, fertilizing both crops increased yield only slightly on one type of soil and that quality was consistently reduced by fertilization of both crops.

K:Ca balance is of concern. Both calcium and boron are important in the control of hidden damage in the seeds. Adequate available calcium minimizes plumule malformation and damage during seed development and therefore plays a direct, major role in seed yield and quality. Boron affects the entire growth and development of the plant. Not only is it essential to control discoloration of cotyledons, but adequate available boron fosters a concentrated flowering period and a crop that matures uniformly; boron deficiency may extend the flowering period and lead to malformed foliage. Prompt application of boron as soon as deficiency symptoms appear will minimize crop losses.

Specific fertilizer recommendations for the peanut plant are based on detailed soil tests and on observations of the crop as it grows. As a rule, the primary plant foods are not applied. But, although the peanut crop may not require the application of fertilizers, it does remove elements from the soil—elements that must be replaced if soil productivity is to be maintained (Table 21-3).

HARVESTING
Since the end of World War II, peanuts have been harvested mechanically, with few exceptions. The windrow method is frequently practiced. A digger-shaker lifts the entire plant from the soil (remember, the peanuts, or pods, are underground) and shakes some of the adhering soil from the roots and pegs. The plants are allowed to dry for two or three days, and then they are windrowed with a windrow-shaker—a process that also removes more soil. If it rains while the crop is in the windrow, peanuts, like other crops including hay, must be turned to avoid heating and spoiling. When the foliage is dry, the pods are removed by combining (Figure 21-3). The pods are then dried and stored. Because prolonged drying in full sunlight may reduce seed quality, pods are first dried to about 20 percent moisture in full sunlight, and then to about 8 or 9 percent moisture in dryers using forced air. Shelled seed should be stored at a moisture level of 5 percent or less in a room with a 60-percent relative humidity, and the seed moisture content should be maintained at 5 percent or less. If the moisture content is higher, the shelled nuts may become rancid in less than two months. For that reason, peanuts are usually left in their shells for prolonged storage.

Peanuts are harvested in the late summer or fall while the tops are still growing, before

Table 21-3
Effects of different rates of three essential minerals on development of peanut plants.

Element	Rate	Days until flowering	Total flowers per plant	Total pegs per plant	Percentage of flowers forming pegs
Ca	Low	44.6	19.2	2.9	15
Ca	Medium	46.3	17.3	1.6	8
Ca	High	47.3	16.0	0.8	4
Mg	Low	46.2	15.8	1.8	10
Mg	Medium	47.0	16.2	1.3	7
Mg	High	44.9	20.5	2.2	10
K	Low	46.6	17.0	1.6	8
K	Medium	44.7	18.7	1.7	9
K	High	46.8	16.8	1.9	10

Source: J. J. Nicholaides and F. R. Cox. Effect of mineral nutrition on chemical composition and early reproductive development of Virginia type peanuts. *Agron. J.* 62(1970): 262–264.

Note: None of the elements affected days until flowering. Increasing magnesium had a positive effect on number of flowers; increasing calcium had a negative effect on total number of pegs and on the percentage of flowers forming pegs.

the first killing frost. The plants are dug when seeds are plump, or full. The grower must be able to recognize the characteristics of mature pods in terms of color and texture in order to select the proper harvest date. Premature harvesting leads to an excessively high proportion of immature pods and shrunken or shriveled seeds. Late harvesting leads to disease problems, trouble with late-summer weeds, and the possibility of losses from fall rains.

ROTATIONS

As noted earlier, peanuts, although rarely fertilized, make severe demands on the soil. A crop of 1 ton (900 kg) of pods and 4 tons (3,600 kg) of hay removes as much as 150 pounds of nitrogen, 25 pounds of phosphorus, 100 pounds of potassium, and nearly 100 pounds of calcium from the soil per acre (168 kg/ha of N, 28 kg/ha of P_2O_5, 112 kg/ha of K_2O, and nearly 112 kg/ha of Ca). Remember, when peanuts are harvested, nearly the whole plant is removed from the soil; thus, little, if any, of the minerals is returned to the soil as stubble. (Note also that peanut harvesting leaves the soil bare and vulnerable to erosion.)

For the sake of soil conservation and pest control, peanuts should not be grown on the same field in consecutive years. Peanuts are frequently included in rotations with cotton, tobacco, corn, and, in some areas, pasture. Tobacco normally does not follow peanuts directly in a rotation because peanuts leave the soil too rich in available nitrogen. Corn, because of its high fertilizer requirements, and because it is not infected by the pathogens that attack peanuts, is an ideal crop to precede peanuts in a rotation. Winter rye, oats, and crimson clover may be seeded as a cover and green-manure crop immediately after peanuts are harvested.

Diseases and Pests

DISEASES

In the major peanut producing areas of the United States, viral and bacterial diseases of peanuts are of minor importance, but, on a worldwide scale, both cause significant losses. Peanuts are also subject to damage from a number of pathogenic fungi. Several species of fungus cause seed rots. The pathogens are

Figure 21-3
Windrowed peanuts are combined to separate the pods from the vines. [USDA photograph.]

usually saprophytic. They live in the soil on crop residues and enter the peanut seed through wounds or damage to the seed coat. Conditions that are unfavorable for germination—cool temperatures and water-logged soils—favor seed rots. Control of seed rots is achieved by seed treatment with an acceptable fungicide; the use of sound, large seed, which reduces the chance of infection by pathogens; and proper preparation of the seedbed and planting at an appropriate time (when the soil is adequately warm), which foster rapid germination and vigorous seedlings that resist seed rot. Damping-off and seedling blights—two diseases caused by several species of fungus—can be managed through cultural practices that foster vigorous seedling growth and the choice of well-adapted cultivars.

Leaf spot, caused by fungi of the genus *Cercospora*, had been a serious problem until plant breeders developed cultivars resistant to the disease. As leaf spot progresses, lesions are formed on leaves. Yields are reduced as a result of the loss of leaf (photosynthetic) tissue, changes in the plant's water use and water-use efficiency, and the parasitic nature of the fungus.

Southern blight, or stem rot, caused by *Sclerolium rolfsi,* is a serious disease and may be the factor limiting peanut production in some areas. Disease symptoms may appear at any time during the growing season, but the disease is most likely to affect the crop as it nears maturity in the late summer or early fall. The pathogen attacks the plant near the soil surface. When established it causes wilting. At times, infection may be limited to only one or a few branches, but, if it reaches the central stem, it kills the entire plant. Although the disease causes wilting symptoms, it destroys succulent cortex tissue rather than affecting vascular tissue directly. The pathogen can live as a saprophyte in the soil and it attacks many plant species. Grasses are not susceptible to southern blight; thus, if volunteer peanut plants are eliminated, crop rotation with grasses is effective for disease control. This is indicated by the fact that, if peanuts follow corn in a rotation sequence, southern blight is less prevalent than if peanuts follow peanuts. At one time it was believed that southern blight was more of a problem in bunch-type peanuts than in runners; current evidence shows that both types are susceptible. The development and use of disease-resistant cultivars appears to be the key to controlling this serious disease.

Pod breakdown, caused by species of both *Pythium* and *Rhizoctonia*, is a disease that affects the pods only when they are in the soil. As the name suggests, it leads to the malformation or breakdown of fruits and seeds. Fertilizer management plays a major role in the control of pod breakdown; high levels of calcium tend to reduce damage from the disease, whereas excessive amounts of both potassium and magnesium sulfates (K_2SO_4 and $MgSO_4$) promote the occurrence of pod breakdown. (The potassium response might be, in part, the role excessive potassium plays in limiting calcium uptake.)

In addition to the serious losses from fungal diseases (and the relatively minor losses from bacterial and viral diseases), peanuts are damaged by several types of nematodes. Yield reductions due to nematode infestations may be as high as 25 percent. Three types of nematodes cause losses: the sting nematode, the peanut root-knot nematode (both of which are rare), and the northern root-knot nematode (which is quite widely spread). Nematodes are controlled through rotating to nonsusceptible crops (such as grasses, including corn and other cereals) for three to five years and through soil fumigation.

As is the case with most field crops, disease prevention is more effective than disease cure. Prevention of peanut diseases includes the development and use of disease-resistant cultivars, seed treatment, and soil treatment, including fumigation for nematode control. Also, proper seedbed preparation and management practices that promote rapid germination and vigorous seedlings (planting at the proper time and using high-quality seed) are extremely important. Finally, crop rotation and the practice of burying crop residues are valuable techniques for disease prevention.

WEEDS

Weeds are a serious problem in peanuts, and weed control is an important consideration in seedbed preparation. Later, weed control is difficult because the developing plants are easily damaged by mechanical cultivation.

In the last fifteen years, chemical weed control has become an increasingly widespread practice, and currently more than 90 percent of the U.S. peanut acreage is treated with some type of herbicide. Use of both pre- and postemergence sprays is relatively common, and preemergence sprays are applied to more than half of the total acreage. (Postemergence sprays used alone are of less value in controlling the major weeds.)

The use of herbicides minimizes cultivation and the consequent damage to the peanut plant and may be responsible for the recent remarkable increase in crop yields. The control of grassy weeds is difficult even with the available herbicides, and some shallow inter-row cultivation is necessary for weed control. Cultivation is usually done with power-driven hoes.

The major weeds affecting peanuts are pigweeds, crabgrasses, nutsedges, cocklebur, panicums, morningglories, beggarweeds, and sickle pods. Cocklebur and beggarweeds have become increasing problems in recent years, but control practices have been developed that reduce the damage caused by sandburs to the extent that this weed is no longer a major pest. Late-summer and early-fall weeds may delay harvesting.

INSECTS

The peanut crop is subject to damage from a number of insects, most of which attack other crops too. Both the larvae and adults of the white-fringed beetle feed on peanut foliage and can severely injure the crop. The use of approved insecticides and rotation to non-susceptible crops, generally small grains, are effective in the field control of the pest. The optimum fertilization of a crop that precedes peanuts in rotation, such as cotton and corn, insures their vigorous growth, which, in turn, inhibits development of the insect and limits its movement to the following peanut crop. Field management of cocklebur, an alternate host for the white-fringed beetle, is another important means of controlling this pest.

Other insects causing significant harm to peanuts are the tobacco thrips (which feed on seedlings), the velvet bean caterpillar, the potato leafhopper, the southern corn root worm, and the larvae of the spotted cucumber beetle. Mites are also a serious pest at times. The proper identification of insects and prompt application of appropriate insecticides when insect populations reach potentially dangerous levels are essential to minimize serious crop losses. Insecticides must be applied with expreme care to avoid contaminating the crop.

Crop Utilization

Peanuts, either in the pod or shelled, are a popular confection and party snack (Figure 21-4); processed peanuts are a highly nutritious food.

As the crop comes from the field (before drying), 15 pounds (6.8 kg) of pods yield about 10 pounds (4.5 kg) of cleaned nuts. The nuts (seeds) are about 50 percent oil and from 25 to 30 percent high-quality protein; processing a ton (about 900 kg) of unshelled peanuts yields more than 500 pounds (227 kg) of oil and about 800 pounds (363 kg) of high-protein meal.

Peanut butter, a popular and highly nutritious product of peanuts, is made mainly from runner types. The nuts are hulled and blanched to remove the papery seed coat and then crushed, homogenized, and flavored with appropriate seasonings.

Peanut oil is a high-quality, nondrying oil that is used in cooking. It competes with other vegetable oils, such as corn and safflower oil, in the edible-oil market. After the oil is extracted, peanuts are milled into a fine flour for baking. This flour is high in protein and is frequently mixed with wheat flour. Coarse-milled peanuts, called grits, are a popular food in parts of the South.

Today, most peanuts are produced for human consumption, and very few are produced as animal feed. Thirty years ago, "hogging" the peanut crop—letting swine graze the crop in the fields by uprooting the plants and eating the pods—was a fairly common practice. Some peanuts are still used for hay; frequently, the crop residue after threshing is baled and fed to livestock. Whole plants, including pods and nuts, may be as high as 12 percent protein, but the protein level drops to about 7 percent after threshing.

In a period of growing population and worldwide food shortages—especially shortages of high-protein foods—peanuts may become increasingly valuable because of their excellent nutritional value, though currently they are not a major crop.

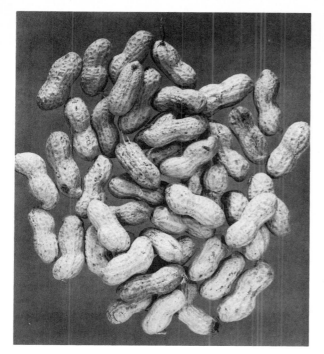

Figure 21-4
Holland Jumbo peanuts in the pod and shelled. [USDA photograph.]

Selected References

The American Peanut Research and Education Association. 1973. *Peanuts: culture and use.* Stone Printing Co.

Development and use of defatted peanut flours, meals, and grits. 1972. *Auburn Univ., Agr. Exp. Sta. Bull.* 431, April.

Effects of deep turning and non-dirting cultivation on bunch and runner peanuts. 1963. *Auburn Univ., Agr. Exp. Sta. Bull.* 344, April.

An evaluation of a mechanized system of peanut production in North Carolina. 1961. *N. Carolina State Coll., Agr. Exp. Sta. Bull.* 413, August.

Hughes, H. D., and D. S. Metcalfe. 1972. *Crop production,* 3d ed. Macmillan.

Study Questions

1. Describe the morphological features peculiar to the peanut plant.
2. What management factors have the greatest effect on peanut yield and quality?
3. What factors should be considered in the fertilizer management of peanuts? in crop rotations with peanuts?
4. What are the major pests of peanuts?

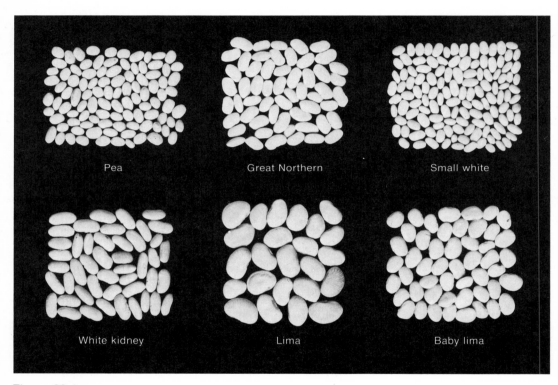

Figure 22-1
Principal types of white beans. [USDA photograph.]

Twenty-Two

Field Beans and Peas

Origin and History of Culture

Field beans There is evidence that various types of field beans, or dry edible beans, evolved in Mexico and South America, where they were domesticated and cultivated more than four thousand, and perhaps as long as six thousand, years ago. Today, field beans, including lima beans, continue to be an important source of protein in parts of Central and South America, where diets lack adequate amounts of animal protein.

The United States produces various market classes of beans in fourteen states on a total of about 1.5 million acres (607,500 ha). The leading producers are California, Idaho, and Colorado; other states producing sizeable quantities of field beans are Michigan, New York, and Nebraska. In addition to the United States, the leading bean-producing nations include Brazil, Mexico, Yugoslavia, and Italy.

Field peas The field pea (*Pisum arvenses*) and the canning pea (*P. sativum*) apparently evolved in the Mediterranean area and in Central Asia. Cultivation of field peas (dry peas) can be traced to Swiss lake dwellers of the Bronze Age (about 3000 to 1100 B.C.). Field peas were brought to North America by early colonists and were valued as livestock feed, as food, and as a soil-building green-manure crop.

Although a considerable amount of field peas is produced in the United States and Canada, they are a less significant crop than other legumes in North America. The major pea-producing areas in the United States are in Washington, Oregon, Idaho, Montana, Colorado, Wisconsin, and Michigan. Saskatchewan is Canada's leading field-pea producing province.

The USSR is one of the world's leading producers of peas, and peas are an important crop in central and northern Europe.

Botanical Characteristics

Field beans All field beans are members of the legume family (Leguminosae). There is some confusion in popular terminology concerning peas and beans because they may be grown as a vegetable (fresh) or as a field crop (dry). For our purpose, all field beans, including lima beans, are classified in the genus *Phaseolus*. White beans (navy or pea, Great Northern, and small white beans), pinto beans, and kidney beans are all classified as *P. vulgaris*. These types of beans constitute the market classes of *P. vulgaris* (Figure 22-1) and each market class includes a number of cultivars. Both standard, or large, lima beans and baby limas are classified as *P. lunatus*.

In spite of some variations, field beans have certain vegetative traits in common: most field beans are herbaceous annuals, although, under tropical conditions, some beans (such as large limas) may behave as short-lived perennials. All field beans germinate in an epigeal manner; that is, the cotyledons rise above the soil surface. However, even within a market class, field beans vary considerably with respect to growth habit. Bush types, common to red kidney beans and baby limas, are determinate in growth habit. Vine, or trailing, types, typical of Great Northern and pinto beans as well as of most large lima bean cultivars, are indeterminate in growth habit.

All field beans have pinnately compound, trifoliolate leaves. Their flowers, typical of the legume family, have been described earlier (see Chapter 19). In bush types, flowers are borne in terminal racemes and, in vine types, in axillary racemes. The fruit is the typical pod with several seeds per pod.

Field peas Field peas, too, are members of the legume family, and they are classified in the genus *Pisum*. The pea plant is an herbaceous annual. Peas germinate in a hypogeal fashion; the cotyledons are not raised above the soil surface, as they are with field beans. Plants develop a taproot system. Stems range from 2 to 4 feet (50.8–101.6 cm) in length and are quite succulent. Leaves are typically pinnately compound, but the apical leaflet is modified into a split, or double, tendril. A large pair of stipules, or leaflike bracts, is found at the base of the petiole of each leaf; these bracts are so large that they can be mistaken for sessile leaves. They usually wither and die before the plant matures.

Peas have a typical legume flower. Flowers are commonly borne singly in leaf axils, although cultivars with pairs of flowers may be preferable to those with single flowers. Flowers vary in color from white to reddish purple. The flowers of processing peas (*P. sativum*) are usually white. Plants are long-day in photoperiodic response and flower indeterminately throughout the growing seasons. The fruit is a typical pod, or legume, and contains from four to nine seeds. Seeds vary in shape from round to angular to very rough and in color from green-yellow to grey and brown. Processing peas are uniformly light green.

Environmental Requirements

Field beans The environmental requirements for the five major market classes of field beans—White (pea or navy), Pinto, Great Northern, Kidney, and Lima—are similar enough to be discussed together. However, there are several marked differences between various classes, which should be noted.

Beans of all market classes are warm-season crops; the optimum temperature for all of them is in the lower range for warm-season crops—about 24°C. All, except kidney beans, require a minimum frost-free period ranging from 120 to 130 days; a minimum of 140 days is needed for kidney bean production. In general, high temperatures during flowering (85°–90°F, or 29°–32°C) cause flower blasting (dropping of buds and flowers), which reduces yields. Baby limas tolerate higher temperatures than most beans and are produced in warmer regions. Large lima beans, instead, are adapted to a cooler, foggy climate, and,

Table 22-1
Water use and frequency of irrigation of field beans at different stages of development and in different soils.

Stage of development	Root zone depth (feet)	Water use (inches/day)	Days between irrigations*		
			Sandy	Loam	Clay
From two to three weeks after emergence	1.0	0.075	7–12	12–17	17–23
Early bloom	2.0	0.125	7–13	13–18	18–23
Pod development	2.5	0.250	4–8	8–10	10–13
Pod maturing	3.0	0.200	6–11	11–15	15–20

Source: J. S. Robins and O. W. Howe. *Irrigating dry beans in the West.* USDA leaflet 499(1961).
*The time between irrigations on any one soil is largely determined by climatic conditions.

because of this peculiar climatic adaptation, they are produced primarily in a narrow band along the Pacific Ocean north of Los Angeles.

Although they are warm-season crops, field beans do not require excessive amounts of moisture; depending on the soil and climatic factors, from 12 to 24 inches of moisture (30.5–61.0 cm) is an adequate amount. Various types of beans differ markedly in their patterns of root growth, and this must be considered in scheduling irrigations. In general, vine types have much more extensive root systems than bush types and therefore require less frequent, but deeper, irrigations (more water per irrigation). Because vine-type plants may produce more total dry matter (roots, stems, and leaves, as well as seeds), they may require more total moisture. Adequate moisture early in the season is essential for all types of beans. In regions with limited spring rain, a preirrigation of 3 to 8 inches (7.6–20.3 cm) is frequently recommended. Because field beans do not tolerate crusted soils or emerge well through a crust, they should be planted into moist soil and should not be irrigated until after they have emerged. Thus, preirrigation is preferable to "irrigating the crop up." Water use and needs change as bean plants grow and develop (Table 22-1). Along the California coast (a large lima-producing area), preirrigation plus one more irrigation may be adequate for the entire season. In warmer areas, three or four irrigations of 3 to 6 inches (7.6–15.2 cm) each are usually required (Table 22-2). More irrigations may be needed for red kidney beans, which are very shallow rooted. All types of field beans require adequate moisture throughout the pod-filling stage (during and immediately after flowering); the soil should not fall below 60 percent field capacity to insure proper moisture availability. Dry weather is desirable for maturing the crop and during harvesting.

In addition to interrupting harvest schedules, late rain may discolor the beans and lower their grade and market value. Sprinkler irrigation is, in general, not recommended for bean production, and, in some states, beans

Table 22-2
Effect of number of irrigations on yield of Great Northern beans grown in western Nebraska.

Number of irrigations	Yield (bu/acre)		
	1954	1955	Average
0	26.9	12.9*	19.9
1	36.5	42.5	39.5
2	39.7	47.2	43.5
3	43.5	54.6	49.1
4	44.1	50.6	47.3
5	43.3	60.7	52.0

Source: O. W. Howe and H. F. Rhodes. Irrigation of Great Northern field beans in western Nebraska. *Nebraska, Univ., Agr. Exp. Sta. Bull.* SB 459(1961).
*In 1955, water for preirrigation was not available, and the subsoil was dry; yield of unirrigated beans in 1955 was less than half of yield in 1954.

produced with sprinkler irrigation are not eligible for certification.

Field beans are adapted to a wide range of soils: in humid areas, they are produced on acidic soils; in the relatively dry West, on neutral to slightly basic (alkaline) soils. A well-drained loamy soil is ideal for field bean production.

Field peas Field peas are a cool-season crop; moderate temperatures are essential throughout the growing season for successful production. Although light frosts can damage the plants during flowering, even moderately warm temperatures (more than 30°C) are undesirable generally. Moisture requirements are variable: more than 20 inches (50.8 cm) of rain is desirable, but field peas can be produced successfully in cool, semiarid regions. Peas are most sensitive to moisture stress at flowering. The water-use efficiency of the pea plant is highest in the absence of moisture stress.

Field peas are best adapted to well-drained, clay-loam soils. They tolerate a moderate soil pH range: the optimum pH is 6.5, but moderate acidity (pH as low as 5.5) is tolerated. In some areas, liming to raise the soil pH to 6.0 may be necessary for profitable yields.

Production Practices

SEEDBED PREPARATION AND SEEDING

Field beans Seedbed preparation and seeding of field beans follows the same general pattern as that for any spring-planted row crop. In the fall, crop residue is incorporated into the soil, and the field is left in a suitable condition for maximum storage of winter rain. In the spring, a deep but firm seedbed is prepared, and the field is leveled and readied for preirrigation. After preirrigation, the seed-

Figure 22-2
Typical bean field, Sacramento Valley, California.

bed is given a final harrowing followed by furrowing-out for irrigation before planting. The seedbed should be deep and firm, but to minimize crusting, it should stay a little cloddy, not finely pulverized.

Field beans are usually planted in rows spaced from 22 to 24 inches (55.9–61.0 cm) apart (Figure 22-2). Row spacing may be as close as 18 inches (45.7 cm), but, if planting large lima beans, rows may be 30 inches (76.2 cm) or more apart. Spacing of seeds for bush types averages four seeds per foot, or a plant every 3 inches (7.6 cm). For vine types, spacing may increase to 6 or even 12 inches (15.2–30.5 cm) between plants. Planting rates for bush types with average spacing are about 75 pounds of seed, or about 90,000 plants per acre (85 kg or nearly 101,000 plants/ha) (Table 22-3). Obviously, the number of plants is reduced for the widely spaced vine types. Also, the seeding rate in pounds per acre is reduced with the use of small-seeded types and increased with the use of large-seeded types.

Planting depth ranges from 1 to 3 inches (2.5–7.6 cm), with a maximum depth of 4 inches (10.2 cm). Seed must be placed into

Table 22-3
Effects of different row spacings, plant spacings, and plant population densities on yield of red kidney beans grown at Davis, California, in 1970.

Rows per bed	Distance between beds (in inches)	Distance between plants (in inches)	Plants per acre	Yield (100 sacks/acre)
2	40	3	104,648	27.95
1	30	6	34,848	26.60
2	30	3	139,392	25.62
2	40	6	52,324	25.46
1	30	3	69,696	25.46
2	30	6	69,696	25.41
2	40	9	34,010	24.52
2	40	12	26,162	24.48
2	30	9	45,302	24.17
2	30	12	34,848	23.07
1	30	12	17,424	22.87
1	30	9	22,651	22.24

Source: Unpublished data courtesy of Carl Tucker, University of California at Davis.

moist soil; therefore, the final seedbed preparation must conserve soil moisture, yet avoid causing crust formation.

Field beans are planted with specially designed four-, six-, or even eight-row planters. Both cotton and corn planters can be equipped with bean plates and used efficiently. Either vertical or horizontal-plate planters are used for all beans except large limas, which require cup planters.

Time of planting for field beans depends on soil temperature and on their market class. Field beans are planted later than corn; planting should be delayed until the soil temperature is about 65°F (18.5°C) (Table 22-4). Planting dates for both limas and small whites, although produced in different regions, are in early May; red kidneys are planted in mid-June and July; and pintos are usually planted in mid-June (Table 22-5). The appropriate planting date is governed by at least five factors: (1) proper soil temperature (neither too cool nor too warm), (2) probability of rains, which may lead to soil crusting and restrict seedling emergence, (3) possibility of high temperatures later in the season, which cause blossom drop, (4) length of growing season (fall wind and rain damage should be avoided); and (5) place of the bean crop in the total cropping system. Many types of beans tolerate a moderate range of planting dates and still produce well except in areas where the likelihood of high temperatures or early-fall rains and wind makes early planting a necessity.

As is the case for all crops, fertilizer applications for field beans should be based on the results of appropriate soil tests. Although all field beans belong to the legume family and should fix nitrogen when inoculated, the addition of nitrogen is required in many instances. Relatively light applications of nitrogen, from 20 to 40, or as high as 60, pounds per acre (22.4–44.8 to 67.2 kg/ha), are generally adequate in humid regions of the East, and little or no additional nitrogen is required if field beans follow another legume. Nitrogen applications in the West vary widely with the soil, the preceding crop, and fertilizer management. Both phosphorus and potassium (in the form of P_2O_5 and K_2O) are frequently applied to field beans; the rates of each range

Table 22-4
Effect of soil temperature on percentage and speed of germination of lima beans in California.

Soil temperature (°F)	Percentage of germination	Days to emergence (avg)
48	0	—
55	2	31.0
61	52	28.0
68	82	17.6
77	80	6.5
85	88	6.7
95	2	9.5

Source: R. W. Allard. Production of dry edible lima beans in California. *Calif., Univ., Agr. Exp. Sta. Ext. Serv. Circ.* 423(1953).

Table 22-5
Planting dates for six market classes of field beans in California.

Market class	Planting dates*		
	Earliest	Normal	Latest
Standard Lima Beans	4/20	5/1–5/10	6/1
Baby Lima Beans	4/25	5/7–5/15	6/10
Small Whites	4/20	5/5–5/15	6/10
Red Kidneys	6/10	6/20–7/1	6/15
Garbanzos	February	March	April
Horse Beans	January	February	March

Source: R. W. Allard and F. L. Smith. Dry edible bean production in California. *Calif., Univ., Agr. Exp. Sta. Ext. Serv. Circ.* 436(1953).
*Planting dates are closely related to soil temperatures.

from as low as 40 pounds per acre to as high as 140 or more pounds per acre (44.8–156.8 kg/ha). Fertilizers may be applied either before or during planting. Phosphorus is banded with the seed, but nitrogen and potassium should not be. In some limited regions, certain micronutrients, such as zinc, may be deficient; a zinc deficiency has been remedied by foliar applications of a zinc sulfate solution. A major benefit of applying zinc is that it encourages early maturity of the crop.

Field peas Seedbed preparation for field peas usually starts with a fall plowing. In northern regions, field peas are planted as early as possible in the spring; in the United States, planting frequently starts in mid-April; in Canada, planting begins early in May and continues through the third week of the month. In the U.S. coastal and southern areas, where the crop must mature before damaging high temperatures occur, fall planting (mid-September to mid-October) is the rule. Planting is commonly done with a standard grain drill. Row spacings may vary from 7 to 9 or even 11 inches (22.9–27.9 cm). The desired stand is from three to four plants per square foot (plants 7–10 cm apart), or about 175,000 plants per acre (432,250 plants/ha). Seeding rates range from about 110 pounds of seed per acre (123 kg/ha) for small-seeded types to nearly 180 pounds per acre (202 kg/ha) for large-seeded types.

Unless peas have been grown recently, the seed should be inoculated with an appropriate strain of *Rhizobium* to insure effective nodulation and nitrogen fixation for satisfactory yields. Like most legumes, peas frequently respond to phosphorus fertilizers. General fertilizer recommendations cannot be made; fertilizer applications should be based on the results of appropriate soil tests.

HARVESTING AND ROTATIONS

Field beans Field-bean harvesting is a fall operation. It starts in September and, in some regions, may continue well into November. The crop is ready to harvest when 75 percent of the pods are dry and yellow. Some types

Figure 22-3
Combine harvesting navy beans in Michigan. Beans are cut and windrowed to dry several days before threshing. [Photograph by J. C. Allen and Son.]

tend to bloom and set pods indefinitely, or until the first frost; this makes the identification of the 75-percent-ripe-pod stage more difficult. Weather during harvesting is critical; rain tends not only to delay harvest, but to discolor beans and lower grades. Even fog or heavy dew delays harvest because pods and vines become tough and leathery, which makes threshing inefficient. Hot, dry weather, on the other hand, leads to excessive shattering of pods and consequent losses in threshing.

During harvesting, field beans are cut with a knife 3 to 4 inches (7.6–10.2 cm) below the soil surface, windrowed to dry, and then threshed with combines. Combines used for harvesting beans are similar in operation to grain combines, but the cylinders are modified to handle more vegetative material and to open pods without damaging the seeds. Frequently, bean combines have two cylinders rather than the single cylinder typical of the grain combines. Pick-up and feed devices, as well as screens and fanning equipment, are usually different (Figure 22-3).

Field-bean yields of 2,000 to 3,000 pounds of seed per acre (2,250–3360 kg/ha) or more are common, although yields vary with market class, geographical location, and year.

Where beans are grown, they are generally a prime cash crop and hold the key place in crop rotation. In the East, rotation away from beans is frequently done to control soil-borne pathogens; in the West, this practice is not as common (see the following section on pests). In the large lima-producing regions, limas are grown continuously on an annual cropping basis; a bean-cereal-bean rotation is also common. In areas where beans cannot be produced frequently, a bean-cereal-sugar beet-bean rotation is employed. In some areas, beans are grown for three to five years alter-nated with three or four years of alfalfa. Beans are also rotated with hay crops and potatoes.

Field peas Field peas are ripe and ready for harvesting when the pods are mature and dry. Frequently, plants are cut and windrowed with a special cutter and then threshed with a standard grain combine, but for more than 60 percent of the field peas, harvesting is done in one operation with a grain combine. Combine adjustment is critical to avoid damaging peas, thus lowering the market value of the crop.

The use of chemical desiccants to hasten the drying of plants and to facilitate harvesting has become increasingly popular in recent years. Chemical desiccants also dry weeds and thereby reduce excessive vegetative materials that foul harvesting equipment. Sulfuric acid solutions and several commercial herbicides have been used as desiccants. The use of desiccants, however, may contaminate the crop residue (the haulm) and make it unfit for use as livestock feed.

Field peas are a good, and fairly common, annual hay crop. Harvesting should be done when pods are well formed but green.

In the North and Northwest, field peas are frequently used as a green-manure crop in a cropping sequence. When grown for seed, field peas can replace fallow in a common cereal-fallow rotation, and, when market conditions are favorable, they may be grown instead of a cereal. Field peas may occasionally be inserted as a third crop in a small grain and corn rotation. Usually, when field peas are produced, they are considered a minor crop, and the rotation is developed around more important crops, such as corn or small grain. However, if peas are produced for processing (canning or freezing), they may hold a priority place in the rotation. In response to consumer demand for convenience foods, more peas are being harvested for the fresh-frozen vegetable processor.

Diseases and Pests

DISEASES

Field Beans Throughout the United States, field beans are attacked by at least forty diseases caused by fungi, bacteria, or viruses. The most severe disease problems and the greatest losses occur in the humid regions, although significant disease losses also occur in the irrigated, semiarid areas.

There are three major groups of diseases caused by various species of fungus. The first, seed and seedling rots, either prevent germination or kill seedlings. These rots are best controlled by seed treatment and by farming practices that insure good germination, such as waiting until the soil warms, planting into moisture, and limiting soil crusting. A second group of diseases caused by fungi is root rots. There is little effective chemical control for this group; pathogens persist in the soil, and only long-term rotation is truly effective. Finally, there are foliar blights caused by fungi. These diseases are serious in humid regions but are rarely a problem in semiarid areas, even with irrigation. Bean anthracnose is one of the more serious of the foliar blights caused by fungi. Because the fungus overwinters in the seed, one very effective method of control is planting disease-free seed. Also, because the disease is not prevalent in semiarid areas, seed from these areas is disease-free and is used for planting in humid regions. For example, red kidney bean seed planted in New York is produced in California to insure freedom from seed-borne bean anthracnose.

Beans are also attacked by several bacteria that cause foliar blights, which lead to reduced yields and shrunken seeds. Control is achieved mainly through planting disease-free seed that is produced in the semiarid environment of the intermountain region of the Pacific coast.

Common bean mosaic—a seed-borne virus, which is also transmitted by aphids from diseased to healthy plants—may cause losses in all areas. The only effective control is through the use of resistant cultivars.

Field peas The greatest losses from diseases of field peas occur in the humid southern areas where peas are grown mainly as green-manure crops. The black or brown stem disease is caused by the two fungi *Ascochyta pinodella* and *Mycosphaerella pinodes*. Infection causes severe degeneration of stem tissue and plants may die before seed is matured. The pathogens persist in or on seed and in the soil, and control is achieved by planting disease-free seed and by means of crop rotations that exclude field peas and other susceptible species for three or four years.

Symptoms of leaf blotch, caused by the fungus *Septoria pisi*, appear on seedling leaves of fall-sown peas. Control of this common disease is achieved through crop rotation. Field peas are also subject to attack by powdery mildew caused by the fungus *Erysiphe polygoni*. Symptoms appear in the late spring, and damage is minimal in crops grown for green manure. There is no known practical control for powdery mildew.

Field peas are susceptible to root rot and to wilt diseases caused by various fungi, as well as to a bacterial blight disease caused by *Pseudomonas pisi*. Peas suffer from no serious viral diseases. Root-knot nematodes may be a problem in the extreme South and in some coastal areas.

WEEDS

Field beans Weeds are a serious problem in field-bean production. Control starts with seedbed preparation and continues with cultivations throughout the growing season. Cultivations that are either too deep or too close to the plant can damage the tops (shoots) or the roots; therefore cultivating requires great caution. Cultivation is, of course, more restricted in vine types of beans than in bush types.

Herbicides, too, play an important role in weed control: more than a third of the total U.S. acreage is treated with some type of herbicide. Preemergence herbicides are most common and are used on about 67 percent of the treated acreage; postemergence sprays are used on less than 30 percent of the treated acreage; and combinations of preemergence and postemergence sprays on about 10 percent. Despite the high cost of labor, some hand hoeing is still used for weed control.

Specific weeds in field beans vary from region to region, but the ten most important weeds, or groups of weeds, in the United States in order of importance are: pigweeds, crabgrasses, lambsquarters, nutsedges, ragweeds, foxtails, nightshades, thistles, barnyardgrass, and kochia. Although the number of acres with specific weeds is decreasing in some areas, the number of acres infested with all the major weeds is increasing in others. Progress is being made, but, generally, the increases in infested acres exceed the decreases.

Field peas Weed problems in field peas are similar to those described for field beans. Because field peas are planted in narrowly spaced rows, weed control depends on seedbed preparation and herbicides, plus the use of weed-free seed. Cultivation during the growing season is impossible for all practical purposes.

INSECTS

Field beans Field beans are subject to attack from a number of insect pests in the field and in storage. Pest control in the field depends on the type of insect, the stage of its life cycle, and the severity of the infestation. Both chlorinated hydrocarbons and organic phosphates have been effective in controlling field pests. However, because field beans are consumed more or less directly by humans (they are not processed, except for packaging), the use of these pesticides has been greatly restricted. New, nonpersistent insecticides are now being explored.

The wireworm (the larva of the click beetle), an insect pest that cuts seedlings off at, or slightly below, the soil surface, can be controlled with insecticides and by rotation if infestations become too damaging. Other insect pests include lygus bugs, which cause damage to developing pods, and leaf miners (a larva form), which destroy foliage, as do the larvae of the Mexican bean beetle and the seed corn maggot. Red spiders, although not insects, also feed on and damage field beans. From time to time, other insects such as the potato leafhopper, white-fringed beetle, and various cutworms also cause damage to field beans.

The bean weevil is a serious pest that attacks stored beans. It is controlled through the use of weevil-free seed for planting and through sanitation and fumigation at the elevator. For this pest, as is generally the case, an ounce of prevention—the use of "clean" seed—is worth a pound of cure!

The root-knot nematode is a highly destructive and persistent pest that survives in the soil for long periods. The only effective control is long-term rotations from susceptible to nonsusceptible crops, such as cereals. (Field fumigation is effective but expensive.)

Field peas The pea weevil is the most serious of the insect pests that attack field peas. The female pea weevil deposits eggs in the developing pod, and, when the eggs hatch, the larvae feed on the seed (embryo). Naturally, the presence of feeding larvae lowers the nutritional value (and the market value) of the seed. Control consists of fumigation of seeds immediately after harvesting, and, occasionally, it may require field spraying to destroy females before eggs are deposited.

Other insect pests include the pea aphid and the pea moth. The pea aphid can be controlled culturally in the seed-producing region of western Oregon by delaying planting until after mid-October when insects are not present to infest plants. It can also be controlled by means of field applications of rotenone dust formulations.

Crop Utilization

The various types, or classes, of field beans require minimum processing for human consumption. They are eaten boiled, baked, or fried and are included in many soups; they are also used as a substitute for meat. The relatively high protein content of dry beans and the relative ease with which they can be stored make them a valuable and dependable food. Of course, beans are also frozen and canned and are available in many prepackaged convenience meals.

Field peas are grown primarily as a green-manure crop. When the crop is harvested for seed, it is most commonly used as split peas in various types of soups. Field peas are high in protein (as much as 24 percent) and may become increasingly important as both a food and a feed crop in the future.

Other Edible Legumes

In addition to field beans and peas, and to soybeans and peanuts (see Chapters 20 and 22), there are many other legumes among which are lentils, that are important as both food and feed crops in various areas of North America. The pink bean is a significant crop in California as are chickpeas, or garbanzo beans, throughout the Southwest. Other valuable legumes include mung beans and broad beans (also called horsebeans). The faba bean has become increasingly popular as a protein source for livestock feed in Canada, and it has significant potential as an edible legume. Faba beans (classified as *Vicia faba*) are small-seeded relatives of the horse bean and are adapted to moist areas and cool growing conditions. Although available cultivars mature relatively late in the season, that drawback can be offset by early seeding because seedlings are quite frost tolerant. Faba beans fit well into many standard cereal crop rotations; faba-bean production uses standard cereal-planting and harvesting equipment. Beans, peas, and the various other edible legumes offer a major protein source in a world where many cannot afford the luxury of animal protein.

Selected References

Allard, R. W. 1953. Production of dry edible lima beans in California. *Calif., Univ., Agr. Exp. Sta. Ext. Serv. Circ.* 423.

Allard, R. W., and F. L. Smith. 1954. Dry edible bean production in California. *Calif., Univ., Agr. Exp. Sta. Ext. Serv. Circ.* 436.

Howe, O. W., and H. F. Rhodes. 1961. Irrigation of Great Northern field beans in western Nebraska. *Nebraska, Univ., Agr. Exp. Sta. Bull.* SB 459.

Martin, J. H., and W. H. Leonard. 1967. *Principles of field crop production.* 2d ed. Macmillan.

Robins, J. S., and O. W. Howe. 1961. *Irrigating dry beans in the West.* USDA leaflet 499.

Study Questions

1. Why are field beans increasingly important to mankind?
2. Describe the morphological differences between bush- and vine-type beans and discuss how these differences affect production practices.
3. What are the production factors that have the greatest effect on the quality of field-bean crops? of field-pea crops?
4. What are the major limiting factors for producing the various classes of field beans and peas? How can these factors be managed under "field conditions"?
5. What are some of the major differences between the production of field beans (or peas) and the production of cereals (including corn)?

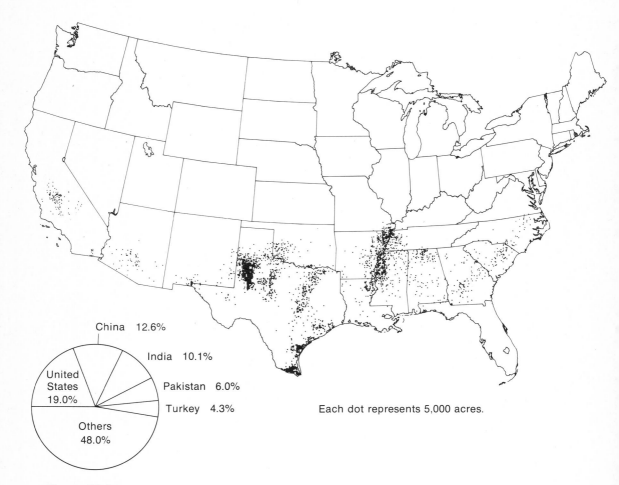

Figure 23-1
Major cotton-producing regions in the United States and contributions to world production. [Data from U.S. Department of Commerce and the United Nations.]

Twenty-Three   Cotton

Cotton has been an important crop in the United States since the colonial period. With the development of the cotton gin in 1794, it became increasingly important during the nineteenth century. The economic dependency on cotton production in the southern United States led to the evolution of what could be called a "cotton society," and the economic and social problems that developed in part precipitated the Civil War.

Since World War II, modern technology has supplied the means for developing numerous synthetic (so-called miracle) fibers. Nevertheless, cotton continues to be in great demand both at home and abroad and is a leading cash crop in the United States.

Origin and History of Culture

Cotton is reported to have evolved in several different locations, but its origin is unknown. There is evidence, however, that, in the Old World, some types originated in Indochina and tropical Africa; in the New World, different types originated in Central and South America. The types of cotton currently grown in the United States (Sea Island, American-Egyptian, and Upland) are of New World origin and are known by the characteristics of the lint (fiber) they yield.

The importance of cotton is reported in man's earliest written records, which trace cotton to Peru in 2500 B.C., to Pakistan 500 years earlier in 3000 B.C., and to Mexico as early as 5000 B.C. Records show that cotton was used for clothing in India about 2000 B.C. Early explorers from Europe reported seeing "wool trees" in India. Alexander the Great is generally credited with having introduced Indian cotton into Europe in the middle of the fourth century B.C. Evidence of cotton production dating back more than 800 years has been unearthed in Arizona in the form of cotton fabrics in Indian ruins.

Cotton was brought to Virginia in 1607 and played a major role in the agricultural and sociological development of the United

States. Cotton culture in the South constituted a monoculture, an ecological phenomenon that has been severely criticized by environmentalists in recent years.

Currently, the distribution of cotton production is changing markedly. Although the cotton belt still exists in the South, production has been moved westward, and California and Arizona are now leading cotton-producing states (Figure 23-1). Several factors contributed to the westward migration of cotton. As the nation expanded to the West, settlers carried the crop with them and cotton proved to be ideally suited to some of the new locations. Of greater importance in the migration of cotton was the practice of growing cotton exclusively until the soil would no longer sustain a crop and then moving to another location; cotton culture was thus forced west of the Mississippi River. Finally, it was moved westward to escape the ravages of the notorious boll weevil, the major insect pest of cotton. With the development of extensive irrigation-water supply systems, cotton culture was moved into Arizona and the southern part of central California, where climatic conditions are ideal, insect pests are not serious, and fields are nearly level, thus favoring irrigation and mechanical picking.

Of all of the major crops produced in the United States, cotton has been the subject of more writing and more regulations than any other crop. Federal control of the marketing and production of cotton has been in effect since 1929. Price restrictions and import-export quotas were established for cotton before similar restrictions were placed on other crops. Domestically, cotton production has been restricted by the establishment of legally recognized "single variety districts." Because the spinning industry depends on an adequate supply of fiber of uniform quality, some districts that have established reputations for producing fiber of a specific quality that is in demand have legally restricted the number and types of cultivars that can be grown there in order to maintain their favorable status in the marketplace. Such districts require that all producers grow only specific cultivars of the crop.

In the past ten years, the United States has produced about 25 percent of the world's cotton on about 15 percent of the cotton acreage. Obviously, U.S. cotton yields are significantly greater than the world average. Cotton is produced in nineteen states, including the relative newcomers, Arizona and California. Worldwide, both Mexico and Brazil produce relatively large amounts of cotton, as do the United Arab Republic, India, Pakistan, Turkey, China (mainland), the Sudan, and the USSR.

Cotton production, from planting through harvesting and processing, has become highly mechanized. The mechanization of processing has been improving continuously since the invention of the cotton gin in 1794, but many other earlier inventions contributed to the growth in significance of cotton. Inventions such as the flying shuttle (in 1733) and the spinning Jenny (in 1764) helped to create the demand for cotton by the spinning and weaving industry. The major industrial developments were made first in England and were matched and surpassed later in the United States.

Mechanization in the production of the crop is quite recent—the mechanical harvester has come into common use in the two decades since 1955, and hand labor is still not a thing of the past. Since the end of World War II, however, manual labor in cotton production has been drastically reduced through the development of mechanical pickers. Other major advances have centered on the development of pesticides and chemical defoliants and the extensive use of aircraft in applying chemicals.

Botanical Characteristics

Cotton is the only agriculturally important member of the mallow family (*Malvaceae*). Two species in the genus *Gossypium* are cultivated: *G. hirsutum* and *G. barbadense*. Upland cotton (*G. hirsutum*) has relatively short fibers, from 0.75 inch (2 cm) to 1.25 inches (3 cm) in length; Sea Island and Egyptian types (both classified as *G. barbadense*) have longer fibers, from 1.5 to 2.0 inches (2-5 cm). Cultivars of both species vary with respect to many traits, including fiber length and strength.

Generally, cotton is classified as an annual; in the tropics, however, plants of *G. barbadense* may be short-lived perennials. Plants grow to heights of 2 to 5 feet (60-120 cm) and develop a deep taproot and a widely branching secondary root system. Many branches are formed from axillary buds on the upper part of the main stem. Leaves are large, hairy, and simple with three, five, or seven lobes. They are borne on petioles with stipules and are palmately veined. Two buds are present in, or adjacent to, the leaf axils—a vegetative bud in the axil itself and an extra axillary flower bud that produces a fruiting branch. Thus, cotton is indeterminate in growth habit.

Flowers are borne on alternate sides of the fruiting branches, and they are typically complete and perfect (Figure 23-2). Three leaflike bracts are found at the base of each flower. The calyx is composed of five large sepals, and the corolla of five large, showy petals. Petal color varies from white to purple and, depending on the cultivar, a reddish purple dot may be present at the base of each petal. Stamens are fused in a double row around the pistil. The pistil has a compound ovary and is composed of three to five carpels. The fruit is a typical capsule with about nine seeds per locule, or as many as forty-five seeds per fruit.

Figure 23-2
An open cotton flower. [Courtesy of University of California Agr. Ext. Serv.]

The immature, unopened bud is called a square. Normally, each plant sheds as much as 40 percent of its squares. The mature pistil (ovary, or fruit) is the cotton boll. Cotton fibers (lint), which yield the cotton staple, are long, single cells produced on the seed coat. The lint starts to grow at about the time that the flower opens and continues to elongate for about eighteen days. After this period, the fiber continues to grow in thickness for about another three weeks. Cotton fibers vary in length from 0.25 to 2.00 inches (0.6-5.1 cm); their length is from 1,000 to 3,000 times their thickness. Adequate moisture is essential for the plant to develop fibers of maximum length. When the fibers are developing, adverse conditions such as drought, disease, and insect damage cause the plant to develop low-quality fibers.

Environmental Requirements

Cotton is a warm-season crop; it is not produced in regions in which the mean annual temperature is below 60°F (16°C). Cardinal temperatures for germination and seedling growth are: minimum 60°F, optimum 93°F, and maximum 102°F (16°, 34°, and 39°C, respectively). Optimum temperatures during flowering range from 91° to 97°F (33°–36°C). Below optimum temperatures in general favor the development of nonreproductive (vegetative) branches; however, the development of reproductive (flower-bearing) branches is favored by cooler days and warm nights. In addition to requiring high temperatures, cotton requires a lengthy frost-free period: approximately 180 days is the minimum, and a frost-free period of more than 200 days is common in many cotton-producing regions. Frost during the seedling stage of development reduces yields by stunting growth and delaying fruiting.

Cotton requires moderate to fairly large amounts of water. The minimum is about 20 inches (50 cm) of rainfall per year; the optimum for economical yields is more than 60 inches (150 cm) distributed throughout the growing season. Irrigation requirements vary from 24 to 42 inches (61–107 cm). Excessive moisture from mid- to late season—whether from rainfall or from irrigation—delays maturation of the crop and interrupts harvesting. Early, heavy rainfalls or excessive irrigation encourages the development of an undesirable shallow root system. Excessive moisture at any time tends to leach essential minerals from the soil. Moisture stress reduces total vegetative growth.

Cotton is day-neutral in photoperiodic response and flowers indeterminately until the fall frosts. However, the crop is sensitive to light quantity; cotton requires high light intensity during most of the growing season for efficient growth and economic yields. Production in areas where 50 percent or more of the days during the growing season are classified as half-cloudy is risky; if more than 60 percent of the days are half-cloudy, cotton is not produced.

In short, ideal climatic conditions for cotton production are a warm spring with frequent showers, a hot summer with adequate moisture, and a long dry autumn.

Cotton can be produced on a wide array of soils. The optimum soil pH for cotton is from 5.8 to 6.5, but the crop is grown successfully on more acidic soils with a pH as low as 5.2, and on alkaline soils with a pH above 8.0. Generally, soils that are high in organic matter produce higher cotton yields; thus, green-manure crops are often included in cotton rotations. Loamy soils (clay-sand mixtures) or alluvial soils of the Mississippi Delta and southern Central Valley in California are ideal for cotton; heavy, clay soils with poor drainage tend to delay maturity.

Production Practices

SEEDBED PREPARATION AND SEEDING

Initial seedbed preparation for cotton is somewhat similar to that described for corn. Initial cultivation is done either immediately after harvesting or as early in the spring as field conditions will allow, depending on the preceding crop in rotation. The primary objective is to work crop residues into the soil. Plowing and chiseling hard pan, if necessary, are completed before final seedbed preparation.

In the East cotton is planted on beds; in the West, it is planted flat with furrows between the rows or on W-shaped beds, in which seeds are planted in ridges with furrows between the ridges. (Some hill planting similar to that employed with corn is still being done but is not efficient if current equipment and production practices are used.) Row spacing varies from 36 to 42 inches (91–107cm); seed spacing in the row is from 3 to 8 inches (7–20 cm). After emergence, the rows may be thinned to one plant every 3 to 12 inches (7–

Table 23-1
The effects of row width and plant population density on the percentage of lint and on yield of seed cotton.

Row width (cm)	Plants seeded per hectare								
	128,000	256,000	Mean	128,000	256,000	Mean	128,000	256,000	Mean
	Plants per hectare			Lint (%)			Yield of seed cotton (kg/ha)		
25.4	107,702	209,384	158,544	38.6	38.6	38.6	4,401	4,300	4,350
50.8	123,757	214,067	168,913	38.8	38.8	38.8	4,065	4,062	4,064
76.2	108,371	200,069	154,220	38.4	38.6	38.5	3,693	3,894	3,794
101.6	111,049	214,428	162,738	38.9	38.5	38.7	3,855	3,787	3,821
Average	112,720	209,487	161,104	38.7	38.7	38.7	4,004	4,011	4,007

Source: B. S. Hawkins, and H. A. Peacock. Influence of row width and population density on yield and fiber characteristics of cotton. Agron. J. 65(1973):47–51.
Note: The yield of seed cotton was not significantly affected by seeding rate; rather, it was affected by row width. The percentage of lint was not affected by either row width or seeding rate.

30 cm). Satisfactory yields are obtained if there are two to eight plants per foot (30 cm) of row, or, in nonirrigated semiarid regions, one plant per foot (30 cm) of row. In the past, thinning was done by hand; currently, mechanical thinning using various types of special cultivators and flame thinning using burners are common practices.

Plant populations may be as low as 20,000 and as high as 50,000 plants per acre (49,400–123,500 plants/ha) depending on row spacing and plant spacing (Table 23-1). Seeding rates generally range from 8 to 20 pounds of acid-delinted seed per acre (9–22 kg/ha) where adequate moisture is available to as low as 7 pounds per acre (8 kg/ha) in nonirrigated semiarid areas. Acid-delinted seed is generally recommended because it is easier to handle and feeds more readily and more accurately through precision planters. In recent years, attempts have been made to produce cotton in very close rows. In some areas, row spacing as close as 9 inches (23 cm) has not reduced yields significantly, and, because less weed control is required, production costs have been reduced by more than 25 percent. Dense planting results in taller plants but less total vegetative growth, fewer and shorter internodes in reproductive branches, the formation of fruits on higher branches, and the production of smaller but more numerous bolls per acre.

As is true for most crops, the seeding depth for cotton should be regulated to insure rapid, uniform germination and seedling emergence. Seeding depth depends on a number of factors, including soil type, available moisture, soil temperature, and expected seedling vigor (which, of course, is related directly to the quality of the seed planted). In areas where cotton is planted early, seeding depths are generally from 1.0 to 1.5 inches (2.5–4.0 cm) to insure that seeds are placed in warm soil; of course, moisture must be available at these shallow depths, and irrigation may be necessary. In relatively dry areas, seeding depths are as much as 3.0 inches (7.6 cm) to insure that seeds are placed into moist soil. Factors that lead to slow germination and emergence (such as cool temperatures, inadequate available moisture, and seeding at too great a depth) may encourage diseases and lead to poor stands and reduced yields. Seedling vigor, and plant vigor in general, is related to the balance of available moisture and temperature.

The planting date for cotton varies with the area of the country; it always follows the date of the last killing frost, frequently by as much as two weeks. Ideally, planting should

Table 23-2
Effects of planting dates on cotton lint yields (in kilograms per hectare) in Texas.

Year	Planting date (approximate)*					Average
	4/1	6/1	6/10	6/20	6/30	
1960	1,132	1,001	779	493	277	736
1961	1,122	901	826	575	286	742
1962	1,039	1,009	908	562	218	747
1964	894	753	528	485	330	598
1965	1,200	873	714	353	160	660
Average	1,077	907	751	494	254	

Source: J. D. Bilbro and L. L. Ray. Effect of planting date on the yield and fiber properties of three cotton cultivars. *Agron. J.* 65(1973):606–609.
Note: Highest yields every year came from earliest planting.
*Actual (optimum) date depends on the specific location.

be delayed until the soil temperature reaches at least 60°F (16°C). This occurs in early March along the Gulf Coast and in Arizona and after mid-April in most of the cotton belt and in California (Table 23-2). After early May, little commercial cotton is planted in the United States.

Planting equipment for cotton is functionally similar to that used for corn; horizontal-plate planters that plant four, six, or eight rows are commonly used (Figure 23-3). Frequently, two or more field operations are carried out during planting: beds are cultivated, preemergence herbicides and/or fertilizers are banded into the seedbed, the seed is placed in the soil, and the bed is rolled.

FERTILIZERS
Like all crops, cotton requires a balanced supply of minerals for efficient growth and high yields. Cotton has a high demand for nitrogen early in its life—from about two months to four months (75–125 days) after planting—with the highest demand for nitrogen coming about 90 days after planting. The common fertilization rate for nitrogen is from 125 to 150 pounds per acre (140–150 kg/ha), but rates of up to 400 pounds of nitrogen per acre (450 kg/ha) have resulted in economical increases in yield. (As is the case with other crops, overfertilization with nitrogen leads to undesirable rank growth and weak plants.)

Harvesting, Processing, and Rotations

Cotton harvesting starts in midfall (October) and continues through midwinter. Cool weather, and even frost, speeds up harvesting by accelerating defoliation; rains delay harvesting and may reduce crop quality (Figure 23-4).

In no aspect of crop production are the contributions of agricultural engineers more evident than in the mechanization of harvesting and processing cotton. The harvesting process used to start fairly early in the season with topping the plants—cutting off the apical buds of either the main or all stems to prevent rank growth and thus reduce lodging. Plants would generally be cut to about 48 inches. At present, however, topping is not commonly practiced. Plants are now defoliated with a contact herbicide before picking so that leaves do not foul mechanical pickers. Defoliation also improves the efficiency of hand picking. Defoliants are applied when a minimum of 60 percent of the bolls are open. Certain conditions, such as drought or frosts, may cause plants to drop their leaves before harvest and make the use of chemical defoliants unnecessary.

Mechanical harvesters are used on nearly half of the cotton planted in the United States (and on about 90 percent of the California crop). Harvesters may be one- or two-row

Figure 23-3
Seeding cotton with row-crop planters. Note the three hoppers for each row, which permit simultaneous application of fertilizers and pesticides. [Courtesy of Deere and Company.]

Figure 23-4
Heavily fruited cotton plants in Arkansas. After leaves have been shed, the crop is ready to harvest. [USDA photograph.]

Figure 23-5
A modern two-row cotton picker. [Courtesy of Deere and Company.]

types and either act as strippers, which remove all bolls (both mature and immature), or as the more complex pickers, which utilize a series of rotating steel spindles to pull the lint (and seed on which it is formed) from the boll (Figure 23-5). Strippers can be used effectively only with cultivars that mature the majority of bolls uniformly and that have bolls that do not open widely or do not tend to "drop" seed from their locules.

Processing cotton starts at the gin. If the cotton from the field—called seed cotton—is wet, it may be artificially dried before ginning. First, trash and immature bolls are removed in the ginning process. Next, the fibers are separated from the seed with a circular saw that catches and cuts the fibers from seed held on a screen or on slated ribs above the blade. The fibers (lint) are brushed from the blade and blown to a condenser. Finally, the lint is baled. The standard bale is about 40 cubic feet (1.1 m^3) and, including tying material, weighs about 500 pounds (227 kg).

Cotton seed removed by ginning is not wasted. The seed can be processed to yield a valuable oil, and the remaining meal can be used as a feed. Seed for planting is given an acid bath to remove the final stubble of lint—a process called acid delinting.

Average yields in the United States are about 500 pounds of lint per acre (560 kg/ha) and nearly 800 pounds of seed per acre (896 kg/ha). About 1200 to 1500 pounds (540–680 kg) of machine-picked cotton yield one 500-pound bale of lint. Because stripped or hand-picked cotton contains more trash and immature bolls, more of it is required to produce a bale.

In many arid and semiarid areas, as well as in some highly productive cotton areas in the Delta, cotton follows cotton in a continuous cropping rotation. In the Southeast and Midsouth, cotton is rotated with peanuts, tobacco, corn, and soybeans. Rotations in these regions may also include winter legumes, such as one of the clovers, and green-manure crops to increase soil organic matter. In the cotton belt, too, cotton is rotated with peanuts and tobacco. In the areas where cotton is produced, it is usually the major crop, and rotations are developed around it.

Diseases and Pests

DISEASES

Cotton is susceptible to a number of serious diseases, and crop losses vary from year to year and from location to location. Fungal diseases pose a major threat, and every year a group of fungi that cause sore shin, damping off, or rotting of the stems and roots of young plants cause serious losses. Control of these diseases can be attained, in part, by means of seed treatment with approved fungicides and by delaying planting until the soil is warm enough to insure rapid germination and vigorous seedling growth.

Cotton is also attacked by fungi that cause root-rot diseases. Partial control of these diseases can be attained through management practices that foster vigorous plants, such as the selection of the proper planting date, appropriate irrigation, and balanced fertilizer applications. Because the pathogens persist in the soil and in and on crop residues, control is also achieved by rotating to a nonsusceptible crop (such as one of the cereals) for two or more years, and by deep plowing to bury crop residues.

Fusarium wilt is another serious fungus-caused disease. The organism blocks the xylem, causing the plant to wilt and display rather typical drought symptoms even when adequate moisture is available. Control is best accomplished through the use of disease-resistant cultivars, although proper levels of available potassium in the soil also help reduce losses. Verticillium wilt, another fungal disease of cotton, is best controlled through the use of resistant cultivars. Many cultivars are susceptible to other fungus-caused diseases, such as boll and fiber rots and at least one rust disease.

Bacteria cause one major disease of cotton —angular leaf spot, or bacterial blight. The organisms overwinter in crop residue and are also carried on the seed. Control is achieved through seed treatment with appropriate fungicides and by rotation to nonsusceptible crops for one or more seasons. To date, no serious viral diseases of cotton have been identified.

WEEDS

Weeds are serious and costly pests in cotton; chemical weed control alone (disregarding yield losses) may cost the grower from five to twenty dollars per acre ($12–$50/ha). Herbicides are applied to more than 90 percent of the cotton acreage in the United States, and their use is increasing in some areas. The application of appropriate combinations of both preemergence and postemergence sprays provides the best weed control; preemergence herbicides alone are reasonably effective, whereas the use of only postemergence compounds is less effective. The use of herbicides has not subsided in cotton-producing states; in some, it has increased. In spite of the heavy use of herbicides, the persistence of chemicals treated acreage; therefore, persistence problems are apparently not serious. is reported for less than 10 percent of the

Cotton is extremely sensitive to herbicides; thus, chemical weed control during the growing season must be carefully managed. Weed control is an integral part of seedbed preparation, and the use of preplanting and pre-emergence herbicides (substituted ureas, toluidines, and carbamates) is effective. After emergence, herbicides can be applied to cotton after the plants have grown more than 3 inches (8 cm) in height but before they flower. Rates and frequencies of application must be carefully regulated. Postemergence herbicides include certain aromatic oils, arsonates, and substituted ureas. Flaming, which is also used for thinning the crop, is an effective method of weed control, but the burners must be adjusted, positioned, and guarded to avoid damaging the young plants. In addition, cultivation for weed control is practiced during the early part of the growing season.

Major weed pests of cotton include pigweeds, crabgrasses, nutsedges, johnsongrass, cockleburs, morningglories, and plants in the *Sida* complex. Plants in the *Sida* complex are the only weeds that have become relatively more serious since 1965, whereas barnyardgrass has diminished in importance in the same period.

INSECTS

Crop damage and losses from insects depend on the location, the year, and many other factors. Beyond doubt, the best-known and most destructive insect pest of cotton is the boll weevil. However, because the boll weevil is a serious menace to cotton only in areas with more than 25 inches (63 cm) of annual precipitation, it is of less concern to growers in the semiarid West. The adult female punctures squares (or bolls) and deposits eggs inside, damaging or even destroying the bolls in the process. When the eggs hatch, the larvae eat the lint inside the boll. This pest may be effectively controlled through use of chemicals on the adults to prevent egg laying. Although organic phosphate and chlorinated hydrocarbon insecticides have been effective in controlling the boll weevil, their use has been restricted. New insecticides are now being tested and seem to be promising (Figure 23-6). Because the boll weevil lives on cotton residue, deep plowing to destroy the residue is also helpful in controlling this dangerous pest.

Like the boll weevil, both the bollworm (also called corn earworm) and the pink bollworm destroy the contents of the boll. Adults of both these bollworms can be controlled with appropriate, approved insecticides; pink bollworms, which overwinter in cottonseed, can be killed through heat treatment of the seed. Spread of this pest is being checked through quarantine of seed from infected areas. Leafworms and webworms feed on foliage and retard boll development; aphids, lygus bugs, and red spiders (which are not true insects) reduce yields and crop quality both by feeding on the foliage and by damaging the plant when laying eggs.

Regardless of the insecticides used, effective control depends on frequent, sometimes daily, field inspections to determine when insect populations have reached critical size requiring control and when the pests have reached a stage of development at which chemical control will be effective. Control of certain insects requires inspection of adjacent crops; for example, lygus bugs seem to prefer to feed on alfalfa, but, after the alfalfa is harvested, they frequently migrate to adjacent cotton fields.

Several species of nematodes cause severe damage to cotton: the root-knot nematode, the reniform nematode, the meadow nematode, and the sting nematode. The only effective control of these pests, other than long-term

Figure 23-6
Spraying for boll weevils on a cotton field near Raymondville, Texas. [USDA photograph.]

rotations away from cotton to nonsusceptible crops such as cereals, is field fumigation, which is extremely expensive. Fumigants, such as DBCP (dibromochloropropane), can be injected into the soil at planting.

Crop Utilization

Although cotton is produced mainly for fiber, two by-products—cotton seed oil and cotton seed meal, which is the seed material remaining after the oil is extracted—should be considered in crop utilization. It is also important to note that, in some years, the United States exports more than 30 percent of the fiber it produces, in addition to processed cotton in the form of fabrics and finished garments.

Cotton as it comes from the gin is graded according to fiber length, strength, and fineness. The color of the fiber and the presence of trash also affect grade. Cotton with long, strong, fine fibers is of premium grade in the spinning industry. Fiber length is determined rapidly by a fibrograph, a device that has replaced hand measurements of samples of cotton. Strength is determined by a stilometer, and fineness is measured by the flow of a gas through a "plug" of cotton with an instrument called a micronaire. Cotton produced under unfavorable conditions tends to be weak and its fibers are thin-walled. Yarn that is spun from such cotton is full of knots and snags (it is "nippy"). These defects lower the quality of the yarn; therefore, such cotton is deemed to be of inferior quality.

The quality of the lint depends on the type of cotton, the cultivar, and the environmental conditions under which the crop was grown. Cultivars of the Sea Island and American-Egyptian types tend to yield finer yarn than do Upland types. Shorter fibers and linters (the fuzz of short fibers remaining on the seed after it has been ginned) are used in producing

Twenty-Four   Tobacco

Origin and History of Culture

Tobacco originated in the Western Hemisphere, and the types of tobacco currently being cultivated evolved in Mexico and Central America. In pre-Columbian times, Indians throughout the Americas grew tobacco, and the custom of smoking tobacco as a part of their religious rites has been traced as far back as 100 A.D. In Cuba in 1492, Columbus and his crew were the first Europeans to see the curious sight of tobacco being smoked. In addition to smoking tobacco, the natives ground the leaves into a fine powder, or snuff, and inhaled it through a Y-shaped, cane tube called a "tobago." The term was modified by the Spaniards and eventually entered the English language as "tobacco."

The Spaniards made the first commercial plantings of tobacco in the West Indies and shortly thereafter, in 1535, tobacco was grown in Spain. By midcentury, it had become popular in Europe, with Spain being the leading tobacco producer until 1575 when Portugal began to produce large quantities. The production of tobacco then spread to colonies in the Dutch East Indies and to colonial Virginia, where it became one of the early major crops.

In Europe, tobacco was used as snuff for medicinal purposes at first, but the habit of smoking it for pleasure spread rapidly during the sixteenth and seventeenth centuries in spite of political and religious bans on its use. Even today, in some cultures, the use of tobacco is prohibited by custom.

In recent years, federal regulations have restricted cigarette advertising and have required that a health warning be clearly printed in advertisements and on cigarette packages. Nevertheless, the production and use of tobacco products have not declined appreciably. Tobacco remains a leading cash crop in North America, although it cannot be considered a major crop on the basis of acreage planted (Figure 24-1).

Botanical Characteristics

Tobacco, *Nicotiana tabacum,* is a member of the nightshade family (Solanaceae). The genus designation honors Jean Nicot, the French ambassador to Portugal who suggested the medicinal merits of snuff to Catherine de' Medici, then queen of France (c. 1561).

Tobacco is a warm-season, herbaceous annual grown primarily for its leaves. A single plant may bear as many as 25 leaves (Figure 24-2). Leaves are simple and may be as long as 2 feet (60 cm) and 1 foot (30 cm) wide, and their surfaces are covered with sticky hairs. Tobacco leaves have relatively large cells; mitosis ceases when the leaf reaches about 15 percent of its mature size, and subsequent growth is a result of cell enlargement, cell-wall development, and the accumulation of materials in cells. The leaves are borne alternately on an erect stem, which is usually unbranched. If branches arise, the "suckers" are removed in commercial crops. Plants vary in height from 4 to 6 feet (120–180 cm).

The stem terminates in a raceme type of inflorescence, which may have more than 150 individual flowers. Flowers of the tobacco plant are complete and perfect. The calyx is composed of five sepals. The corolla has five petals that are fused into a floral tube, or funnel. The petals are commonly pink but may vary from white to red, depending on the cultivar.

Each flower has five stamens, one of which is markedly extended. The pistil is compound and matures into a capsule type of dehiscent fruit. Tobacco seed is exceedingly small; a single fruit may contain as many as 8,000 seeds, and there are about 5 million seeds per pound (11,000 seeds/g).

Tobacco plants generally form a shallow, branched root system. The majority of the root system is frequently confined to the upper two to three feet (60–90 cm) of the soil.

There are several other species of New World origin. None are commercially important, except perhaps in plant-breeding programs and as ornamental plants.

Environmental Requirements

Although tobacco can be grown in a wide range of environments—from the southern United States to parts of southern Canada—it is best adapted to the humid southeastern United States. The major tobacco-producing states are North Carolina, Virginia, and Kentucky. Significant amounts of tobacco are also produced in Florida, Georgia, South Carolina, Maryland, and Ohio. Some tobacco is grown in Pennsylvania and in limited areas of Wisconsin; in the past, tobacco has also been produced in the western United States. Limited quantities of tobacco are grown in the Quebec-Ontario region of Canda; 90 percent of the flue-cured tobacco grown in Canada is produced in Norfolk County, Ontario. Pipe tobacco and tobaccos for cigar filler and binder are produced in Quebec.

Tobacco seeds require a high temperature for germination; the minimum is about 70°F (21°C) and temperatures between 80° and 90°F (27–32°C) are desirable for rapid, uniform germination and for control of some root-rot diseases. After the seedling stage, the crop requires a growing season of about four months to mature leaves. The tobacco plant grows and matures most rapidly when average temperatures are about 80°F (25°C), but leaf quality is poor. Average temperatures in the mid-70s (about 24°C) are optimum for high yields and quality. Flue-cured tobaccos thrive when daytime temperatures are in the range of 70° to 90°F (21°–32°C).

Tobacco requires from 40 to 45 inches (100–115 cm) of annual precipitation for successful production. Moisture stress lowers leaf

Figure 24-2
A tobacco plant ready for harvesting. [USDA photograph.]

quality and leads to undesirably high levels of nicotine and to low levels of potassium and soluble carbohydrates, which are essential for superior burning and ash quality. Relatively high humidity is desirable in the fall to foster optimum curing of the leaves after harvesting.

Soil conditions play a major role in determining leaf quality. Tobacco is adapted to moderately acidic soils, with a pH ranging from 5.5 to 6.5. Alkaline conditions may favor some diseases. Excessive acidity is also undesirable; when soil pH is below 5.5, plants may absorb excessive amounts of aluminum and manganese and insufficient amounts of calcium, magnesium, and phosphorus. An imbalance of these elements tends to reduce leaf quality.

Tobacco plants grown on sandy loams that have relatively low moisture-holding capacities and that are low in soluble (available) minerals produce leaves that are light in color, are finely textured, and have a weak aroma. Plants grown in heavy, clay soils yield smaller, darker leaves with a stronger aroma. Highly fertilized clay-loam soils are not suitable because they favor lush growth that leads to a high nicotine content and a low potassium content.

Because the tobacco plant has a shallow root system, it can be grown on shallow soils. Adequate moisture must be available, however, and water-table levels must be more than 3 feet below the soil surface. When the crop is irrigated, water quality may affect leaf quality; for example, excessively high concentrations of iron and manganese in irrigation water have caused leaf discoloration and spotting and thus reduced the quality (and value) of cigar-wrapper tobacco.

Tobacco is day-neutral in photoperiodic response. Frequently, plants are clipped to remove flowers—a process that promotes leaf size and quality. Tobacco that is to be used for cigar wrappers is frequently grown under artificial shade; reducing light intensity favors the development of uniformly thin, uniformly colored leaves. Coarse burlap is frequently used for shading material.

Tobacco is susceptible to damage from air pollution. High concentrations of ozone are known to cause a condition called weather fleck—discolored margins and spots that lower leaf quality. Weather fleck is an increasing problem in or near metropolitan areas, but plant breeders have developed cultivars that are resistant to ozone damage. In addition, tobacco is extremely sensitive to chlorine in the soil or in fertilizers.

Production Practices

Tobacco is a specialty crop that requires an unusual amount of hand labor. Tobacco fields are planted with established seedlings, not seeded directly. Successful tobacco production requires that an adequate supply of seedlings be available for transplanting into the field at the right time.

SEEDBED PREPARATION AND SEEDING

First, the seedlings must be grown. The first step in production is to prepare the seedbed in a cold frame. Space requirements for seeding in the cold frame depend on the type of tobacco produced and range from about 35 square yards per acre (about 80m^2/ha) for cigar-wrapper tobacco to more than 150 square yards per acre (340 m^2/ha) for burley types. Cold frames are usually 1 to 3 meters wide and as long as is necessary for the type of tobacco being grown and for the size of the projected crop. Before seeding, the soil is cultivated, raked, and smoothed to form a firm, fine seedbed. Seedbeds are sterilized for insect, weed, and disease control before seeding; sterilization methods include steaming and the use of a wide array of chemicals, such as methyl bromide and calcium cyanamide. Sterilization may precede seeding by as much as three months.

Fertilizer management starts in the cold-frame seedbed. Fertilizers are applied at relatively high rates: from 0.5 to 1.5 pounds per square yard (250–750 g/m²). Fertilizers are as much as 8 percent nitrogen, 12 percent phosphorus (in the form of P_2O_5), and 12 percent potassium (in the form of K_2O). Types of fertilizers and exact amounts depend on soil tests and the type of tobacco to be grown; higher rates are used for flue-cured tobacco, and lower rates for fire-cured types.

To give a uniform distribution of seed, the cold frame is seeded as many as three times. Seeding rates are set to give from two to four seedlings per square inch, or about one seedling per square centimeter. Between 1 and 2 ounces (28–56 g) of seed is required to seed an area of 100 square yards (83.6 m²), which will supply enough seedlings for as much as 6 acres (2.4 ha). As stated earlier, tobacco seed is extremely small; there are more than 300,000 seeds per ounce (11,000 seeds/g). For flue-cured tobacco, the seeding rate may be as low as 0.125 to 0.166 ounce (3–5 g) per 100 square yards. This rate gives about 40 seedlings per square foot (or 430 seedlings/m²). The seed is mixed with a carrier, or spreader (sand or other inert material), to promote uniform distribution. After seeding, the seedbed is rolled or given a light irrigation to press the seeds into the soil.

Immediately after planting, the beds are covered with glass, cheesecloth, nylon (or other synthetic material). In Virginia, a combination of plastic film over cheesecloth, suspended flat only 4 inches above the soil surface, has given excellent results in fostering vigorous seedlings. This method of covering keeps soil temperatures more than 10°F (6°C) higher than the minimum temperatures that can be maintained under cheesecloth alone or under combinations of coverings that are suspended higher (24 inches, or 61 cm) above the soil. In North Carolina, the best results have been obtained by covering the bed with a thin layer of wheat-straw mulch immediately after seeding, and placing the cover directly on the mulch. The higher temperatures resulting from covering the seedbeds promote rapid seedling growth and development; thus, seedlings are ready to transplant earlier. Higher temperatures also reduce some root-rot problems.

TRANSPLANTING

From six to twelve weeks after sowing, seedlings are as much as 6 inches (15 cm) tall with four to six leaves and are ready for transplanting to the field (Figure 24-3). Transplanting starts in April in low areas of the South and continues into June in northern areas and at higher altitudes.

Fields are plowed in the fall or spring to turn under cover crops or weedy growth. A seedbed with good tilth is prepared before transplanting. The seedlings are transplanted by hand or with machines. Unless the soil is already damp, seedlings are watered individually after transplanting. In high rainfall areas, plantings may be made on beds.

Field spacings depend on the location and type of tobacco grown. Distance between rows varies from less than 3 feet to more than 4 feet (90–120 cm). The distance between plants in a row varies from 1 to 2 feet (30–60 cm). These row and plant spacings provide from 4 to 12 square feet (0.4–1.1m²) per plant. Plant population density ranges from 18,000 plants per acre (45,000 plants/ha) to as low as 3,600 plants per acre (9,000 plants/ha). Higher densities tend to promote growth of smaller, thinner leaves, which is desirable (Table 24-1). Specific planting rates are determined by the type of tobacco, moisture availability, and fertilizing practices. High rates require more moisture and extra fertilizer. Cigar-wrapper tobacco plants grown under artificial shade

Figure 24-3
Transplanting tobacco seedlings. Contour planting protects the soil from erosion. [USDA, SCS photograph.]

are usually spaced closely to minimize the costs of shading material.

FERTILIZERS
When the seedlings are transplanted, fertilizers should be banded from 3 to 4 inches (7.5–10 cm) on both sides of the seedling and about 2 inches (5 cm) below the soil surface for best results. This practice prevents fertilizer injury to the seedlings. Fertilizers are also applied as side-dressing (Figure 24-4).

Because leaf quality is so markedly affected by the availability and concentration of numerous elements, fertilizer management is of critical importance in tobacco production.

A yield of 900 pounds of leaves per acre (1,000 kg/ha) removes approximately 80 pounds of nitrogen (90 kg/ha), 20 pounds of phosphorus pentoxide (22 kg/ha), and 110 pounds of potassium oxide (120 kg/ha), 70 pounds of calcium (78 kg/ha), 10 pounds of magnesium (11 kg/ha), and 4 pounds of sulfur (4.5 kg/ha).

The potassium-calcium ratio in the leaves affects rate of burning and ash quality—both important quality factors—and should be 1:1. Phosphorus deficiency leads to dark green, lower-quality leaves. Excessive nitrogen increases the nicotine content of the leaves. Nitrate forms of nitrogen are more desirable

Table 24-1
Effects of plant populations on leaf area index and on seasonal yield for Type 41 Pennsylvania broadleaf tobacco

Date of sampling	Leaf area index* by plant population		
	10,500	14,600	20,900
1965			
July 6	0.02	0.02	0.03
July 19	0.03	0.30	0.38
Aug. 2	2.10	1.78	2.04
Aug. 16	4.60	3.60	3.52
Aug. 28	5.23	4.15	3.82
1966			
July 20	0.51	0.47	0.44
Aug. 3	1.20	1.08	1.02
Aug. 19 BT†	3.34	2.89	2.44
Aug. 19 AT†	2.44	2.32	2.19
Sept. 8	2.53	2.26	2.14
Seasonal yield (kg/ha)	2773	3478	3813

Source: J. O. Yocum and G. W. McKee. Yield and leaf area of Type 41 Pennsylvania broadleaf tobacco as affected by variety and plant population. *Agron. J.* 62(1970):377–380.

Note: As plant population increases, leaf area per plant decreases, but total yield increases. No change in quality was observed with change in plant populations in this experiment.

*Computed by multiplying mean leaf area per plant (m²) by plant population per plot and dividing by surface area per plot (m²).

†BT stands for before topping and AT for after topping.

Table 24-2
Effects of form of nitrogen applied on yield and quality of tobacco.

Nitrogen applied (%)		Yield (kg/ha)	No. 1 grade (%)
NH₄ form	NO₃ form		
100	0	1472	37
75	25	1518	45
50	50	1562	54
25	75	1563	56
0	100	1570	53

Source: F. M. Rhoades. A comparison of ammonium and nitrate nitrogen for cigar wrapper tobacco. *Agron. J.* 64(1972): 209–210.

for high yield and leaf quality than are ammonia forms (Table 24-2). Because chlorine lowers leaf quality, fertilizers with excess chlorine (e.g., KCl) should be avoided.

Lower fertilizer rates generally increase leaf quality but may decrease crop yield. Thus, management decisions regarding fertilizer application rates must consider the impact that higher yields may have on quality. In Virginia, high yields of high quality, and consequent high crop value, have been attained with applications of as much as 2,000 pounds per acre (2,240 kg/ha) of a 4-8-12 commercial fertilizer, compared with applications of 750 pounds per acre (840 kg/ha) of the same fertilizer. Elsewhere, highest yields were produced with about 240 pounds of nitrogen per acre (270 kg/ha), but highest quality was reached with 160 pounds of nitrogen per acre (180 kg/ha). Similar results have been obtained with phosphorus (in the form of P_2O_5) and potassium (as K_2O). Fertilizer applications for flue-cured tobacco average about 50 to 80 pounds per acre for nitrogen (56–90 kg/ha), 60 to 90 pounds per acre (67–100 kg/ha) for phosphorus, and 100 to 140 pounds per acre (110–157 kg/ha) for potassium.

Leaf quality is enhanced by topping and suckering the plants. Topping is the removal of the inflorescence to allow the upper (younger) leaves to mature and develop more body. When plants are topped, apical dominance—an auxin-regulated phenomenon—is broken, and lateral buds, called suckers, develop. Sucker growth can be controlled by hand clipping, by chemicals, such as maleic hydrazide, and by some oils that are very effective and require very little labor.

HARVESTING

Tobacco is ready to harvest from 90 to 120 days after transplanting, when leaves turn light green to slightly yellow. Some classes of tobacco, such as flue-cured and cigar-wrapper, are handpicked. Hand harvesting,

Figure 24-4
Thirty-five day-old tobacco plants being cultivated for weed control and side-dressed with potassium and nitrogen fertilizer. [USDA photograph.]

or priming, requires several pickings, and the harvest period for prime leaves may last for three to six weeks. Picking starts at the lowermost leaves, which mature first, and continues until the uppermost leaves, which are larger and heavier, have been harvested. The number of leaves per plant depends on the plant type and on the number of leaves discarded in topping. Fire-cured types may have as few as eight leaves to be picked (bottom leaves are frequently discarded in addition to the leaves discarded in topping), and shade-grown tobacco (burley types and others) may have as many as twenty leaves.

In other classes of tobacco, such as burley, the stalks are cut, and the entire plants are harvested. This method is less time-consuming and requires far less labor, but reduces crop quality and yields.

Because so much hand labor is required, an individual grower usually plants less than 10 acres (4 ha) per year. Even this amount is profitable because of the crop's high cash value. Mechanical harvesters are now available that enable individual growers to manage up to 100 acres (40 ha) successfully. In areas where tobacco is adapted, it may be grown continuously. This is not, however, recommended practice, and a winter cover crop is often included as a soil conservation measure.

Diseases and Pests

DISEASES

Tobacco is subject to a variety of diseases caused by fungi, bacteria, and viruses. Blue mold (downy mildew) is a fungal disease caused by *Peronospora tabacina*. The overwintering spores are found in the soil of cold frames and are the source of primary inoculum for seedlings. Leaf symptoms vary with the time of infection: in young plants, the leaves tend to be very erect; in slightly older plants, the leaves develop round, yellow (chlorotic) areas. Following these early symptoms, a mass of whitish or violet fungal growth appears. Early detection is important to minimize spreading of the disease. Blue mold is controlled by regular use of approved fungicides and by sterilizing soil prior to planting.

Black root rot is caused by the fungus *Thielaviopsis basicola*. Symptoms include black lesions and, frequently, chlorotic areas and stunted growth. Ultimately, the disease causes root decay. Control measures include soil sterilization and use of disease-resistant cultivars. Black root rot is not a problem when seedbed soil temperatures are above 80°F (27°C).

Black shank, caused by the fungus *Phytophthora parasitica nicotianae*, produces root and stem rot. It is the most serious disease of flue-cured tobacco in the Southeast. Black shank is controlled by crop rotation and by the application of appropriate chemicals.

Other root and stem rots caused by fungi include southern stem rot, caused by *Sclerotium rolfsii*, and fusarium wilt, caused by *Fusarium oxysporum*. Both of these diseases can be controlled through soil sterilization and crop rotation, and there are some cultivars that are resistant to fusarium wilt.

Tobacco is susceptible to at least three significant diseases caused by bacteria; wildfire, or bacterial leaf spot; blackfire, or angular leaf spot; and bacterial wilt. The first symptoms of wildfire, caused by *Pseudomonas tabaci*, are lemon-yellow leaf spots with dead centers. The disease can spread rapidly throughout a field, and it is best controlled with appropriate sprays, including antibiotics, and by use of disease-resistant cultivars. Similar methods are used to control blackfire, a disease caused by the bacterium *Pseudomonas angulata*. In this disease, leaf lesions have a characteristic angular shape. Blackfire is of significance predominantly in Virginia and Kentucky. Bacterial wilt is caused by *Pseudomonas solanacearum*, which penetrates the host plant through wounds in the roots. The pathogen causes blockage of the vascular (xylem) tissue, and, typically, infected plants are wilted. Disease control consists of rotation to nonsusceptible crops (cotton, corn, small grains) and the use of resistant cultivars.

The most serious of the viral diseases causing tobacco crop losses is tobacco mosaic virus (TMV), also called calico (or walloon) disease. The virus (and the disease) is present wherever tobacco is grown. The virus infects other members of the nightshade family, and it can persist in dead material, including commercial tobacco made from infected plants, for longer than twenty-five years. Disease symptoms vary with the strain of virus, but generally the leaves of infected plants display a typical green, greenish yellow, or yellow mosaic pattern. The magnitude of crop loss depends on when the plants become infected; if they are infected as seedlings, yields may be reduced nearly 40 percent, and the value of the crop may be lowered as much as 60 percent. The virus is transmitted by insects (aphids) and through wounds by contact of healthy leaves with infected plants or even by workers who smoke or have handled infected tobacco or products made from infected tobacco.

Control of tobacco mosaic virus disease is complex and includes sterilizing soil and shade materials, destroying weedy virus hosts, practicing clean culture (destroying tobacco refuse around beds), and washing hands with disinfectants as well as refraining from using any tobacco product while working with plants. Certainly, the adage "an ounce of prevention is worth a pound of cure" is valid for this disease. Control also includes the use of a few virus-resistant cultivars.

Other viral diseases include tobacco etch and ring spot. Viral diseases are sometimes controlled through the field management of insect vectors.

Root-knot nematodes also attack tobacco.

Control is through soil fumigation and crop rotation. Broom-rape is a parasite of tobacco plants.

WEEDS

Weeds can be a problem both in the cold frames (seedbeds) and in field plantings. Weed control in the cold frames starts with seedbed preparation and sterilization of the soil before planting. Herbicides are not generally used for weed control in tobacco, although their use is increasing. In the United States, about 10 percent of the total acreage, and more than 50 percent of the flue-cured tobacco acreage, is treated with an herbicide. Postemergence sprays are used on about 70 percent of the treated acres; combinations of preemergence and postemergence sprays are used on less than 1 percent of the treated acres (i.e., less than 500 acres, or 200 hectares).

Crabgrass is the most frequently reported weedy pest, followed by pigweeds and lambsquarters. Other weeds include bermudagrass, carpetgrass, cocklebur, nutsedge, fall panicum, Florida pursley, and ragweeds. With the exception of a slight increase in nutsedges, levels of infestation for most weeds in tobacco have been constant in recent years.

INSECTS

Tobacco plants are subject to damage from insects in cold frames, in the field, and in storage. The most damaging insects to the growing plants are the hornworms, or greenworms. The tobacco hornworm, *Protoparce sexta*, and the tomato hornworm, *P. quinquemaculata*, eat the entire leaves of plants in the field. The tobacco flea beetle, *Epitrix hirtipennis*, too, attacks plants in the field, feeding on the leaves and leaving holes, or punctures. The tobacco budworm, *Chloridea virescens*, feeds on new, unopened buds, often chewing holes in several unopened leaves at a time. This insect is a pest in the southern tobacco-growing areas.

Figure 24-5
Insecticides are applied to tobacco with specially designed (sometimes homemade) boom sprayers. [USDA, SCS photograph.]

All insect pests in the field can be controlled with appropriate, approved insecticides (Figure 24-5). Insect management also requires the destruction of crop residues, weed control, and, generally, sanitary production practices.

Tobacco in storage may be attacked by the tobacco (cigarette) beetle, *Lasioderma serricorne,* and by larvae of the tobacco moth, *Ephestia elutella.* Crop value is reduced by losses from feeding and by the presence of insect contaminants. These pests can be controlled effectively by fumigating storage areas.

Remember, insecticides are dangerous; any chemical pest-control program must be carefully regulated to avoid harm to humans and livestock and to prevent damage to, or contamination of, the crop.

Crop Utilization

The ultimate use of tobacco depends to a large extent on how it is cured. During curing, the leaves dry; chlorophyll decomposes until the leaves lose all green color; nitrogen compounds are changed and ammonia is released; starches are hydrolized; and sugars are respired. The grain of the cured leaf is the result of the crystalization of mineral salts.

During curing, tobacco leaves lose more than 85 percent of their weight as a result of both dehydration and respiration. Dry-weight loss, mostly from respiration, may be as high as 12 percent. After curing, the leaves are allowed to regain moisture until they have a moisture content of 24 to 32 percent of total weight. Increasing moisture makes the leaves more pliable and reduces damage in handling.

Curing is done in barns that are designed to hold the tobacco leaves in tiers. For flue-curing, the barns are sealed, and heat is circulated through a closed system of ducts, or flues. Flue-curing hastens the early steps of leaf processing (dehydration and chlorophyll destruction) so that the cured leaves retain a bright yellow color. In the first stage of flue-curing, the barn is allowed to heat to 100°F (38°C) in 48 hours and the humidity is held at 65 to 85 percent. This is called the "yellowing stage," in which the leaves are cured to the desired color. Next, the temperature is gradually increased to 130°F (55°C) for a 40-hour period. This is the color-setting stage—the most important in the curing process. Drying is completed in about 40 hours as the temperature is raised to more than 160°F (71°C), and the humidity is reduced to less than 10 percent. The leaves are killed during this period. After four or five days, the heating is stopped, and the leaves are allowed to regain moisture. The entire process must be done with care; improper curing leads to discolored, low-quality leaves.

Air-curing is done in airtight barns with ventilators for regulating temperature and humidity. Sometimes, artificial heat is used (gas burners, stoves, or charcoal fires) to lower the relative humidity; high relative humidity at temperatures of 60°F (16°C) or less lowers leaf quality. Air-curing requires as long as two months, and air-cured tobacco is generally darker than flue-cured tobacco.

Fire-curing, like flue-curing, uses artificial heat to dry the leaves, but, unlike flue-curing, the fires are open. Usually, leaves are allowed to yellow and wilt in the barn from 3 to 5 days before fires are started. After that period, and until the yellowing is completed, fires are controlled to keep the temperature between 90° and 95°F (32°–35°C). After yellowing is completed, the temperature is raised to 130°F (55°C) for three to five days to cure the tobacco. After curing is completed, the leaves are allowed to regain moisture. Materials burned in the open fires flavor the leaves, giving the tobacco a characteristic flavor and odor. For example, burning hardwood during the curing of chewing tobacco is responsible for its distinctive character.

Nearly all the tobacco produced in the United States and Canada is processed for smoking as cigarette, cigar, and pipe tobacco; a small amount, for chewing tobacco. The U.S. Department of Agriculture recognizes twenty-seven types of tobacco, which are grouped into seven classes on the basis of curing methods and use. The various types are grown in fairly well defined regions. Flue-cured tobacco comprises about 60 percent of all the tobacco grown in the United States and consists of five types. It is used chiefly for cigarettes, although some is also used for pipe tobacco and chewing tobacco.

Air-cured types comprise about 35 percent of the domestic crop and are used for cigarette, pipe, and chewing tobaccos, as well as for cigar wrappers and binders. Air-cured tobaccos comprise four classes: light air-cured and dark air-cured (class 3, a and b), air-cured cigar filler, air-cured cigar binder, and air-cured cigar wrapper. Each class consists of several types. Fire-cured tobacco (class 2) comprises 5 percent of the domestic crop. Like fire-cured tobacco, class 7 tobaccos, which include *La Perique* tobaccos—a very pungent type—constitutes a small part of the total domestic crop.

In recent years, serious questions have been raised regarding the effects of smoking on health. In spite of current medical evidence and government restrictions on the advertising and sale of cigarettes, tobacco products are still in demand, and, in spite of relatively small acreages, tobacco continues to be a crop of significant economic importance (Figure 24-6).

Figure 24-6
A floor of tobacco ready for sale. [USDA photograph.]

Selected References

Ignatieff, V., and H. J. Page, eds. 1958. *Efficient use of fertilizers*. Food and Agriculture Organization of the United Nations.

McKee, C. G., and D. O. Street. 1963. Effects of fertilizer rate and method of application and plant spacing on yield, quality, and value of Maryland tobacco. *Maryland, Univ., Agr. Exp. Sta. Bull.* A-126.

LaPrade, J. L., J. C. Petty, and W. H. Wells. 1964. Use of plastic film in production of tobacco seedlings. *Virginia Agr. Exp. Sta. Bull.* 167.

Study Questions

1. What are the environmental factors that most affect tobacco yield and quality?
2. What are the major differences between the production practices for tobacco and those for other major crops?
3. What is topping and what is the botanical basis for its importance?
4. How is tobacco cured? Describe the differences in the techniques and effects of flue-curing, air-curing, and fire-curing.

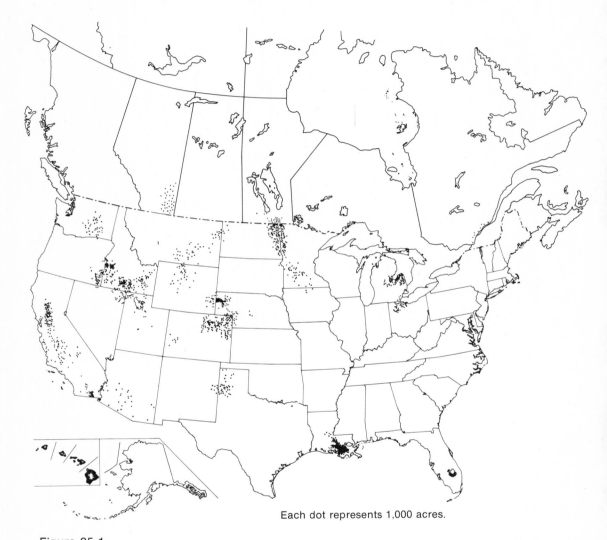

Each dot represents 1,000 acres.

Figure 25-1
Major sugar-beet– and sugarcane–producing regions in North America. Sugarcane production is confined to Hawaii and the southern United States. [Data from U.S. Department of Commerce and *Canada yearbook,* 1972.]

Twenty-Five

Sugar Beets and Sugarcane

Origin and History of Culture

The written history of sugar can be traced to pre-Christian Sanskrit, and linguists have tied the Sanskrit word *sarkara* to the root of the word *candy*. Apparently, man has had a sweet tooth historically! There are two major sources of sugar: sugarcane and sugar beets. In North America, approximately two-thirds of the sugar consumed comes from cane; the other third comes from beets (Figure 25-1). In Europe, sugar beets are the primary source of refined sugar. Other, relatively minor, sources of sugar are maple sugar and syrup and sugar from some sorghums (see Chapter 14). Historically, plant nectars and honey have been important sources, but today they are relatively unimportant.

Sugar beets The sugar beet belongs to the same species as the familiar red, or garden, beet. It originated in what is known as the Mediterranean center of origin. The beet has been cultivated for thousands of years in one form or another, but its potential as a source of sugar remained unnoticed, or at least unreported, until the middle of the eighteenth century, when a German chemist discovered that sugars found in sugar beets were identical with sugars found in sugarcane. Early attempts at processing beets for sugar failed because the sugar content was much too low. In an effort to eliminate France's dependency on foreign sources of sugar, Napoleon directed that the sugar beet be developed, and eventually French agriculturalists were successful in raising a strain of beets that could be processed to yield profitable amounts of sugar.

The first serious attempt at producing sugar beets in the United States came in 1832 in Massachusetts—an effort that met with little, if any, success. However, some forty years later, production of sugar beets was successfully started in California. Today, sugar beets are grown commercially in several areas of North America. In the past five years, Canada has produced from 1.0 to 1.3 million tons of sugar beets per year; in the same period, the United States has produced an average of about 25 million tons per year.

Since the earliest days of successful sugar-beet production, growers have been dependent on processing companies, with respect to selling their crops. The limited capacities of processing plants, more than federal regulations, determined the acreage to be planted, field management, time of harvesting and tons harvested, and purchase price. In many, if not all, sugar-beet producing areas of the United States, a grower will not seed sugar beets until he has a binding purchase contract for his crop with a processing company or cooperative.

Sugarcane Records of sugarcane, or its ancestors, exist in the earliest Indian writings, dating back to at least 1000 B.C., but its center of origin has not been established. Both northeastern India and some of the Polynesian islands have been suggested as centers, and current evidence, based on existing wild ancestors, indicates that northeastern India is the probable center of origin. Sugarcane was introduced into China by early travelers or by nomads sometime between 1800 and 1700 B.C. From China it was brought to the Philippine Islands, to Java, and perhaps to Hawaii.

Early in the fifteenth century, Portuguese and Spanish explorers carried the crop to the Azores and the Cape Verde Islands, and, in 1493, Columbus brought it to Santo Domingo, where it reportedly thrived. Sugarcane as a crop was not introduced into the United States until 1751, when Jesuit priests brought it to Louisiana, but small refineries in New York had been processing imported sugarcane since 1690. The importation of sugarcane was a minor issue in the Revolutionary War: the Sugar Act of 1764, which taxed sugar imported by the American colonies, was among the grievances that the colonists had against the British crown.

Sugarcane became a significant crop in the United States after the development of larger processing facilities. However, its production was essentially stopped during the Civil War and was not resumed until nearly twenty years later in Louisiana.

Botanical Characteristics

Sugar beets From a botanical point of view, the name "sugar beet" is misleading because sugar beets and red (garden) beets are the same species. They belong to the goosefoot family (Chenopodiaceae), which includes such crops as the mangel (or mangel-wurzel, a very close relative of beets) and chard, as well as a major weed pest, pigweed. Both the garden beet and the sugar beet are classified as *Beta vulgaris;* the sugar beet has also been classified as *B. saccharifera*—a designation totally lacking any botanical justification.

The sugar beet is biennial in habit, but it is farmed as an annual for its sugar. In the first year, it produces a dense canopy of leaves and a large root in which the sugar is stored (Figure 25-2). In the second year, the reserve sugar is utilized by the plant to produce a flowering stalk, flowers, and seeds.

The beet "root" is divided into three regions: the top, which is a convex cone from which leaves are borne in a dense spiral; the neck (which is in fact the hypocotyl), a narrow zone but the broadest part of the "root"; and the fleshy root, which terminates in a long taproot. There are vertical grooves along opposite sides of the root from which secondary roots arise.

Figure 25-2
Sugar beet at the end of its first season of growth. [USDA photograph.]

The leaves are simple and fairly smooth in outline, with a large blade and petiole. Flowers are typical of the goosefoot family: they are perfect, but incomplete. The calyx is five-parted, and the corolla is absent. There are five stamens and three styles leading to a compound pistil. Flowers are surrounded by leaflike bracts and are borne in a branched, paniclelike spike. The sugar beet has a typical aggregate fruit that yields a seedball with two or more viable seeds, or germs. The entire fruit, not individual seed, is used in field planting. Multiple germs make it difficult to obtain uniform stands with appropriate spacing between plants in a row. Uneven stands with crowded seedlings must be thinned to achieve optimum root growth. Thinning is both time-consuming and expensive; to minimize thinning, monogerm seed, which produces only one seedling (germ) per seed, is often used. Monogerm seed is produced by mechanically treating the aggregate fruit before planting to remove or damage one of the germs so that the seedball contains only a single, viable germ. Plant breeders have also successfully developed some sugar-beet cultivars that naturally produce monogerm fruits, or "seed."

Sugarcane Sugarcane is a robust tropical perennial grass (Figure 25-3). The high-quality (noble) canes are classified as *Saccharum officinarum*. Many of the commonly grown cultivars are hybrids between *S. officinarum* and closely related species, especially *S. spontaneum* and *S. barberi*. These two species are used as parents in plant-breeding programs (see Chapter 32) to develop resistance to various diseases in cultivars of the noble canes.

Plants vary in height from 15 to 20 feet (4.6–6.1 m) or more. The stems are from 1 to 3 inches (2.5–7.6 cm) in diameter, and they are solid. Internodes vary in length, generally from 2 to 6 inches (5–15 cm). A leaf is formed at each node; the blades are as much as 6 feet (2 m) long, somewhat lance-shaped, and usually quite erect, although they may nod, or droop, at their tips. The occurrence of long leaves from each node and the fact that sugarcane tillers freely give the mature plants an excessively leafy appearance.

Sugarcane has a panicle type of inflorescence. The inflorescence, called an arrow, is from less than 1 foot to more than 2 feet (30–61 cm) in length. Spikelets occur in pairs—one sessile and the other on a short stalk, or pedicel— and they usually have two florets, one (and sometimes both) of which is sterile. Pollen sterility is quite common. Long, silky tufts of hairs are formed at the base of each spikelet.

Sugarcane seed is commonly of low viability, but this is of little consequence in field planting because the crop is propagated vegetatively by pieces of stem. Pieces of stem that have buds (eyes) at each node and that develop adventitious roots are used for field plantings. Following harvest, regrowth occurs, and this is harvested as a ratoon crop in several subsequent years. Seed is used almost exclusively in plant-breeding projects to produce new, superior cultivars—either new genotypes or new interspecific hybrids.

Environmental Requirements

Sugar beets Sugar beets are grown in the United States mainly in the western half of the country. They are grown at sea level in California, in the mountain valleys of Colorado, and in Montana and Idaho; they are also produced in Michigan, in Ohio, in western Minnesota, and, to a lesser extent, in Wisconsin and Illinois. Most of the sugar beets produced in Canada are grown in the three prairie provinces: Alberta, Saskatchewan, and Manitoba.

Figure 25-3
Sugarcane field in Louisiana showing typical, robust growth. [USDA, SCS photograph.]

To understand the optimum environmental conditions for sugar-beet production, it is essential to bear in mind the fact that the sugar beet is a biennial that is farmed as an annual. Therefore, the environmental factors of concern to sugar-beet growers are those that affect root growth and sugar accumulation during the first year. (Seed production has special requirements that will not be discussed here.)

The optimum soil temperature for sugar-beet germination is about 60°F (15°C); thus, beets must be considered a warm-season crop. For growth and sugar accumulation, a twenty-four hour average temperature of about 70°F (21°C) is desirable. High average temperatures—near, or above, 90°F (32°C)—retard sugar accumulation but favor rapid growth; thus, higher average temperatures for short periods are not wholly undesirable. Cool nights are not desirable during most of the growing season.

Temperatures below 25°F (−4°C) kill beet seedlings. Although older beets can tolerate temperatures below freezing, temperatures below 27°F (−3°C) frequently destroy foliage. Regrowth of foliage following cold damage consumes sugar stored in the roots and thereby reduces yields. In some instances, lower temperatures cause plants to bolt—that is, to produce flowers and seed in the first year rather than developing storage roots. The propensity to bolt is inherited, and bolting-prone cultivars should not be grown in areas where low temperatures are common during the growing season. In the fall, before harvest, lower temperatures—that is, in the mid-60s, or about 18°C—favor the conversion of starch into sugar and the accumulation of sugars in the root.

The amount, distribution, and management of moisture are critical factors for efficient sugar-beet production. Irrigation is necessary in regions where annual precipitation is less than 18 inches (46 cm) or where precipitation may not be appropriately distributed. In the United States, sugar beets are grown as a nonirrigated (dryland) crop only in North Dakota, Minnesota, and parts of Iowa; they are grown under irrigation in other areas.

Although light quantity is rarely a limiting factor for sugar accumulation, inadequate light reduces total plant growth under field conditions. Thus, weed control, particularly in the seedling stage, is critical, and planting rates must be carefully managed. High light intensity is essential for efficient utilization of minerals by the plants; light also tends to "sterilize" plants and aids in the control of some diseases. Sugar beets are classified as day-neutral for root development, and as long-day for seed production.

Sugar beets are best adapted to fertile, loamy soils. Sandy soils do not retain adequate moisture between irrigations, and heavy clay soils cause problems in harvesting. In general, sugar beets are best adapted to slightly acidic soils, although they are also successfully grown on slightly alkaline soils; once established, the plants tolerate moderate alkalinity.

Sugarcane Sugarcane is more than a warm-season crop; it is a tropical or subtropical crop. It is best adapted to areas where the minimum mean monthly temperature is 70°F (21°C) or above. However, sugarcane does not tolerate excessively high temperatures. The noble canes (*S. officinarum*) have a narrow range of tolerance and are harmed by temperatures above 100°F (38°C). The best root growth in sugarcane occurs when soil temperatures are between 70° and 80°F (21°–27° C); root growth is retarded when soil temperatures are below 70°F (21°C); and it is essentially stopped at 50°F (10°C) or below. Foliage is harmed when air temperatures fall below 53°F (11°–12°C). Minimum temperature for "germination" of the vegetative seed pieces (pieces of stem used for vegetative propagation) is 70°F or slightly below (about 18°–20°C).

Stem growth is directly related to temperature. In tropical or semitropical areas, plants may grow more than 12 feet (3.7 m) per year and reach maturity (maximum sugar content in the stems) in twenty-four months. In Hawaii, where the average February temperature is about 70°F (21°C), stems grow about 6 inches (15 cm) in that month. In July, when the average temperature is nearly 78°F (26°C), stems grow slightly more than 2 feet (60 cm). The difference in growth during these two months is due directly to the differences in average monthly temperatures. In Louisiana, where the growing season is shorter than in tropical or subtropical areas, the sugarcane crop matures in seven to eight months, and the yields are proportionately lower.

Sugarcane is a short-day plant, and in tropical and subtropical areas, plants flower in response to short days. Normally plants do not flower in Louisiana because days are too long, but they do flower in Florida, which is further south. Plants may be topped to prevent or delay flowering. In addition to reacting to light duration, sugarcane is sensitive to light quantity. Like cotton and soybeans, sugarcane requires bright, cloudless days (i.e., high light intensities) for maximum yields. Studies in Hawaii have shown that both stalk yield and the percentage of sugar are reduced when light intensity is decreased by cloudy conditions.

For high yields, sugarcane requires from 45 to 50 inches (115–130 cm) of precipitation per year. Moisture must be available and uniformly distributed throughout the year, but in warmer periods, when growth rates are high, water needs are proportionately higher. Irrigation is good insurance against insufficient precipitation, and it may be essential.

In Louisiana and Florida, sugarcane is grown on heavy clay loams, and on muck or peaty soils; some soils are more than 75 percent organic matter. In Hawaii, the crop is grown on inorganic soils derived from volcanic material. Some of these soils are infertile and require heavy applications of primary plant foods for economic yields. Sugarcane does not tolerate salinity; the ratoon regrowth crop is even less tolerant of salts than the "seeded" crop. Saline conditions in the soil that give soil conductivity readings of more than 5 mmhos generally reduce the percentage of polysaccharides in the stems and thus cause reduced sugar yields.

Considering its staggering growth, the sugarcane plant is unusually sensitive to environmental conditions, and its basic range of adaptation has not been greatly modified through plant breeding. In contrast, sugar beets are well adapted to many temperate, and even cool, regions in which sugarcane would be totally unadapted, such as the prairie provinces of Canada.

Production Practices

Sugar beets Seedbed preparation for both fall and spring planting starts in the late summer or in the fall. Fields are plowed to a depth of 8 to 12 inches (20–30 cm). Deeper plowing is of questionable value for beet production, although it does promote the incorporation and decomposition of crop residues. Before planting, fields are leveled for surface irrigation (an unnecessary step if sprinklers are used), and a mellow, but firm, seedbed is prepared by discing, harrowing, and, if needed, rolling. The seedbed should be prepared in such a way that crusting is minimized. If beets are planted flat (as they generally are in the United States), furrows for irrigation are opened between rows; if they are planted on elevated beds, shaping and finishing the beds is the final step in seedbed preparation.

Row spacing usually ranges from 18 to 22 inches (45–56 cm). Depth of seeding varies from 1.0 to 1.5 inches (2.5–3.8 cm). Seeding at depths of 2.0 inches (5 cm) or more reduces emergence and is not recommended.

As stated earlier, the sugar beet has an aggregate fruit; thus, the "seed," or seedball, may have more than a single embryo and may produce more than one seedling. Because multiple seedlings are undesirable, most plantings are made with monogerm seed. Planting rates of monogerm seed range from 5 to 8 pounds per acre (5.6–9.0 kg/ha)—a planting rate that gives a field seedling stand of one or two seedlings per foot (30 cm). Seeding is done with special four- to six-row planters that operate similarly to corn and bean planters.

Dates of planting sugar beets depend on soil temperatures and, consequently, differ widely with location: in central California, seeding is done during the winter months and in early spring (February and March); in southern California, beets are planted in the fall; and, in most other regions, sugar beets are planted from late April through May.

Figure 25-4
Field of sugar beets near Davis, California, in early summer.

As the plants develop, the stand must be thinned to insure optimum growth and maximum yield. Plants should be spaced 10 to 12 inches (25–30 cm) apart in the row (Figure 25-4). Closer spacing leads to competition for water and minerals in the soil and to shading (competition for light) in the tops. Competition lowers yields by reducing root size. Usually, the normal seeding rates produce stands with more than one seedling per foot of row; this is either done intentionally to insure an adequate stand or is the unintentional result of using multigerm seed, which produces more than a single seedling. Thinning is generally done after the plants have developed from eight to ten leaves. In the past, much of the thinning was done by hand—an expensive and time-consuming procedure. In the last twenty years, various precision thinners have been developed. These machines operate either down the row, by cultivating and skipping measured distances, or across the row, by using flexible toothed or knife weeders that are electronically or mechanically activated. Down-the-row thinners are particularly effective in thinning the uniform stands resulting from the use of monogerm seeds. Flame thinning (also done in cotton production) is effective if done properly and on time. The thinning operation is also an important method of controlling early weed growth.

Because the amount and distribution of

Table 25-1
Total soil moisture (cm) in 183 cm of soil in response to different sugar-beet population densities and fertilizer applications.

Date	Density (plants/ha)		Nitrogen fertilizer (kg/ha)			
	26,400	70,000	0	56	112	224
6/5	87.7	87.0	88.7	86.7	86.1	87.9
7/2	86.4	85.9	87.6	86.8	84.3	85.9
7/13	82.3	79.7	82.6	80.7	79.6	81.1
7/27	74.6	72.0	74.9	72.2	73.6	72.5
8/10	66.9	63.3	66.6	64.7	64.4	64.7
8/24	61.2	56.3	60.3	58.9	57.8	57.9
9/8	56.6	51.6	55.9	53.6	54.0	52.9
9/19	56.2	51.2	55.2	53.3	53.3	52.9

Source: J. T. Moraghan. Water use by sugar beets in a semiarid environment as influenced by population and nitrogen fertilizer. *Agron. J.* 64(1972):759–762.
Note: In this study, fertilizers had less effect on the amount of water used than did plant density.

moisture is critical for efficient beet production, scheduling of irrigations is extremely important. The frequency of irrigations and the amount of water applied depend on weather conditions and the stage of plant development. Young plants have a very sparse root system and require frequent, light irrigations. After the crop is established, irrigations of 2 to 6 inches (5–15 cm) may be required at ten- to fourteen-day intervals throughout the growing season. The last irrigation is scheduled to allow the plants to mature and to insure, at the same time, that the ground is dry enough to harvest but not enough to impede digging the beets. Total irrigation requirements range from 12 to 36 inches (30–90 cm), depending on natural precipitation in the area.

High-yielding sugar beets make a heavy demand on minerals in the soil. Even a moderate yield of, say, 20 tons per acre (45 MT/ha) removes about 80 pounds per acre (89.6 kg/ha) of nitrogen, 30 pounds (33.6 kg/ha) of phosphorus (P_2O_5), and, surprisingly, more than 100 pounds (112.0 kg/ha) of potassium (K_2O).

The use of commercial fertilizers has increased dramatically in the past thirty years, and the trend continues. (Manuring is still practiced primarily in the Great Plains, but supplemental commercial fertilizers, notably phosphorus, are also used.)

Both the timing of fertilizer application and the placement of fertilizer are critical. Sugar beets require a well-balanced supply of minerals throughout their life cycle for maximum growth; available minerals, especially nitrogen, affect plant growth and water use (Table 25-1). However, available soil nitrogen should be depleted from ten days to two weeks before harvest; nitrogen deficiency at this time forces the plant to reduce vegetative (top) growth and, at the same time, forces it to convert the starch in its root into sugar. To insure depletion of nitrogen, applications of fertilizer are carefully controlled, and samples of plant tissue (frequently petiole) are chemically analyzed to determine the nitrogen levels in the plant.

Fertilizer application rates may be as high as 120 pounds of nitrogen per acre (134.4

kg/ha), more than 150 pounds (168.0 kg/ha) of phosphorus (P_2O_5), and 100 pounds (112.0 kg/ha) or more of potassium (K_2O). In some areas—especially those with muck soils—fertilizers containing both copper and boron are added. Fertilizers can be applied before or during seeding, or as a topdressing during the growing season. Fertilizers applied during seeding operations should be banded below and to the side of the seed because sugar-beet seedlings are very sensitive to high concentrations of fertilizers. As much as half or more of the total recommended nitrogen may be applied during seeding; the remainder is applied in irrigation water. In this way, exact amounts can be applied to insure maximum yield increases and the timely depletion of nitrogen.

The harvesting of sugar-beet roots is highly mechanized and consists of three distinct steps: topping the plant, lifting the root, and loading. In addition, transportation from the field to a processing plant or a railroad loading point is an integral part of harvesting.

Topping, as the name implies, is the process of removing the leaves and crown (the upper part of the root) from the main part of the root. If tops are to be saved (e.g., used for feed), topping is done with knives that cut the beets just below the soil surface; if tops are discarded, various types of beaters are used. Uniform topping is essential to avoid excessive losses of beets, and processors may specify how the beets are to be topped.

Lifting and loading may be done as a single operation or in two separate operations (Figure 25-5). Mechanical lifters, or diggers, loosen the soil with disclike openers and then pick up the beets with spiked wheels. If loading is part of the operation, the beets are knocked from the spiked wheel onto a conveyor belt and dumped directly into a truck; otherwise they are piled in the field.

In most areas, except parts of California and Arizona, the sugar beet harvest covers a relatively short period, from September through October; harvesting wet fields is difficult, and lifting beets from frozen soil is essentially impossible. Because of the short harvest period, processing plants do not operate year round. Limited processing capacity dictates how many tons of beets can be processed. To insure prompt processing, most beets are grown on contract with a sugar company, and the time of harvest and volume of beets are specified in the contract (Figure 25-6). The price paid by processors depends on the sugar content of the beets, which ranges from 12 to 18 percent by weight. Yields, in terms of tons per acre, have been increasing in the past ten years. In California, yields of nearly 30 tons per acre (67 MT/ha) are not uncommon; elsewhere, about 20 tons per acre (45 MT/ha) is a common yield.

Climatic conditions in California and parts of Arizona allow beets to be harvested during the fall and winter. Beets that have not been removed by the time the winter rains come are allowed to overwinter in the ground and are harvested as early as possible in the spring. The mild temperatures in these areas minimize bolting, and there is little, if any, spoilage from freezing. Overwintering the beets lowers the tonnage because of the respiration of the plants during the winter and increases the percentage of sugar because in the cool weather starches are converted into sugar. Thus, the yield in terms of tons of sugar per acre is not significantly reduced. Processing throughout the year permits sugar plants to operate longer and more efficiently.

Sugar beets are a prime cash crop and are usually given a priority position in crop rotations. Rotation away from sugar beets is done to control diseases and root-knot nematodes. Rarely can sugar beets be grown continuously; usually, they can be grown only once in four or five years. Rotations include cereals, beans, vegetables (such as tomatoes and lettuce),

Figure 25-5
Modern sugar-beet harvester that digs topped beets, lifts them from the soil, and conveys them to a truck. [Courtesy of Deere and Company.]

Figure 25-6
A sugar-beet dump that is owned and operated by the processing company. Crops are brought to such dumps from the field and then hauled to the processing plant. The capacity of the processing plant dictates the volume of beets accepted in a given period.

corn, grain sorghum, and forage legumes (both alfalfa and clover). For pest control and soil-fertility management, sugar beets should not immediately follow legumes.

Sugarcane Cultural practices for sugarcane production are dictated by the perennial nature of the crop, its robust growth, and the fact that it is reproduced vegetatively. Because it is a perennial, seedbeds are not prepared every year. In Louisiana, sugarcane fields may be harvested for three years, then rotated to a green-manure crop, and then returned to cane; a grower may schedule plantings so that approximately a third of his land is replanted annually with cane. In Florida, where the crop is grown on soils that are extremely high in organic matter, a green-manure crop is not included in the rotation; otherwise, cultural practices are similar to those employed in Louisiana. In Hawaii, sugarcane represents a virtual monoculture; there is essentially no rotation to other crops. The cane is harvested two or three times in six or eight years, and then the fields are replanted.

Where a green-manure crop is included in the rotation, seedbed preparation starts with incorporating crop residue into the soil. Sugarcane may be planted flat (as in Florida) or on beds (as in Louisiana). Beds range from 1 to 2 feet (30–61 cm) in width, and rows (or beds from center to center) are spaced from 3 to 7 feet (90–210 cm) apart (Table 25-2). In Hawaii, seed pieces are normally placed in furrows spaced 5 feet (1.5 m) apart or less. The

Table 25-2
Effects of row spacings on yields of three types of sugarcane cultivars at Cairo, Georgia.

Cultivar	Yield (kg/ha) by row spacing (cm)					Average
	107	122	137	152	168	
C.P. 29-116	59.9	47.1	48.3	49.9	43.8	43.8
C.P. 36-111	55.9	57.7	59.7	51.3	45.1	53.9
C.P. 52-48	107.8	104.8	110.3	94.4	90.2	101.5
Average	74.5	69.9	72.8	65.2	59.7	

Source: K. C. Freeman, Influence of row spacings on yield and quality of sugarcane in Georgia, *Agron. J.* 60(1968):424–425.
Note: In general, close spacing favored higher yield. Marked exceptions are represented by 137-cm spacings of C.P. 36-111 and C.P. 52-48.

wider spacings are more common where growing seasons are longer, and more crops are harvested before a field is replanted; more space is needed as the plants tiller and spread in the course of several years.

Sugarcane is propagated vegetatively. Planting is done by opening furrows and inserting sections of stalks—from 2 to 3 feet (60-90 cm) in length—end-to-end in the furrow. As many as three rows of these pieces of stalk are placed in each furrow. The seed pieces are covered to a depth of 2 to 3 inches (5-20 cm); covering may be done by turning a shallow furrow on each side of the planted row. For frost protection, the stubble from which the ratoon crop grows may be covered with soil. In the early spring, the excess crop stubble is scraped off with a harrow or other implement so that the soil warms more rapidly, which favors rapid and early growth.

Time of planting varies with the location. In Hawaii, cane can be planted any month of the year, but the rate of growth depends on temperature; winter plantings grow relatively little until April. In Florida and Louisiana, plantings are made from August to mid-November. In the northern regions where sugarcane is adapted, cane may be planted in the spring. In general, plantings are scheduled so as to avoid interfering with harvesting operations in other cane fields.

In Hawaii, the crop matures and is harvested in about two years following seeding. The plants regrow as a ratoon crop and reach maturity and are harvested again in about another two years—a process that is repeated two or three times before replanting. In the rest of the United States, the crop matures in seven or eight months. After harvesting, the plants are more or less dormant during the cool winter months, and regrowth is initiated in the spring with the onset of warm weather. Frequently, two ratoon crops are harvested in successive years, before fields are replanted. Harvesting on the mainland usually starts in October, when cool weather or light frosts stop plant growth. In Hawaii, sugarcane may be harvested throughout the year because growth-stopping frosts essentially never occur. However, most harvesting operations cease during the wet period (from December to February) to avoid damage to the fields and significant reductions in sugar yield. This pause in the harvesting operation also allows time to repair the processing mills.

Sugarcane responds both to fertilizers and to innate soil fertility as reflected by the cation-exchange capacity of the surface foot

Table 25-3
Response of sugarcane grown in Mississippi to different rates of fertilizer: 5% N, 4.3% P_2O_5, and 13.4% K_2O.

Fertilizer (kg/ha)	Tons of cane per hectare			
	Plant cane	First year stubble	Second year stubble	Average
449	74.4	68.8	50.0	64.4
673	78.2	72.0	53.8	68.0
898	80.0	76.0	57.2	7.1
1,122	79.8	78.2	56.9	71.6
1,346	80.3	78.0	58.3	72.2
Average	78.5	74.6	55.2	—

Source: C. Freeman. Fertilizer rates on sugarcane for syrup production. *Agron. J.* 64(1972):639–640.
Note: Yields increased directly with the amount of fertilizer applied in this test.

of soil. On all but the highly organic soils typical of cane production areas in Florida, yields increase with the application of nitrogen fertilizers in amounts ranging from 50 to 100 pounds of nitrogen per acre (56–112 kg/ha) per year. Nitrogen may be supplied in a dry, in a liquid, or in an anhydrous form. As is done with sugar beets, the nitrogen may be applied in irrigation water. The maximum uptake of essential minerals occurs during the warmest part of the growing season during periods of maximum rate of growth. In Louisiana, for example, from 75 to 85 percent of the minerals are taken up during June, July, and August—the warmest months of the growing season. Cane yields of 30 tons per acre (67,000 kg/ha) remove about 4.3 pounds (1.9 kg) of nitrogen, 1.7 pounds (0.7 kg) of phosphorus (P_2O_5), and 6.7 pounds (3.0 kg) of potassium (K_2O) per ton on a whole-plant basis or 129 pounds (144.5 kg/ha) of nitrogen, 51 pounds (57.1 kg/ha) of phosphorus (P_2O_5), and 200 pounds (224.0 kg/ha) of potassium (K_2O) per acre. Higher yields in Hawaii—as much as 150 tons per acre, or 118 MT per hectare—remove proportionately more. Depending on the amount of plant residue left in the fields, more than half of the minerals taken up by the plants may be removed from the field. Fertilizer applications must be made to replace these minerals and to insure adequate amounts of essential minerals during the periods of peak growth. In addition to nitrogen, applications of both phosphorus and potassium may increase yields (Table 25-3). Much of the highly organic (muck) soil on which sugarcane is produced in Florida is deficient in at least three micronutrients: manganese, copper, and zinc. Yields on muck soils increase with applications of sulfate salts of these elements; as much as 60 pounds of manganese sulfate per acre (67.2 kg/ha) and 25 pounds (28.0 kg/ha) of copper and zinc sulfates may be applied. In Hawaii, fertilizer requirements are based not only on soil tests, but also on the results of plant tissue tests that determine specifically which elements the plants require.

Sugarcane yields tend to increase directly as the cation exchange capacity of the surface foot of the soil increases. Yields as low as 25 metric tons per acre are produced in soils

Figure 25-7
Experimental sugarcane harvester for recumbent cane developed in Florida by USDA engineers. [USDA photograph.]

with a cation exchange capacity of 3 milliequivalents and increase to more than 100 metric tons per acre in soils with a cation exchange value of 8 milliequivalents. (Cation-exchange capacity is a measure of soil fertility.) Although sugarcane, compared with most crops, is quite tolerant of aluminum, even moderate amounts may reduce cane yields and, in general, cane yields increase as exchangeable aluminum decreases. Yields of 100 metric tons of cane per acre are produced in soils with exchangeable aluminum values of less than 2 milliequivalents. (Note: Such yields may result from a number of factors other than the quantity of exchangeable aluminum.)

Sugarcane harvesting is becoming increasingly mechanized (Figure 25-7). Harvesting consists of stripping the leaves from the plants, topping the plants, cutting the stalks, and, finally, transporting the cut stalks to the processing plant (Figure 25-8). Topping may be done by hand or with special machines. Stripping the leaves is done by burning, which destroys rank weedy growth in the process. In Florida and in Hawaii, the standing crop is burned. In Louisiana, generally the crop is first topped, cut, and windrowed; then, after drying several days, windrows are burned to remove leaves. Burning the crop during harvesting is an operation unique to sugarcane production. This procedure does not damage the ratoon crop.

Diseases and Pests

DISEASES

Sugar beets Black root, a disease caused by a group of fungi, either kills germinating seeds and young seedlings or weakens those plants that survive to the extent that they become ready prey for other diseases and some insects. Black root causes reduced or uneven

Figure 25-8
Loading sugarcane for transport to a processing plant. [USDA photograph.]

stands and, consequently, lower yields. Control is possible through long-term rotations to cereal crops and through cultural practices that encourage rapid and vigorous germination and seedling growth, such as supplying balanced nutrition, irrigating properly, and insuring adequate drainage (overirrigating promotes the disease). Seed treatment with appropriate fungicides is also helpful.

Cercospora leaf spot, a widely prevalent fungus-caused disease that affects leaves and results in serious yield reductions, is best controlled through the use of resistant cultivars. Other fungal diseases include root and crown rots and downy mildew.

One of the major viral diseases of sugar beets is curly top; the diseased plants display curled, misshaped leaves. The virus lives on weedy members of the goosefoot family (e.g., pigweeds) and is transmitted to beets by insect vectors—the beet leafhopper or the whitefly. Control is achieved exclusively through the use of disease-resistant cultivars. There are several other viral diseases that affect sugar beets, including yellows diseases and mosaic diseases.

Sugar beets are also susceptible to mineral-deficiency diseases and to other nonparasitic diseases.

Sugarcane Diseases caused by both fungi and viruses have severely affected sugarcane production in the United States and, currently, loss from diseases may exceed 20 percent of the total crop. In the mid-1920s, disease epidemics virtually eliminated commercial crops in Florida and Louisiana. The development of disease-resistant cultivars, many of them from interspecific hybrids, has made possible the continued production of sugarcane in these states. More than sixty diseases affect sugarcane throughout the world, but relatively few of them are of significance in the United States.

All the viral diseases that cause losses in sugarcane are either transmitted by specific insect vectors or carried in infected seed pieces. Mosaic disease was a major contributing factor in the decline of sugarcane production in Louisiana in the first quarter of the twentieth century and generally caused significant crop losses until the 1920s, when disease-resistant cultivars were developed. Other viral diseases carried by insect vectors are the fiji disease, which causes severe stunting, and the chlorotic streak, transmitted by leafhoppers. Chlorotic streak reduces germination of seed pieces and reduces ratooning. Like other viral diseases, fiji disease and chlorotic streak are best controlled through the use of resistant cultivars.

Hot-water treatments have been developed to control seed-borne viruses. Exposure to a temperature of 125.6°F (52.4°C) for twenty minutes has proved to be effective and does not damage the seed. Hot-air treatments using electric ovens have been effective in controlling the "seed borne" virus that causes ratoon stunting disease. Treatments of seed pieces for eight hours at 136° to 138°F (58°–59°C) have afforded satisfactory disease control. With control, ratoon crop yields increase from 5 to 10 percent and sugar content increases more than 25 percent.

Several sugarcane diseases are caused by various fungi. Seed pieces are ready targets for many rot-causing organisms, partly because they are cut before planting and thus have wounds that afford avenues of penetration for pathogens and partly because they are rich in carbohydrates. Red rot—caused by *Collectotrichum falcatum,* which is carried on or in the seed piece—is a major disease that is favored by low temperatures and high moisture conditions. Pineapple rot, which causes seed rotting quite similar to that resulting from red rot, is favored by dry conditions. Both diseases are best controlled through the use of disease-resistant cultivars, although cultural practices that favor rapid germination —such as insuring adequate available moisture (and drainage, if necessary) and planting into warm soil—also help. Pineapple rot is caused by *Ceratostomella paradoxa,* one of several species of fungi causing sugarcane seed rots. Seed rots are also caused by various species of *Fusarium* and *Phytophthora.*

Fungi also cause a variety of sugarcane root-rot diseases, the most common of which is pythium root rot, which reduces tillering and stunts plant growth. Symptoms of the disease include severe yellowing of the leaves and wilting, as well as watery lesions in the roots and a reduction of secondary root development and tillering. If the disease is severe, plants die. The only control is through the use of resistant cultivars.

Root rot diseases are more serious on heavy, clay soils; a fine soil texture coupled with a high water table favors disease development. Cool temperatures also favor root rots.

WEEDS

Sugar beets Weeds are a serious problem in sugar-beet production. Weed control starts with seedbed preparation. In addition, early weed growth is controlled, in part, during thinning operations. Because sugar beets are planted in relatively widely spaced rows, it is possible to cultivate for weed control in the spring and early summer. Herbicides also play a major role in the control of weeds growing in sugar beets. More than 60 percent of the 1.5 million acres (0.6 million hectares) of sugar beets harvested annually in the United States is treated with herbicides. Soil-incorporated herbicides are sometimes applied when the land is being prepared for sugar-beet production in the fall. Preemergence, postemergence, and combinations of pre- and postemergence sprays are used, but most areas are treated with only preemergence sprays and relatively few acres receive both sprays. The use of herbicides is increasing in most areas. Some problems of herbicide persistence in the soil have been reported, particularly in the Northeast, but in general, it is not a major problem.

The most common weeds in sugar-beet fields are pigweeds, lambsquarters, barnyardgrasses, mustards, kochias, and wild oats. The list of major weeds has remained unchanged in the past five years.

Sugarcane Weed control using herbicides is practiced on about 95 percent of the U.S. sugarcane crop. Postemergence sprays are used most frequently, but both preemergence and postemergence sprays are used on more than 25 percent of the area treated. More than 100,000 acres (40,500 ha) are treated annually with only a preemergence spray. Weed control is part of seedbed preparation, but cultivation for weed control during the growing season is impractical. Some weed control results from burning the mature crop before cutting.

The major weeds are warm-season grasses that are not serious pests for other crops: alexandergrass, guineagrass, napiergrass, and paragrass. Other weed pests include crabgrasses, panicums, johnsongrass, three-lobe morningglory, and wingleaf passionflower. Paragrass, crabgrass, and wingleaf passionflower are pests in Hawaii; the others are pests in Florida and Louisiana.

INSECTS

Sugar beets A variety of insects attack sugar beets, but most of the major pests can be controlled with organic phosphate and carbamate insecticides. Most of the crop losses result from insects feeding on the foliage of mature plants, but sometimes seedlings are eaten at the soil surface. Leading insect pests include beet armyworms, beet webworms, beet leafhoppers, and grasshoppers.

The root-knot nematode; a microscopic eelworm that persists in the soil, is a serious pest of sugar beets and can cause a total loss of stands. Control is best accomplished through soil fumigation or through long-term rotation away from sugar beets, and all other susceptible crops, to cereals. To avoid infesting clean fields, soil from beet dumps should not be brought into them, and equipment used in infested fields should be cleaned before it is moved to clean ones.

Sugarcane Sugarcane is attacked by several common insect pests, but the most important are the sugarcane borer and the sugarcane beetle. The sugarcane borer attacks the shoot apex (growing point), causing deadheart, and it may kill the tops of older plants. The boring of this insect reduces sugar content and makes the plant subject to rot.

Other insect pests include mealybugs, termites, chinch bugs, and three corn pests—the cornstalk borer, the southwestern cornborer, and the lesser cornstalk borer. The larvae of the click beetle—called wireworms—are also sugarcane pests. Losses due to all insects may be as much as 15 percent of the total U.S. crop. Chlorinated hydrocarbon insecticides were effective in controlling insect pests in the past, but they gave rise to persistence problems. New, nonpersistent compounds have not been fully evaluated, but they seem to be promising.

Crop Utilization

The major product of both sugar beets and sugarcane is refined sugar. The sugars obtained from processing the two crops are indistinguishable, and differences in consumption of sugar from sugar beets and sugar from sugarcane are attributable entirely to the type of crop grown or processed locally. For example, in Europe sugar beets are the primary source of refined sugar.

Processing sugar beets starts with washing and slicing the roots and cooking them to extract their juices. The solids are then removed, and the juice is purified and crystalized. The residual root solids are dried and shredded for use as livestock feed. Molasses, another by-product of sugar-beet processing, is also used as livestock feed and in the production of alcohol and other products. Beet tops and crowns, removed in the harvesting process, are often baled or ensiled for use as livestock feed.

Processing sugarcane starts with crushing the cane with heavy, grooved rollers to extract the juice. After this first step, the process is similar to that used for sugar beets; the juice is boiled, filtered, and crystallized. One by-product is molasses; another is a fibrous stalk residue called boagasse. Because of its high fiber content, boagasse is not suitable for livestock feed. Instead, it is used in the manufacture of some plastics, as well as some paper products, including building boards. It also serves as livestock and poultry bedding.

Selected References

Golden, L. E., and R. Recaud. 1963. The nitrogen, phosphorus, and potassium content of sugar cane in Louisiana. *Louisiana State Univ. Bull.* 574 (October).

Johnson, Russell T., J. T. Alexander, G. E. Rush, and G. R. Hawkes, eds. 1971. *Advances in sugar beet production.* Iowa State University Press.

Lill, J. G. 1964. Sugar beet culture in the north central states. *U.S. Dep. Agr., Farmers' Bull.* 2060.

Van Hook, Andrew. 1949. *Sugar.* Ronald Press.

Study Questions

1. Describe the major botanical features of (a) sugar beets and (b) sugarcane.
2. What are the major factors that limit the distribution of sugar beets? of sugarcane?
3. What field-management practices have the greatest effect on (a) sugar-beet yields and (b) sugarcane yields?

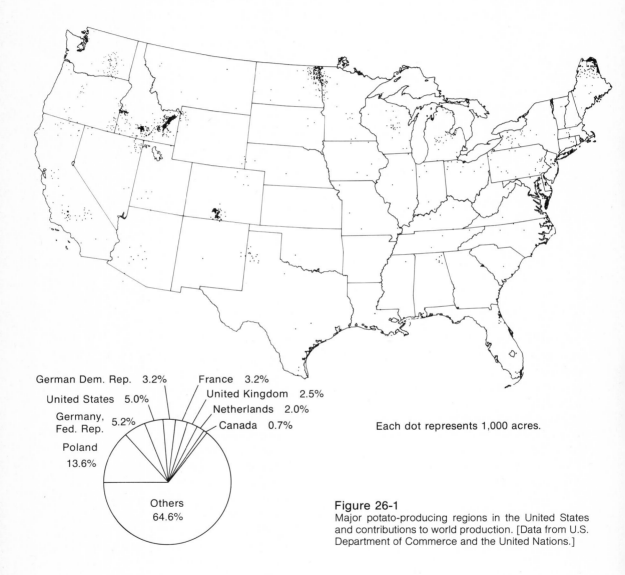

Figure 26-1
Major potato-producing regions in the United States and contributions to world production. [Data from U.S. Department of Commerce and the United Nations.]

Twenty-Six Potatoes

Origin and History of Culture

Like field beans and corn, potatoes originated in the New World. The probable center of origin is in South America in the central Andes. Evidence indicates that potatoes were cultivated for centuries by South American Indians, whereas they are relative latecomers to European agriculture; it is believed that Spanish explorers brought the potato from Peru to Spain in the middle of the sixteenth century. Following its introduction into European agriculture, the potato became an important food crop in northern Europe. Ireland became so dependent on the potato that, when a serious disease wiped out much of the crop there in 1845–1846, a famine resulted, causing mass starvation. This famine was one of the reasons for the large scale emigration to the United States in the middle of the nineteenth century. The human suffering caused by the failure of the potato crop in Ireland is a tragic example of the danger of any society becoming overly dependent on a single source of food.

Although the potato originated in the New World, it did not become an important crop in North America until after it was reintroduced from Europe sometime in the first half of the seventeenth century. Since then, it has become a major food crop. Potatoes have been a major crop in Peru for more than 2,000 years. Like other important commodities in primitive cultures, they were also significant in spiritual practices.

Of the major crops grown throughout

the world, potatoes rank second only to cereal crops in importance. They are grown in Canada and in every state in the United States, where the major regions of production are widely distributed: in the west, Idaho, California, Washington, and Colorado; in the east, Maine and New York; and in the central states, Minnesota and North Dakota (Figure 26-1). Although potatoes are produced on many small farms, commercial production is confined to a relatively few very large operations.

Botanical Characteristics

The potato, *Solanum tuberosum,* is a member of the night-shade family (Solanaceae), which also contains tobacco and tomatoes. Potatoes are herbaceous annual plants grown for their edible tubers. (Recall that tubers are the enlarged, fleshy tips of underground stems; they are storage organs and are high in carbohydrates.) The eyes of a potato are buds from which lateral branches will grow. Each potato (or tuber) has at least three buds that can produce normal vegetative branches. It is easy to distinguish the spiral arrangements of the buds (eyes). The tuber generally has a somewhat pithy core, which is surrounded by a mass of parenchyma tissue, in which large amounts of starch are stored. The parenchyma is surrounded by an outer vascular ring and an external layer of cambium tissue, which has pigments that give potatoes a variety of skin colors. The outermost layer of cells—the skin, or periderm—is frequently coated with suberin and has the ability to form cork tissue and thereby close wounds. Cells of the periderm contain chloroplasts, which are responsible for making the potato turn green when exposed to light.

Vegetative stems (or branches) of the potato plant are usually quite thick and erect and are greatly branched (Figure 26-2). They range from 1 to 2 feet (30–61 cm) in length. Leaves are compound; they have a terminal leaflet and from two to four pairs of fairly large, oblong, pointed leaflets and two or more small leaflets at the base. There are stipules at the base of the petiole.

Flowers are borne in compound (branched) terminal cymes; thus, the plant is determinate in growth habit. Individual flowers are complete and perfect. The corolla consists of five petals, which vary in color from white to rose or purple. There are five stamens and a single style; the pistil is compound and the ovary develops into a berry-type fruit (like a tomato), but the fruit is only an inch or less in diameter.

True seed is not used for commercial plantings. Because potatoes can reproduce vegetatively, pieces of tuber (called seed potatoes) that include one or more eyes are planted to produce the crop. However, true seed must be produced and used in plant-breeding programs. (Another major crop that is seeded with a vegetative part of the plant rather than seed is sugar cane; see Chapter 25.)

Environmental Requirements

Because potatoes are grown for their tubers rather than for their seed, the environmental conditions discussed here are those suitable for tuber production and not necessarily for seed production.

Potatoes are classified as a cool-season crop. An average soil temperature of 75°F (24°C) in July is optimum for tuber growth. Satisfactory tuber growth occurs if soil temperatures are in the mid-60s (about 17°–19°C), but tuber development virtually stops if temperatures rise above 85°F (30°C). At higher

Figure 26-2
Typical potato field in midsummer.

temperatures, the respiration rate increases, and the carbohydrates produced by photosynthesis are consumed rather than stored in the tuber. In effect, the rate of respiration exceeds the rate of photosynthesis, and you might say that the plant is spending more than it is saving. Because of this "spendthrift" characteristic of potatoes in high temperatures, in the South they are produced only in the early spring.

Although potatoes are a cool-season crop, freezing temperatures are damaging to both seedlings and older plants. The plants are more cold tolerant during the middle of the growing season; however, tubers freeze at about 28°F (-2°C). Long days with cool temperatures favor flowering and seed production.

Potatoes require a uniform supply of water throughout the growing season, but dry weather is desirable for harvesting. Adequate moisture is essential during seedling growth and during the period of tuber formation to minimize the development of malformed tubers. The total water requirement (precipitation plus irrigation) is from 12 to 24 acre-inches (30–61 cm/ha). Because most roots are between 12 and 25 inches (30–64 cm) below the soil surface, frequent light irrigations, or frequent light rains, are essential (Figure 26-3). As many as six irrigations may be required for optimum yields. Fields should be irrigated when they approach 60 percent of field capacity (much the same as for beans).

Potatoes can be produced on a wide range of soils. They are well suited to acidic soils (pH from 4.5 to 5.5). In fact, acidic conditions tend to limit the potato scab disease, which is favored by more basic soils, and liming to raise the pH is not recommended because

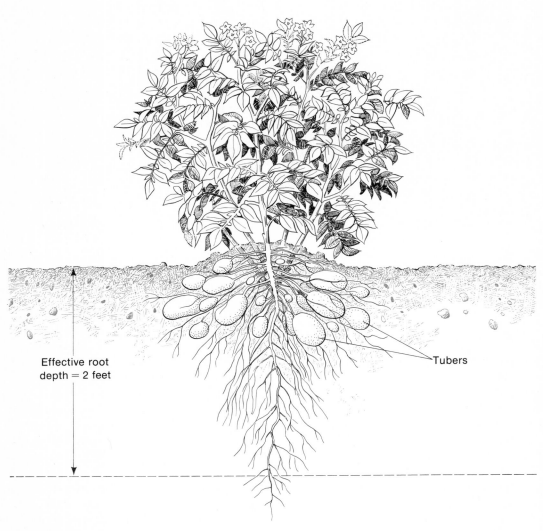

Figure 26-3
Pattern of root growth of a potato plant.

alkalinity favors the disease. Potatoes can also be successfully produced on neutral and even on basic soils. Light soils are preferred because they tend to promote more uniform temperatures and make harvesting the crop easier, but the crop can be produced on peat soils. Heavy clay soils may make digging tubers difficult.

Production Practices

SEEDBED PREPARATION AND SEEDING

Potatoes are planted in the spring. Fall operations are designed primarily to incorporate crop residue. Spring seedbed preparation entails plowing to a depth of 5 to 7 inches (13–18 cm). (Deeper plowing does not increase yields and is not generally recommended.) Subsoiling is not commonly practiced, but periodic chiseling to depths of 12 to 16 inches (30.1 to 40.6 cm) may be done on heavy soils.

As mentioned earlier, potatoes differ from other crops (except sweet potatoes and sugar cane) in that true seed is not used for field planting. Instead, pieces of tuber that include at least one eye (or bud), called *seed potatoes*, are planted. Thus, potatoes are reproduced asexually, or vegetatively. Most seed-potato pieces weigh between about 1.5 and 2.0 ounces (43–57 g), or a quarter of a well-shaped 5-to-8 ounce (142–227 g) tuber. In some cases, whole small tubers, from 1.5 to 2.0 inches (4–5 cm) in diameter, are planted. Care must be exercised in cutting seed pieces. Cleanliness is essential to minimize the spread of seed-borne diseases and to insure vigorous seedlings. Sometimes seed potatoes are treated for disease control—mainly the control of pathogens carried on the surface of the seed piece. However, treatment is not effective in controlling viruses, soil-borne pathogens, or pathogens inside the seed piece; such control is better accomplished by planting disease-free (e.g., certified) seed potatoes. If seed treatment is required, special care must be taken to avoid damaging the seed piece with the treatment solutions.

If whole, small potatoes are used for seed, losses from disease are reduced because there is no large cut surface to offer access to pathogens. An additional benefit of planting with whole potatoes is slightly higher yields. This is true even when cut seed pieces are comparable in size to the whole seed potato.

Fresh potatoes are dormant for three to five months after harvesting. Sometimes, seed potatoes are taken from storage and allowed to sprout before planting. If very fresh potatoes are used for seed, it may be necessary to treat the potatoes in order to break or end this dormant period. The treatment consists of applying one of three compounds: ethylene chlorohydrin, potassium thiocyanate, or sodium thiocyanate.

Seed pieces are planted in rows and spaced from 6 to 10 inches (15–25 cm), or slightly more, apart; rows are spaced from 34 to 36 inches (86–91 cm) apart. Closer row spacing leads to smaller, but more uniform, tubers. This is a general result of competition among plants for water, essential minerals, and space as the tubers develop. Depending on location, time of planting, and available moisture, plant populations may be as low as 11,000 and as high as 30,000 seed pieces per acre (27,000–74,000/ha). Russet types and spreading types are seeded at the lighter rates.

Planting depth varies between 3 and 5 inches (8–13 cm); the depth increases for light soils, dry conditions, and later planting dates. In northern areas, planting is done as soon as fields can be worked and the soil temperature exceeds 40°F (4°–5°C); these conditions usually occur within 10 days before the average date of the last killing frost—early to mid-May. In the South, where the soil does not freeze to seeding depths, planting may be done as much as six weeks before the last killing frost (in Georgia, this would be in mid-January). Although the planting date for potatoes does not follow Hopkins' Bioclimatic Law precisely, a rule of thumb used in the mid-Atlantic region is to delay the planting date one day for every 8 miles (13 km) from south to north (about 0.125° latitude).

Potato production can be separated into three groups, according to when crops are harvested. The early crop, harvested in winter or spring, is grown mainly in the south—in Florida, Alabama, North and South Carolina, Louisiana, Tennessee, Texas, and southern California. This crop is less than 15 percent of the total U.S. crop. The intermediate crop, harvested in early and late summer, is less than 20 percent of the total U.S. potato crop. This crop is produced in the east and southeast—in Virginia, New Jersey, Kentucky, New York (including Long Island), Missouri, Maryland, and Delaware. The major crop is the late crop. This fall-harvested crop is produced in Maine, but production extends to the Great Lakes. All potatoes in the western United States are harvested as part of the late crop. This crop comprises about 70 percent of the total U.S. crop and nearly all of the Canadian crop.

In small operations, plantings are made by dropping seed pieces into furrows; in larger commercial operations, special mechanical planters are used.

FERTILIZERS
Fertilizer requirements depend on the location, on the preceding crop in the rotation, and on whether the crop is irrigated. Under dryland conditions (no irrigation), if potatoes do not follow a legume immediately, from 80 to 100 pounds of nitrogen per acre (90–112 kg/ha) may be applied (Table 26-1). Under irrigation, rates of more than 250 pounds of nitrogen per acre (280 kg/ha) have proved to be economical. Applications of phosphorus (P_2O_5) ranging from 30 to 130 pounds per acre (34–146 kg/ha) are common, but rates as high as 300 pounds per acre (336 kg/ha) are not unknown. Potatoes require a sizeable amount of potassium (K_2O): about 0.1 pound (45 g) per bushel of tubers, or between 100 and 150 pounds per acre (112–183 kg/ha). Where deficiencies exist, magnesium may also be required. For highest yields and quality, minerals must be continuously available to growing plants, but excesses must be avoided. Excessive amounts of nitrogen lead to hollow heart, a breakdown of tissue in the center of the tuber that reduces quality. Excessive nitrogen, particularly in relation to available potassium, reduces the percentage of solids in the tuber. (The percentage of solids is an important quality trait in potatoes and is measured by determining the specific gravity of the tuber in a laboratory test.) Excessive phosphorus in relation to potassium also may lower the specific gravity of a tuber. Micronutrient balance is of major importance in quality and yield, particularly the sulfur: chlorine ratio. High sulfur levels are favored.

Fertilizers are frequently applied during planting; they are banded below and 2 inches on either side of the seed pieces. Fertilizer placement is critical because seed pieces are sensitive to fertilizer salts, and exposed (cut) edges are easily burned.

HARVESTING, STORAGE, AND ROTATIONS
Potatoes are harvested from 3 to 4 months after planting. Harvesting of very small plantings is done by cultivating next to the rows to loosen the plants, turning the tubers to the soil surface, and gathering the tubers by hand. In large, commercial fields, digging and gathering are done in a single operation with special equipment (Figure 26-4). About two weeks before digging, vegetation is destroyed

Table 26-1
Response of potatoes to nitrogen and phosphorus.

Fertilizer (lb/acre)	Yield (cwt/acre)		Quality (percentage of Grade 1)	
	1963	1964	1963	1964
Nitrogen				
100	331	541	68	70
200	437	644	69	67
300	498	673	66	62
400	527	686	63	59
Phosphorus (P_2O_5)				
0	425	585	70	65
133	448	649	67	65
267	454	652	66	64
400	466	658	64	64

Source: R. Kunkel. Results of the 1963–1964 fertilizer trials with Russet Burbank potatoes. *Proc. 4th Ann. Wash. State Potato Conf.*, 1965.
Note: An average of 200 lb/acre of P_2O_5 and K_2O was applied with all rates of nitrogen, and 200 lb/acre of nitrogen and K_2O with all rates of P_2O_5. Nitrogen increased yields but reduced quality slightly; P_2O_5 had less effect on yield and also lowered quality.

by cultivating so that the tubers develop a firm skin. Care must be exercised to avoid bruising or cutting the tubers. In North America, potatoes are harvested somewhere every month of the year, sometimes before the plants are mature. Premature harvesting leads to lower yields (if left in the ground, the tuber continues to enlarge until the plant dies) and to lower quality. But these losses are offset by the higher prices that out-of-season potatoes command.

Storage conditions are critical; most potatoes are stored in cellars where both temperature and humidity are maintained fairly constant. Storing potatoes at 65°F (18°C) with the relative humidity near 85 percent for two weeks after harvesting favors the healing of wounds resulting from harvest operations. After two weeks, tubers can be stored safely at 36° to 38°F (2°–3°C) with the relative humidity near 90 percent.

Potato rotations vary regionally. In Maine, potatoes may be grown in a short rotation with a green manure crop. Potatoes may be produced on the same field for three out of every five years. Longer rotations in which potatoes are grown only once in 5 to 7 years are common and may be required for insect and disease control. Potatoes can be rotated with a number of other crops, depending on the region, the soil, the potential crop pests, and the available moisture. Potatoes do particularly well following legumes in a rotation. In some regions, if crops are grown without irrigation, a cereal crop immediately preceding potatoes may dry the soil to too great a depth and thus lower potato yields.

Diseases and Pests

DISEASES

Potatoes are susceptible to a number of diseases caused by fungi, bacteria, and viruses—all of which lower yields and reduce crop quality.

Figure 26-4
A combine digging potatoes, discarding soil, vines, and weeds, and dumping potatoes into a truck by means of a rod-chain conveyer. [USDA photograph.]

There are at least four major diseases caused by fungi. The first, fusarium (eumartii) wilt, is a soil-borne disease that is expressed as a tuber rot. The fungus is carried in the soil, on seed potatoes, and on seed-potato bags. The disease can be controlled in part by not using contaminated potato bags, by not transferring soil from infected fields, and by using certified seed potatoes. Crop rotation is also effective; fusarium wilt can be controlled by growing potatoes only once every five years and seeding nonsusceptible crops in the other years.

Potato scab is a soil-borne fungal disease that causes lesions (scabs) on the surface of the tuber, which lower crop-quality ratings. Growing nonsusceptible russet types of potatoes is a major method of controlling potato scab. Planting into raw organic matter that favors the disease should be avoided, and maintaining a proper nutrient balance in the soil is helpful for control. Rotations that use potatoes only once every five years are also effective.

Rhizoctonia canker—also called black scurf—results in the rotting of roots and stems and leads to weak plants, poor stands, and, of course, reduced yields. Control of this soil-borne fungal disease is achieved through cultural practices that foster rapid, vigorous seedling growth. Cultivations, including seedbed preparation, should minimize soil crusting; excessive irrigation should be avoided; and planting should be delayed until temperatures are above 55°F (13°C).

Early blight is a foliar disease caused by a

fungus. It is a serious problem only where sprinkler irrigation is used. Control methods include rotation away from potatoes and, if the disease is severe, application of fungicides in the field.

The serious bacterial diseases infecting potatoes include blackleg, verticillium wilt, and ring rot. Blackleg is controlled by using sound seed-potato pieces (the bacteria enter only through wounds in the seed piece). Verticillium wilt disrupts water movement in the plant causing it to display droughtlike symptoms. Aside from clean farming practices and attention to optimum irrigation and fertilizer applications that favor healthy plants, there is no effective way to control verticillium wilt. Ring rot affects both foliage and tubers, causing internal rotting of the tuber. If the disease is severe, serious yield losses can occur. Control is accomplished by growing resistant cultivars and by using certified, disease-free seed potatoes.

There are several important viral diseases of potatoes, and more are being identified every year. Currently, control of all viral diseases is restricted to the use of disease-free seed. Leaf roll, characterized by an upward roll of the leaf margins, may cause yield losses as high as 30 percent. Spindle tuber causes the development of long, thin tubers that reduce the value of the crop. Rugose mosaic causes mottled and rolled leaves and malformed tubers. Other viral diseases are mild mosaic, latent mosaic, and purple-top wilt.

WEEDS

Because potatoes are grown virtually from coast to coast, they are subject to attack from a wide array of weeds that vary in importance from location to location. Of the ten most prevalent weeds or groups of weeds, lambsquarters and pigweeds are the most important in terms of acres infested; the others are barnyardgrass, crabgrass, foxtails, kochia, nightshades, nutsedges, quackgrass, and ragweeds. The number of acres infested with one or more of these weeds has increased, at least slightly, in the past ten years. Although, currently, nutsedges do not infest a markedly large area, they have been spreading rapidly, and they may eventually become an even more serious pest.

Weed control in potatoes, as in most row crops, is achieved through a series of steps: seedbed preparation, cultivation during the growing season, and the use of herbicides. More than 30 percent of the U.S. potato acreage is treated with some type of herbicide; both pre- and postemergence sprays are available. About 90 percent of the treated acreage receives preemergence herbicides, about 10 percent is treated with a postemergence spray, and less than 5 percent is sprayed with both pre- and postemergence herbicides.

Cultivation for weed control during the growing season is an important part of potato production, and sometimes the size of the available cultivation equipment dictates the row spacings. As the plants develop and spread during the growing season, cultivation must be done with care in order to avoid damaging the roots and the growing tubers.

INSECTS

A number of insects cause reductions of potato yields. Chemical control of insect pests is effective, but caution must be used to meet all restrictions and regulations. Both the young and the adults of the Colorado potato beetle feed on, and defoliate, plants. Insecticides are the major method of control when infestations become severe. Wireworms (larvae of click beetles) cut seedlings off at the soil surface and cause stand losses and yield reductions. Some insecticides are effective, as is crop rotation to alfalfa. The same control methods are used on white grubs, which can be a serious problem in areas with highly organic soils. The use of insecticides against wireworms and white grubs also controls leather jackets (crane fly larvae). Other insect pests include the potato tuber worm, potato maggot, potato leafhopper, and the seed corn maggot.

Crop Utilization

The potato is a source of carbohydrates no matter how it is consumed. On a dry weight basis, the tuber ranges from 65 to 80 percent starch. The utilization of potatoes has changed markedly in the past twenty-five years in response to the market for modern conveniences.

As a general rule, potatoes are easy to store. Storage should be cool (36°–40°F, or 2°–5°C) with a high relative humidity (85–90 percent). Because potatoes store well, they have been a dependable food supply. Throughout the world, fresh (unprocessed) potatoes are prepared in a variety of ways: boiled, baked, and fried.

Since the end of World War II there has been an increasing demand for various "convenience types" of processed potatoes. The earliest form, which predates World War II, is the ordinary potato chip. This was followed by the frozen french fry. The advent of the home freezer contributed greatly to the demand for frozen potato products. Tuber quality plays a major role in the processing of potatoes. For both potato chips and french fries, the most critical factor is the percentage of solids, or specific gravity. Both require tubers that have a high specific gravity, that is, above 1.066 (the specific gravity of water is 1.000). A number of common by-products of the frozen french fry industry are seen in the supermarket freezer: diced, mashed, hashbrown, and au gratin potatoes, as well as potato puffs and potato cakes are but a few examples. Recently, an entirely new industry centered on dehydrated potatoes has emerged.

In addition to the more recent frozen and dehydrated conveniences, potatoes are milled into potato flour. Some potatoes are also fermented for alcohol by the distilling industry. Finally, although the potato is an important energy source for the human diet, some potatoes are fed to livestock.

There are marked regional differences with respect to tuber quality for processing. In general, potatoes produced in the western United States (e.g., in Idaho) have a higher percentage of solids than those produced in the east and are therefore in greater demand by the processors of frozen products. The frozen french fry industry originated in Idaho, partly in response to the availability of potatoes of acceptable quality. The potato certainly must be considered a crop of antiquity that has changed its face in the light of modern technology.

Selected References

Jensen, M. C., and J. E. Muddleton. 1965. *Proc. 4th Ann. Wash. State Potato Conf.*

Kunkel, R. 1965. Results of the 1963–1964 fertilizer trials with Russet Burbank potatoes. *Proc. 4th Ann. Wash. State Potato Conf.*

Martin, J. H., and W. H. Leonard. 1967. *Principles of field crop production.* 2d ed. Macmillan.

Ohms, Richard E. 1962. Producing the Idaho potato. *Idaho, Univ., Agr. Ext. Serv. Bull.* 367.

Study Questions

1. What are the most important irrigation and fertilization practices in potato production?
2. How does the field management of potatoes differ from that employed for most other crops?
3. What factors affect the quality of seed potatoes?
4. Describe the growth and development of a potato plant.

Figure 27-1
Safflower heads: (left) after flowering; (center) flowering; and (right) before opening. Note also sessile leaves with sharp spines. [USDA photograph.]

Twenty-Seven

Safflower, Rapeseed, Flax, and Sunflower for Oil

Origin and History of Culture

Mankind's use of vegetable oils—oils derived from plants—began before recorded history. It has been suggested that seed-oil crops may have been discovered by accident when "weeds," from which the vegetable oils come, were harvested along with cereal crops. Unlike cereal crops and some edible legumes, the cultivation of seed-oil crops has not played a significant role in mankind's cultural development, although several important species of oil crop plants are grown throughout the world.

Vegetable oils—the main product of seed-oil plants being grown today—can be classified according to their uses; edible oils, lubricating oils, and drying oils.

Safflower Safflower, which is known to have been cultivated for thousands of years, reportedly originated in northeastern India. Throughout history, safflower has been used primarily as the source of a dye to color a variety of cloths, including silk. Egyptians used safflower dye to tint the wrappings of mummies more than 3,500 years ago, and the dye is still being used in some parts of the world.

The value of safflower as a food crop—more precisely as an edible-oil crop—was recognized in India more than 100 years ago. In the 1930s, attempts were made to produce safflower in the United States, but they failed. As a result of the development of cultivars having a high oil content, safflower became a favored crop in some areas in the 1950s.

Although these cultivars were developed in Nebraska, California is now the leading safflower producing state. This relatively new crop is grown on lands that were rotated out of rice production as a result of federal acreage limitations on rice.

Rapeseed The origin and early culture of rapeseed is obscure. There are two species commonly referred to as rape: *Brassica napus* (Argentina rape) and *B. campestris* (turnip rape). The former is believed to have evolved in Europe or in northwestern Africa, and the latter in Europe or Asia. Because there are many interspecific hybrids, tracing the evolution of these species is extremely difficult.

The earliest written records of rapeseed (2000 to 1000 B.C.) are found in India. Apparently rapeseed was introduced into China from Korea more than 2,000 years ago, and traces of rapeseed have been identified in Bronze Age (c. 3500 B.C.) settlements in Germany. The crop spread through Europe in the Middle Ages and was used primarily for lamp oil. Because olive oil is generally considered superior for cooking, rapeseed never became a significant crop in areas where olive trees are grown.

Today, rapeseed is the most important oil crop grown in Canada, and in terms of acres cultivated, it is the most rapidly expanding crop. The commercial production of rapeseed in Canada started in 1942, and by 1970 more than 5 million acres (2 million ha) were planted to rapeseed. The oil was in demand as a superior lubricating oil for marine engines. Today, rapeseed that is low in erucic acid is grown for edible oil; oil from rapeseed that is high in erucic acid is used in the potash mining industry.

Flax The history of flax is nearly as old as the history of civilization. Egyptian records trace the culture of flax back 5,000 years to the Bronze Age. The long history of flax culture can be attributed in part to the fact that the plant evolved in the Mediterranean area where cereals were first cultivated, and in part to the fact that the flax plant has served several different and important functions. Fibers from the stems of the plants were (and still are) used to make linen—the fabric that helped preserve Egyptian mummies. The seed was eaten directly or in a milled form, much as cereals. Oil was also extracted from the seed, and the drying properties of linseed oil (the oil from flax seed) were recorded in Roman writings (c. 230 A.D.). Linseed oil is still used as the base of many high-quality paints, although it has been replaced by modern synthetic compounds in some products.

Flax has been grown for both its oil and its fiber in the United States and Canada since the latter part of the eighteenth century. During World War I, linseed oil was in great demand for marine paints, and linseed cake, or meal (an oil by-product), was valued as an animal feed in Europe. In recent years, little or no flax has been produced as a source of fiber.

Sunflower Sunflower is a native of the New World, and it is believed to have been revered by ancient sun worshipers in Peru. Apparently, the crop was introduced into Spain from Central America by early explorers and merchants; there are no records of its cultivation as a crop until the middle of the sixteenth century. Sunflower was brought to New England from Europe in the colonial period. In the past, sunflower seeds were processed for cosmetics, notably hair oil, and the plant served as a food crop in Central America. Today, sunflower oil is valued as a quality edible oil. In Canada, sunflower is grown as an oil crop on approximately 250,000

Figure 27-2
Safflower "seed," which is actually a simple, dry fruit called an achene. [USDA photograph.]

acres (100,000 ha), an acreage that gives some indication of its importance as a source of edible oil.

Botanical Characteristics

Safflower Safflower is a member of the thistle, or composite, family (Compositae); it is classified as *Carthamus tinctorius*—a species name that reflects its early use as a source of dye, or tincture. Safflower is an annual with stems ranging from 2 to 5 feet (60–150 cm) in height, depending on the cultivar and planting date. The stems branch prolifically near the top and are determinate in growth habit.

The leaves are sessile (without petioles), toothed along their margins, and usually armored with sharp spines. (Plant breeders have been trying to develop spineless cultivars, but, currently, none are in commercial production.

Each stem terminates in a typical head type of inflorescence (Figure 27-1). These heads vary in size from 0.5 to 1.5 inches (1.3–3.5 cm) in diameter. Flowers are generally complete and perfect with conspicuous white, yellow, or red-orange petals. (The petals are the source of dyes produced from safflower.) The fruit is a smooth, shiny white achene, very angular and nearly wedge shaped, less than 0.5 inch (1.3 cm) in length (Figure 27-2).

Safflower is grown mainly for its oil. Whole seed is from 20 to 35 percent oil of high quality both for human consumption and for commercial use in making paints and soap. In recent years, safflower oil has been used in salad oils, margarine, and similar products because it is unsaturated; unsaturated fats have been reported to be desirable for human consumption. The unsaturated nature of safflower oil makes it a drying oil, which is desirable for making paints.

Rapeseed Rape is a member of the mustard family (Cruciferae). Although there are two distinct species of rape, *Brassica napus* and *B. campestris,* their morphological features are so similar that a general description fits both. Rapeseed is a herbaceous annual. Plants vary in height from 1 to 3 feet (30–90 cm): *B. napus* from 2.5 to 3 feet, and *B. campestris* from 1 to 3 feet. The stems are heavily branched; the leaves are from 4 to 12 inches (10–30 cm) long and are borne on relatively short petioles. The leaf-blade margins are lobed, or pinnatified (Figure 27-3). The roots of the rapeseed plant are hard and they branch several inches below the soil surface.

The flowers, which are variable in color, are complete and perfect and are borne on elongated racemes. As is typical of flowers in the mustard family, they have four sepals and four petals, which open into the square, or cross-like, pattern that gives the family its name—Cruciferae, or "cross bearing." Each flower has six stamens; the filaments of four are of equal length, and the filaments of the others are markedly shorter. The pistil is compound and the ovary matures into a two-celled fruit (a silique), which tends to dehisce at maturity. Each fruit contains from fifteen to forty small seeds. At maturity, the seeds have a thin band of endosperm surrounding the embryo. The seed coat, or testa, is normally

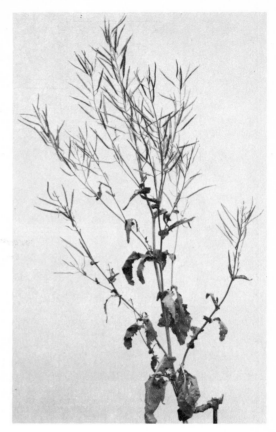

Figure 27-3
A typical rapeseed plant.

yellow but may vary from light yellow to black.

Flax Flax, *Linum usitatissimum,* is a herbaceous annual and belongs to the Linaceae family. Plants range in height from 1 to 4 feet (30–120 cm), depending on the type of cultivar, environmental conditions, and the seeding rate, which also affects branching. Cultivars grown for seed (oil) production are usually shorter than those grown for fiber production (Figure 27-4). Stems are narrow and may branch from the base. With the heavy seeding rates used for fiber production, plants tend to be tall and have only a single stem. The leaves are short and narrow; they are alternate on the stem and are sessile. Plants form a poorly branched taproot system that is fairly shallow—2 feet (61 cm) deep.

Flowers, which are usually either white or blue, are complete and perfect with five sepals, five petals of equal size, and five stamens. The pistil is compound and the ovary matures into a capsule housing from four to ten cells, each of which may contain two seeds. The seeds are flat, shiny, and relatively small. They range in color from brown and yellow to green. The blue flowers are borne on terminal cymes; on clear, warm days, individual flowers open at sunrise, and the petals usually fall before noon. Some cultivars of flax have been developed for ornamental purposes.

Sunflower Sunflower, *Helianthus annuus*, is a member of the thistle family, Compositae, which also includes safflower. Sunflower plants are herbaceous annuals that grow from 4 to 20 feet (1-6 m) tall, usually with only a single, hair-covered stem that may be more than an inch in diameter. Leaves are as long as 12 inches (30 cm) and are borne on petioles arranged alternately on the stem.

Sunflower has a typical head type of inflorescence (Figure 27-5). The ray flowers, located around the margin of the head, are sterile; the corollas of these flowers are the petals of the sunflower. The disc flowers, located in the central part of the head, are fertile (perfect) but incomplete. The ovary of the disc flower develops into an achene, the same type of fruit as that of the safflower.

The sunflower has a peculiar response to light. In the morning, the head is facing eastward, toward the rising sun. During the day,

Figure 27-4
Flax plants showing (left and right) tall growth and no branching and (center) shorter growth with branching. The shorter, branching habit is favored for oil production. [USDA photograph.]

Figure 27-5
Typical sunflower head showing ray flowers and disc flowers.

the head (or the stem tissue supporting it) turns so that the head follows the path of the sun. At sunset, the head is facing west. During the night, the head rotates toward the east, and the cycle is repeated. The heliotropic response of the sunflower is different from the usual phototropic reaction of plants in that sunflowers anticipate the appearance of the sun. Sunflower leaves display the same type of heliotropic response, and, in fact, they may trigger it: if the leaves are removed, the head no longer follows the sun. The plant ceases its heliotropic behavior at, or shortly after, anthesis. As a result, as much as 90 percent of the heads are facing east at maturity. This phenomenon simplifies harvesting. By planting sunflowers in north-south rows, the vast majority of the mature heads, which end up facing east, may be handpicked from trucks moving along one side of the row.

Environmental Requirements

Safflower Safflower is considerd a cool-season crop; minimum temperature for germination is near 40°F (5°C) and optimum temperature is 60°F (about 16°C). Seedlings will tolerate subfreezing temperatures ranging between 10° and 20°F ($-7°$ to $-17°C$) without irreparable damage, but older plants are less cold tolerant. (At, or after, the bud stage, temperatures below freezing damage plants and reduce yields.) Spring-planted safflower requires a frost-free period of about 120 days; safflower planted in the fall or winter requires a frost-free period of 200 days—the same period required by cotton.

A minimum of 16 to 18 acre-inches (41–46 cm/ha) of moisture is required for safflower production. In arid regions, the water requirement may be as high as 44 acre-inches (112 cm/ha). Both excessive moisture and high humidity are very damaging in that they favor the development of diseases in safflower. Thus, proper irrigation management is critical.

Safflower has very deep roots and can utilize moisture deep in the soil profile. In the major safflower-producing areas, most growth occurs after the end of the rainy season; therefore, irrigation and proper management of moisture stored in the root zone are very important.

Irrigation must be regulated to avoid standing water on the crop; to prevent standing water around the crowns, safflower is planted on elevated beds. On heavy soils, which retain large quantities of water, safflower may receive only a preirrigation; in some areas the crop is artificially subirrigated. In order to rigidly regulate the amount of water, sprinkler irrigation is sometimes used.

Safflower is adapted to deep, fertile soils with a high water-holding capacity and a neutral pH. Good drainage is essential. (On poor soils, barley may be a more productive crop than saflower.)

Rapeseed Rapeseed is a cool-season crop; it is well adapted to southern Manitoba and the northern districts of the prairie provinces of Canada. As is true of wheat, there are both winter and spring types of rapeseed (in Canada the spring types are the most commonly planted). Rapeseed may be pelleted and sown in the fall. Pelleting the seed protects it during the winter months. (The pelleting material prevents water uptake by the seeds until freezing and thawing cause it to decompose.) The pelleted seed germinates with the onset of spring (frequently before normal spring planting), and fall-sown crops are in full flower by the time spring-sown crops are only 6 to 9 inches (15–23 cm) tall; thus, sowing with pelleted seed allows maximum use of a short growing season.

Rapeseed is not as drought tolerant as wheat; successful production requires an annual precipitation of 15 to 18 inches (38–46 cm) or a crop-fallow rotation. The crop also does not tolerate flooding well; excessive water contributes to some disease problems.

Rapeseed can be produced in regions with a minimal growing season. Some rapeseed cultivars mature in as little as 90 days; thus, rapeseed is a safer crop to grow in areas with a short frost-free period than safflower, which requires about 120 days to mature. The crop tolerates frosts in the seedling stage better than in later stages of development. In general, cultivars of turnip rape (*B. campestris*) tolerate both frosts and drought conditions better than those of Argentine rape (*B. napus*), but turnip rape gives lower yields. Turnip rape is grown in southern Manitoba, where the growing season is marginal; Argentine types are favored instead in the parkland area of Saskatchewan. Most cultivars of rapeseed tolerate moderately warm temperatures (80°–90°F, or 27°–33°C) during and after flowering. Rapeseed is long-day in photoperiodic response.

Rapeseed tolerates a wide range of soil conditions. Successful crops have been produced on acid soils with a pH as low as 3.8 and on slightly alkaline soils with a pH of 7.8. Plants tolerate moderate salinity reasonably well. In general, rapeseed is more productive on the poorer soils than are cereals.

Flax Flax is definitely a cool-season crop. Seedlings will stand temperatures as low as 28°F (−4°C); after the plants are established, but before fruits are formed, flax may tolerate temperatures as low as 15°F (−9°C) without serious damage. After flowering, below-freezing temperatures may reduce yield and oil quality, and high temperatures (above 90°F, or 32°C) lower yields and the oil content of the seeds. Moderate temperatures (70°–80°F, or 21°–27°C) are ideal. Cultivars grown for fiber require cool, moist conditions during the growing season, but, after flowering, dry weather is essential to cure the plants.

From 18 to 30 inches (46–76 cm) of precipitation is required for successful production. Because plants have a shallow, sparse root system, distribution of precipitation throughout the growing season is critical. (Under drought conditions, a deeper root system may develop.) Flax yields are generally higher on loam, silt-loam, and clay-loam soils than on sandy soils; the heavier soils have a higher moisture-holding capacity, which favors flax plant growth.

Flax, like other cool-season crops, is long-day in photoperiodic response. The plants flower indeterminately until stopped by fall frosts.

The major flax-producing regions are along the western borders of Minnesota and Wisconsin, in the eastern parts of North and South Dakota, and in the adjacent regions of Canada. Some flax is also produced in Texas and California.

Sunflower Sunflower is generally considered a warm-season crop. Seedlings tolerate frosts moderately well until they reach the four- to six-leaf stage of development; from then, through the remaining period of vegetative growth and during flowering, the plants are sensitive to low temperatures. However, they once again tolerate frosts without significant damage during the seed-ripening period. In general, sunflowers require a minimum frost-free period of about 120 days to mature (similar to that required by safflower), although cultivars differ with respect to the minimum frost-free period in which they will mature.

Sunflower is grown on a wide range of soils and tolerates a moderate pH range and some salinity. The crop is grown under irrigation and is adapted to dryland production in regions that have more than 20 inches (51 cm) of annual precipitation. Sunflower is day-neutral in photoperiodic response.

Sunflowers vary markedly in yield with the region in which they are grown. Seed yields of 1,800 pounds per acre (2,000 kg/ha) are not uncommon in California; yields of 1,000 pounds per acre (1,100 kg/ha) or less are typical of the corn-belt states, such as Missouri and Iowa.

Sunflower has been grown as a silage crop in northern areas or at high elevations because the seedlings are more frost tolerant than corn seedlings.

Sunflower is adapted to the northern two-thirds of the United States, and production extends into Canada's prairie provinces. It is also grown in Texas and California on a commercial scale. (Wild sunflower is common in the Great Plains and is the state flower of Kansas.) The USSR is the world's leading producer of sunflower seed, followed by Rumania, Bulgaria, and Hungary. Sunflower is also grown as a crop in South America and in Africa.

Production Practices

SEEDBED PREPARATION AND SEEDING

Safflower Initial seedbed preparation for safflower is similar to that described for wheat and barley (see Chapters 16 and 17). Fall plowing is recommended. Fields must be leveled, prepared for preirrigation, and then harrowed to avoid a cloddy seedbed. The subsoil should be firm, but good drainage must be insured. Following preirrigation, or after the spring rains, cultivations should be planned in such a way as to minimize water loss.

In nonirrigated crop production, row (or

Table 27-1
Effects of planting date on safflower water use and seed yield in a crop grown near Lancaster, California.

Planting date	Total water use (inches)	Seed yield (lb/acre)
1/16	36.0	1,885
2/13	35.5	1,636
3/27	31.6	1,346

Source: R. E. Luebs, D. M. Yermanos, A. E. Laage, and W. D. Burge. Effect of planting date on seed yield, oil content, water requirement of safflower. *Agron J.* 57(1965):162–164.

bed) spacings commonly range from 18 to 24 inches (46–61 cm); rows are rarely spaced more than 30 inches (76 cm) apart. Under irrigation, row spacings are closer—from 6 to 12 inches (15–30 cm). The best seeding rate is about eight seeds per foot (30 cm) of row. Seeding rates range from about 15 to 20 pounds of seed per acre (22–34 kg/ha) on nonirrigated lands, but may be as high as 30 pounds per acre (34 kg/ha) on irrigated lands.

Most safflower is planted with a grain drill; drill-box openings are plugged to obtain the desired row spacings. Planting depth varies between 1 and 2 inches (2.5–5.0 cm), and the seed should be placed into moist soil. Seeding for spring-planted crops begins in mid-February (or earlier) and continues well into March, but water-use efficiency decreases with later plantings (Table 27-1). Plantings can be made even into June; but the plants mature later, are smaller, and yields are reduced. Relatively little safflower is planted in the fall.

Safflower responds favorably to fertilizers (Table 27-2). For dryland culture, applications ranging from 20 to 50 pounds of nitrogen per acre (22–56 kg/ha) are common. If the crop is irrigated, as much as 150 pounds of nitrogen per acre (168 kg/ha) may be applied. Rates

Table 27-2
Effects of various fertilizer combinations and row spacings on safflower yields in North Dakota averaged over three years (1962–1964).

Fertilizer applied (kg/ha)			Yield (kg/acre) by row spacing (cm)			Average yield for each fertilizer
N	P	K	15	53	91	
0	0	0 (check)	1,437	1,335	1,168	1,313
0	7.3	0	1,465	1,466	1,279	1,403
0	14.6	0	1,561	1,494	1,312	1,456
0	21.8	0	1,559	1,518	1,306	1,461
11.2	14.6	0	1,634	1,506	1,319	1,486
33.6	14.6	0	1,455	1,558	1,328	1,414
11.2	14.6	14.0	1,519	1,562	1,416	1,499
Average yield for each spacing			1,519	1,477	1,304	

Source: B. K. Hoag, J. C. Zubriski, and G. N. Guszler. Effect of fertilizer treatment and row spacing on yield, quality, and physiological response of safflower. *Agron. J.* 60(1968):198–200.
Note: All fertilizer combinations exceeded the check and higher yields were obtained with narrower row spacings; the difference between 15- and 53-cm spacings is not statistically significant.

of 30 to 60 pounds (34–67 kg/ha) of P_2O_5 are also common. The application of other elements varies too widely to permit any specification here. As is true for all crops, applications of fertilizers should be governed by the results of soil tests.

Rapeseed Seedbed preparation and seeding practices for rapeseed production are quite similar to those described for spring wheat (see Chapter 16). Initial seedbed preparation consists of a fall or early-spring plowing to incorporate crop residue; final seedbed preparation consists of tillage for control of early-growing weeds and the establishment of a firm seedbed. Moisture conservation is essential in all stages of seedbed preparation. If the crop is grown under irrigation, a preirrigation and a single, late-spring irrigation are usually sufficient.

Seeding is done in mid-to-late May, frequently after spring wheat has been planted. Although row spacings may vary from 9 to 14 inches (23–36 cm), spacings between 12 and 14 inches (30–36 cm) are the most common. Planting is done with a standard grain drill; the use of press wheels on the drill is strongly recommended to insure a firm seedbed and maximum contact between seed and soil. To foster rapid, uniform germination, seed should be planted into moist soil. Seeding depth varies from less than 1.0 inch to 1.5 inches (2.5–4.0 cm). Seeding rate varies with seed weight; rates are higher with larger-seeded, spring types and lower with the smaller-seeded, winter types. Seeding rates for Argentine rape range from 5 to 7 pounds per acre (6–8 kg/ha) and, for turnip rape, from 4 to 6 pounds per acre (5–7 kg/ha). In areas with plentiful moisture, large-seeded types may be planted at a rate of as much as 13 pounds per acre (15 kg/ha). Fall-planted (not pelleted) rapeseed is sown before winter wheat, about mid-August. Plants must be from 3 to 6 inches (8–15 cm) tall to survive the winter.

Rapeseed requires more nitrogen and phosphorus than either wheat or flax. The plant makes the greatest demand on minerals before flowering; thus, late-season side-dressings of essential minerals are of questionable value. Yields of 3,600 pounds of seed per acre (4,032 kg/ha) remove about 120 pounds (134 kg/ha) of nitrogen, 50 pounds (56 kg/ha) of phosphorus (P_2O_5), and more than 60 pounds (67 kg/ha) of potassium (K_2O). Yields frequently increase with the application of nitrogen fertilizer, as well as with the application of both nitrogen and phosphorus (Table 27-3). Compared with many other crops, rapeseed plants apparently have a high capability of extracting phosphorus from the soil; therefore, in many soils, applications of phosphorus fertilizers alone have not markedly increased yields.

Seeds of some cultivars of rapeseed are sensitive to phosphorus fertilizers; phosphorus fertilizers should, therefore, be banded to the side of the seed, not placed with the seed. Nitrogen should also be banded to the side of the seed, unless application rates are very low—that is, less than 10 pounds per acre (11 kg/ha). The same practice should be employed with all potash fertilizers.

Specific fertilizer recommendations depend on the results of soil tests. Application of nitrogen fertilizer is recommended for soils with less than 85 pounds of nitrate nitrogen available per acre (95 kg/ha). Potassium (P_2O_5) may be required on sandy soils, frequently at rates of 30 to 60 pounds per acre (34–67 kg/ha). Sulfur may also be required.

Flax Seedbed preparation for flax is quite similar to that described for wheat and barley (see Chapters 16 and 17); a firm seedbed is desirable. Extra care must be exercised in weed control because, in general, flax does not compete well with weeds. (The root system of flax is sparce, which makes the plant a poor competitor for water and minerals; the shoots are fine and thus do not shade weeds.)

Seeding is done with a standard grain drill. Because seeds are small, planting depths are usually 1 inch (2.5 cm) or less. Seed should be treated with an approved fungicide for protection against seed and seedling rots. In areas of relatively high precipitation, or where the crop is irrigated, seeding rates range from 40 to 60 pounds per acre (45–67 kg/ha). In most semiarid regions, seeding rates are from 25 to 30 pounds per acre (28–34 kg/ha). Row spacings are the same as those for wheat and barley. Narrower row-spacings do not increase yields markedly providing that good weed control is practiced both in seedbed preparation and by the use of herbicides. However, in cases in which good weed control is not practiced, decreasing row spacing from 12 to 3 or 4 inches (30 to 8 or 10 cm) has increased yields. (The narrower row spacings afford the flax plants a competitive advantage, but they are not recommended in lieu of effective weed control.) In Canada and in the northern parts of the United States, flax is planted in mid-May, after spring wheat is seeded. In California and Texas, where the

Table 27-3
Responses of turnip rape (grown on summer fallow at Scott and Saskatoon, Saskatchewan) to different seeding rates and to the application of 40 pounds per acre of 11-48-0 fertilizer.

Seeding rate (lb/acre)	Seed yield (lb/acre)	
	Fertilized	Unfertilized
2	1,145	941
4	1,130	980
6	1,099	948
8	1,134	991
10	1,139	929
Average	1,129	958

Source: R. K. Downey, A. J. Klasse, and J. McAnsh. *Rapeseed, Canada's "Cinderella" crop*, 3d ed. (Rapeseed Association of Canada, 1974).
Note: Seeding rates did not affect yield nearly as much as fertilizer.

Table 27-4
Performance of five sunflower cultivars in western Canada.

Cultivar	Days to blooming	Days to maturing	Seed yield (lb/acre)	Oil (%)
Krasnodarets	78	121	1,150	43.3
Peredovik	85	126	1,240	43.9
Sputnik	86	128	1,270	47.5
Commander	83	124	1,180	29.3
Valley	86	127	1,490	37.8

Source: R. K. Downey, A. J. Klassen, and J. McAnsh, *Rapeseed, Canada's "Cinderella" crop.* 3d ed. (Rapeseed Association of Canada, 1974).

crop is grown as a winter annual, flax is planted in October or later.

Flax responds to applications both of nitrogen and of nitrogen and phosphorus; on stubble, only nitrogen may be required. (In Manitoba, applications of 40 to 60 pounds per acre (45–67 kg/ha) of nitrogen have increased yields.) Sulfur may be required on some of the gray forest soils; potassium may be required on some sandy soils. Flax seedlings are sensitive to fertilizer salts; therefore, fertilizers should not be drilled with the seed. Broadcasting of fertilizer has proved to be effective.

Sunflower Seedbed preparation for sunflower production is quite similar to that described for corn (see Chapter 14). Tilling the soil to incorporate crop residue is done in the fall or early spring; the final seedbed is prepared early in the spring. Cultural methods should insure the conservation of soil moisture. Sunflower is planted with a corn (row-crop) planter equipped with special sunflower plates. Rows are commonly 3 feet (1 m) apart; plants are spaced 18 inches (46 cm) apart. Seeding rates that yield 70,000 plants per acre (173,000 plants/ha) are usually recommended. As row spacing is decreased, yield per acre increases, but both yield per plant and seed size decrease.

Time of planting is a major factor in determining sunflower yields. Early plantings (late April in Minnesota, and March in Georgia) result in the highest yields. In Canada, the highest yields are obtained from pelleted seeds sown in the fall. Because sunflowers require such a short growing season (100 days or less in some areas), the crop is frequently planted later, but then yields are reduced. Because the moisture and temperature conditions early in the season favor sunflower growth, crops that mature in midsummer give the highest yields—larger seeds and higher oil content (Table 27-4).

HARVESTING AND ROTATIONS

Safflower Depending on the region and the planting date, safflower harvest may start as early as mid-August and continue into November. Most safflower crops are combined directly with a grain combine; the modifications are minor, and usually consist of adjusting concave clearance to avoid damaging the seeds and reducing cylinder speed. Yields of as much as 4,000 pounds of seed per acre (4,500 kg/ha) are not uncommon for crops grown in optimum environmental conditions.

Safflower fits well into rotations with cereals and with a number of row crops, such as sugar beets and tomatoes, but row crops gen-

Figure 27-6
A shock of mature flax plants ready to be threshed. [USDA photograph.]

erally occupy the priority position in the rotation because safflower crops have a relatively lower value. Safflower fits well into rotations with rice because the soil on which rice is produced generally has a high moisture-holding capacity that minimizes the need for irrigation. (However, subsoil hard pan, which is often found in rice-producing areas, offers poor drainage, which may lead to problems in safflower production.)

Rapeseed Rapeseed is harvested in late August and September. Because of the high oil content, rapeseed, like seeds of other oil crops, must be quite dry to minimize spoilage, and harvesting of rapeseed starts when the moisture content of the seed is 8 percent or less. Usually rapeseed is combined directly. Shattering of the seeds can be a serious problem; to reduce shattering, a ground-driven, not power-driven, reel is used. Combine cylinder speed must be lower than that used to harvest wheat and barley, and fans and sieves must be adjusted to reduce tailing losses. If large quantities of green, weedy material are picked up with the crop, the seed should be cleaned before storing to keep moisture levels acceptably low.

Flax Flax is ready to harvest when 75 percent or more of the capsules are dry (Figure 27-6, on the preceding page). Because of the plant's indeterminate flowering habit, it is difficult to establish the optimum harvesting time. Early harvesting results in many green (immature) fruits; delayed harvesting may result in losses from shattering and, consequently, reduced yields. Flax can be combined directly with a grain combine, but, if shattering is a serious problem, the crop is windrowed and then threshed with a pickup combine. Because flax seed is smaller and lighter than small-seeded grains, combines must be adjusted to minimize tailing losses.

Sunflower Sunflower is ready to harvest when the backs of the heads are yellow or yellow brown. Dwarf types are combined directly; the heads of standard types are frequently cut by hand and then threshed. Because sunflowers are not a major crop, production is not as fully mechanized as it is with many other crops.

Diseases and Pests

DISEASES

Safflower Phytophthora root rot, a fungal disease, may damage the crowns and roots of safflower plants, in which case the plants are either killed or weakened, and crop yields are reduced. Losses from this disease are minimized by rigidly controlling irrigation and by avoiding standing water on the crop.

Verticillium wilt, a fungal disease, affects safflower in addition to cotton; it causes the wilting of plants, which leads to reduced yields. The pathogen is carried by the seed; thus, the use of disease-free seed is strongly recommended as a method of control. Rotation away from susceptible crops is also effective in verticillium-wilt control. Excessive nitrogen in the soil leads to lush, rank vegetative growth—a condition that favors the development of verticillium wilt; thus, careful regulation of nitrogen application is helpful in controlling this disease.

Safflower is susceptible to a number of other diseases, including rusts and fungal head rots, bacterial blights, and, possibly, viral diseases (viral diseases have not been well studied in safflower). Like other crops, safflower is subject to physiological diseases and abnormalities, including deficiencies and toxemias.

Rapeseed Rapeseed plants are susceptible to a number of fungal diseases. Many of the pathogens also attack wild relatives (e.g., mustard), making disease control through crop rotation and cultural practices difficult and complex. Basal stem rot (late root rot) is caused by several different fungi (one of the pathogens is seed-borne). The typical symptom is a clearly defined, brown oval lession—usually less than 1 inch (2.5 cm) long—at the base of the stem. The disease is also found on cultivated mustard, safflower, flax, sunflower, and several weeds.

Stag head, or white rust, can be a serious fungal disease in rape crops. Symptoms include creamy-white pustules that appear on the undersides of leaves in late spring or early summer. As the plant matures, the pathogen causes individual flowers, or entire stems, to develop into swollen brown spiny structures. No seed is produced by the infected parts. Instead, overwintering fungal spores are produced. Spores may overwinter on seed or in the soil and in crop residue. The best control is through the use of disease-resistant cultivars: all common cultivars of Argentine rape (*B. napus*) are resistant to staghead; cultivars of turnip rape (*B. campestris*) are susceptible.

Sclerotinia stem rot, a fungal disease that affects rapeseed as well as forage legumes (alfalfa and clovers) and mustard, causes bleaching of parts of stems, or even of the entire plant. Severely infected plants mature early; their stems become filled with overwintering spores (sclerotia) and tend to shred easily. The disease is controlled by rotating to cereals (including fallow) and cleaning seeds to remove sclerotia. Control of alternate hosts also reduces disease losses.

Alternaria black spot, also caused by a fungus, is recognized by spots of varying size on leaves, stems, and pods; leaf lesions are usually brown or gray; those on stems and pods are black. The pathogen is seed-borne, and control is mainly through the use of disease-free seed and seed treatment. (Seed produced in drier areas is usually relatively free of the pathogen.)

Ring spot and blackleg are two more fungal diseases that cause significant losses in rapeseed. Rapeseed is not susceptible to any serious bacterial diseases. However, the crop is susceptible to aster yellows, a viral disease transmitted by the gray leafhoppers. Infected plants develop round seedless bladders rather than seed-filled pods. Control includes early seeding and control of alternate hosts for the virus. Separating the crop from flax, which is highly susceptible to aster yellows, is strongly recommended.

Flax Flax may be infected with aster yellows, a viral disease transmitted by the gray leafhopper. Bacterial diseases are relatively insignificant in flax, but the crop is subject to a variety of fungal diseases. Two major fungal diseases are rust, caused by *Melampsora lini*, and fusarium wilt, caused by *Fusarium lini*. Both of these diseases are best controlled by the use of disease-resistant cultivars. Boron deficiency may increase susceptibility to fusarium wilt; adequate available calcium tends to minimize the disease. In addition, flax is subject to several seed and seedling decay, or rot, diseases. Pathogens are carried on the seed or they overwinter on flax stubble, and seed treatment with an approved fungicide is highly recommended. Flax is also susceptible to heat canker; diseased seedlings are girdled at ground level, and they collapse and die. Plants that are only partly infected may continue to grow, but their stems may break, causing the plants to die. Cultural practices designed to reduce temperatures at the crown of the plant are useful in controlling heat canker; early dense plantings and planting flax in north-south rows, to foster shading, are recommended. In general, disease losses are reduced if flax does not follow itself immediately in a rotation, and if stubble in adjacent fields is plowed under before flax seedlings emerge.

Sunflowers Four major fungal diseases cause losses in sunflowers. Sunflower rust (caused by *Puccinia helianthi*) infects plants in both Canada and the United States. The disease can be controlled by dusting the plants with sulfur and through crop rotation. A stem-rot disease (or wilt) caused by a fungus (*Sclerotinia sclerotiorum*) occurs in Minnesota. Verticillium wilt and both powdery and downy mildew also cause sunflower crop losses.

Genetic resistance to some diseases has been identified in European cultivars and in wild sunflowers; however, it has not yet been fully developed in commercial cultivars.

WEEDS

Safflower Weeds, which are always a problem in safflower production, are particularly damaging pests during the early stages of crop growth. Safflower is a broad-leafed species and is susceptible to herbicide damage;

hence, weed control is critical in seedbed preparation, and most chemical weed control is in the form of preplanting or preemergence herbicides. In some regions where the crop is planted with wide row spacings, cultivation for weed control is also practiced.

In California, the types of weed pests that attack safflower are typical of those that infest barley—watergrass, lambsquarters, pigweed, and wild oats, to name but a few. These four weeds are serious pests wherever safflower is grown; other weeds of importance vary from region to region.

Rapeseed Weed control—an important part of rapeseed production— starts with seedbed preparation and the use of clean, weed-free seed and continues with the use of herbicides. Major weeds include barnyardgrass, green foxtail, wild oats, wild buckwheat, lambsquarters, and redroot and prostrate pigweeds—all of which can be controlled, in part, with preplanting herbicides.

The presence of seeds from weeds such as wild mustard and cleavers may affect the value of the crop; mustard seed in the rapeseed lowers the grade of the crop, and the presence of cleavers seed may make the rapeseed crop valueless. The seed of cleavers is quite similar to rapeseed and is therefore difficult to remove in cleaning. Rapeseed may be rejected if it contains cleavers seed.

Flax As stated earlier, the growth habit of flax makes it a poor competitor with weeds, and many weeds may become serious pests. Extra care for weed control is essential in seedbed preparation, and the use of clean, weed-free seed is highly recommended. Although narrow row spacings and heavy seeding rates may help limit weed growth, the use of standard seeding rates, coupled with chemical weed control, is more effective.

Postemergence sprays should be applied after plants are 2 inches (5 cm) tall and until just before buds are formed. Early weed control is the most effective for reducing losses due to weeds; moreover, young flax plants are less susceptible to permanent, yield-reducing damage from herbicides than the older plants, but they still require careful regulation of amounts of herbicides and volumes of water applied to minimize crop damage. Environmental conditions play a significant role in determining the effectiveness of herbicides and the damage they do to the plants. For example, high light intensities and lower temperatures, 65°F (8°C) rather than, say, 85°F (30°C), minimize damage when 2,4-D is applied. Among the weedy pests that infest flax are a number of mustards (boll, haresear, Indian, tumble, wild, and tall wormseed), stinkweeds, lambsquarters, cocklebur, flaxweed, kochia, pigweed, ragweeds, shepherdspurse, thistles, buckwheats, green foxtail, and the seemingly ever-present wild oat. Heavy weed infestations may reduce flaxseed yields by as much as 70 percent.

INSECTS

Safflower More than fifty different insects are reported to attack safflower; however, insects do not cause significant safflower crop losses, and little, if any, insect control is practiced.

Rapeseed Rapeseed is attacked by six major insect pests, all of which can be managed with appropriate insecticides. Before control measures are initiated, the pest must be identified and the potential damage estimated; sprays should be used only if significant losses seem likely, and, of course, all precautions given on the labels of pesticides should be taken.

Cutworms, a larval form of butterflies,

cut the stems of seedlings at the soil surface. Flea beetles, which migrate to the crop from stubble fields of rapeseed or mustard, pit seedling leaves, scallop leaf margins, or devour entire leaves. Similar damage to leaves is caused by the red turnip beetle. Diamondback moths attack young and maturing pods and may also eat leaves if the infestation is severe. The bertha armyworm eats or cuts off the leaves, flowers, and fruits; the damage is usually seen in late June and in July and August. The beet webworm also eats leaves and flowers; it may also eat stem tissue. This insect feeds on weeds such as lambsquarters and Russian thistle, and it sometimes migrates to rapeseed fields from weeds.

Flax Cutworms, beet worms, and bertha armyworms—all pests in rapeseed crops—also attack flax; the damage they cause in flax is similar to that described for rapeseed, and control practices are essentially the same. Flax is also attacked by the flax bollworm and by the potato aphid. The bollworm infects immature bolls (fruits), and the larvae eat the developing seed. The potato aphid punctures the leaves by feeding and covers the plants with honeydew, a sticky exudate generally seen in July and August.

Sunflower Sunflowers are attacked by several insects including wireworms, cutworms, thrips, and aphids—all of them able to cause significant crop losses. At times, grasshoppers also damage the crop and, in Canada, the thistle caterpillar may cause crop damage. The sunflower headbeetle can be a problem, but cultivars resistant to this insect have been developed. Generally, insects that pose a serious potential threat to sunflowers can be controlled effectively with insecticides. However, because the insect pests of sunflower also attack a wide spectrum of other crops and weeds, the destruction of eggs and overwintering insects is difficult and often ineffective.

Crop Utilization

The seed of all four crops is processed to extract the oil. The various extraction methods employed can be broadly classified either as pressure methods (using a mechanical press) or as chemical extraction methods (using solvents). The quantity of oil extracted depends on the method employed; although pressure methods are still employed, in the new processing facilities there is a general trend toward using solvent methods.

The drying capacity of an oil may determine how the oil can be used; it results from the presence of unsaturated bonds in oil molecules that can cause the oil to absorb oxygen. The drying capacity, or the degree of saturation of the oil, is measured by determining the number of grams of iodine that can be absorbed by 100 grams of oil. The number of grams of iodine absorbed is the iodine number; drying oils have an iodine number of 177 or more; edible oils have low iodine numbers.

Safflower Safflower is grown nearly exclusively for its oil; the "seed" (which is actually the achene) is usually from 24 to 36 percent oil. Safflower oil is a semidrying type; its iodine number ranges from 140 to 150. The oil is used as a cooking oil and in the manufacture of mayonnaise and margarine. Although the issue is still being debated, there is some medical evidence to suggest that foods containing oils with high levels of unsaturated fatty acids are more healthful than foods containing saturated fatty acids; consequently safflower oil has become increasingly popular in recent years.

Safflower oil is also used in soaps and in the manufacture of paints and varnishes. The meal that remains after the oil has been extracted is high in protein and is a valuable livestock feed supplement. The whole "seed" (actually the fruit, or achene) contains from 18 to 24 percent protein; the true seed (called the "meat") contains from 28 to 50 percent protein.

Rapeseed Rapeseed is grown for its edible oil, commonly called colza oil. Colza oil contains erucic acid, a highly undesirable, and possibly dangerous, compound. The tendency to produce erucic acid is inherited, and rapeseed cultivars that are free of this trait have been developed. Use of these cultivars is strongly recommended, unless the processor buying the crop specifies that the presence of erucic acid is no handicap.

Rapeseed yields as much as 2,500 pounds of seed per acre (2,800 kg/ha). Oil content varies from 30 to nearly 50 percent, depending on the cultivar and on the environmental conditions under which the crop is grown. At maturity, the majority of the oil comes from the embryo (the testa contributes only about 6 percent of the total oil, and the thin endosperm contributes virtually no oil).

In addition to its use in foods, rapeseed oil is used in soaps and as a high-temperature lubricant for jet engines. The meal that remains after the oil has been extracted contains as much as 37 percent protein; between 60 and 70 percent of the protein is digestible, and it is suitable for human consumption, as well as being a good livestock feed supplement.

Flax Flax is the source of linseed oil, a drying oil that is the base of many paints. Although the majority of linseed oil goes to the paint industry, some is used in plastics. Flaxseed is as high as 45 percent oil; a bushel of flax (56 lb, or 25 kg) will yield as much as 19 pounds (9 kg) of oil. The cake, or meal, remaining after the oil has been extracted is a valuable livestock feed supplement; it is as much as 35 percent protein, 85 percent of which is digestible.

Flax is no longer grown commercially for fiber in the United States. However, flax straw or stubble is used by the paper industry in the production of cigarette paper.

Sunflower Sunflower oil is also edible, and it is used much like other edible oils. In addition, sunflower seed is sold as a confection and is included in many commercial bird feeds. In areas where corn production may be marginal because of cool temperatures, sunflower is also grown as a silage crop.

The potential of sunflower as an oil crop has not been fully developed. Because the plant matures in a short time, it is quite well adapted to many northern regions, where it is not now a significant crop. Eventually, sunflower may become increasingly important in human diets. In most areas, seed yields are from 900 to 1,000 pounds per acre (1,008–1,120 kg/ha); in California, yields may be as high as 1,800 pounds per acre (2,016 kg/ha).

Other Oil Crops

The seeds of many major crops, such as corn, cotton, peanuts, and soybeans, are processed for oil—sometimes as a by-product. Other crops grown for oil either are of minor significance or require production practices that differ so markedly from place to place that it is not possible to discuss them in any detail here. Such crops include various types of mustards, castor beans, crambe, and sesame.

In the past, we discovered that many weeds were, in fact, the source of useful oils; in the future, we may discover many more valuable oils in the plants we now dismiss as mere weeds.

Selected References

Appelqvist, L. A., and R. Ohlson, eds. *Rapeseed.* Elsivier.

Downey, R. K., A. J. Klassen, and J. McAnsh. 1974. *Rapeseed, Canada's "Cinderella" crop.* 3d ed. Rapeseed Association of Canada.

Dunham, R. S., R. G. Robinson, and R. N. Andersen. 1968. Crop rotation and associated tillage practices for controlling annual weeds in flax and reducing the weed seed population of the soil. *Minn., Univ., Agr. Exp. Sta. Tech. Bull. 230.*

Eastman, Whitney. 1958. *The history of the linseed oil industry in the United States.* T. S. Denison & Company, Inc.

Knowles, P. F., and M. D. Miller. 1965. Safflower. *Calif., Univ., Agr. Exp. Sta. Ext. Circ. 532.*

Martin, J. H., and W. H. Leonard. 1967. *Principles of field crop production.* 2d ed. Macmillan.

Study Questions

1. Describe the most important environmental factors and production practices for safflower, rapeseed, flax, and sunflower. Describe how environmental and production factors can be managed for each of these crops.
2. List the major uses of oil for each crop discussed in this chapter. Why, in some instances, might the oil of one species be preferred to that of another?
3. Describe the vegetative and floral characteristics for each of the four crops discussed here.

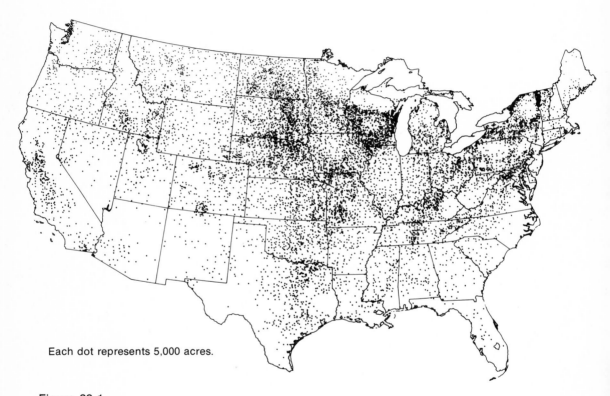

Each dot represents 5,000 acres.

Figure 28-1
Land in the United States from which hay was cut, excluding sorghum hay, in 1969. Forage as hay is an important crop in every state. A similar pattern of land use is found in Canada, especially in the southern provinces. [Data from U.S. Department of Commerce.]

Twenty-Eight   Forages and Man

Introduction

Forages played an important role in mankind's history long before animals were first domesticated; the grazing lands of the world were also the hunting lands of man. Later, when it was discovered that green forage could be stored as cured hay to feed domesticated animals in winter and during droughts and inclement weather, a long step was taken in the direction of developing a stabilized agriculture, and it is a stabilized agriculture that has formed the base upon which most civilizations have grown.

Even our civilization today depends heavily on the presence of forages. In the United States, the importance of forages is demonstrated by the broad scope of their culture: more than half of the U.S. land area is under cover of forages—more than one billion acres is used to produce these crops (Figure 28-1). In Canada, between 10 and 15 percent of the occupied, improved farmland and vast areas of unoccupied land are under cover of forages. In the United States, more than one-third of all nutrients consumed by ruminant livestock comes from pastures and rangelands; in Canada, approximately two-thirds does. Much grazing land is unimproved; the application of modern technology and agricultural management techniques could significantly increase the production capability of the world's grazing lands and so increase the world's food-producing capacity.

Forages on croplands and rangelands serve many important functions. Forage plants cover the soil protecting it from the weather, thus decreasing erosion and the consequent loss of nutrients; forages can enrich the soil and thereby increase crop yields on the lands where they are grown; forages provide economical, good-quality feed for livestock in the form of pasture, hay, and silage (Figure 28-2).

Conservationists have long recognized the value of forages in reducing runoff from intensive rains and erosion of the soil by both wind and water. A good forage cover serves as a canopy that reduces wind velocity and protects the soil surface from wind erosion. It intercepts raindrops and gently distributes the water over the soil surface where it will be more likely to soak as far in as the root zone rather than run off as surface water. Forage residue, whether green manure plowed into the soil or the exposed stubble from hay or meadow, tends to have a beneficial effect in controlling runoff and erosion from plowed surfaces. Exposed residue on the

Figure 28-2
In northern areas, hay produced during the growing season is essential for livestock feed during the winter months. Here, livestock producers must either grow or purchase large quantities of hay.

surface of plowed fields allows greater infiltration of rain and less soil is lost as a result of wind and water erosion. Forages are also used in waterways, where they stabilize the soil in the channel bed, protecting it from the erosive action of rapidly moving water.

We should always keep in mind that soil is a basic resource; sound soil management practices on most farms and ranches include the proper use of forages. Forages in a crop rotation can greatly affect the physical condition of the soil (in addition to reducing erosion and promoting water infiltration), because the benefits provided by forage plants do not stop at the soil surface. Sod crops, particularly grass, tend to improve soil tilth; the presence of organic matter in the soil, which comes from forage plants, tends to improve aggregation of soil particles; the network of roots that is formed by sod crops not only stabilizes the soil but tends to improve soil aeration, and forages with taproots may penetrate several feet into the soil, thus opening channels for the deep percolation of water.

Finally, forages provide the major raw materials used in the production of meat, milk, and other animal products (Figure 28-3). Animals are often the "harvesters" of forage, and they are the most important forage processors because it is livestock, not man, that converts this crop into a form that is of value to the producer and of benefit to the consumer.

Effective production of row and cereal crops is essential if we are to have the quantities and varieties of foods and fibers that our civilization demands. One way in which the productivity of our tilled lands can be insured is by maintaining a proper combination of forages and row crops in rotation.

Certainly forage crops are important, and volumes have been written about these crops and their production. Here, we shall confine ourselves to outlining some of the desirable characteristics that one should look for when selecting a forage crop; in following chapters, we shall briefly describe some of the most important forage legumes and grasses and dis-

Figure 28-3
Beef cattle grazing coastal bermudagrass in a highly productive pasture near Tefton, Georgia. [USDA photograph.]

cuss forage management practices as they relate to rangeland and pasture, as well as production of hay and silage.

Forage Characteristics

Many attributes must be considered when evaluating various alternative forage crops. No single attribute can be considered exclusively. It is the combination of all attributes that determines whether a forage crop has the characteristics needed to do a particular job for a particular area.

YIELD

The quantity of forage produced must certainly be one of the leading criteria for determining a crop's desirability. The components of yield are the stems, petioles, leaves, and flowers; the measure of yield is usually expressed as kilograms of dry matter per hectare or as pounds of dry matter per acre by the researcher, whereas, in practice, hay measured in the field is assumed to be at about 12 percent moisture (100 lb, or 45 kg, of hay is assumed to contain about 12 lb, or 5 kg, of water), and silage is generally measured in terms of tons of forage ensiled per acre, or hectare (including the water content). The production of a given forage should be figured for the entire growing season and not for any one cutting; frequently a forage will have a large first cutting and little or no regrowth for subsequent harvests. The distribution of forage throughout the growing season is another, often very important, consideration.

QUALITY

The factors that are considered in determining forage quality generally relate to the nutritive value of the crop. Because a high percentage of the nutritive value of most forages is found in the leaves, leafiness is generally a very desirable characteristic. For example, much of the protein found in legumes—which are generally considered a quality forage, in great part because of their high protein content—comes from their leaves. Palatability of the forage is another factor in determining quality. Livestock find some species much more palatable than others, usually because of respective fiber content, steminess, leafiness, and level of maturity, or age, at harvest. Young shoots are generally more palatable (and nutritious) than mature stalks. Palatability may also be related to chemical composition; the presence of alkaloids or volatile chemical components, for example, often reduces the animal's acceptance of the plant. The chemical composition of the forage plant itself may substantially affect intake, digestibility, and subsequent livestock production, whether milk, meat, or wool production.

There are trade-offs that must be considered in evaluating the characteristics of forage quantity and quality. To get the highest quality, one would select only the new shoots of the forage plant, but this would result in a tremendous sacrifice of yield. By the same token, there is a point at which projected increases in yield would not offset the loss in quality that results from the growing quantity of dry matter characteristic of older plants. The point at which the most favorable quantity:quality ratio is obtained differs from species to species, and there are many other factors that have to be considered in determining the most opportune time to harvest, as we shall see later.

It should be noted that the quality of forage may be affected by the way in which the crop is handled. A hay crop that is initially of excellent quality can become very low in quality if it rains at the time of harvest or if the forage becomes wet and moldy in storage.

DROUGHT TOLERANCE

Except in irrigated areas, most forage-producing areas eventually experience drought conditions to a greater or lesser extent. The ability of forage plants to tolerate drought conditions without serious consequences for quantity and quality of yield varies greatly from species to species. Ladino clover, for example, does not withstand drought conditions for very long; crested wheatgrass, on the other hand, can withstand long periods of drought without being killed. But regardless of the species, drought conditions reduce yields. When selecting a forage for a given area, one should consider the probability of drought in the area and then weigh the characteristics of yield, quality, and drought tolerance (among others) to determine the forage crop that will best meet the needs of that particular area.

WET-SITE TOLERANCE

A very wet site, or a site that is frequently flooded, requires a forage cover that will tolerate "having its feet wet" much of the time. Bogs and swampy areas, sidehill seeps, subirrigated areas, lake and farm pond shores, and other wet sites require particular species of forage plants. A species like reed canarygrass tolerates wet sites so well that it is often considered a weed in these areas, yet it can be a productive forage crop if managed properly. Legumes such as strawberry clover may also produce valuable stands on wet sites. (In general, however, the legumes are not as tolerant of "wet feet" as are the grasses, and they do not do well on peat soils, which may remain frozen until late spring.) Flooding tolerance of several forage species is shown in Table 28-1.

RESISTANCE TO INSECTS AND DISEASES

Many pesticides have been banned and, in many cases, there are no chemicals available that can effectively control insects and diseases. Through breeding, some forage cultivars have been developed that can tolerate certain insects and resist particular diseases. (In some cases, insects and diseases are very species specific.) Through further breeding and selection of resistant and tolerant species and cultivars, losses from insects and diseases can be reduced and in some cases eliminated.

SEASONAL DISTRIBUTION

One desirable characteristic of pasture forage is its ability to sustain production for a long period, preferably throughout the growing season. A number of species have a large burst of growth early in the spring and produce very little in mid- and late summer; other species produce only in mid- and late summer; only a few have a relatively even, sustained growth over the full season. It is possible to work out a pasture calendar that will serve as a management guide in determining which forage species should be planted in order to be available for pasturing at a given time. With proper management of various species or combinations of species, forage can be made available in different pastures throughout the growing season.

STAND ESTABLISHMENT

Forage must be relatively easy to establish if it is to have widespread acceptance among producers. Seedbed preparation and seeding practices, the importance of using quality seed, and other factors affecting stand establishment are discussed in Chapter 12. Birdsfoot trefoil is an excellent forage, but it has not

Table 28-1
Flooding tolerance of several forage species.

Crop	Days of flooding tolerated
Red clover	0–7
Sweetclover	7–14
Alfalfa	7–14
Alsike clover	7–14
Strawberry clover	7–21
Intermediate wheatgrass	21–28
Meadow fescue	21–42
Timothy	21–56
Reed canarygrass	35–56
Slender wheatgrass	35–56
Bromegrass	35–56

Source: From *Crop adaptation and distribution* by C. P. Wilsie. W. H. Freeman and Company. Copyright © 1962.

received widespread acceptance because it is relatively difficult to establish unless some exacting criteria are met, such as the appropriate application of large quantities of *Rhizobium* inoculum, adequate phosphorus, and some starter nitrogen. Alfalfa, on the other hand, is relatively easy to establish, and, because it has many other desirable characteristics, it is widely accepted in many forage-producing areas.

SEED PRODUCTION

The ability of a given forage cultivar or species to produce seed has a great bearing on its desirability. The price of seed is directly related to seed-producing ability—an obvious example of the law of supply and demand. There are some exceptions to this rule, however: for example, when a new cultivar or a new species is released for commercial production, or when growing conditions and other factors have caused the seed from a good seed-producing crop to be temporarily in short supply. An interesting example of a factor that can affect seed prices is supplied by birdsfoot trefoil. Among the many desirable characteristics of this forage plant is the fact that it is an excellent seed producer. However, it is indeterminate in its flowering habit, and it will flower and set seed all summer; also, as the pods mature and dehisce, the seeds may be thrown 3 or 4 feet (about 1 m) from the plant. The uneven ripening of the seed and the shattering characteristic make it difficult to determine the best time to harvest seed. Consequently, it is very difficult to harvest birdsfoot trefoil seed, and the price of the seed goes up accordingly. On the other hand, crested wheatgrass, grown in many of the dryland areas of the western United States, is a profuse and relatively reliable seed producer. The species is broadly established, and the price of seed is relatively low.

COMPATIBILITY WITH OTHER SPECIES
Many management systems require a combination of forage species to best meet particular needs. Some pasture mixtures may have as many as five or six species in one plant community, all of which must be compatible. Grasses and legumes are frequently grown together because of their compatibility and because, when the two species are grown together, they produce higher yields than when either species is grown alone (see Figure 28-4). Some species will not grow well with others because they are too aggressive or, conversely, because they are not competitive enough. Competition for light, water, and nutrients is an important factor governing the compatibility of various species. A tall-growing species may completely dominate a shorter-growing companion species. Differences in basic growth habits of shoots (e.g., vine types versus bush types) and of roots (e.g., taproots versus fibrous roots) affect a plant's ability to compete for water and nutrients and play a significant role in determining the compatibility of species.

WINTERHARDINESS
In northern regions of the United States, as well as in Canada and Alaska, it is essential that perennial forages be relatively winter-hardy. The degree of winterhardiness demanded depends on the location, the length of time that winter conditions persist, the severity of the winter (in terms of wind and of temperature and its variability), and other environmental factors. Most forage legumes that are winter-hardy have a different growth pattern than nonhardy legumes. Environmental conditions characteristic of winter's onset trigger metabolic mechanisms within the plant that change the pattern of growth and prepare the plant for winter. Some cultivars lack these triggering mechanisms and continue their normal growth; plants that continue to grow actively as winter approaches do not survive. (Factors related to winterhardiness are discussed in Chapter 8.) Plant breeders have been able to improve the winterhardiness of a number of forage cultivars that have other desirable agronomic characteristics, and research is continuing to develop even greater resistance or tolerance to winter conditions.

POISON CONTENT
A highly productive, highly nutritious forage is of little value if it contains compounds that are either toxic or lethal to livestock. Poisonous plants, such as lupine and larkspur, that grow among desirable forages constitute a real nuisance. Also, some forage plants are poisonous under certain environmental conditions. For example, certain cultivars of sudangrass are particularly dangerous when young or when they have regrown after a drought or frost, because they may contain toxic levels of prussic acid. Plant breeders have attempted to develop cultivars that are low in prussic acid, but currently the grazing of sudangrass must still be done with great caution.

The accumulation of nitrate-nitrogen in forage can be toxic to livestock. It can reduce the rate of weight gain, cause abortion, and, in extreme cases, cause death. Nitrate poisoning in oats or other cereals being grown for forage is most likely to happen under stress conditions such as drought with high levels of

Figure 28-4
Grass-legume mixtures in test plots at Montana State University near Bozeman, Montana. Various combinations of different grasses and legumes are evaluated to determine compatible species for highest forage yields.

soil nitrogen and low levels of phosphorus and potassium. However, it is not confined to these species.

FLEXIBILITY IN USE

To meet the needs of the producer, a forage should be flexible in use. Clearly, a forage that can be used for pasture, hay, or silage—depending on the specific needs of the producer in a given season—is more valuable than a forage that has only one or two uses. Flexibility gives the forage producer the option of selecting the most advantageous use and helps take some of the risk out of farming or ranching.

Selected References

Canada. Ministry of Industry, Trade and Commerce. 1972. *Canada Yearbook, Statistics.*

Heath, M. E., D. S. Metcalfe, and R. E. Barnes. 1973. *Forages.* 3d ed. Iowa State University Press.

Smith, Dale, 1962. *Forage management in the north.* Wm. C. Brown.

Study Questions

1. How do forages contribute to a stabilized agricultural economy?
2. Why do conservationists place such value on forage crops?
3. What characteristics or combination of characteristics are most important for forage crops in your area?
4. What are the most desirable characteristics of forage crops grown on a dryland western range compared with those needed in a grain-forage rotation in Illinois or Iowa?

Twenty-Nine Forage Legumes

The family Leguminosae comprises nearly 500 genera and some 11,000 species; almost 4,000 species are found in the United States and Canada. Clearly, we will not be able to discuss the nature and uses of each species in detail; of necessity, the discussion in this chapter will treat briefly only some of the more significant forage legumes.

Legumes are generally flexible in use; many can be used as hay, pasture, green manure, nectar, and human food. An anlysis of most forage legumes shows that their crude protein content varies between 12 and 20 percent. This high protein content, along with other desirable constituents, makes legumes very valuable forages. In general, legumes are known as soil-building plants; their capacity for working in association with certain bacteria to take nitrogen from the air and make it available for plant growth makes them unique among the forage species. Legumes and bacteria of the genus *Rhizobium* have a symbiotic relationship. Unless the proper bacteria are present in sufficient numbers, the protein content and quantity of forage yield may be drastically reduced. Also, the general quality of the forage yield is affected because legumes themselves need and use the nitrogen provided by their symbiotic relationship with the bacteria. Thus, it is extremely important that sufficient quantities of the proper inoculant be supplied when legumes are seeded. (See Chapter 12 for a more detailed discussion of legume inoculation.)

Origin and History of Culture

Most of the legumes commonly grown in the United States and Canada have their origin in southwest Asia and in the European Mediterranean area referred to by Vavilov as the Near-Eastern Center and the Mediterranean Center, respectively. Alfalfa, the queen of forages, is believed to have evolved in southwest Asia; wild forms of alfalfa and related species are currently found in central Asia and as far north as Siberia. Reference to alfalfa as a feed for horses and other animals was made in Greece as early as 490 B.C. Later, alfalfa spread into Italy and the rest of Europe, and eventually Spanish explorers took it to Central and South America. Alfalfa from Chile was introduced on the West Coast of the United States in the 1850s, during the California Gold Rush. It became a very popular crop and soon spread east—through Utah, to the midwestern states, and finally to the eastern seaboard. In New Zealand, Australia, and many European countries, alfalfa is frequently called lucerne.

Red clover is believed to have originated in southeastern Europe; references to what is believed to have been red clover were made as early as the third and fourth centuries A.D. Red clover spread throughout Europe into Russia; it was carried to the United States by early European colonists and soon was grown in most of the colonies. Today, red clover is widely distributed throughout the world, and it is an important forage in the United States, Canada, Australia, and New Zealand, as well as in many countries in Europe and Central Asia.

References to the production of alsike clover in Europe were made in the eighteenth and nineteenth centuries, but its origins are obscure.

Sweetclover is believed to have had its origin in temperate Europe and Asia. It was introduced into North America in the early eighteenth century; this widely adapted crop was first produced in Virginia and soon found its way to nearly all parts of the United States and Canada.

White clover probably had its origin in the eastern Mediterranean or in southwestern Asia; various forms of white clover are found throughout the world, and it is one of the most widely distributed legumes. It is believed that British colonists brought white-clover seed with them when they settled on the Atlantic coast. Because of its broad adaptation and the ease with which it spreads, white clover soon covered the whole area colonized by the early settlers and eventually moved across the continent with them. Because it seemed to have been found everywhere the white man settled, the Indians named it "whiteman's foot."

The origin of birdsfoot trefoil is somewhat obscure; there are no reliable references to this species until the mid-seventeenth century. Currently, it is found throughout much of Europe, and it was presumably carried to the United States by the early immigrants.

Most of the economically important vetches, which are similar in habit to the legumes discussed above, are native to Europe and the adjacent Asian countries.

Lespedeza, however, does not fit the pattern of the other legumes mentioned. Korean and Kobe lespedezas, both native to eastern Asia, have been introduced into the United States relatively recently. They are warm-season plants and grow best during the warm summer months. Kudzu, another exception that was introduced from Japan in 1876, was not used very much in the United States until the early 1900s. Soybeans were brought to the United States from China (probably in the late eighteenth or early nineteenth century) as a forage crop. The soybean was later developed into an oil seed crop—a use that far overshadows its function as a forage (see Chapter 20).

The impact of these introduced legumes on the development of agriculture in the United States and Canada must not be underestimated. They have not only increased the production of forage but also improved the productivity of the land.

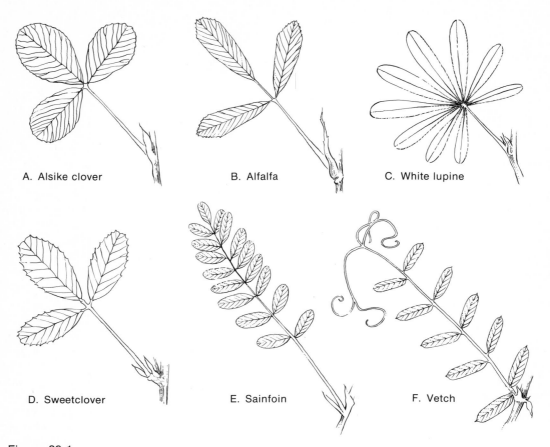

Figure 29-1
Compound leaves of common forage legumes: A, B, and D are trifoliolate; A and C are palmately compound and differ in number of leaflets; B, D, E, and F are pinnately compound with different numbers of leaflets (note the tendrils instead of terminal leaflets in F and the absence of conspicuous stipules). Leaflet margins also may differ (compare A and C with B and D).

Botanical Characteristics

ROOTS

Plants of the Leguminosae family have a number of characteristics in common. Legumes characteristically have a taproot system with lateral roots branching from one main root. The root system is the site of the unique characteristic of legumes discussed earlier—the ability to fix nitrogen. Bacteria of the genus *Rhizobium*, which are symbiants and are responsible for the nitrogen-fixing ability of legumes, live in the roots; they enter through root hairs and spread through the cortical tissues. The plant reacts to this invasion by forming gall-like root nodules. The nitrogen-fixing bacteria are found in these nodules. Nodules are formed on the roots of most species of forage legumes, providing that the right nitrogen-fixing bacteria are present.

The Leguminosae is the only plant family of economic significance that hosts bacteria capable of fixing atmospheric nitrogen, making it available for plant use. Efforts to find bacteria that will infect grass plants in a similar manner have produced some significant research discoveries but, as yet, no developments of economic importance.

STEMS

The stems of legumes vary greatly from species to species in length, size, branching characteristics (which range from simple to profuse) and structure; some stems are woody, others are delicately succulent. Rhizomes (modified underground stems) are not generally found in legumes, except for some strains of alfalfa. Stolons (horizontal stems above the soil surface) are found in some species, such as white clover.

LEAVES

The leaves of legumes vary from simple to compound. The most typical leaves are the trifoliolate and the pinnately compound types (Figure 29-1). The clovers are examples of legumes with trifoliolate compound leaves; vetch and sainfoin are examples of legumes with pinnately compound leaves. The typical compound leaf of legumes is composed of either two or three leaflets, a petiole, two stipules at the base of the petiole, and, in some cases, such as field peas, tendrils. Some legume plants, such as sainfoin and vetches, have many leaflets (from twelve to fifteen are not uncommon), which, in general, are characteristically flat and oblong. Leaflets of legume plants have netted venation.

FLOWERS

Legume flowers typically consist of a calyx (made up of five sepals that are united at the base); a corolla, which consists of five petals (one standard, two wings, and two petals united on one side to form the keel); ten stamens (nine united into a tube and one separate); and a pistil, which is simple and enclosed within the stamen tube.

Major Legumes

Alfalfa For many years, the most popular single forage species in the United States and Canada has been alfalfa (*Medicago sativa*). Because it is a perennial, it is ideal for long rotations. One of the most significant qualities of alfalfa is its yield, which is greater than that of other legumes and of most grasses in a given environment. In most northern areas of the United States and in Canada, where there is adequate rainfall or where the crop is irrigated, from two to three cuttings per year can be expected; total yields range from 3 to 6 tons per acre (6.7–14.1 MT/ha), depending on the soil fertility and weather conditions. In southern areas, when the crop is irrigated, six or seven cuttings per year may be obtained, with yields of as much as 11 tons per acre (24.6 MT/ha).

Alfalfa has a very extensive root system; root penetration is usually from 6 to 8 feet (2–3 m), but under favorable conditions, its taproot has been known to penetrate to depths of 20 to 30 feet (6–9 m). This deep root system and the ability to use water held at great depths in the soil enable alfalfa to tolerate droughty conditions that some other species would not survive.

Alfalfa grows well in association with other forages. The common alfalfa mixtures include smooth bromegrass or orchardgrass and, at times, other legumes, such as red clover and 'Ladino' clover. Proper management is necessary to maintain a desirable balance among the species in a stand. For instance, cutting too frequently may weaken the legumes and allow the grasses to dominate the stand; improper fertilizer application can affect the balance of species (e.g., phosphorus favors legumes; nitrogen favors grasses).

Alfalfa is relatively easy to establish, but it is important that a fine, firm seedbed be prepared, and that seeding depth be carefully controlled. The seeds are relatively small; therefore, they require a shallow seeding depth. Seeding rates vary according to row spacing, but usually from 10 to 12 pounds per acre (11–13 kg/ha) is considered adequate.

In general, alfalfa is a highly nutritious feed, palatable to livestock. It does, however, have one drawback: it frequently causes bloat when pastured by ruminants, such as cattle and sheep. Growing alfalfa with grasses will reduce the incidence of bloat, but careful management is still extremely important when livestock are being grazed on alfalfa (as well as a number of other legumes).

Alfalfa is easy to cure and makes an excellent hay crop. Because most of the nutrients—vitamins, minerals, and protein—are located in the leaves, it is essential that as many of the leaves as possible be retained when alfalfa is harvested as hay. The stage of maturity at which alfalfa is cut greatly affects the quality of the hay produced. To get the highest possible quality, one would cut the new shoots soon after they emerge from the crown; the hay would be extremely nutritious and palatable, but there would be very little of it. On the other hand, to get the greatest yield, one would wait to cut the plants until they were relatively mature (after flowering, but before the leaves have dropped); however, the nutritive value of the forage would be relatively low by then.

Determining the most advantageous time to cut alfalfa involves not only trade-offs between quality and quantity of yield, but also other factors; one of the most important is a variable over which one has little, if any, control—weather. It has been found that cutting at about the one-tenth bloom stage (when one-tenth of all the flowers are open) gives the best combination of palatability, protein content, feeding value, and yield. However, weather conditions at the time that the crop reaches the one-tenth bloom stage may make it necessary to select an earlier or later cutting date. For example, if when the forage has been cut and is lying on the ground or in the windrow it is subjected to a series of heavy rains, it loses its nutritive value and quality very rapidly; many vitamins and minerals may be leached away, and leaves are lost because of shattering and falling.

It has been determined that the time of the last cutting in the fall can affect the ability of the plants to survive the winter. In northern areas, it is generally advisable to make the last cutting at least four weeks before the average date of the first killing frost. This gives the plants time to recover and produce an adequate store of food in the roots, which allows them to overwinter successfully. In southern areas, where winterhardiness is not a relevant characteristic, the date of the last cutting is not a significant factor, and six or more cuttings with high yields may be obtained each year, rather than the two or three cuttings that are common in the north.

The quality of alfalfa hay depends not only on when the hay is cut, but also on how the hay is handled after it has been cut. In general, the shorter the interval between the time the hay is cut and the time it is put in storage, the higher its quality. Exposure to direct sunlight literally bleaches out the vitamins in the hay; other weather conditions may also reduce quality. (For this reason, crimpers are often used to hasten the curing process.) It is advisable not to put the hay into storage if it is too wet; storing hay that is too wet promotes microbial (bacterial and, to some extent, fungal) activity, causing the generation of heat, which can result in spontaneous combustion and subsequent damage from fire. The occurrence of mold is another possible problem that results from storing overly wet hay, but this problem may be prevented in commercial practice by the use of organic acids, which act as preservatives.

Alfalfa is subject to a few diseases and to serious damage from insect pests. Bacterial wilt is one of the most serious alfalfa diseases and, through the years, has infected stands in many areas of the United States and Canada. Cultivars that are susceptible to bacterial wilt are short-lived; affected plants are stunted, and the disease is usually lethal. Wilt-resistant cultivars are available and are effective in preventing losses from bacterial wilt. Most other diseases affecting alfalfa cause damage to the leaves, thus reducing the quality and quantity of the crop. Alfalfa yellows, a viral disease transmitted by the leafhopper *Empoasca fabae,* is a serious problem in the eastern half of the United States. The disease causes a yellowing of foliage and results in a loss of carotene, a loss of leaves, and, consequently, reduced feed value of the crop. As a general rule, however, alfalfa is subject to relatively little damage caused by disease.

In recent years, the alfalfa weevil has caused considerable crop losses in large areas of the United States and Canada. Attempts at chemical control of this insect have had some degree of success, and plant breeders are currently attempting to develop cultivars that are resistant to the alfalfa weevil. The green pea aphid and the spotted alfalfa aphid—most prevalent in the southern part of the United States—can cause a great deal of damage. Lygus bugs are also significant pests in some areas and can cause extensive damage in seed fields. Stem and root-knot nematodes cause damage in some cultivars; others (e.g., 'Lahontan') are resistant to stem nematodes.

A number of cultivars of alfalfa are available, and selection of the best one for any given environment depends on the weather, climate, altitude, latitude, and use to be made of the crop. Cultivars differ significantly in their regions of adaptation, resistance to diseases and insects, growth characteristics, and morphological characteristics.

Alfalfa cultivars are divided into five general groups that will be discussed separately.

Common alfalfas. The alfalfas in this group include winter-hardy and nonwinter-hardy types and both slow- and fast-recovery types, depending on their origin. Winter-hardy types tolerate lower temperatures because they become dormant during the cold season; they generally grow more slowly and recover more slowly after harvesting than the nonhardy types. Nonhardy types do not become dormant. Generally, they have purple flowers; most of them have developed through natural selection. When purchasing common alfalfa cultivars, it is important to know the origin of the seed; strains of common alfalfa produced for many years in the southern part of the United States are not adapted to the northern part of the United States and Canada, and vice versa.

Variegated alfalfas. Cultivars in this group are, in general, very winter-hardy, and, as a result, they are relatively slow in recovery after being cut. Their flowers display a variety of colors including white, yellow, and purple. Most of the recommended cultivars being grown in the central and northern United States and in Canada come from this group. Many of the variegated cultivars, which include 'Ladak,' 'Vernal,' and 'Ranger,' have been developed by plant breeders who have successfully introduced qualities of superior production and winterhardiness.

Turkistan alfalfas. Extreme winterhardiness (and fall dormancy) is characteristic of the Turkistan group. Turkistan cultivars store large amounts of carbohydrates before fall dormancy, which allows these plants to resist

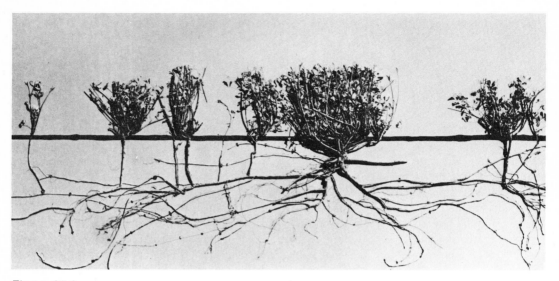

Figure 29-2
Rhizomatous alfalfa, showing five plants that have developed along the lateral root system extending from the larger "parent" plant. [USDA photograph.]

cold injury to vital tissues and supplies energy throughout the winter dormancy period, as well as that required for spring growth until plants are self-sustaining through photosynthesis. Cultivars in this group are slow-growing and recover slowly after harvesting; as a result, they are generally lower in yield than cultivars of the variegated and common groups. Cultivars in this group are relatively insignificant from the standpoint of use, but they are extremely important from the standpoint of their contribution to the development of new cultivars: germ plasm from the Turkistan group has been introduced into some of the variegated cultivars to improve their winterhardiness, wilt resistance, and drought resistance.

Nonhardy alfalfas. Cultivars in the nonhardy group are more upright in growth than the other groups and do not go dormant in the late fall. Nonhardy cultivars have rapid growth and very rapid recovery after cutting, and their nondormant growth habit results in more cuttings and much higher yields. However, because of their growth habit, they do not harden for winter and thus are not able to survive in northern climates. Instead, these cultivars are well adapted to southern areas, where winterhardiness is not a factor; in northern areas, they are used mainly as single-season crops such as green manure.

Rhizomatous alfalfas. Although most alfalfa cultivars are characterized by a heavy, aggressive taproot that penetrates deeply into the soil, there are some strains—the rhizomatous alfalfas—that develop from a crown at the soil surface and spread into the surrounding area (Figure 29-2). In some cases, rhizomatous plants have been known to spread to as much as 15 feet (5 m) in diameter. Rhizomatous types of alfalfa have not gained broad acceptance in the United States; for the most part, they do not spread vigorously and their yields are not superior to those of the normal taprooted cultivars. However, in the dryland areas of western Canada, rhizomatous cultivars are now grown on a significant acreage, and seed production suggests that they are becoming the dominant type of alfalfa. Where frost heaving is a problem, they will probably become important in the development of superior alfalfa cultivars. The spreading habit of rhizomatous plants has many advantages and, if the other characteristics of these strains can be improved through plant breeding, they may become increasingly popular in the future.

Red Clover Red clover (*Trifolium pratense*) is a very important legume hay crop in much of the northern part of the United States and southern Canada; in recent years, however, the acreage of red clover has diminished as other species have become more popular. (Red clover is also found in west-central Asia, northern and western Europe, New Zealand, Australia, as well as in several other areas of the world.)

There are three types of red clover. The first type, early-flowering red clover, is usually referred to as medium red clover and is commonly found in the United States and Canada. Medium red clovers produce two hay crops per year. Although the plants are short-lived perennials in their growth habits, they are usually managed as biennials and are best adapted in rotations that include two years of meadow. Recently developed cultivars of the medium red clover type display a more perennial character.

The second type of red clover is the later-flowering type generally referred to as American mammoth red clover. Plants of this type are usually slower-growing and generally flower two or three weeks later than the medium red clovers. They produce only one crop and a small aftermath growth that may be used for pasture. The third type of red clover, wild red clover, is found in England and parts of Europe, but it is unknown in North America.

Most of the clover grown along the Atlantic coast and in the northwestern United States—mostly under irrigation—is medium red clover. It is used mainly for hay but may be found in pasture mixtures, and occasionally it serves as a green-manure crop. With most cultivars, many of the plants die after the second cutting of the second year. If left another year, winter injury, diseases, and infestation of insects, such as the clover root borer, may reduce the stand to the point that it is uneconomical to continue the crop into the third year.

The usual management practice is to take a hay crop at the end of the first season, although, under the right climatic conditions, a light seed crop may be taken at that time. In the second year, if there is sufficient forage and if a seed crop is desired, the first cutting may be taken for hay, and the second cutting may be allowed to go for seed. Red clover is normally cut for hay at the full-bloom stage—the stage at which the crop has the most desirable combination of yield and quality (Figure 29-3).

In Canada, red clover cultivars are generally designated as single cut (American mammoth type) or double cut (medium type). Recently, these two types have been subdivided into two categories—diploid and tetraploid. The tetraploids are generally the more robust, winter-hardy, and long-lived (4 to 5 years). However, seed production with the tetraploids is limited by the need for large numbers of certain bees of the genus *Bombus* to serve as pollinators.

Because the lush, heavy growth, which is usually characteristic of red clover, may lead to bloat when the forage is pastured, red

Figure 29-3
The upper part of a single red clover plant at the fullbloom stage of development. [USDA photograph.]

clover intended for grazing is usually seeded with grass, which reduces the bloat hazard. Extreme care must be exercised when grazing ruminant animals on this crop; swine can graze with no danger of bloat.

In the fall of the first season, red clover should be grazed with care. In northern areas, it should not be pastured after September 1, to allow the plants to build up root reserves that increase the chances of overwintering without injury.

Red clover is not adapted to strongly acidic soils or poorly drained and light sandy soils; a fertile, well-drained soil with high moisture-holding capacity is ideal. Although red clover is not quite as sensitive to acidity as alfalfa, liming is desirable for obtaining highest yields on acidic soils.

Red clover, as is typical of most legumes, has a taproot. The taproot is heavily branched, and a large part of the plant's root system is concentrated in the top 12 inches (30 cm) of soil. Consequently, red clover does not tolerate drought conditions as well as alfalfa, which has a very deep root system.

The generally accepted seeding rates for red clover are from 8 to 10 pounds per acre (9-11) kg/ha) in a pure stand or from 4 to 6 pounds per acre (5-7 kg/ha) when seeded in a grass mixture. Usually, the stand is clipped in August of the first season to reduce competition from weeds. Regrowth of the plant comes from the crown buds at the ground level; to avoid injury of this vital crown tissue, care must be taken not to clip too closely.

There are several diseases that attack red clover. Crown rot, caused by the fungus *Sclerotinia trifoliorum*, is common in southern areas; root rots, caused by various species of *Fusarium*, attack susceptible cultivars mainly in the north. Northern anthracnose and southern anthracnose, formerly serious diseases, are currently less of a problem because of the development of disease-resistant cultivars.

There are other diseases of local significance in some areas. In general, the most effective control for diseases infecting red clover is by the use of resistant cultivars developed through plant breeding.

Major insect pests of red clover include the chalcid fly (*Bruchophagus gibbus*), the clover root curculio (*Sitona hispidula*), and the clover root borer (*Hylastinus obscurus*). Other pests include species of leafhoppers, aphids, and grasshoppers. The use of insect-resistant cultivars and, to some extent, the application of insecticides help control insect pests.

In summary, the desirable characteristics of red clover include its adaptation to a wide variety of climatic conditions (it is equal or superior to alfalfa in this respect), its quick growth, its ease of establishment, and its reasonably high yield. Red clover is a high-quality forage that can be used as hay, silage, or pasture. It can also be used as a soil-improving crop; when properly inoculated, it serves as an economic way of adding nitrogen to the soil as well as protein to the forage.

The main disadvantages of red clover include the fact that it is a very short-lived plant; it is adapted only for short meadow rotations and does not stay in a pasture mix long enough to have the desirable characteristics of being a permanent pasture forage. Also, it is not as high-yielding as alfalfa, although it is superior to many other forages in this respect. Its very lush growth is most conducive to bloat in ruminants, and, finally, the pubescence on leaves and stems characteristic of red clover tends to produce annoyingly dusty hay.

Alsike clover Alsike clover (*Trifolium hybridum*) is managed in much the same way as red clover. It is a perennial, but it is usually treated as a biennial in forage rotations. Alsike seeds are greenish yellow, and they are much smaller than the seeds of either red clover or alfalfa. It is finer-stemmed and has smaller inflorescences than red clover, but the plant does not have the pubescence that is normally found on red clover. Flowers range in color from pinkish white to pink. In the United States, alsike is grown in New England, in the Great Lakes region, and, under irrigation, in the northwestern and intermountain areas.

Because of its decumbent growth habit (Figure 29-4), alsike clover is not as popular as other legumes. It is usually grown with grasses, frequently timothy, which tend to keep the plants more erect.

Alsike is best adapted to cool climates, and stands may be injured by continuous hot weather. It is also adapted to heavy, wet soils and will live through flood conditions for a brief period. Alsike tolerates strongly acidic soils more than does red clover, and it will survive on sites that are too wet for 'Ladino' (a cultivar of white clover). Because the seeds are much smaller, the recommended seeding rates is about one-third to one-half that of red clover—that is, from 3 to 4 pounds per acre (3.5–4.5 kg/ha).

Alsike clover usually will produce only one cutting of hay per year. It is recommended that the crop be cut when the plants reach full bloom. However, because the plant is indeterminate in both growth and flowering habit, identifying the full bloom stage is rather difficult, and the optimum cutting time must be estimated. Because alsike is adapted to wet and acidic soils, it is a valuable species for use on naturally wet meadows or irrigated pastures.

One cannot expect to produce both a hay crop and a seed crop in the same year with alsike clover, and in this sense it resembles mammoth red clover.

Alsike clover is susceptible to about the same diseases and insects that damage red clover; however, it is considered to be resistant to northern and southern anthracnose.

White clover One of the most widely distributed legumes throughout the world is white clover (*Trifolium repens*). Under favorable conditions, white clover is a perennial; however, depending on cultivars or strains, droughts, disease, insect pests, and management practices, plants may sometimes behave like annuals or biennials.

There are three main types of white clover: large, intermediate, and small. The large type, generally grown in northern regions, is typified by the cultivar 'Ladino,' and the more recently developed cultivars 'Pilgrim' and 'Merit.' The large types generally grow from two to four times the size of small, or common, white clover; otherwise, there is little difference between the three types. This difference in size makes greater demands upon the soil nutrients; to get the most out of 'Ladino' and other large types, it is necessary that adequate fertilizer and lime be applied, that proper management be practiced, and that they be grown in association with compatible grass species. The intermediate type of white clover is grown mainly in the South, where it is widely used in permanent pastures; intermediate types blend well with dallisgrass, bermudagrass, and bahiagrass. Small, or common, white clover is found throughout the United States and much of Canada. It grows voluntarily in lawns and pasture areas, but it is not considered a particularly desirable plant in either area; its growth characteristics depend on soil fertility, the herbicides used, and other management practices, but generally it produces low yields.

The white clover plant is prostrate in its growth habit. The seeds are very small, and relatively low seeding rates are required; 2 pounds per acre (2.2 kg/ha) when seeded in a grass mixture, or 4 pounds per acre (4.5 kg/ha)

Figure 29-4
Alsike clover plants in the vegetative stage of development. Note the decumbent growth habit and the taproot system. [USDA photograph.]

in pure stand, will generally give an adequate plant population. Because the seeds are small, shallow seeding is essential. White clover seedlings have a rosette type of leaf growth. The plant develops a small crown, from which creeping fleshy stems (stolons) arise and spread outward. These stolons root at the nodes and give rise to new plants. Because of its spreading habit, a thin stand of white clover seedlings soon develops into a thick stand, providing that competition from weeds is held down.

'Ladino' clover—one of the large types of white clovers—is an extraordinary pasture legume. It is unexcelled in its nutritive value; the protein content varies from 20 to 30 percent. Because of its growth habit, only its leaves and petioles are grazed, which makes it not only a highly palatable forage but one that is highly concentrated in terms of nutritive value. Also, the fact that only the leaves and petioles are grazed enables it to be grazed early without damage to the plant. 'Ladino' clover reseeds itself; it is very aggressive and will compete in a mixture with several grass species; it is superior to other legumes in its tolerance for wet sites and in its ability to recover quickly after grazing. 'Ladino' is a relatively high yielding legume and is excelled only by birdsfoot trefoil in its tolerance of close grazing.

The main disadvantage of 'Ladino' clover is that it is more likely to cause bloat than any other legume. One of the characteristics that makes it a most desirable forage also contributes to causing bloat in ruminants—the fact that the stolons lie on the ground and that only the leaves and petioles are grazed. Apparently, leaves, more than stems, contain the causative factors of bloat (probably protein fractions and a lack of tannins). Other disadvantages include the fact that 'Ladino' is not drought-resistant and is not adapted to areas with low rainfall, except under irrigation. The species is susceptible to winterkill, and only its capacity to replace dead plants by propagating with its stolons allows 'Ladino' to be grown widely in northern regions. Because of its stoloniferous growth habit, it is hard to cure; therefore, it is not considered a good hay crop, although, on occasion, it is included in a mixture that may be used for hay or grazing.

'Ladino' clover makes a good pasture for hogs and poultry; its high concentration of nutrients compared with other forages makes 'Ladino' clover particularly suitable for animals with simple stomachs. In fact, it makes an excellent nutritious salad for humans (although it is not as palatable as lettuce).

Sweetclover One of the most drought-resistant of the common forage legumes is sweetclover (species of the genus *Melilotus*). It has a wide range of adaptation and is found throughout the United States and much of Canada; its drought-resistant characteristic makes it a satisfactory crop in the Great Plains. It is best suited to fertile, adequately drained soils that are relatively high in organic matter. Sweetclover's broad range of adaptation is limited by a lack of tolerance for acidic soils; it is unadapted to soils with a pH below 6.0 and, thus, may require liming. Most of the sweetclover grown in North America belongs to one of two species: the yellow-flowered (*M. officinalis*) and the white-flowered (*M. alba*). Yellow-flowered sweetclovers are biennial in growth habit; white-flowered sweetclovers include both annual and biennial strains.

Yellow sweetclover flowers from about ten to fourteen days earlier than the white sweetclover. It has finer stems, is leafier, has less top growth, and is somewhat shorter (Figure 29-5). Yellow sweetclover is less sensi-

Figure 29-5
A single sweetclover plant in the vegetative stage of development. Note the fine stems, numerous leaves, and erect growth habit. [USDA photograph.]

tive to adverse conditions, such as drought and competition from companion crops and weeds, than alfalfa; it is also more common than white sweetclover in the northern Great Plains.

The biennial strain of white sweetclover is favored in the corn belt because it matures later and produces more hay or pasture; also, its ability to improve the soil matches, if not exceeds, that of yellow sweetclover.

Yellow and white sweetclovers have about the same number of stems per plant, and the size of their root systems is about equal. Both are generally considered to be winter-hardy, although, as a whole, sweetclovers are somewhat susceptible to winter injury. Annual white clover, such as the cultivar 'Hubam,' does not overwinter. It has a large top growth and small root development and is an excellent green-manure crop. 'Hubam' may be planted with small grains; after the grain crop has been removed, the sweetclover is allowed to grow up, and it is plowed under as a green-manure crop in the fall. When properly inoculated, sweetclover is unexcelled in its ability to fix and utilize atmospheric nitrogen.

Cultural practices needed to successfully establish a stand of sweetclover are practically the same as those required for alfalfa. If the soil pH is below 6.0, lime should be applied well in advance of seeding. Sweetclover has a high incidence of hard seed, and the seed should be scarified before planting; from 10 to 15 pounds of scarified seed per acre (11–17 kg/ha) is recommended in a pure stand. Care must be taken when selecting chemicals to control weeds and insects; as with all legumes, sweetclover is susceptible to 2,4-D and several other selective herbicides.

Unlike alfalfa and red clover, sweetclover produces only a single main stem in the first year. If the stem is cut off, the new growth must come from axillary buds on the part of the stem that remains (there is no new growth from the crown, as in alfalfa). If too much of the stem is removed when sweetclover is clipped during the first year, a severe reduction in the size of the root will result. This may reduce the ability of the plant to overwinter and will most certainly reduce the vigor of the plant during the second year, because the stem serves as a food storage area for the overwintering plant.

Sweetclover has a high content of coumarin, a substance that seems to reduce palatability to the point that livestock may refuse to eat the crop, especially if they have a choice. But the presence of coumarin may have even more serious consequences. Sweetclover that has not been properly cured before storage may contain sufficient moisture to cause heating and spoilage—processes that transform coumarin into a toxic substance called dicoumarol. Dicoumarol reduces the clotting ability of the blood, and animals that have eaten spoiled hay containing this substance have been known to bleed to death from small wounds or from internal hemorrhaging. Consequently, it is extremely important that sweetclover hay be carefully and completely cured before it is stored and that the hay be managed in such a way that it does not heat and spoil. Plant breeders have developed low-coumarin-content cultivars of sweetclover, but such cultivars have not been particularly successful. The cultivar 'Cumino,' for example, is low in coumarin but more susceptible to insect pests (blister beetles, in particular, have devastated seedling stands of this cultivar). The application of a fatty acid mixture (propionic and acetic acids) will prevent mold growth and heating in hay. Preventing the formation of mold should prevent the possibility of "bleeding disease."

In summary, sweetclover is an outstanding pasture crop, and it has no equal in its soil-improving ability. It is a good seed producer, and seed is relatively inexpensive; its drought

Figure 29-6
A close-up view of birdsfoot trefoil showing the prolific growth of a fine-stemmed, leafy, slightly decumbent plant. [USDA photograph.]

tolerance allows it to be grown in a wide array of climatic conditions; it can easily fit into established cropping systems (but is best for short meadow rotations). The biennial varieties are winter-hardy and can stand relatively wet conditions as well as drought; it will grow on soils that are too alkaline for alfalfa or most other cultivated crops.

Some disadvantages of sweetclover are its low tolerance for acidic soils; its tendency, when hay is improperly cured, to produce dicoumarol, which causes "bleeding disease" in livestock; and its high percentage of hard seed, which requires that seed be scarified before seeding.

The main diseases attacking sweetclover are root rot and black stem. It is also subject to damage from the sweetclover weevil.

Birdsfoot trefoil (broadleaf type) Broadleaf birdsfoot trefoil (*Lotus corniculatus*) is a long-lived, highly branched, taprooted, yellow-flowered legume adapted to the northeastern and northcentral parts of the United States and to parts of southeastern and central Canada (Figure 29-6). It is also grown under irrigation in the West, but it is not suited to the humid South. Birdsfoot trefoil grows well with most grasses and produces good mid-to-late summer pasture. It is a relatively deep-

Table 29-1
Dry-matter yields of sward and birdsfoot trefoil component under different management systems.

Type of management	Total sward yield (kg/ha)	Birdsfoot trefoil component (kg/ha)
1969		
Pasture	5,076	3,120
Pasture-reseed	5,078	3,355
Hay	5,773	4,472
Hay-reseed	5,060	3,243
1970		
Pasture	5,652	4,207
Pasture-reseed	4,335	3,286
Hay	6,336	5,040
Hay-reseed	5,426	4,154
1971*		
Pasture	4,595	2,372
Pasture-reseed	5,513	2,796
Hay	4,819	2,282
Hay-reseed	6,511	3,550
1972*		
Pasture	2,643	1,853
Pasture-reseed	3,597	1,326
Hay	3,471	1,200
Hay-reseed	4,663	2,061

Source: T. H. Taylor, N. C. Templeton, Jr., and J. W. Wyles. Management effects on persistence and productivity of birdsfoot trefoil (*Lotus corniculatus* L.). *Agron. J.* 65(1973):646–648.

*Neither pasture-reseed nor hay-reseed plots were permitted to set seed in this year. Note the decline in the birdsfoot trefoil component in 1972, when reseeding had not been permitted in 1971. Data for 1969, 1970, and 1971 show that good stands can be maintained through natural reseeding.

rooted crop—root depths are between those of red clover and those of alfalfa. It is not as drought tolerant as alfalfa and sweetclover, but it is more drought tolerant than 'Ladino' (white clover).

Once established and given proper management, birdsfoot trefoil is a very long-lived perennial (Table 29-1). Some stands in the eastern United States have remained established for more than thirty years.

Although there are winter-hardy cultivars that will withstand moderately severe winter conditions, such as 'Empire' and 'Leo,' many cultivars are not winter-hardy, and selecting a cultivar for a given climatic condition must be done with care.

If there is adequate available moisture, birdsfoot trefoil will produce very well during mid- and late summer—the time of year when often there is the greatest need for forage. Most grasses in the northern and central United States produce best in the spring and early summer; therefore, birdsfoot trefoil is a very valuable forage for balancing the forage available during the pasture season.

This forage legume tolerates a wide range of soil pH; it has been grown successfully on relatively alkaline soils in the West and on acidic soils in the East. It is susceptible to few diseases or insect pests.

Some cultivars of birdsfoot trefoil survive close grazing better than other legumes; because of their prostrate growth habit, the grazing animal harvests mainly the leaves and petioles, but relatively few stems, allowing the plant to regenerate itself more readily. It also withstands trampling. However both of these abuses—close grazing and trampling—hinder optimum production.

One of the most significant advantages of birdsfoot trefoil is that, unlike most other legumes, it does not appear to cause bloat. Repeated experiments, including intense pasturing of animals in lush, moist stands, have produced no incidents of bloating. This removes one of the biggest risks common to most other legumes.

Although birdsfoot trefoil is not a generally recommended hay crop, it can be cut for hay if conditions warrant. A delay in cutting birdsfoot trefoil for hay does not result in as serious a loss in quality as is the case with alfalfa and other legumes; plants retain their leaves, and the stems neither lose their feed value as rapidly nor become as unpalatable as alfalfa.

The disadvantages of birdsfoot trefoil are numerous. Stand establishment is difficult; the plant is very slow-growing when it is becoming established. It is a weak competitor in the seedling stage; therefore, it should not be seeded with a companion crop, and weed competition must be held to a minimum, or the stand will fail.

Birdsfoot trefoil requires careful inoculation. Because the specific bacteria that inhabit the roots of birdsfoot trefoil are not normally found in the soil in most regions, inoculation becomes especially important in order to obtain high yields. Not only does this species require high rates of inoculum, but the bacteria must be specific for birdsfoot trefoil. Moreover, because there is a unique interaction between a particular cultivar of birdsfoot trefoil and a specific strain of *Rhizobium*, it is probably necessary to select bacteria strains that are compatible with specific cultivars. In Canada and the northern United States, where early and rapid stand establishment is an important factor, strains of *Rhizobium* that will infect plants at low temperatures (50°–64°F, or 12°–18°C) are generally required.

Most cultivars of trefoil produce little forage in the early spring; they produce most heavily in midsummer. Some Russian introductions, together with suitable *Rhizobium* inoculation, will initiate growth before alfalfa, red clover, and sainfoin. In general, however, trefoil initiates growth a little later in the spring and tends to reduce growth a little earlier in the fall than other legumes.

Trefoil does not make as good a hay crop as other legume species. It is weak-stemmed and tends to lodge. When cut for hay, the stems are usually severed, requiring the plants to regenerate both stems and leaves; thus, recovery after harvest is relatively slow. Trefoil is difficult to cure because it is fine-stemmed and leafy; however, when properly cured, the hay is of relatively good quality.

Another disadvantage of birdsfoot trefoil relates to seed production. Flowering and seed set continue for a six-to-eight-week period. Some of the pods are ripe and dehisce while others are still being formed. Depending on the humidity and other weather conditions, from 20 to 60 percent of the seed may be lost because of shattering; accordingly, the cost of trefoil seed is relatively high.

In summary, birdsfoot trefoil is very difficult to establish, but, once established, it makes an extraordinary pasture legume that will persist for years if well managed. It is highly nutritious and palatable, and it does not cause bloat—one of the few legumes that does not present this hazard (others are lespedeza, sainfoin, and soybeans). Birdsfoot trefoil produces most heavily in mid- and late summer—the period when grasses have become less productive. It does not make as good a hay crop as alfalfa and many other legumes.

Lespedeza The lespedezas, which are grown primarily in the southeastern states, are warm-season legumes that grow well only during the warm summer months. Because they are grown primarily on soils that are low in fertility, the yields are generally lower than those of other legumes.

The two most important and widely grown species, *Lespedeza stipulacea* (Korean) and *Lespedeza striata* (Kobe), are annuals. The only perennial species of any significance from the standpoint of acreage is *Lespedeza cuneata,* referred to as sericea lespedeza.

Lespedeza is valued and used mainly as a pasture crop. Because of its warm-season characteristic, the forage it produces is usually not available until late spring or early summer. Its period of growth comes at about the end of the most productive time for most cool-season grasses, and, therefore, it extends the carrying capacity of the pasture complex through midsummer when the productivity of most grasses is tapering off.

Lespedeza has a number of advantages. It tolerates acidic, infertile soils but does respond well to fertilizer. Annual types are easy to establish (sometimes they are seeded on the snow in winter or early spring, and the seed germinates as the snow melts), and they reseed themselves each year, so that the stands persist as if the plants were perennial. Lespedeza resists heat and drought conditions and provides good summer grazing. It is also a good seed yielder; thus, seed is relatively inexpensive. The crop seems to be relatively tolerant of or resistant to most diseases and insect pests. Lespedeza can also be used for hay; it is easy to cure, and the hay is palatable and nutritious and has one definite advantage in that it does not cause bloat.

Lespedeza has several disadvantages. The yield of forage, whether it be pasture or hay, is relatively low, ranging from 1,000 to 3,000 pounds per acre (1,120–3,360 kg/ha) depending on the soil and available water. Unless early supplemental pasture is available, the characteristic of late-spring growth initiation can be a disadvantage. Lespedeza does not compete well with weeds and does not tolerate wet soil sites.

Seeding rates ranging from 25 to 30 pounds per acre (28–34 kg/ha) of good-quality seed are suggested to obtain a good stand for hay and pasture or for a seed crop in the first season. To insure a full stand in the second season, from 10 to 15 pounds per acre (11–17 kg/ha) may be seeded in the second year to supplement natural reseeding.

Sainfoin Sainfoin (*Onobrychis viciaefolia*) is an ancient European forage crop that has not been popular in North America until recently. It is now being grown on significant acreages in the western United States and in Canada. It was unpopular previously because of its poor tolerance for acidic soils and its lack of tolerance for high soil-moisture conditions. Its coarse stem suggests low palatability, and the finer-stemmed crops such as alfalfa have been preferred by many growers. There are several reasons for the emerging popularity of sainfoin on dryland sites in the West: it is frequently higher-yielding than alfalfa under dryland conditions; it grows well in association with crested wheatgrass, intermediate wheatgrass, basin wildrye, and other dryland grass species; it is alfalfa's equal in digestibility and, in spite of its apparently coarse stems, excels alfalfa in palatability; it matures earlier than alfalfa and holds its leaves better as it advances in maturity; and it does not cause bloat.

Table 29-2

Average yields in metric tons per hectare of mixtures containing 'Eski' and 'Remont' sainfoin in each of four years. (Average of each cultivar grown alone and in mixtures with each of two grasses and four legumes.)

Year	First harvest		Second harvest		Total	
	'Eski'	'Remont'	'Eski'	'Remont'	'Eski'	'Remont'
1968	3.67	3.27	1.68	2.64	5.35	5.91
1969	5.53	5.11	3.92	4.48	9.45	9.59
1970	6.45	6.05	2.55	3.04	9.00	9.09
1971	5.85	5.76	1.92	2.39	7.77	8.15

Source: C. S. Cooper. Establishment, hay yield, and persistence of two sainfoin growth types seeded alone and with low-growing grasses and legumes. *Agron. J.* 64(1972):379–381.

Given proper management of the stubble, sainfoin has shown excellent regrowth after grazing. At Lethbridge, Alberta, and at Outlook, Saskatchewan, from 600 to 800 pounds of beef per acre (672–896 kg/ha) has been produced from sainfoin pastures.

Sainfoin is relatively winter-hardy. (The cultivar 'Melrose' is significantly more winter-hardy than 'Eski,' and, therefore, it is better adapted in Canada.) Sainfoin also exhibits considerable tolerance for saline soils and is, therefore, adapted to areas where many other legumes do not persist. Under dryland conditions, this forage will persist for many years; stands at Swift Current, Saskatchewan, have been established for more than twelve years. The longevity of sainfoin grown under irrigation is less than that of alfalfa.

Sainfoin has few insect problems, and its resistance to the alfalfa weevil has increased its popularity significantly. When grown under irrigation, sainfoin is subject to serious root and crown rot diseases; the irrigated crop is probably only productive for the seeding year plus two subsequent years.

As is true of all species of forage legumes, sainfoin has a number of disadvantages. It will not tolerate acidic soils or wet sites. When harvested for hay, it has less regrowth after the first cutting than alfalfa and several other legumes. The cultivar 'Remont' has a more even yield distribution than 'Eski,' (Table 29-2). Currently available cultivars are adapted mainly as a single-cut hay; the aftermath growth can be pastured.

Because most inocula appear to be ineffective, inoculation of the seed does not increase its yield; consequently, it may be necessary to fertilize sainfoin with nitrogen as well as the other elements. Canadian researchers have developed some strains of *Rhizobium* that appear to be effective in promoting the nitrogen-fixing capability of sainfoin, particularly when inoculated at high rates.

It is not difficult to establish a good stand of sainfoin. For best germination and seedling development, the seedbed should be firm with moisture near the surface. As with all forages, a weed-free seedbed is desirable. Row spacing depends on the amount of available moisture and the purpose for which the crop is to be used. For hay, a solid seeding with row spacing between 6 and 12 inches (15–30 cm) is usually best; for seed production, wider row spacing may be better. When there is little available moisture, a row spacing greater than 12 inches (30 cm) is desirable. The recommended seeding rate on dryland for row spacings ranging from 6 to 12 inches (15–30 cm) is 26 pounds per acre (29 kg/ha); for 24-inch (60 cm) row spacing, 13 pounds per acre (15 kg/ha); for 36-inch (90 cm) row spacing, 9

Figure 29-7
Sainfoin at full-bloom stage of development, ready to be harvested for hay. Note that at this stage at least one flower on each raceme is open.

pounds per acre (10 kg/ha). The heavy seeding rates are a reflection of the pods, which are not separated from the seeds for planting purposes (a single-seeded pod is planted).

When sainfoin is harvested for hay, it should be cut at the full-bloom stage—the point at which quality and quantity are at a combined maximum (Figure 29-7). To determine the bloom stage, the inflorescence (a raceme) on each stem should be examined; if there is at least one open blossom on each stem, the plant is considered to be in full bloom.

Other Legumes

Vetches There are several legume species that are referred to as vetch. The genus *Vicia* includes, among others, common vetch and hairy vetch; the genus *Astragalus* includes several species referred to as milk vetch; and the genus *Coronilla* includes the species known as crown vetch. Species of the genus *Vicia* are grown mainly in the southeastern United States and to some extent on the West Coast; species of *Astragalus* and *Coronilla* are scattered throughout the United States and Canada. Vetch is not a crop of broad significance, but some species are important in certain areas. Because of their matting type of growth, which helps protect the soil from erosion, vetches are sometimes used as a cover crop. Vetches of the genus *Vicia* are generally grown as winter annuals in regions having mild winter temperatures. Cicer milk vetch (genus *Astragalus*) is a long-lived perennial adapted to dryland conditions such as those that exist in the Great Plains and the Northwest. It is slow to become established and spreads rather slowly (by rhizomes), but, once established, cicer milk vetch is good for livestock grazing. (Some of the native species of *Astragalus* are poisonous, and caution should be exercised when this vetch is grazed.) Crown vetch (genus *Coronilla*) is also a long-lived perennial that spreads by rhizomes; it is adapted generally throughout the humid north, and it is a good soil-conserving plant but recovers very slowly after being harvested for hay.

Faba beans Faba beans (*Vicia faba* minor) are emerging as a significant crop in Canada. Although it is grown primarily for the bean, which is a good source of protein and energy, the faba bean plant can also be made into silage. The plants should be cut for silage just after the bottom pods turn black, and they should be ensiled after very little wilting. (The silage gradually becomes very dark in color.) Faba beans do not compete well with other plants; however, they may be planted in strips alternating with barley (at the respectively appropriate planting dates) and harvested across the strips for a mixed silage of barley and faba beans.

Field peas Field peas (*Pisum arvense*) are used as a hay and pasture crop in combination with oats or other small grains. They require a temperate or cool climate and do best where high temperatures during the growing season are uncommon. Field peas are used as livestock feed mainly in the northwestern states and, in limited quantities, in other northern areas.

Soybeans The soybean (*Glycine max*) was brought to the United States from China during the early colonial period. Until the 1940s, most of the soybeans produced in this country were grown for hay. Recently, nearly all of the soybeans have been harvested for their seed (see Chapter 20).

Cultivars of soybeans are very day-length sensitive, so there are hundreds of cultivars varying greatly in their area of adaptation. When grown for forage, this crop is used mainly for hay; it is cut when the pods are well developed, but before the leaves have turned yellow and begin to fall. The hay is stemmy and coarse and is slow to cure in the field because of the plant's thick, pubescent stems. The pubescence also results in a relatively dusty hay.

Strawberry clover The pink-to-white inflorescence that resembles a strawberry is characteristic of strawberry clover (*Trifolium fragiferum*). It is grown mainly in temperate climates on wet sites with saline or alkaline soil conditions. Strawberry clover is especially well adapted to areas that are flooded for several weeks at a time; it is one of the few known forage legume species that will survive prolonged flooding. It is used mainly in the northwestern United States in areas that are too wet, or on lands that are too alkaline, for other legumes.

Black medic Black medic (*Medicago lupulina*) is a pasture legume that is sometimes considered a weed. It is an annual that reseeds itself. Its stems are small and leafy, and the species has a relatively low productivity. Black medic is not adapted to acidic soils and needs a reasonable amount of moisture and a moderate level of fertility to flourish. It is broadly dispersed throughout much of the temperate United States and Canada.

Kudzu Kudzu (*Pueraria thunbergiana*) is adapted mainly in the humid South and in other areas that are not subject to winterkill. It is a long-lived perennial that can be used for both pasture and hay. Its main value, however, is for controlling rapid erosion, which produces deep gullies. Kudzu grows as a coarse vine with long runners, or stems; it can root at each node on this stem, and thus it is able to spread rapidly. Once established, kudzu persists for many years. It has become a pest in some areas, where its persistent growth covers trees and shrubs, thereby weakening and, in some cases, killing them.

Legumes have been an important source of good quality livestock feed for many years, but the recent, growing demand for high-protein food and animal feed is making these high-protein-content plants increasingly valuable. The ability of legumes to fix atmospheric nitrogen not only to meet their own growth needs but also to make nitrogen available to other plants is a feature that may take on added significance in the future. Although it is possible to transform nitrogen into a form available to plants by burning fossil fuels, the process is expensive and relatively inefficient; moreover, the fossil fuels themselves show signs of becoming scarce. Thus, in a world of dwindling food supplies and shrinking energy sources, the legume plant could well become a vital resource.

Selected References

Grass. 1946. *Yearbook Agr.* USDA.

Heath, M. E., D. S. Metcalf, and R. E. Barnes. 1973. *Forages.* 3d ed. Iowa State University Press.

Isley, Duane. 1951. The Leguminosae of the north central United States: Loteae and Trifolieae. *Iowa State Coll. J. Sci.* 25:439–82.

Smith. Dale. 1962. *Forage management in the north.* Wm. C. Brown.

Study Questions

1. What is the unique soil-building property of legumes?
2. Why is legume forage often considered superior to grass forage?
3. How are origins of legumes reflected in their adaptation characteristics?
4. Describe the typical root, stem, leaf, and flower characteristics of forage legumes.
5. Contrast the advantages and disadvantages of two forage legume species adapted to your area.
6. List the major legumes that do not cause bloat in ruminant livestock.

Thirty   Forage Grasses

The grass family, Gramineae, is one of the principal families of forages along with the legumes. Hitchcock's *Manual of the Grasses of the United States* shows that there are about six hundred genera and five thousand species of grasses. More than two-thirds of the cultivated forage cropland includes species of grasses, and grasses are a major part of most native rangelands.

Botanical Characteristics

ROOTS
The plants of the grass family have many characteristics that are common to most or all of the various species. Grasses, in general, have fibrous root systems composed of extensively branched roots that are fairly uniform in size. The extensiveness of the root system varies from species to species. The depth of the root system may vary from a few inches to several feet, but, as a rule, it is not as deep as many of the taprooted legumes. The pattern of root development of various species determines many of their uses and limitations.

Grasses have both seminal and adventitious root systems. The seminal root system forms when the seed germinates and is found at the depth at which the seed is planted. The adventitious root system—the plant's permanent root system—arises above the seminal roots as the plant develops; the usefulness of the seminal root system is generally limited once the seedling is established.

STEMS
The stems of grasses are made up of nodes and internodes. Nodes are the solid enlarged joints on the grass stem; internodes are the

hollow cylindrical parts of the stem and vary in length depending on the age of the tissue and the species of plant. Various species of grasses have modified stems, either rhizomes or stolons.

LEAVES

All grasses are monocots; their single cotyledon is called the scutellum—a modified leaf that is formed at the first node of the embryonic stem. A special cone-shaped leaf, the coleoptile, is formed at the second node. The coleoptile is wrapped around the culm enclosing the upper part of the stem and protects the delicate leaves and the growing point during emergence. As the plant grows and the culm elongates, the leaves and growing point are pushed through the tip of the coleoptile. True leaves—the first of which originates at the third node—consist of a sheath and blade and at the collar may have a ligule and auricles (see Chapter 3). Under favorable conditions, a single grass plant may have several culms or stems. The majority of culms arise from axillary buds at the lower internodes; thus, grass plants frequently develop a crownlike mass of culms and leaves near the soil surface. Stems, or culms, that form from axillary buds are referred to as tillers, and the initial stage of tiller development is sometimes referred to as stooling.

INFLORESCENCE

The inflorescence of the grass species is generally one of two types: spike or panicle (see Figure 4-4 on page 66). The spike type of inflorescence, typified by wheatgrasses, is made up of spikelets, which are sessile. The panicle type, typified by smooth bromegrass, is a loosely branched inflorescence with spikelets borne on pedicels. A third type of inflorescence is called a panicle of racemes. These inflorescences are branched; a single branch is definitely a raceme, and the entire inflorescence is a reduced panicle. Dallisgrass exemplifies this somewhat unusual inflorescence.

Classification of Grasses

There are many practical ways of classifying grasses; one common method is to distinguish between bunch types and sod-forming types. Bunchgrasses do not form stolons or rhizomes; sod formers do. Grasses that are used as forages have been arbitrarily classified, mainly on the basis of their adaptation and response to temperate and subtropical environments, into two other main categories: cool-season and warm-season grasses. There are a number of characteristics that separate the members of these two broad groups of grasses.

COOL-SEASON GRASSES

The cool-season perennial grasses had their origin in northern Europe and northeastern Asia. As a rule, they are relatively winterhardy; they need a cool, moist climate or a cool climate with irrigation for optimum growth (they are not particularly drought resistant).

Cool-season perennials also require cool temperatures and long days (or short nights) for floral induction (i.e., most are long-day in photoperiodic response). Generally, they produce one large hay crop and some aftermath growth. When grown for seed, they usually produce only one crop of seed a year plus some aftermath vegetative growth. Examples of cool-season perennial grasses include smooth bromegrass, timothy, orchardgrass, fescues, reed canarygrass, bluegrass, and redtop.

Some cool-season annual grasses, such as oats and other cereals, are also used for forage. Although they differ from the perennial grasses in origin (which is thought to be the

Near East), their characteristics are generally similar.

WARM-SEASON GRASSES

Most of the warm-season grasses originated in Africa, east Asia, and South America; a few are native to North America. As a rule, they are not winter-hardy; they require warm temperatures for optimum growth. They are relatively tolerant of heat and are drought-resistant. Warm-season grasses require short, warm days for floral induction.

The warm-season perennial grasses can produce several crops of hay and several crops of seed each year. Some examples of warm-season perennials are bermudagrass, dallisgrass, bahiagrass, carpetgrass, and johnsongrass. Annual warm-season grasses, such as sudangrass, have most of the characteristics of perennials, except for longevity.

Grasses of the Humid or Irrigated North

Smooth bromegrass One of the most extensively grown cool-season grasses in northern areas is smooth bromegrass (*Bromus inermius*). As alfalfa is sometimes referred to as the "queen of the forages," smooth bromegrass might be referred to as the "king of the forages." Smooth brome has some readily identifiable characteristics: it has an open-panicle type of inflorescence (Figure 30-1), no auricles, a solid leaf sheath, and the leaf blade has a W-shaped constriction. It spreads by rhizomes (and thus tends to fill in empty spaces in the stand).

Smooth brome has good drought resistance, high yield, and good palatability. It makes a highly nutritious pasture or hay, and it grows well with other forages. This forage is also a good seed producer, and the price of seed is usually quite reasonable.

Smooth bromegrass does have some disadvantages, however. Because the seed is paper-thin, it is difficult to get an even distribution of seed when planting smooth brome. It requires a highly fertile soil for good production (if the soil is low in nitrogen, smooth brome sometimes becomes sod-bound and unproductive). It does not tolerate poorly drained sites, and it is susceptible to a few leaf diseases. During the first year, the seedling growth is somewhat weak, but once plants are established, they persist for several years.

Orchardgrass Another popular, long-lived cool-season forage grass is orchardgrass, *Dactylis glomerata*. This palatable bunchgrass has numerous, broad, folded basal leaves and a tall seed stalk with a compact panicle that reaches well above the upper leaves of the mature plant (Figure 30-1). Orchardgrass is known for its high yield (especially when grown with adequate nitrogen) and particularly for its productive aftermath growth (Table 30-1). It is nutritious and palatable if cut early; as is true of other grasses, it loses palatability and some nutritional value as it advances toward seed maturity. Orchardgrass does not spread by rhizomes; but plants will, in some cases, fill in a sparse stand by reseeding. It is characterized by early spring growth, quick recovery after cutting, ease of establishment, rapid germination, and compatability with legumes in pasture mixtures.

Orchardgrass is not as winter-hardy or as drought-tolerant as smooth bromegrass and some of the other cool-season grasses; however, heavy applications of nitrogen fertilizer combined with adequate phosphorus and potassium promote winterhardiness and plant longevity. It has been known to spread as a weed if not properly managed. Grazing or mowing regularly tends to keep the plants in

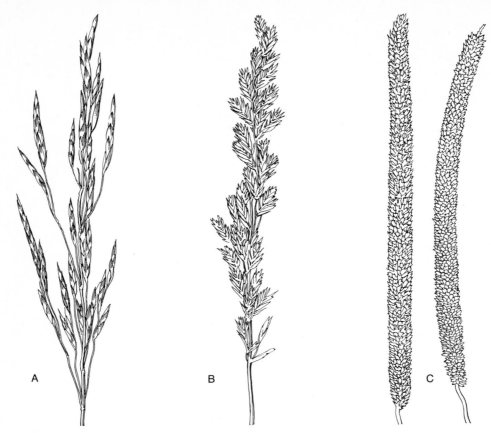

Figure 30-1
A. Typical smooth bromegrass plants at the flowering stage; note the loose panicles. B. Typical orchardgrass plants at the flowering stage; note the dense panicles. C. Typical inflorescence of timothy. Although the inflorescences appear to be spikes, they are in fact dense panicles.

a vegetative state—an important consideration when managing orchardgrass because it loses palatability more than some other grasses with the approach of seed maturity.

Fescues There are several species of fescue with a variety of characteristics: annuals and perennials, fine-leaved and coarse-leaved, and creeping or erect in growth habit. The two most important species in the United States and Canada are tall fescue (*Festuca arundinacea*) and meadow fescue (*F. elatior*), both perennials. Tall fescue grows under a wider range of conditions than meadow fescue, and it is the more extensively grown. The fescues are less palatable than most other grass species. Acceptance, intake, and animal performance seem to be related to the alkaloid content, namely, perloline and perlolidine. (Attempts are being made to breed fescue cultivars with a lower alkaloid content.) Some of the newer cultivars of tall fescue are more productive than orchardgrass and bromegrass and have been found to be relatively acceptable to cattle and sheep. Fescues are relatively tolerant of shade; they are able to

Table 30-1
Effects of nitrogen fertilizers on dry-matter yields
(in lb/acre) of orchardgrass and bromegrass grown at
Ames, Iowa.

Nitrogen applied (lb/acre)	1958		1959	
	Orchard	Brome	Orchard	Brome
0	1,260	1,905	827	1,345
30	2,864	3,420	2,113	2,275
60	3,867	4,133	2,755	2,800
120	5,890	5,133	3,617	3,941
60 + 60*	6,906	5,524	3,975	3,835
240	8,877	7,140	5,640	5,516
240†	8,459	7,149	1,239	1,492

Source: L. P. Carter and J. M. Scholl, Effectiveness of inorganic nitrogen as a replacement for legumes grown in association with forage grasses, *Agron J.* 54(1962):161–163.
Note: Orchardgrass gave a greater response to nitrogen fertilizer at the higher rates than bromegrass.
*Split application: 60 pounds in the early spring, 60 pounds after the first harvest.
†In this trial, 240 pounds applied in 1958, none in 1959.

survive long periods of flooding during the winter, and they are also tolerant of alkaline soil conditions.

Fescues form a very dense sod (root system) and are thus useful in waterways, lanes, around water tanks and barn lots, and on athletic and recreational fields.

The tall fescue plant is a deep-rooted, tufted, long-lived perennial that grows to a height of 4 to 5 feet (1.2–1.5 m). It has a panicle-type inflorescence. A significant morphological distinction between tall fescue and meadow fescue is seen in the auricles: tall fescue has very obvious hairs on the auricles, whereas the auricles of meadow fescue are "bald." The two species also have different chromosome numbers: tall fescue is 2N = 42, and meadow fescue is 2N = 14. Meadow fescue is significantly more winter-hardy than tall fescue and, therefore, is used more throughout Canada.

Reed canarygrass Reed canarygrass (*Phalaris arundinacea*) is found in much of the northern half of the United States and in southern Canada. It tolerates wet sites and is frequently found along irrigation ditches, bogs, and poorly drained sites. (Its persistence in these sites has sometimes caused it to be classified as a weed.) Reed canarygrass spreads by rhizomes and is especially aggressive and quick-spreading on wet sites, making it a good forage for gully control. It is a tall, coarse, sod-forming perennial that reaches a height of 4 to 7 feet (1–2 m) and is usually found growing in clumps. In solid stand, it forms a dense sod and is relatively high in forage production. When grazed, it must be kept from reaching maximum maturity or it loses palatability.

If properly managed, reed canarygrass will produce good-quality and highly nutritious pasture forage. It has been used successfully in grass waterways and in areas subject to overflow from farm ponds. (It is also useful around the edges of farm ponds to stabilize areas frequently trampled by livestock.) This species has a tremendous underground root system. In a research trial, canarygrass produced 17,000 pounds of roots per acre (19,000 kg/ha) compared with 11,000 pounds per acre (12,300 kg/ha) for smooth bromegrass and 3,500 pounds per acre (4,000 kg/ha) for alfalfa. The heavy sod helps the plant withstand running water. When grown in a gully, the grass tends to slow down the rate of water flow, which allows silt particles to settle out. The grass continues to grow vigorously through the silt forming new shoots, which in turn continue the effect of slowing down the flow of water and eventually serve to heal the gully.

Reed canarygrass responds to high rates of nitrogen fertilizer and heavy applications of animal waste. In Canada, yields of 7 tons per acre (15.7 MT/ha) of dry matter and more than 2,000 lbs per acre (2,240 kg/ha) of crude protein have been obtained from heavily fertilized stands.

Timothy Timothy (*Phleum pratense*) is a bunchgrass with erect stems that grow to a length of 20 to 40 inches (51–102 cm) and a dense cylindrical panicle inflorescence (Figure 30-1). This short-lived perennial is unusual among the grasses in that one of the lower internodes remains relatively short and enlarges to form a haplocorm. The haplocorm looks like a bulb and serves as a storage organ until the plants bloom; buds on the haplocorm then give rise to new growth, which propagates the plant from year to year. Although the individual timothy shoots are biennial in nature, new shoots develop vegetatively each year from the haplocorm, and the plant thereby maintains itself as a perennial.

Timothy stands are easy to establish and produce either pasture or hay in the first year; the hay is bright and clean and known to be well suited for horses. The plants are very palatable both when grazed and as hay. The seed is relatively inexpensive, and it is of comparable size and about as easy to spread as alfalfa and clover seed. Timothy grows well in association with most legumes. It is a very winter-hardy crop, and it is attacked by very few diseases.

One disadvantage of timothy is its lack of drought resistance—probably a result of the relatively shallow root system that forms each year from the new, vegetatively produced plants. It cannot withstand very close grazing for a long period, or the haplocorm will not form. The plant's shallow root system makes timothy easily uprooted by grazing animals. Injury from frost heaving is common also. Timothy does not yield as well as bromegrass and orchardgrass.

Bluegrass The bluegrasses, various species of the genus *Poa*, are broadly distributed throughout the temperate and cooler regions of the world. Kentucky bluegrass (*Poa pratensis*) is the most extensively grown species in the United States. It is a rhizomatous plant that produces a dense sod. The leaves occur in small tufts and are usually from 4 to 12 inches (10–30 cm) long. In late spring, the plant produces an erect culm from 15 to 25 inches (38–64 cm) tall; an open-panicle type of inflorescence forms on this culm. Apomixis is common in bluegrass; at least 90 percent of the seed is formed without fertilization.

Kentucky bluegrass is the most prominent and widespread of the cool-season grasses; about 90 percent of the acreage has been spontaneously established. Given favorable spring moisture and temperature conditions, Kentucky bluegrass is hard to beat as far as early productivity is concerned, and its early-spring growth is highly palatable and nutritious. However, Kentucky bluegrass produces a rather low annual yield of forage, and does not make a good hay crop. It tolerates close grazing, and the stand still persists even when it is subject to grazing abuse late in the fall. It grows well with legumes, such as birdsfoot trefoil.

Some disadvantages of Kentucky bluegrass are that it is not resistant to heat and drought, and it usually goes dormant in midsummer. Because some 80 percent of the roots are in the top 6 to 8 inches (15–20 cm) of soil, it is a poor producer under drought conditions.

Improved cultivars of Kentucky bluegrass are extremely popular in the northern and central United States for use in lawns; they will withstand close mowing and form a nice turf. (Cultivars that are resistant to many of the persistent diseases are available.) When seeding Kentucky bluegrass in a lawn or a pasture, it is very important that the seed be planted shallowly, because it requires light for good germination, particularly if the seed is freshly harvested. (This requirement for light becomes less critical as the time between seed harvesting and planting increases.)

Redtop and the bentgrasses Redtop (*Agrostis alba*), a forage grass that is occasionally put in lawn mixtures to thicken the turf, is not widely used. It grows well on acid soils that are low in fertility and can persist on wet soils. The bentgrasses, various species of the genus *Agrostis,* probably got their name from the characteristic growth of the individual plants: the stems tend to follow the ground surface for a short distance and then bend upward. Bentgrasses are smaller than redtop and have finer leaves. Rigorous management is necessary to maintain a good stand; watering and fertilizing must be carefully handled. When well managed, bentgrasses produce an extraordinary turf, which is widely used on golf-course putting greens.

Warm-season Perennial Grasses of the Humid South

Johnsongrass Johnsongrass (*Sorghum halepense*) is a very aggressive, rhizomatous, persistent, palatable, and nutritious grass. Its aggressiveness has sometimes led to its being classifiied as a weed in the southeastern United States. It is classified as a noxious weed in Canada where a johnsongrass hybrid was introduced into British Columbia and other western provinces a few years ago.

Johnsongrass usually grows to a height of 3 to 6 feet (1–2 m). The density of the stand depends on soil fertility, moisture, and available space. The blade of the leaf is extraordinarily broad, approximately 1 inch (2.5 cm); the inflorescence is an open panicle and the plants are highly cross-fertile (Figure 30-2). It differs from all of the other sorghum species in that it has a perennial habit and spreads by rhizomes. Johnsongrass is used extensively in the Southeast for hay and pasture. Forage-type sorghums and sorghum-sudangrass hybrids are grown to some extent in southern Canada and the corn-belt states.

Carpetgrass Carpetgrass (*Axonopus affinis*) is found mainly along the coastal plains from Virginia to Mexico. It is a low-growing sod former that spreads by stolons and by reseeding itself. Carpetgrass is distinguished by compressed, two-edged creeping stems and blunt leaf tips. Seed stalks grow approximately 9 to 10 inches (23–25 cm) high and usually have three branches. Carpetgrass is a profuse producer of seeds. It is adapted to soils of low fertility where moisture is available in the root zone most of the year.

Pangolagrass Pangolagrass (*Digitaria decumbens*), sometimes referred to as pangola digitgrass, is best adapted to Florida and the fertile, moist soils along the Gulf Coast. It grows best on well-drained, fertile soils that are not too acid. Pangolagrass is a creeping perennial that spreads by stolons; the plants grow from 2 to 4 feet (60–120 cm) high. Seed viability is very low; consequently, seeding rates must be increased accordingly. Pangolagrass is used principally as a pasture forage

Figure 30-2
A clump of johnsongrass showing open panicles and numerous, wide leaves. [USDA photograph.]

Figure 30-3
A bermudagrass plant showing the stems, with adventitious roots, arising from both rhizomes and stolons.

but can be used for hay or silage also; it is a highly productive and important forage in its area of adaptation.

Bermudagrass Bermudagrass (*Cynodon dactylon*) is adapted to tropical and subtropical regions throughout the world, and it is extensively grown in the southern United States. It grows well on soils that are fertile and relatively high in clay content. It is not winter-hardy, and it thrives where the weather is hot. Bermudagrass is a long-lived perennial that is capable of spreading rapidly; it propagates itself by stolons, rhizomes, and seed (Figure 30-3). The erect panicle inflorescence is usually from 6 to 12 inches (15–30 cm) high. Leaves are bluish green and relatively short and flat.

Bermudagrass is used principally for pastures and lawns and occasionally for hay. Coastal bermudagrass seems to be the most popular of the improved cultivars; it will tolerate frosts, makes a better growth in the autumn, and is more drought-resistant than common bermudagrass and other grasses that are commonly grown in the South.

Dallisgrass Dallisgrass (*Paspalum dilatatum*) is generally adapted to the cotton belt. This perennial grows in clumps to a height of

2 to 4 feet (60–120 cm) and produces many leafy stems from a basal clump of leaves. The seeds which are produced on erect, leafless seed stalks, are oval and hairy. The inflorescence is unusual in that a single branch is a raceme that bears flowers in two rows, but the entire unit could be referred to technically as a panicle.

Dallisgrass begins growth earlier in the spring than most of the other perennial grasses of the humid South. It grows continuously during the warm weather, can withstand a moderate frost, and is the last of the grasses in this region to become dormant in the fall; consequently, dallisgrass produces forage for a longer time span than the other grasses in the area and is thus becoming one of the more popular species of the Southeast. Because dallisgrass has a bunch-type growth, which is somewhat less competitive than the sod-forming types, it grows well in association with legumes.

Dallisgrass does not withstand close grazing or frequent cutting as well as some of the other species; timely mowing is recommended to keep the grass in a vegetative state of development and, therefore, palatable.

Bahiagrass Bahiagrass (*Paspalum notatum*) is a deep-rooted rhizomatous perennial, which forms a dense sod and is very competitive once established. It is adapted to Florida and to the areas along the Gulf Coast. The culms of this plant arise from thick, short, woody rhizomes and reach a height of 1 to 2 feet (30–60 cm), depending on soil fertility and moisture. Bahiagrass appears to be well adapted to sandy soils and produces better on these soils than most other species in the same region. The extensive and deep root system of bahiagrass also makes it adapted to relatively dry soils. Bahiagrass is used mainly as a pasture forage on relatively dry, low-fertility soils. It ranks between carpetgrass and bermudagrass in productivity and nutritive value.

Other warm-season grasses There are a number of other perennial warm-season grasses adapted to the humid South, including vaseygrass (which is especially adapted to wet sites), rescuegrass, natalgrass, paragrass, and centipedegrass, among others.

Range and Dryland Grasses

It is not possible to discuss all the native and introduced species of forage grasses grown on the rangelands of the western United States and in low-rainfall areas of the Great Plains. The following discussion is confined to a brief description of some of the principal forages used in these areas.

Crested wheatgrass Plants of crested wheatgrass are classified as belonging to one of two species—*Agropyron cristatum* or *A. desertorum,* depending on ploidy (*A. cristatum* is diploid; *A. desertorum* is tetraploid). Both species are vigorous perennial bunchgrasses with deep, extensive root systems (Figure 30-4). Both were introduced into the United States from Siberia in 1889, but they were not grown extensively until after 1920.

Crested wheatgrass is well adapted to the cool, dry areas of the northern Great Plains and the prairie provinces of Canada. It also grows extensively in the intermountain region of the United States and at high elevations in the Rocky Mountain states. Much of the marginal and submarginal dryland farm acreage was seeded to crested wheatgrass in the early 1900s in a revegetation (conservation) program after an unsuccessful attempt had been made to produce crops. Crested wheatgrass is extremely drought-tolerant, winter-hardy, and able to withstand close grazing. It produces

Figure 30-4
Typical crested wheatgrass plants at the flowering stage; note very characteristic, loose spikes.

good forage early in the spring, before most native species have grown sufficiently to be grazed; it produces little during the hot, dry part of the summer. Crested wheatgrass forage is palatable and nutritious, and it is relished as hay or pasture by all classes of livestock; however, both palatability and nutritional value decline rapidly as the plants approach maturity.

Western wheatgrass Western wheatgrass (*Agropyron smithii*) is a native, drought-resistant, sod-forming perennial that spreads by means of rhizomes. It is very winter-hardy and well adapted to areas with severe winters. It is also adapted to a wide range of soil conditions but is most abundant on lowlands and, to some extent, on benchland and hillsides.

Western wheatgrass plants are rigid and upright, usually growing to a height of 1 to 3 feet (30–90 cm). The stems, leaves, and spikes have a characteristic bluish color. The leaves are less than 0.25 inch (6 mm) wide and from 2 to 10 inches (5–25 cm) long; when dry, they roll into a wirelike form, making the plant appear to have very little foliage. The spikes, which are composed of closely crowded spikelets, are from 2.5 to 5.0 inches (6–12 cm) long. Western wheatgrass is relished by livestock both for grazing and as hay.

The seed is relatively slow in germinating; seedlings are slow in becoming established. Because the seed is hard to harvest, its price is relatively high.

Intermediate wheatgrass It is believed that intermediate wheatgrass (*Agropyron intermedium*) is native to the Caspian region of Russia. It was introduced into the United States in the early 1900s but was not used extensively until the 1930s. Intermediate wheatgrass is adapted mainly to the northern and central Great Plains and the Pacific Northwest range and dryland areas. This sod-forming plant has a very extensive root system, which is frequently associated with drought-resistant species, but it is only moderately tolerant of drought. It is a perennial grass; it grows as high as 4 feet (1.2 m), and it is a good seed producer. Intermediate wheatgrass spreads to some extent by means of rhizomes; the plant is only moderately tolerant of alkaline soil conditions.

As is the case with many of the cool-season grasses, intermediate wheatgrass does not produce well during the hot, dry midsummer period. It matures in mid-to-late summer and will grow again in the autumn, when temperatures drop; under favorable conditions, the leaves will stay green until frost. Intermediate wheatgrass makes an excellent pasture or hay crop; it has excellent palatability if harvested before the plants mature. It grows very compatibly with alfalfa; together, they form one of the highest-producing hay crop mixtures in areas where intermediate wheatgrass is well adapted.

Pubescent wheatgrass Pubescent wheatgrass (*Agropyron trichophorum*), unlike intermediate wheatgrass, has pubescent spikes and seeds; in other respects, the two grasses have similar morphological characteristics. Pubescent wheatgrass, however, has the advantage of being better adapted to low moisture conditions, higher elevations, lower fertility, and alkaline soils.

The plant does not generally produce seed at elevations above 6,000 feet (1,828 m). The seed of this species is relatively large, and a fairly high seeding rate—usually 8 pounds per acre (9 kg/ha) or more—is required. Because pubescent wheatgrass has large seeds and vigorous seedling growth, it has a low susceptibility to root rot and seedling blight.

Tall wheatgrass Tall wheatgrass (*Agropyron elongatum*), a native of southern Europe and Asia Minor, where it grows on saline meadows and seashores, is a coarse, tufted perennial. It was introduced into North America in the early 1900s, and it has become valued in the northern Great Plains and intermountain plateaus for its adapation to alkaline soils and its resistance to flooding. It develops a good growth in the spring and autumn and remains green well into the summer. Tall wheatgrass produces a very palatable hay if it is cut before bloom. Seed production is generally good, but because the seeds take a long time to ripen, seed production at higher elevations (or in the far north) is not recommended.

Tall wheatgrass is a perennial bunchgrass with erect blue-green foliage. This species develops an abundance of long basal and culm leaves; the leaf blades are firm, flat or loosely rolled, and they are prominently nerved. The plant has an erect, elongated spike; the spikelets do not overlap. Tall wheatgrass matures late—about thirty days after crested wheatgrass.

Slender wheatgrass Slender wheatgrass (*Agropyron trachycaulum*) is a native perennial bunchgrass that develops numerous erect, stiff, flowering culms, from 1.5 to 3.0 feet (45–90 cm) tall, and a long, slender spike. The spikelets grow close to the rachis and are awnless. The basal leaves are moderately abundant, often have a bluish tinge, and reach a length of 2 to 10 inches (5–25 cm).

This species compares favorably with other wheatgrasses in productivity; the seed germinates well, and the seedlings are very vigorous. It is grazed by all classes of livestock and also serves as a good winter forage.

Slender wheatgrass is widely distributed in southwestern Canada and in the northern and western areas of the United States, but it is most prevalent in the Rocky mountain and intermountain regions. It is relatively tolerant of alkaline soil conditions, but it is less drought-resistant than western wheatgrass and crested wheatgrass. The plants are very winter-hardy and adapted to areas with very severe winters; in Canada, the species frequently behaves like a short-lived perennial (3–5 years).

Thickspike wheatgrass Thickspike wheatgrass (*Agropyron dasystachyum*) is a vigorous, sod-forming native grass that spreads by rhizomes and forms a tight sod under dryland conditions. The plant is low-growing with many light green, fine leaves and is a good seed producer.

Thickspike wheatgrass is recommended primarily for use as a ground cover on disturbed areas, roadsides, airports, recreation areas, and construction sites, where it is often necessary to stabilize the soil; it can also be used as a range grass but is not as productive as some other species. This species is adapted mainly to the northern Rocky Mountain region and the adjacent Great Plains.

Bluebunch wheatgrass Bluebunch wheatgrass (*Agropyron spicatum*) is a native perennial bunchgrass. The plant, which is green to bluish green in color, attains a height of 1 to 3 feet (30–90 cm) and develops narrow spikes from 3 to 6 inches (7.5–15 cm) long. Florets bear a conspicuous geniculate awn; as the seed reaches maturity, the awn bends abruptly away from the spike.

Bluebunch wheatgrass is an excellent range grass while green. It produces abundant, palatable forage that stays green longer than other grasses in the same area. It will not withstand heavy grazing in spring and early summer; thus, overgrazing will cause the stand to reduce in vigor and ultimately disappear. This species is found mainly in the Pacific Northwest and in the intermountain region.

Russian wildrye Russian wildrye (*Elymus junceus*) was introduced into the United States from Siberia in 1927 but was not used extensively until the early 1940s. Today, it is found mainly in the northern Great Plains in the United States and in western Canada. It is adapted at elevations as high as 8,500 to 10,000 feet (2,600–3,000 m).

It is a large bunchgrass with a dense cluster of basal leaves and erect naked culms about 3 feet (1 m) tall. The leaves, which are soft and lax, are from 6 to 18 inches (15–45 cm) long and as wide as 0.25 inch (6 mm). The inflorescence is a dense, erect spike with seeds that shatter readily when mature. The roots are fibrous and may penetrate to a depth of 8 to 10 feet (2.5–3.0 m).

Russian wildrye is a cool-season plant but has a longer summer season than most cool-season, dryland grasses. It is one of the most winter-hardy of the grasses introduced into North America, and it is quite drought-resistant. It grows best on fertile loam soils; it does poorly on low-fertility soils, and it tends to become sod-bound. As Russian wildrye grows older, stands usually become thin (some agronomists attribute this reduction in stands to a shortage of available nitrogen in the soil). Although this species is generally considered to be tolerant of moderately saline or alkaline soil conditions, this tolerance varies considerably with the specific cultivar. Russian wildrye tends to crowd out legumes in a mixture, because of its vigorous root system and heavy growth of basal leaves.

Russian wildrye is used for both grazing and hay, and it is particularly useful for late-summer and winter grazing. It is grown in Western Canada for pasture. Under favorable conditions, this grass will recover exceptionally rapidly. It also produces an early spring growth that is very palatable. In pasture trials, cattle and sheep have produced excellent weight gains on this grass.

Green needlegrass Green needlegrass (*Stipa viridula*) is a perennial bunchgrass that is native to the central and northern Great Plains. Since the late 1930s, it has been recognized as a valuable forage grass because of its seedling vigor, rapid recovery after defoliation, high yield, nutritional value, and palatability. The plant is bright green with abundant basal leaves. Culms grow to a height of 1 to 3 feet (30–90 cm) and produce a panicle type of inflorescence whose short branches make it look almost like a spike. The inflorescence is from 4 to 8 inches (10–20 cm) long. The plant produces abundant seed with long twisted and bent awns. The awns must be removed before seeding.

One of the major reasons why green needlegrass is not used more extensively is its low germination rate in the seeding year. The seed may germinate in subsequent years, but the value of late germinating seeds in establishing a stand is questionable.

The bluestems There are many species of these warm-season grasses; in areas where they are adapted, they start growing late in the spring and produce most of their growth during the warm summer months. Two species of bluestem are native to North America: big bluestem (*Andropogon furcatus*) and little bluestem (*A. scoparius*). Big bluestem, which is native to the central states and the eastern edge of the Great Plains, will grow to heights of 4 to 6 feet (1–2 m) when adequate moisture and essential nutrients are available. It is a sod-forming grass with a deep, extensive root system. Tall bluestem is an excellent forage; it also makes excellent hay if cut before it becomes too stemmy and seed heads start to form.

Little bluestem is shorter (as might be expected); it only grows to heights of 2 to 4 feet (60–120 cm). It is native to the drier areas of the Great Plains—the area west of where big bluestem is found—and it is usually considered less palatable than big bluestem.

Seeding the bluestems is very difficult; their fuzzy seeds will not flow through a seeder, and the seeds are very hard to process and clean.

Buffalograss Buffalograss (*Buchloë dactyloides*), which is found mainly in the central and southern Great Plains, will tolerate alkaline soil conditions and is relatively drought-resistant. The species is a perennial, and the plants spread by stolons. It is a low-growing grass; plants are usually no more than 2 to 6 inches (5–15 cm) high. The foliage is grayish green. Because of its low-growing habit and stoloniferous growth, the plant tolerates heavy grazing. Buffalograss is used mainly as pasture; it produces a highly palatable and nutritious forage. This forage is especially useful because it maintains its feed value in a dry, cured state during the winter months and, therefore, makes good winter grazing.

This species has an unusual characteristic: the plants are dioecious, bearing either male or female flowers, but not both. The seeds are borne in a hard burr, which makes seed harvesting very difficult.

Grama grasses Among the species of grama grasses, there are two of great significance: sideoats grama (*Bouteloua curtipendula*) and blue grama (*B. gracilis*). Both species are warm-season grasses adapted to the Great Plains area. Sideoats grama is a perennial bunch-type grass; it has a distinctive flowering stalk with short, dangling, purplish spikes on culms that grow to a height of 2 to 3 feet (60–90 cm). It is found in areas having a little more rainfall than the regions in which blue grama is adapted.

Blue grama is smaller and has finer growth than sideoats grama; it is also considerably more drought tolerant and is sometimes found growing in association with buffalograss and other dryland range species. Blue grama is a bunch-type grass that tolerates alkaline soils. It does not spread by stolons or rhizomes. The plants are relatively short—from 6 to 12 inches (15–30 cm) at maturity.

Both sideoats grama and blue grama are relatively palatable and retain their nutritional value into the winter months.

This discussion of forage grasses has covered only the species of major importance in the United States and Canada; there are many others that may have significance in certain locales.

The forage grasses contribute significantly to the economic well-being of people around the world; grasses constitute about 75 percent of the cultivated forages. It is imperative that livestock producers be familiar with the advantages and disadvantages of the available grass species, the techniques of grass management and the best use for the various species.

Selected References

Gould, Frank W. 1968. *Grass systematics* McGraw-Hill.

Grass. 1948. *Yearbook Agr.* USDA.

Heath, M. E., D. S. Metcalfe, and R. E. Barnes. 1973. *Forages.* 3d ed. Iowa State University Press.

Hitchcock, A. S. 1951. *Manual of the grasses of the United States.* rev. ed. USDA Misc. Publ. 200.

Study Questions

1. What type of root system do grasses have? Compare the root system of grasses with those of legumes.
2. Describe the characteristics of the two major types of grass inflorescence.
3. Differentiate the characteristics of bunch-type and sod-forming grasses, and of cool-season and warm-season grasses.
4. Contrast the advantages and disadvantages of two cool-season grasses; of two warm-season grasses; of two dryland or range grasses.

Thirty-One Forage Management and Utilization

An integral part of the agricultural production system in both the United States and Canada is the production of forage crops. Forages had not received major attention in the management of farms and ranches until recent years when their real value and potential were recognized. Much of the land that is now being used for forage production could be managed in such a way as to produce economically from two to four times as much forage as it is now producing. A successful forage producer must have an understanding of the adaptation and limitations of the various species and their growth habits, of cultural practices, of pest control practices, of harvesting and storage methods, and of the marketing and utilization of forages.

Forage management varies from region to region (i.e., the western range versus the midwestern pasture) and according to the way in which the forage will be used (i.e., pasture, hay, or silage) and the general management scheme applied (i.e., permanent, rotation, or temporary pasture). These factors, together with an understanding of the general principles of plant growth, reproduction, and utilization, collectively applied, to arrive at the most advantageous forage production practices are referred to as forage management.

The Western Range

Rangelands cover more than 700 million acres (284 million hectares) in the United States, mostly in the western states (Figure 31-1). The objective in managing these rangelands is to obtain maximum livestock production while conserving the land resource. Their primary use is for livestock grazing, but they are also

Figure 31-1
Open range of the intermountain region in the western United States. [Photograph courtesy of J. E. Taylor.]

used for wildlife grazing, for recreation, for their timber, and for mining, and they serve as watersheds.

The western range lies west of an irregular line running north and south through the Great Plains from North Dakota to Texas. The rangelands in this area surround many irrigated valleys and dryland farming regions. In the southern part, year-around grazing is possible; however, in the north, the grazing season is limited because of the severe winters.

Irrigated pastureland in the western range, exclusive of irrigated pastures that are rotated with crops, has an area of only about 4 million acres (1.6 million hectares). About 65 percent of the irrigated pastureland is in the mountain regions, 31 percent in the coastal states, and only about 4 percent in the Great Plains region. The term "range" refers to a type of vegetation that is a native climax or that can be developed from climax by natural or induced succession. In contrast, the term "pasture" has generally come to mean tame pasture or forage areas that have been seeded with introduced species.

In the eleven western states, more than 50 percent of the rangeland is federally owned, and this land may be grazed by obtaining a permit from the U.S. Forest Service or from the Bureau of Reclamation. Recreational use of the western rangelands is increasing, and this may eventually create many problems for the range operator. However, many ranch owners may be able to capitalize on this trend by charging fees for hunting and vacationing.

There is little, if any, difference between the management of a range crop and that of a more intensively cultivated crop. Both consist of overcoming difficulties inherent in factors related to climate, soil, plants, and harvesting. The economic return from rangeland is almost exclusively measured in terms of livestock production. Therefore, in managing the western range, it is natural that attention be focused on animals and livestock production because animals convert forage into a product that can be used by man.

Range conditions are extremely variable, and there is no simple formula for range improvement. A knowledge of the effects of geological formation, topography, drainage, slope, aspect, soil textures, soil horizons, soil pH, climate, and elevation on plant adaptation, growth, and reproduction is necessary to develop sound range management practices. A well-managed range maintains a high proportion of useful, palatable, and nutritious forage species in the plant community. On poorly managed ranges, it may be necessary to replace an undesirable plant community with more desirable species. It is necessary for a range manager to know the species adapted to his particular environment, their nutrient requirements, growing habits, and how they will respond to the management practices employed. Undesirable plants on the range can be controlled by various means, such as fire, mechanical treatment, chemical treatment, biological control, and competition—namely by seeding adapted species in cleared areas to help establish the desired species.

Because livestock are known to be very selective grazers, it is necessary to control the grazing system utilized. The grazing pressure exerted on desirable species can result in the complete elimination of these species from the plant community. Deferred grazing, rotation grazing, seasonal stocking, and combinations of these grazing systems have been used to protect desired species. Fences may be needed to control livestock movement. The placement of salt and supplemental feed as well as water sources also are used to control livestock distribution and movement.

Money spent on range improvement programs can often return great dividends, depending on the circumstances. Before embarking on such a program, a thorough study of alternative programs should be made that takes into consideration the investment requirement, the income expected, and the time required for improvement to take place.

Pastures

In contrast to the native rangelands, pasturelands are normally considered to be those areas seeded with improved or introduced species. Thus defined, the pastureland of the United States and Canada are found primarily in the central and eastern parts of these countries, but irrigated pastureland and irrigated cropland in rotation with pastureland are found in the western parts. Pastures have been classified in three categories: permanent pastures, rotation pastures, and supplemental, or emergency, pastures.

PERMANENT PASTURES

A permanent pasture is one maintained indefinitely for grazing purposes (Figure 31-2). It is usually made up of perennials and of annual species that are able to propagate themselves by reseeding. Permanent pastures may be established by seeding or they may

Figure 31-2
Permanent pasture is an important source of forage for nearly all types of livestock production. [USDA photograph.]

become established through the spreading of aggressive species that compete with and crowd out the other species present. Once a permanent pasture is established, it is seldom, if ever, plowed for use in crop rotations.

In many areas where permanent pastures are being grown, this is the only feasible agricultural use that can be made of the land. Often the land is too steep or too wet or inaccessible for the large equipment required for economic cultivation.

Fertilizing and liming Many permanent pastures are less productive than others because they lack some of the basic nutrients required for plant growth. One way of restoring productivity to these pastures is to fertilize. In many cases, nitrogen is the most limiting factor, but other nutrients may be limiting also. A soil test will give some indication of the fertilizer needs for any given pasture. In the eastern United States where rainfall is relatively high, pasture soils may be acidic. The addition of lime to such soils will often reduce the acidity, thus increasing the productivity. Fertilizing and liming not only increase the yield of vegetation, but also increase the proportion of desired pasture species and improve the stand density.

Reseeding Reseeding as a method of improving permanent pastures has proved to be a valuable technique. It consists of the preparation of a seedbed and the seeding of desirable species at the right time of year to get a good stand. Reseeding to desirable species has been found not only to extend the grazing period, but also to make the pasture more productive. Overseeding or seeding on top of existing vegetation without preparing a seedbed has not generally been successful because of poor germination and stand establishment.

Renovation The most effective way of improving poor-quality permanent pastures has been found to be complete renovation rather than fertilization or reseeding. Renovation includes a number of steps and requires a greater investment of time and money. The first step is to destroy the existing vegetation and to prepare a seedbed. The next step is to take a soil test to determine the fertilizer and lime requirements. The recommended fertilizer and lime should be applied before seeding. The last step is to select desirable species and seed at the recommended rate and depth (see Chapter 12).

Experimental attempts have been successfully made to renovate by using chemicals to destroy existing vegetation and reseeding directly with a grassland drill, but the cost and availability of these chemicals may make this method impractical.

Seeding a well-adapted grass species, together with appropriate fertilization and adequate water, will probably give the highest yields of forage. However, mixtures of grasses and legumes frequently yield more total digestible nutrients per acre. Seeding mixtures may improve the stand because not all species are affected by the same hazards to stand establishment. Adding a legume improves the forage quality, especially the protein content. Grass-legume mixtures generally yield more, and higher-quality, forage than legumes alone. The grasses in a mixture have a higher protein content than grasses grown alone (with the exception of grasses heavily fertilized with nitrogen). Weed control in mixtures seems to be less of a problem. Legumes suffer less damage from heaving due to frost if planted in mixtures with grasses, and animals grazing on such mixtures have fewer bloat problems. Overall, the grass-legume mixture is frequently the best management decision.

It is advisable to delay the grazing of pastures during the first year until the plants have become established. The new seeding is especially susceptible to weeds, which compete for light, moisture, and nutrients. If weeds become a problem, mowing may be advisable to prevent weeds from crowding out and destroying new seedlings. Like that of rangeland, the grazing of permanent pastures must be properly managed to get the greatest return per acre. Rotational grazing has been found to increase the carrying capacity of pastures and the animal weight gain per acre. Continous grazing requires less labor but results in excessive trampling, overgrazing and damaging more palatable species, and excessive waste.

Good pasture management also includes the proper timing of grazing. In determining the time or season that a pasture can be grazed, stockmen must consider the stage of maturity of the particular species growing in it. For example, in the north central part of the United States, permanent pastures that include bluegrass and clovers are at their peak production in the spring and early summer. Later in the summer, the productivity of these pastures can be expected to be very low and supplemental pastures are then necessary.

Grazing pastures closely in late fall can be harmful to the stand. The energy reserves in the roots of the plants can be depleted to the extent that the perennials will not be able to survive the winter. Most perennials must be allowed a period in which to build up their root reserves in the fall, and perennial forage plants require a large store of food, especially readily available carbohydrates, to be sufficiently hardy to resist winter injury. Large amounts of carbohydrates do not assure winter hardiness, but plants that do not have adequate reserves at the onset of winter are much less likely to survive. Grazing or cutting for hay too frequently or too late in the fall will reduce the carbohydrate reserves; this will retard the hardening process, making the plant more susceptible to winter injury, which results in reduced vigor or in death. Harvest management is closely related to maintaining a good stand of high-yielding forage.

The morphology of vital growing parts, particularly the crown tissue, is also associated with winter survival. In some species the crown is partly covered by soil and litter, which affords some protection against extreme change in temperature and reduces the incidence of injury due to low temperatures. For example, much of the crown tissue of reed canarygrass, sweetclover, and bromegrass is covered by the soil, whereas the crowns of timothy and red clover are usually exposed to the elements and more susceptible to injury.

ROTATION PASTURES

Pastures that are grown at more or less regular intervals with other crops are referred to as rotation pastures. Most rotation pastures are mixtures of grasses and legumes, but if only grasses are used, nitrogen fertilization may be necessary for good production. The species to be planted depends in part on the number of years of forage in a rotation. If a rotation includes a single year of forage, annual crops such as annual sweetclover, sudangrass, millet, small grains, or ryegrass may be used. If forage is included in the rotation for two or more consecutive years, then biennials and perennials such as alfalfa, biennial sweetclover, bromegrass, orchardgrass, timothy, or many others may prove to be the best species.

Because rotation pastures must be reseeded periodically, they are relatively expensive. Even so, most cropland should be occupied by a sod crop at fairly regular intervals. The pasture crop in a rotation helps maintain soil structure, prevents excessive leaching and erosion, and, if legumes are included in the mixture, may increase the nitrogen content of the soil, thus resulting in increased yields of other crops in the rotation. Pasture rotation is also effective in controlling the weeds and diseases that affect many other crops.

Most species used in rotation pastures are tall-growing, deep-rooted, and highly productive. The forage production from rotation pastures can be expected to exceed that from permanent pastures and rangelands. The species suitable for rotation pastures differ from region to region and with different soil types. Those that are commonly used in the northern United States and Canada are alfalfa, 'Ladino' clover, red clover, and such grasses as bromegrass, orchardgrass, timothy, and other cool-season species. In the southern states, dallisgrass and bermudagrass are common, along with white clover and crimson clover.

SUPPLEMENTARY OR TEMPORARY PASTURES

Pastures falling into the supplemental or temporary category must be able to produce a large amount of forage in a short time after seeding. Most of the species used for supplemental or temporary pastures are annuals that serve only in emergencies. An unexpected shortage of pasture can be met economically by additional forage from such crops as oats, winter rye, and rape, or from some warm-season crops like sudangrass, soybeans, and millet. If properly planned, these supplemental pastures can supply forage when other

Figure 31-3
Timothy, cut and windrowed, curing in the sun before being stored as hay. Quality is maintained by careful drying and proper stacking or baling.

pastures are relatively unproductive, thus giving a sustained production of forage throughout the season.

Hay

Dry, green foliage with a moisture content ranging from 12 to 18 percent is referred to as hay (Figure 31-3). Hay can be kept for a relatively long period with little loss of nutritive value if properly stored; therefore, it allows for sustained livestock production in the northern regions in which forage cannot be produced year-around. It constitutes a considerable part of the winter rations for livestock in such regions, supplying much of the energy and essential vitamins, minerals, and proteins that are needed. Hay is especially important in feedlot operations and during periods of heavy snow cover on the range. In contrast to pasture and range, hay can be bought and sold on the open market. Because it can be sold for cash, it allows a manager more flexibility in his operation than does pasture or range, which can be turned into cash only through livestock grazing.

QUALITY

The three main characteristics that determine the quality of hay are leafiness, color, and the presence or absence of foreign material. Good-quality hay also has a pleasant aroma, a pliable texture, high nutritive value, and palatability. Because most of the protein in forage is found in the leaves, every effort should be made to save as many leaves as possible in harvesting and handling hay. Bright green color indicates a high vitamin A, or carotene, content. Conversely, brown, washed-out looking hay is very low in vitamin A. The content of vitamin D in hay is related to the length of time it is exposed to sunlight in the curing process: exposure to sunlight after cutting increases the vitamin D content. However, exposure to sunlight destroys vitamin A. In general, it is cheaper to supplement vitamin D than vitamin A; so it is more economical to limit exposure to sunlight to the length of time necessary to dry the hay sufficiently for safe storage.

Foreign material is classified as injurious and noninjurious. Injurious foreign material includes wire, nails, and other hardware, spiny seed, and poisonous plants. Noninjurious foreign material includes weeds and other matter that lower the quality of the hay but will not injure the animal.

Weather probably affects the quality of hay more than it does that of any other crop. Rainy weather at harvest time can account for the loss of as much as 40 percent of the nutrients in hay.

The stage of maturity at which hay is harvested is very closely associated with quality (Figure 31-4). If plants are allowed to become too mature, they begin to lose nutritive value and tend to become unpalatable. This is due to some extent to the loss of leaves (especially the lower ones) as the plants reach maturity. In determining the proper stage at which to harvest forage for hay, three factors must be taken into account: (1) yield, (2) nutritive value, and (3) the effect of harvesting at any given stage on the stand in subsequent years. Normally the longer one waits to harvest, the higher the yield will be, until after maturity has been reached. As a general rule, the nutritive value of a given volume of forage diminishes as the plant reaches maturity. For example, good-quality native hay in the western mountain states that is cut at the proper time is about 8 percent protein, which is 60 percent digestible. If this hay is permitted to become mature and weathered before cutting, the protein is reduced to about 4

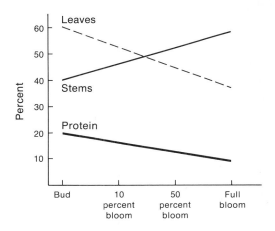

Figure 31-4
The effect of the stage of maturity of alfalfa on the percentage of the plant that consists of leaves, of stem, and of protein. [From *Management of Irrigated Forages in Nevada*. Nevada Agricultural Experiment Station, Reno, Nevada, in cooperation with the Soil Conservation Service.]

percent and is only 40 percent digestible. Another measure of the quality of a forage is its percentage of digestible nutrients, including protein. The digestible-nutrient content of the good-quality hay in the preceding example is 52 percent, whereas that of the mature and weathered hay is 36 percent. The most advantageous stage of development for harvest depends on the effect on the stand in subsequent years. For example, if alfalfa is harvested at the prebloom stage year after year, the stand can be reduced drastically.

GRADING

To facilitate the orderly marketing of hay (especially if it is not possible or practical for a buyer to inspect the product before purchase) the Agricultural Marketing Act of 1946 established the official United States standards for hay and straw. The standards take

Figure 31-5
A modern stack wagon, which mechanizes the handling of long, loose hay and builds high-density stacks, weighing as much as 4 tons, that are shaped to resist wind damage and to shed rain and snow. [Courtesy of Deere and Company.]

into account the various factors that determine quality—namely, leafiness, color, and the presence of foreign material—and establishes grades (i.e., U.S. No. 1, 2, 3 and Sample Grade) for each class of hay. However, the grading of hay has very limited use in the modern marketing system.

STORING AND HARVESTING
In recent years there have been a number of revolutionary changes in the techniques used for storing and harvesting hay. Much of the backbreaking labor that was at one time associated with hay making has now been replaced by mechanization.

There are several ways in which hay can be harvested and stored: as loose hay, in bales, chopped, and as pellets and wafers.

Long, loose hay. Storing hay in this form entails the conventional cutting of hay, windrowing it, loading it onto buckrakes or wagons, and putting it in the mow of a barn or in a stack in the long, loose form. Hay stored in this form takes up a great deal of space, but this method does allow rapid artificial drying; warm air is introduced under the hay and is forced through it to remove excess moisture that may have been present at the time of harvest.

Figure 31-6
A tractor-drawn hay baler picking up cured forage from a windrow and compressing it into bales. [Courtesy of International Harvester.]

Hay is often stacked in the field in the long loose form. Stacking can be done using a farmhand, which moves the hay from the windrow to the stack, or by other means such as the beaverslide stacker that is used in the intermountain regions of the West. Stacking can now be done mechanically as well. Equipment is available that will pick the windrowed hay up, load it on a trailer, form it into a stack, and deposit the stack (Figure 31-5). The stack can then be reloaded on the trailer later and taken to a feedlot. Stacking eliminates the space problem in a hay mow, but it introduces the problem of spoilage due to weathering.

Baling. With this technique, the forage is forced into a tight package, thus reducing the amount of space required for storage (Figure 31-6). The bales can be easily moved and, thus, are suitable for commerce and transport. Recently developed balers have self-loaders, which reduce to a minimum the labor required for handling bales, but they necessitate additional capital investment in the labor-saving equipment.

Field chopping. Field chopping requires an additional investment in equipment but is one step closer to the complete mechanization of the haying operation. The forage is chopped in the field and blown into a wagon. If self-unloading wagons are used, its trip from the field to storage in the mow, stack, or feed bunk is completely mechanized. Forage can be chopped either green or dry. If it is chopped green, it must either be fed to livestock as "green chop," or "soilage," or be artificially dried in storage. Even though field chopping is initally more expensive than baling, it can be economically feasible for large operations. Advantages to feeding livestock soilage are that the animal cannot be as selective in grazing, and all of the forage is utilized by the animal. Soilage feeding also allows for rotational utilization of forage in the field, which results in greater animal gains per acre.

Field pelleting and wafering. A recent development that might have a real future is field pelleting and wafering. This method requires an even greater investment in equipment than does any of the aforementioned methods, but it puts the forage in a very condensed form—an advantage that makes selling and transporting the hay over long distances more likely to be economically feasible. Many feedlots are completely mechanized, and pellets and wafers are more suitable for a mechanized feedlot than chopped, baled, or loose hay. Because of the large investment required, a pelleting or wafering implement would have to be kept in operation most of the growing season to be economically feasible.

Dehydrated Forage

In many hay-producing regions throughout the United States, commercial dehydration plants have been established. These establishments buy forage, usually alfalfa, from farmers and ranchers in the vicinity. This forage is dried and ground to be used primarily as feed additives.

There are a number of advantages to processing alfalfa, or other legumes, by this method. Dehydrating forage preserves more protein, carotene, B-complex vitamins, and essential minerals than do other processing methods.

The production and use of dehydrated alfalfa has increased significantly in recent years. Canada and the United States exported approximately 500,000 tons of "dehy cubes" to Japan and Europe in 1973. Pellets, cubes, and meal are produced for feed for ruminants (Table 31-1). The possibility of a loss of forage due to bad weather is eliminated. Harvesting machines can be operated twenty-four hours a day in almost any weather to harvest the crop when it is ready for processing. Only heavy rains can stop the harvesting operation when the ground becomes too soft to support the equipment.

The dehydration process at present requires large amounts of fossil-fuel energy. The future of dehydration of forage may depend on the availability and the cost of such energy sources.

Silage

Silage is the product formed if chopped green forage is allowed to ferment; fermentation occurs if the plant material is stored in an airtight container—a silo (Figure 31-7). One of the crops more commonly used to make silage is corn, but silage can be made by fermenting any moist hay or pasture crop: grasses, legumes, green cereals, or any combination of these.

Like other methods of handling forage, making silage has its advantages and its disadvantages. Some advantages are:

Table 31-1
Feedlot performance of cattle fed alfalfa cubes, coarsely ground pellets, or finely ground pellets for roughage in 119-day feeding tests.

Form of alfalfa	Number of steers	Average initial weight (lb)	Average daily gain (lb)	Daily feed intake (lb)	Feed/pound of gain (lb)	Carcass grade*	Condemned livers (%)
Cube	119	761	2.71	21.98	8.10	11.3	47.2
Coarsely ground pellet	114	766	2.66	21.97	8.27	11.5	47.9
Finely ground pellet	117	752	2.55	21.95	8.62	11.1	55.7

Source: J. K. Matsushima. Role in feedlot feeding. In *Alfalfa science and technology,* ed. C. H. Hanson. Series in Agronomy, 15 (American Society of Agronomy, 1972).
Note: Alfalfa cubes give the best daily gain and the lowest amount of feed per pound of gain.
*USDA grade: good = 9–11; choice = 12–14.

1. Forage that might have otherwise been ruined by inclement weather can be preserved as a high-quality product.
2. There is less loss of vitamin A in silage than in sun-cured hay.
3. The amount of dry matter conserved is greater than it would be if the same crop were field-cured.
4. A higher percentage of the protein is preserved in silage than in sun-cured hay.
5. Crops can be harvested at an earlier stage of development; thus, their nutritive value is high.
6. Weed seeds are destroyed in the fermentation process.
7. The ensiling process is mechanized so that little physical labor is required.
8. Because of the nature of silage, feeding can be fully mechanized.
9. Crops that cannot be made into hay can be readily ensiled; for example, sorghum, sudangrass, faba beans, sunflowers, and Jerusalem artichokes.
10. Silage will keep for several years, making it a good reserve feed supply.

Some disadvantages are:
1. The high moisture content of silage may result in a reduction in dry-matter consumption by livestock.

Figure 31-7
Silo for the storage of plant materials for livestock feed. Air-tight silos such as the one shown load from the top; preserved feed is removed from the bottom.

2. The storage facility must be able to withstand two to four times as much weight as would be required to store the same volume of hay.
3. To mechanize the process, it is necessary to purchase expensive equipment; so the capital investment is increased.

Fermentation of the chopped forage depends on the action of bacteria, which results in chemical changes in the forage. The plants are still actively respiring after the forage has been cut, and they will continue to respire for quite a while, depending on how they are handled. Aerobic bacteria that are already present on the plant leaves and stems multiply in the presence of oxygen. The aerobic bacterial action plus respiration use plant carbohydrates to produce heat and carbon dioxide.

After the chopped forage is ensiled, the available oxygen is soon used. It is very important that as much air (oxygen) as possible be excluded from the moist, stored forage. If oxygen is not excluded, mold and undesirable acids such as butyric acid will form, and the quality of the forage will deteriorate. Sealing the silage from the air prevents the development of undesirable bacterial growth and putrifaction. In the absence of oxygen, the beneficial anerobic bacteria multiply rapidly, forming mainly lactic acid, but also acetic, propionic, and succinic acids. Ultimately, the amount of acid becomes great enough to cause all the bacteria to be killed; the ensiled material has a pH of about 4.2. The bacteria literally eat themselves to death because the by-product (lactic and other acids) creates an environment that inhibits their growth and thus preserves the silage.

Mold development can be prevented by applying a mixture of volatile fatty acids to the forage at the time of ensiling. A mixture of propionic and acetic acids will prevent heat production and mold growth.

The quality of silage depends on the plant material and on the amount and type of acids that are formed during the fermentation process. Quality also depends on the amount of protein broken down in the fermentation process. The fermentation process is greatly affected by the chemical composition of the plant, the fineness or coarseness of chopping, the speed with which the silo is filled, and the tightness of the air seal. A good-quality silage contains a predominance of lactic acid. If butyric acid or an excessive amount of ammonia form, the silage is likely to be of poor quality.

It is important that the silage be properly distributed and packed in the silo to exclude as much air (oxygen) as possible. Packing can be done by hand or with the aid of mechanical equipment. In bunker or trench silos, it is possible to drive a tractor over the silage with a blade or scoop to distribute it properly. The weight of the equipment does an adequate job of packing. If the silage is well packed, small air leaks in the silo walls or top seal will result in a minimum amount of spoilage.

The amount and type of readily available carbohydrate has a significant bearing on the quality of the silage formed. Grasses (and especially legumes) may not have sufficient readily available carbohydrates for favorable bacterial growth. It may be worthwhile, if not necessary, to supplement with a high carbohydrate concentrate such as molasses, or shelled corn, or barley, or other cereal grain at the time of ensiling. The use of organic acid preservatives to preserve silage has increased in North America and Europe in recent years.

Making good silage may be determined by the way the forage is handled during the first few hours after harvest. Fine chopping, rapid filling, and a good seal will enhance the quality.

The forage may be direct-cut with a forage chopper and ensiled. A disadvantage of this process is that the excess water in the forage may drain away from the silage taking significant quantities of water-soluble nutrients with it. Another method of harvest is to cut the forage, allow it to wilt to about 65 percent moisture (about 4 hours on a warm, sunny day), chop it, and then ensile it. Low-moisture silage (40–60 percent) can be successfully made if it is stored in a gas-tight silo. Limited fermentation takes place in such a silo and preservation is dependent on the exclusion of oxygen.

Silage can be stored in a number of different kinds of structures, providing they allow for the exclusion of air. There are two basic types: upright silos and horizontal silos. Upright silos include the typical tower constructed of wood, concrete, steel, or glass-coated steel. Temporary upright silos can be made from snow fence or heavy galvanized wire fence lined with plastic or waterproof paper.

Gas-tight silos made of steel and equipped with mechanical unloaders are now available. The distribution and trampling of the silage and its moisture content within gas-tight silos are not as critical. Once sealed, the oxygen present within the silo is soon used up by respiration, and the silage is properly preserved.

There are several kinds of horizontal silos including trenches, bunkers, and stacks. A portable feedgate placed at one end of the horizontal silo can make it possible to self-feed.

The trench silo is simply a trench dug in the ground. It should be located so that water will drain away from it and so that it is accessible to machinery. Most trench silos have concrete floors. In some cases, the soil serves as the walls of the silo; in others, a lining of concrete, wood, or plastic can be used.

Horizontal silos that are built at ground level or above the ground with supported walls made of concrete or wood are called bunker silos. Like trench silos, most bunker silos have concrete floors to prevent mud problems at the time the silage is used.

Stack silos have no walls or floors and require no construction at all. A stack silo with minimum waste can be made by using black polyethylene for a ground sheet and a cover with the seams sealed to make them airtight. Polyethylene silo "socks" are available that can be rolled up as they are filled and sealed at the top. A vacuum is then applied to evacuate as much air as possible. Silage made in this way has very little spoilage as long as the seal is maintained. During the winter the silage can be moved to an upright silo for mechanical feeding. This is a common practice in Minnesota and North Dakota.

It is important to try to reduce silage losses as much as possible. Some dry matter is lost in seepage from the silo. In addition, there are losses due to top spoilage in almost any structure because of exposure to air. The larger the area exposed to the air, the greater the loss due to top spoilage. Losses may also be due to respiration of plant material. Bacterial action results in some loss of dry matter; the activity of other microorganisms also causes some loss. These losses can be kept to a minimum by excluding air from the silage as much as possible.

Silage is an excellent way to handle forage if a producer has the proper equipment. There has been a significant increase in the amount of forage being made into silage in recent years.

Selected References

Grass. 1948. *Yearbook Agr.* USDA.

Handbook of Official Hay and Straw Standards. 1958. USDA.

Heath, M. E., D. S. Metcalfe, and R. E. Barnes. 1973. *Forages.* 3d ed. Iowa State University Press.

Making and feeding hay-crop silage. 1962. *U.S. Dep. Agr. Farmers' Bull.* 2186.

Martin, J. H., and W. H. Leonard. 1967. *Principles of field crop production.* 2d ed. Macmillan.

Morrison, F. B. 1956. *Feeds and feeding,* 21st ed. Morrison Publishing Company.

Study Questions

1. Describe the characteristics of well-managed rangeland.
2. Give alternative practices that might be adopted to improve the yield of an unproductive permanent pasture.
3. What are the characteristics of high-quality hay?
4. There are a number of ways of handling forage: loose hay, baling, field chopping, and field pelleting and wafering. What should be considered in deciding which method is best for a given operation?
5. Why is it important to exclude air from silage?

Thirty-Two Progress Through Research

Introduction

In an era of manned space flights, moon walks, orbiting space laboratories, and exploration of outer space, it is tempting to symbolize man's scientific progress by the fiery blast of a superrocket. Without denying the technological achievement of space exploration, the increase in agricultural production, both of crops and of livestock, constitutes the most significant contribution of basic and applied research to human welfare. Even major advances in the medical sciences are of relatively minor significance in the face of growing hunger and malnutrition.

There are a multitude of ways to describe and measure the impact and significance of research on agricultural production. Perhaps the simplest measure of its success is to note that we have survived! In spite of dire predictions of mass starvation, in spite of natural disasters (drought, flood, disease epidemics, and insect infestations), in spite of political conflict among nations, agriculture is feeding mankind. In fact, the quality and quantity of the human diet throughout the world is improving each year. But we should not conclude from this that starvation and malnutrition are no longer major problems confronting national leaders, agriculturalists, and ordinary citizens. That the earth may someday be unable to nourish its inhabitants is still a major cause of concern, and it is the agriculturalists who must find ways in which to increase food production.

There are more than a few pessimists who believe that the struggle to feed mankind is already lost. Those who share this belief apparently feel that the progress made in increasing agricultural productivity has been a matter of happy coincidence, or luck, and now the luck has run out. This is simply not true. In agriculture, as in industry, increased production and increased production efficiency come not by chance, but by the concerted effort of interested, educated workers who actively seek new and better methods of production. Research in agriculture still has the potential to feed mankind. However, the problem is acute in that, as the world's population continues to increase, available land for crop production decreases (see Chapter 10 for a brief discussion of the problems of land-use planning). As recently as fifty years ago, total food production could be increased simply by bringing more land under cultivation, but the resource of available, suitable land for crop production has been used up. The cost of bringing jungle and desert lands under cultivation is great; thus, the challenge is one of producing more from each acre.

Certainly few, if any, researchers in any agricultural field would deny that a little luck helps in a research program, but it is education coupled with dedicated, hard work that gets the job done. Growth in crop production and improvements in efficiency have resulted from research in a wide array of specialized areas. The rural America of the Revolutionary War has become the urban, industrialized America of the space age. Yet, in spite of industrialization, in spite of mass migration from farm to city, North Americans still eat more food of better quality than do people living in most countries. The quantity and quality of the average American's diet are reflected in several facts. More than half of the world's population lives on a near-starvation diet of 1,500 kilocalories or less. In the United States, the average diet is close to 2,200 kilocalories and consists of more protein than most. In Asia, for example, nearly 75 percent of the human diet consists of plant starch and less than 10 percent of animal protein; in contrast, in the United States only about 40 percent of the average diet consists of plant starch and about 35 percent of animal protein. There are many countries in the world that cannot afford the luxury of feeding plants to animals and then eating the animals. Yet, even though the average North American diet is adequate in quantity and quality, there are many children and adults subsisting on an inadequate diet.

The application of the results of research in nearly all phases of plant science (the use of new knowledge) has increased total productivity and production efficiency. Research has made it possible today for the labor of one man to produce food for more than fifty people; at the turn of the century, the ratio was about 1:4. From 1947 to 1951, U.S. crop yields were more than 30 percent higher than those obtained from 1935 to 1939, and yields have continued to increase in the past two decades. Increases in yield and greater production efficiency of major crops in Canada have paralleled the changes in the United States: although there are fluctuations from year to year in per-acre yields that reflect environmental conditions during the growing season, yields of the major field crops have increased from 10 to 20 percent in the brief period from 1967 to 1972 (Table 32-1).

Plant scientists cannot take all of the credit for increased yields and increased production efficiency. The work of agricultural engineers has resulted in improved farm machinery—from new plows and bigger, more efficient tractors to special seeders and mechanical cotton and tomato pickers. Agricultural chemists, working in conjunction with plant and soil specialists, have developed new and better fertilizers, which increase yields dramatically. For example, the availability and use of commercial nitrogen fertilizers in the past thirty years has helped to double, and in some cases triple, the yield of winter wheat

Table 32-1
Yields of five important field crops in Canada from 1967 to 1972.

Year	Crop (bu/acre)				
	Spring wheat	Barley*	Oats	Flax seed	Rapeseed
1967	19.4	31.1	40.9	9.2	15.2
1968	21.8	36.8	48.0	12.9	18.4
1969	26.7	39.7	48.6	12.0	16.6
1970	26.0	41.3	51.3	14.5	17.8
1971	27.0	43.0	53.2	12.7	17.9
1972	24.7	41.4	49.2	13.4	17.5

Source: Statistics Canada, 1973. *Quart. Bull. Agr. Statist.* 66(1). Published by authority of the Minister of Industry, Trade, and Commerce.
*The acreage of barley increased from 8.0 to 12.5 million acres from 1967 to 1972. Total production in both 1971 and 1972 was more than twice the 1967 production of 252,867,000 bushels. Total production increased more rapidly than did the area planted to barley.

grown in Montana, compared with yields obtained using noncommercial sources of nitrogen such as manure or nitrogen fixed in association with legumes. However, these increases were not due solely to the use of commercial fertilizers; the cultural practices for wheat production were modified, as a result of more research, so that the crop could benefit from additional nitrogen. One modification was to rotate wheat with fallow, which helped to insure that the crop would have adequate moisture so that the nitrogen could be efficiently utilized. Another was to improve weed control. A third was to use cultivars with higher yield potential. In addition, the development of methods for producing triple superphosphate increased phosphorus availability for plant use from less than 5 percent with rock phosphate to more than 40 percent in commercial fertilizers. About five percent of the phosphorus fertilizer used in 1930 contained triple superphosphate; more than 30 percent of that used in 1960 contained it; and the increase is continuing. Like the nitrogen in commercial fertilizers, triple superphosphate has contributed to higher yields from crops grown in soils that are deficient in phosphorus.

The cooperative efforts of plant scientists, entomologists, and plant pathologists have led to the development of chemicals for the control of specific weeds, of specific insects, and of serious plant diseases. Although many insecticides have fallen into disfavor in the 1970s, the effectiveness of chlorinated hydrocarbons and organic phosphates in pest control from the 1940s until recent legislation limited their use is unquestionable. Fields of alfalfa that have been saved from the alfalfa weevil, or corn from the European corn borer, and of cotton from other pests are only a few key examples. But these benefits must be weighed against the problems and costs of potential pollution. Currently, agricultural researchers are developing new, nonpersistent, nonpolluting families of effective insecticides. Selective herbicides such as 2,4-D, which allow the chemical control of broadleaf (dicot) weeds growing in cereals, have been similarly effective. Fantastic savings in the costs of control and in reduced losses have been realized since the late 1940s when 2,4-D became widely used. Currently, more refined and more specific herbicides are being developed, which will allow the chemical control of specific weeds growing in a particular crop, such as the wild oats that grow in wheat, or the cheatgrass that grows in peanuts. The development of highly selective herbicides triggered the use of no-tillage or minimum-tillage cropping systems for corn and to a lesser extent for soybeans and other crops. In addition to the conservation of soil that results from minimum tillage, scarce and expensive fuel is conserved through the use of herbicides because fewer cultivations are required for adequate weed control.

Plant breeders have contributed a great deal to the increase in crop productivity. Through the efforts of plant breeders, the adaptation of many crops has been significantly extended and cultivars have been developed that respond positively in terms of yield to more intense field management; for example, the use of fertilizers and irrigation.

To some extent, major progress in crop production has involved all of the diverse disciplines described above. The complex relationships between them can be readily illustrated through the story of hybrid corn. Hybrid corn was developed through research in plant breeding more than fifty years ago, and hybrid corn cultivars were accepted and widely used by the early 1940s. Due to their genetic makeup, hybrid cultivars have the potential for much higher yields than do many older types of cultivars. But, as yields are increased, more minerals are removed from the soil and more fertilizer is required; the expertise of the soil chemist and plant nutritionist is needed to meet this requirement. Like any crop, hybrid corn is subject to damage from weeds, insects, and diseases. Considering the extra cost of buying hybrid corn seed and the investment in additional fertilizer, special chemical and mechanical weed control methods must be employed, as well as insect and disease control steps, so that the yield potential bred into the cultivar can be realized. Special or modified planting, cultivating, and harvesting equipment may be needed to manage the crop properly in the field for highest production. Finally, because of higher yields and greater investments in growing the crop, more care and more sophisticated equipment are called for in harvesting and storing it. Taken together, the contributions from the separate disciplines to the production of hybrid corn have succeeded in doubling the corn yield. Granted, the increase was made possible by the development of hybrid corn by plant breeders, but its potential could not have been realized without research in other disciplines. The development of new and better fertilizers, of pesticides, and of new and more efficient machines are all products of research.

The Scientific Approach to Problem Solving

"Research" and "science" are frequently considered to be synonymous. By definition, they are not, but, in terms of the increase in crop production, they can at least be discussed as a single concept. A science, such as plant science, is defined as "a body of organized knowledge," but in practice this knowledge is used to obtain new knowledge or facts through research.

Research in plant science, as in most other fields, is an organized attack, using the scientific method, on specific problems or groups of problems. The problem that plant scientists have been addressing (and must continue to address) is the prospect of starvation on a large scale. Possible hypotheses for solving this problem might be to increase food supplies, to limit population growth, or to redistribute the food supply so that underfed people receive more of what is currently being produced. To test these hypotheses requires research.

Let's assume for the moment that the hypothesis of limiting population growth is found to be unacceptable as a solution. Let's also assume that reliable data indicate that the redistribution of food will not solve the problem. Obviously, then, the hypothesis to be tested is that of producing more food. How is more food to be produced? Research must answer such critical questions as, How much will yields increase with more fertilizers? What is the maximum yield from adding fertilizer alone? Can new cultivars be developed that will yield enough to help solve the problem? What diseases, insects, and weeds must be controlled to obtain highest yields?

Let's say that the hypothesis of increasing food production to solve the problem of starvation is accepted; it then becomes a theory. When a theory is tested repeatedly and found to be valid, it becomes a natural law. Rarely would a single researcher, or even one group of scientists (e.g., the agricultural research scientists at one university) attempt to study the entire problem of hunger and all possible solutions. As a rule, individual studies would be confined to some manageable facet of the entire problem. The precise area in which any researcher would work depends on his expertise: soils and plant nutrition, pest management, plant breeding, and a host of others.

Statistics as a Tool

In plant science, as well as many other fields of research, statistical methods are used as a major guide in identifying the best hypothesis. Although hypotheses for the solution of the major problem of human starvation could be subjected to statistical analysis, the problem itself is too general for a precise illustration.

Let us look instead at "the systems approach" and the simple statistical analysis of one facet of solving the problem of starvation, plant breeding. Suppose that a plant breeder establishes the hypothesis that through plant breeding higher-yielding cultivars of important crops (such as wheat, rice, and corn) can be developed. Following one or more accepted methods of plant breeding to develop and identify superior genotypes, the plant breeder develops one or more new strains of wheat (or some other crop). He compares the yields obtained from the new strains and then compares these yields with those of the best, currently available cultivars to verify that at least one of the strains he has developed is superior to cultivars already in use (which he considers to be his check). He must also grow his new strains and his check under environmental conditions comparable to those under which a farmer would produce the crop. He then compares the average (or mean) yields of his new strains with the average yield of the check, which he obtains by growing several plots (replications) of each new strain and of the check. The results from such a yield test are given in Table 32-2. Although yields are expressed in bushels per acre, the test was carried out using small plots. Note that the yields from each replication of each of the five new strains and of the check differ slightly (e.g., for the check the yields for the four replications are 26.0, 27.1, 25.8, and 26.5 bushels per acre). The four replications of the check are samples of the same cultivar, or "genetic population," as are the replications of the five experimental strains. Thus, the differences, or variation, in yield within the check, or within any one of the experimental strains, are not due to genetic differences. Even though the entire study was conducted in a relatively small area, the differences (variation) within any of the six strains (including the check) can be

Table 32-2
Yields from five experimental strains of spring wheat developed by a plant breeder and from a currently grown check cultivar.

Replication	Yield (bu/acre)					
	Check	Strain 1	Strain 2	Strain 3	Strain 4	Strain 5
1	26.0	27.2	32.0	35.4	37.2	27.0
2	27.1	27.0	33.2	34.8	36.8	26.1
3	25.8	26.5	31.9	35.0	37.0	26.5
4	26.5	26.9	32.1	35.4	37.4	24.6
Total	105.4	107.6	129.2	140.6	148.4	104.2
Average	26.35	26.83	32.25	35.15	37.10	26.05

attributed to the different environments of the four locations in the experimental area in which the replications of the strains were grown. The variation within each of the six strains is also expressed in part as the variation among the averages (means) of the six strains. It is possible that the differences between the averages of the strains are due only to the effects of the environment in which each of the four replications of each strain was grown; it is also possible that the differences are due to genetic differences between the strains. The plant breeder can determine the probability that the differences between the averages are due to environmental effects. If this probability is small (usually 5 percent or less), he rejects the idea that the differences are due to environmental effects and concludes that they are due in part to his plant-breeding efforts. Further analysis reveals that at least two strains are superior to the check, strains 3 and 4. He concludes, therefore, that the yield of wheat can be increased through plant breeding.

The statistical procedures employed in this case are relatively simple. The general method is the analysis of variance, which serves to partition the total variation among the twenty-four values (four replications each of the six strains) into identifiable causes. The number of identifiable causes depends on the field design or layout of the experiment, but the important comparison is the ratio of the variation among the strain averages to the variation due to environmental effects. The analysis of variance affords estimates of both of these quantities. The ratio of variance is expressed as F; it is known as the F test. When F exceeds a prescribed, theoretical value, a value determined by the F-distribution, the probability that variation among cultivar means is due to environmental factors is 5 percent or less. From this it can be concluded that the differences are "real differences," or that the strains differ by more than environmental effects. The strains are genetically different as a result of plant-breeding efforts.

A modified t-test is used to compare pairs of cultivar means to determine which ones differ or cause the significant variation. This t-test is known as the least significant differences (LSD). The analysis of variance for the data in Table 32-2 is presented in Table 32-3.

The same general method is used to determine the effectiveness of different rates of

Table 32-3
Analysis of variance for data given in Table 32-2.

Source of variation	Mean square	D.F.	F	$F_{0.05}$
Total	—	23	—	
Among block means	0.54	3	1.80	3.29
Among strain means	94.63	5	315.43	2.90
Residual (error)	0.30	15	—	

$$LSD = 2.13 \frac{2(.30)}{4} = .83$$

Note: Field design was a randomized complete block. The F value for variation among strains ($F = 315.43$) is greater than the theoretical value ($F_{0.05} = 3.29$); therefore there is significant variation among the strain means. Any two strain means that differ by the value of LSD or more are deemed to be significantly different. Note that in Table 32-2 Checks, Strain 1, and Strain 5 are not significantly different. Strain 4 is significantly better than all others.

fertilizers or rates or types of pesticides or cultural practices. The use of these statistical procedures is not confined to the agricultural researchers. Any crop producer can carry out simple "on the farm" experiments to help determine the best cultivar or most appropriate planting or fertilizer rate for a given crop. An extensive consideration of statistical methods and the design of experiments is beyond the scope of this discussion; however, many essential concepts are presented in an introductory course in statistics and county agents may be able to provide advice.

Plant Breeding

The significance of plant breeding to human welfare is reflected in many aspects of modern life: improving the quality of the average diet, increasing average yields, and expanding the areas of adaptation of many important crops are but a few examples. In 1970, the importance of plant science received attention worldwide when an American plant breeder was awarded the Nobel Peace Prize. Dr. Norman Borlaug received international recognition for his outstanding leadership in breeding superior cereal crop cultivars.

Plant breeding began when man first recognized superior types of plants and saved seed from these plants to produce more, similar plants. Thus, the early plant breeder selected and preserved superior plant types in preference to inferior types. Part of the superiority was due to heritable (genetic) differences between plants. The basic function of the plant breeder today is still the same, to recognize and to select genetically superior plant types. Plant breeding is more than a simple search for specific genotypes. It includes fundamental research in genetics to determine how important traits are inherited, and it may require the creation of complex hybrids to assemble the combinations of genes wanted in a single plant or population of plants. The final step in plant breeding is selecting superior genotypes. As the need for specific types of plants is recognized and grows (the need for resistance to new diseases, or for expanded adaptation to new environments, or for superior quality, such as higher protein content), the modern plant breeder finds it increasingly difficult to find individual plants with all of the traits that are needed. He may be required to intercross plants to obtain them. In some cases, the genetic factors he needs cannot be found in commonly available plant materials, and he must then look for diverse types of plants in search of the trait he wants. A major source of diverse types of plants is the collection of plants from the center of origin of the species with which the plant breeder is working. Periodically, international teams travel to various centers of origin to collect seed and plant samples of cultivated species, and of their relatives, for use by plant scientists. The seed stocks collected are carefully stored in special laboratories and are available to researchers.

The foundations of modern plant breeding are the basic laws of inheritance first reported by Gregor Mendel in 1865 and rediscovered independently in 1900 by Correns, de Vries, and Tschermak. The applications of Mendel's laws and the use of statistics have led to remarkable success in increasing yields through plant breeding. In the five decades from 1915 to 1964, wheat yields have nearly doubled and corn yields have more than doubled (Table 32-4). A large part of the increased yields must be attributed to the development of higher-yielding cultivars through plant breeding.

Plant breeders have improved a wide array of traits in many crops. Successful efforts have been made to increase seed yield in all of the cereals and forage (vegetative growth) yields in alfalfa, other legumes, and grasses. In addition, resistance to many diseases and some insects has been "bred into" cultivars of many different crops (e.g., in wheat, resistance to certain races of stem rust). Plant-growth characters have been changed to create cultivars that mature early or cultivars that are dwarf in stature, but still have high yield potential. Currently, major efforts are being made to improve the nutritional quality of wheat, corn, and other grains by increasing the total protein content and by increasing the quantity of a specific essential amino acid; for example, lysine in corn.

Table 32-4
Average yield in pounds per acre of three major crops in the United States for five decades.

Period	Wheat	Corn	Soybeans
1915–24	833	1488	—
1925–34	825	1354	805
1935–44	910	1595	1085
1945–54	1027	2103	1199
1955–64	1427	3102	1398

Source: USDA data.

The following example illustrates how a plant breeder selects superior new plant types. Assume that a given gene controls resistance to a specific disease (e.g., leaf rust in wheat) and that allelic forms of this gene are A and a. The genotype aa makes a plant susceptible to the disease. Genotypes Aa and AA are fully resistant to the disease, and A is said to be dominant over a—that is, plants that have either Aa or AA genotypes appear to be the same in their resistance to the disease and thus have the same phenotype. If the two genotypes Aa and AA do not have the same phenotype, if AA is fully resistant and Aa is partly resistant, or intermediate between the aa and AA genotypes, the effect of A is said to be semidominant or nondominant.

In breeding for disease resistance, the plant breeder must first determine if the various plants (or seeds) he has available differ genetically for the trait he wishes to study. This could be determined by growing a number of seedlings from each of several different strains or cultivars of wheat and uniformly exposing all the seedlings to the disease-causing organism. Some might prove to be resistant to the disease and others susceptible.

Then the plant breeder must determine how the resistance is inherited. To do this, he makes hybrids between resistant and susceptible plants by removing the anthers from the florets of one type of plant and placing pollen from the other type on the stigmas of those florets from which the anthers were removed. The anthers are removed to prevent self-pollination. The cross pollination yields hybrid seed; it is heterozygous for rust resistance. If one parent was AA and the other aa, the hybrid (called the F_1) is Aa. Both the AA and the aa types are homozygous. The Aa type is heterozygous. If the heterozygous plant is resistant, like one of the parents, then the plant breeder concludes that resistance is dominant: both the Aa and AA genotypes have the same (resistant) phenotype. If the heterozygous plants are intermediate in phenotype, resistance is nondominant, and, if the heterozygous plants are susceptible to the disease, as is one parent, resistance is recessive. Only homozygous recessive plants are disease-resistant.

Assume for the moment that resistance is dominant. The plant breeder has a number of disease-resistant plants. He wishes to produce a cultivar that is homozygous for disease resistance, but the plants to be used as the seed source may be either homozygous or heterozygous. He must save the seed of only homozygous plants. His job is to determine what plants are homozygous for disease resistance. To determine homozygosity, he grows a sample of seed from each disease-resistant plant and exposes the seedling uniformly to the disease-causing pathogen. The seedlings are the progeny, or offspring, of each plant. The behavior of the seedling in terms of resistance or susceptibility to the disease indicates the genotype of the parent plant on which the seed was produced. If the plant was homozygous resistant, then all of its offspring (the seedlings tested) would also be resistant. If the plant was heterozygous, the ratio of disease-resistant offspring to susceptible offspring would be 3:1. This ratio is the result of chromosomal behavior during meiosis and the pattern in which the sperm and egg nuclei unite at fertilization (review Chapter 4, Figure 4-6, and Table 4-1). In a plant that is Aa, two types of pollen, A and a, can be produced; the same types of megaspores (and ultimately eggs) are also produced. That either type of pollen fertilizes either type of egg is a matter of chance. Thus, the A pollen could fertilize the A egg to yield an AA embryo (seed), or it could just as easily fertilize the a egg to yield an Aa embryo. Similarly, the a pollen could fertilize the A egg to yield an Aa embryo, or it could fertilize an a egg to yield an aa embryo. Because both the AA and Aa types are disease-resistant, the phenotypic ratio of resistant offspring to susceptible offspring derived from the self-pollination of one heterozygous plant is 75:25. If seed from only homozygous resistant plants is used as the foundation for the new cultivar, the new cultivar must be "pure" for resistance.

Now let us assume that the plant breeder is concerned with two traits at the same time: disease resistance and plant height (dwarf versus normal), which is controlled by gene B. The gene controlling plant height is on a different pair of homologous chromosomes: during meiosis, A and B are independent. As a result of studies like those described for determining the inheritance of disease resistance, the plant breeder has determined that normal height is recessive to dwarf—that is, dwarfness is dominant.

In the available plant materials, the plant breeder has two types of homozygous plants: disease-resistant, normal plants, and disease-susceptible, dwarf plants. He seeks to develop a disease-resistant, dwarf plant. The genotype of the disease-resistant, normal plant is $AAbb$, and that of the disease-susceptible, dwarf plant is *aaBB*. He produces a hybrid by cross-pollinating these two types. Its genotype is *AaBb*; it is heterozygous for both *A* and *B*. If the meiotic process is normal, this type of plant will produce four types of pollen grains (sperms) and the same four types of megaspores (eggs): *AB, Ab, aB,* and *ab.* When the heterozygous plant is self-pollinated, seed of nine different genotypes is produced by the heterozygous plant.

The seeds from the heterozygous plant are tested (by planting them) to determine which have the disease-resistant and dwarf traits. All plants that are susceptible to disease, or normal in height, or both are discarded. The remaining plants are progeny tested to determine which plants are homozygous for both disease resistance and dwarfness. Seed from these homozygous plants becomes the foundation of a superior, new cultivar.

This plant-breeding program is quite simple, but, frequently, a plant breeder must make many hybrids in order to combine the traits that are wanted into one superior grouping. In addition, many traits, such as yield, are controlled by many genetic factors, or genes, each one of which has only a slight effect. The effect of the individual factors cannot be measured; therefore, complicated, statistical methods are used to determine which hybrids should be produced and which plants should be retained in the breeding program or discarded from it.

Many crop species are polyploid; that is, rather than having pairs of homologous chromosomes, they may have three or more individual chromosomes of any one type. Chromosomal behavior during meiosis is more complex, and expected genetic ratios—even for simply inherited traits, such as disease resistance—may be distorted, making genetic analysis more difficult.

One of the most commonly cited contributions of plant breeding to increasing crop yields and quality is the development of hybrid corn. It has long been known that the hybrid of two (or more) relatively homozygous parents is more vigorous and higher-yielding than either parent. This phenomenon is referred to as heterosis or hybrid vigor.

Corn is normally cross-pollinated. To produce hybrid corn seed required two major steps: First, lines to be used as parents had to be developed. These lines had to be nearly homozygous and at the same time had to carry the genes that would yield a superior plant. The lines were developed by forcing plants to self-pollinate. Repeated self-pollination leads to homozygosity (see Figure 4-7 on page 72). Unfortunately, self-pollination of a normally cross-pollinated species for several generations frequently leads to a weak plant, but vigor is restored in the hybrid.

Tens of thousands of inbred lines were developed by plant breeders and tested to determine which would yield superior hybrids. Hybrid seed was produced by removing the tassels of plants of one inbred line by hand and allowing pollen from plants of a second inbred line to pollinate the hand-detasseled plants. Hand detasseling is time-consuming and a major expense in producing hybrid seed for commercial plantings.

A genetic factor that caused male sterility, discovered in the 1940s, eliminated the need to detassel by hand. This factor is most unusual in that it is not carried in chromosomal DNA, but is associated with the cytoplasm. Remember, that most of the initial cytoplasm in the embryo is maternal; pollen contributes only a sperm nucleus to the embryo. An inbred plant or line with cytoplasmic male sterility cannot self-pollinate, but seed can be produced on these plants as a result of cross-pollination from a normal male-fertile plant. The seed produced from such crosses has the cytoplasmic male-sterile trait and would be of little value for commercial plantings. However, along with the discovery of cytoplasmic male sterility, another, closely related discovery was made: a fertility-restorer gene was found, and this gene is carried in the nuclear DNA. When pollen carrying the nuclear fertility-restorer gene fertilizes an egg borne on a cytoplasmic male-sterile plant, the embryo, even though its cytoplasm is male-sterile, is male-fertile. The nuclear fertility-restorer gene "overrides" the effects of cytoplasmic male sterility. By carefully selecting male and female parents in corn-breeding programs, cytoplasmic male sterility and/or nuclear fertility-restorer genes can be transferred into superior inbred lines and male-fertile hybrid seed can be produced from crosses between the lines.

Two types of hybrid corn cultivars are grown commercially. A single cross is the hybrid resulting from a cross between two inbred lines. A double cross is the hybrid resulting from a cross between two single crosses. Double-cross cultivars have been very popular, but in recent years there has been a trend to grow single crosses adapted to quite specific environmental conditions.

A pattern of crosses for producing both single and double crosses using the cytoplasmic male-sterile–fertility-restorer system is outlined in Figure 32-1. In both types of crosses, the hybrid seed is produced by growing rows of male-fertile plants interspersed with rows of the cytoplasmic male-sterile plants. Seed is harvested from only the male-sterile plants. This seed must be hybrid because the pollen that fertilized the egg can only have come from the male-fertile plant. The first commercial seed produced by this method was marketed in 1951. For many crops, a grower saves some of the seed he produces for planting the next crop. If hybrid corn, or any other hybrid cultivar, is used, however, this should not be done. The hybrid, as a result of crossing inbred lines, is a uniform heterozygote. The seed that the hybrid produces—that which a grower would ordinarily save from other crops he produces—will reflect extremes in segregation because of intercrossing between plants and will not be uniform. It will be comparable to the F_2 generation resulting from self-pollinating the F_1 plants that were developed from a cross between two pure lines.

Cytoplasmic male sterility has been a boon to hybrid seed production; however, hand detasseling is still practiced and is required in developing new hybrids if the parent plants do not have the necessary sterility. Male-sterility systems have been found in both wheat and barley. If methods to produce hybrid seed can be refined, hybrid cultivars of these crops may become quite common.

In recent years, plant breeders have developed man-made species by hybridizing different species, and even different genera. Although the idea of man-made plant species is not new, triticale, which was developed from crosses between wheat and rye, is the first to have significant potential for alleviating starvation. The development of triticale entailed more than the crossing of two plants; it was derived from crosses between rye (*Secale cereale*), which is a normal diploid species, and one of either of two species of wheat, common wheat (*Triticum aestivum*) or durum wheat (*T. durum*). Both species are polyploids; they have more than two chromosomes of a given type. Common wheat is a hexaploid with twenty-one pairs of chromosomes, which can be divided into three groups of seven pairs of chromosomes each. Each group of chromosomes is called a genome; the number of chromosomes in a genome is the basic (N) number of chromosomes. Durum wheat is a tetraploid with fourteen pairs of chromosomes, which can be divided into two groups of seven pairs of chromosomes each.

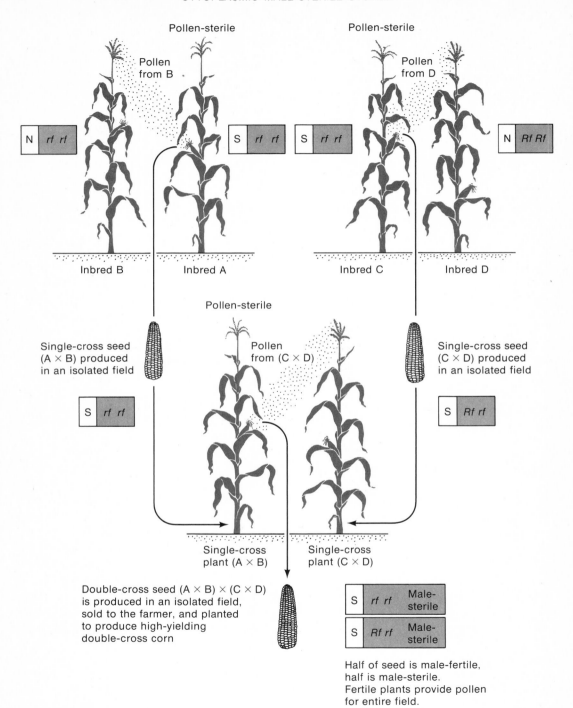

Figure 32-1.
The development of single crosses between pairs of inbred lines of corn and of a double cross using cytoplasmic male sterility and a nuclear fertility-restorer gene (*rf* is sterile, and *Rf* restores fertility). [From *Plant science*, 2d ed., by J. Janick, R. W. Schery, F. W. Woods, and V. W. Ruttan. W. H. Freeman and Company. Copyright © 1974.]

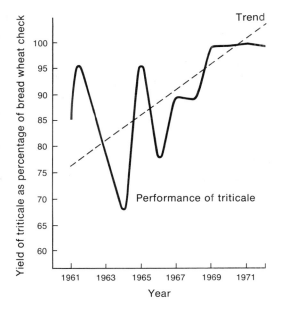

Figure 32-2
Performance of triticale in Canada expressed as a percentage of the performance of adapted types of wheat. Until 1969 yields were low and varied from year to year; since then, they have been stable and close to those of wheat. [Courtesy of E. N. Larter.]

Within the groups of seven pairs, T. aestivum and T. Durum have some chromosomes that are similar.

Gametes produced by rye have seven chromosomes. Those produced by T. aestivum have twenty-one, and by T. durum, fourteen. The embryo formed by a cross between rye and T. aestivum thus has twenty-eight chromosomes, but these are twenty-eight different types of chromosomes, not fourteen pairs. Given that the basic number of chromosomes in a group in either wheat or rye is seven, this hybrid is a tetraploid, but it is not four times a common group of seven chromosomes. If the embryo grows, the mature plant will not be able to reproduce because the twenty-eight chromosomes are not homologous and therefore cannot pair during meiosis. Both micro- and megaspores are nonfunctional, or sterile. This is also true of hybrids produced by crossing rye and T. durum, but the resulting embryo has twenty-one individual chromosomes instead of twenty-eight and is classified as a triploid.

In either case, the problem is to double the chromosome number so that the hybrids have homologous chromosomes and meiosis is normal. This requires the creation of an octoploid ($8 \times 7 = 56$ chromosomes) from the tetraploid or a hexaploid from the triploid ($6 \times 7 = 42$ chromosomes). This has been accomplished, but the job is far from complete. Now that viable, fertile hybrids have been established, plant breeders must develop superior cultivars. The triticale lines derived from crossing rye and T. durum show the most promise. The ultimate goal is to develop cultivars that, like rye, are tolerant of low temperatures and other adverse conditions and that retain the yield and nutritional quality that is characteristic of wheat.

In the early stages of the development of triticale, yields were generally lower and more variable than those of the wheat cultivars that were normally grown in the same region (Figure 32-2). As a result of continued efforts by plant breeders, yields have increased and stabilized since then. Part of this increase and stabilization resulted from the increased fertility of multiple crosses between early triticale types and wheat (Figure 32-3). Currently, high-yielding types that have the characteristics wanted are being developed (Figure 32-4).

Canadian plant breeders have been among the leaders in the development of triticale. This new cereal is an indication of the potential for developing other valuable crop species through plant breeding and through research

Figure 32-3
Spikes of triticale showing increased fertility and increasing similarity to a wheat spike: (A) early, ryelike spike that is quite sterile; (B) intermediate type; and (C) very fertile, wheatlike spike. [Courtesy of E. N. Larter.]

Figure 32-4
Single plant of a very fertile, high-quality triticale cultivar developed in Canada. [Courtesy of E. N. Larter.]

in general to meet the challenge of feeding the world's growing population.

The Future of Crop Production

Research in crop production began when the nomadic hunter-herdsmen turned to farming. Early farmers selected superior plants and preserved seed for future planting. Crop pests were recognized although difficult to control, and conservation measures were taken to retain soil fertility through the use of manure. Considering the long history of agricultural development, it is surprising that the first experiment station, which demonstrated new methods in agriculture, was not established until 1834 by J. B. Bassengault in Alsace, France. Since then research and experimentation in agriculture has received the support of governments in many countries. Private industry has made major contributions as well, not only through the promotion and sale of new and superior materials, but also through independent research and through direct financial support of government researchers.

Thus, the cycle is completed; cooperative research is the road to continued success in increasing agricultural production and production efficiency, and education is the foundation from which this success will arise.

Selected References

Briggs, F. N., and P. F. Knowles. 1967. *Introduction to plant breeding*. Reinhold.

Cochran, W. G., and G. Cox. 1957. *Experimental design*. 2d ed. Wiley.

Study Questions

1. How and why is the use of statistics of importance in plant science?
2. Describe the basic steps and goals of plant breeding.
3. Describe several products of research in plant science that have contributed significantly to human welfare.

Glossary

Abscisic acid A plant-growth regulator, or hormone, that affects dormancy or acts as a growth inhibitor.

Abscission layer The layer of cells at the base of the petioles of leaves of deciduous trees. When these cells die, the leaves fall.

Absorption Taking up water by assimilation or imbibing. The absorption of water by a seed is generally the first step in germination.

Acetic acid An organic acid with the formula CH_3COOH.

Achene A simple, dry indehiscent fruit; the seed is free inside hardened pericarp. This is the characteristic fruit type of the sunflower family.

Acid-delinted seed *See* Acid delinting.

Acid delinting A process to remove the short fibers from seed cotton.

Acidic soil Soil with a pH below 7.0.

Acre Unit of area commonly used in the United States: equals 43,560 square feet. In countries that use the metric system, the basic land measure is a hectare (10,000 square meters). An acre is about 0.40 hectare.

Acre-foot A measure of water volume equal to 43,560 cubic feet, or the amount of water that can cover one acre to a depth of one foot.

Adenosine diphosphate (ADP) A high-energy phosphate compound. ADP is converted into ATP by the addition of phosphate, which is attached to the ADP molecule by a high-energy bond.

Adenosine triphosphate (ATP) A high-energy molecule that is the major source of usable chemical energy for cellular work.

Adsorption Taking up vapor or liquid by a surface on which they remain.

Adventitious Used in reference to buds, roots, or other structures borne on parts of the plant where they are not generally expected to arise.

Adventitious root A root that originates neither as a branch from the pericycle nor as a seminal root. Roots of most grasses are adventitious and support the grass plant during most of its life.

Aftermath growth Regrowth of forage after harvest for hay or seed.

Aggregate fruit A fruit formed from numerous, unfused carpels in a single flower.

Agronomy The art and science of soil management and crop production.

Aleurone Outer layer of cells surrounding the endosperm of a caryopsis.

Alkaline soil Soil with a pH above 7.0.

Alkaloids Any of a group of organic bases occurring in plants and containing carbon, hydrogen, oxygen, and nitrogen.

Alternate Used in reference to an arrangement of leaves or buds in which a bud or leaf grows on one side of a stem at one node and on the other side at the next higher node.

Alternation of generations In plants, a life cycle including a diploid sporophyte generation and a haploid gametophyte generation. During the sporophyte generation, spores are produced by meiosis. These haploid spores develop into the gametophyte, which produces gametes by mitosis.

American Farm Bureau An organization founded in the early twentieth century for the purpose of protecting the economic interests of farmers and ranchers.

Amino acid The fundamental building block of proteins, an organic acid containing nitrogen. There are twenty common amino acids found in living organisms, each having the basic formula $NH_2-CHR-COOH$.

Anabolism Synthetic metabolism: the metabolic processes that form more complex molecules from simpler ones, as in photosynthesis or protein synthesis. *See also* Catabolism.

Anaphase The phase of mitosis that immediately follows metaphase. During anaphase, chromatids separate as a result of the splitting of the centromere, and chromatids move to opposite poles of the nucleus.

Anaphase I Follows metaphase I of meiosis.

Anaphase II Follows metaphase II of meiosis. Chromatids separate at the centromere and move to opposite poles, as in mitosis.

Anatomy The study or science of the shape and structure of organisms.

Angiosperm A plant in which the female gamete is protected within an enclosed ovary. A flowering plant.

Angstrom (Å) One ten-millionth of a meter: 1×10^{-7} meter.

Anion A negatively charged atom or molecule.

Annual A plant that completes its life cycle in a single year or less.

Anther The saclike structure of the stamen in which microspores are produced; usually borne on a filament.

Anthesis The developmental stage at which anthers rupture and pollen is shed.

Anthocyanins A class of water-soluble pigments that account for many of the red to blue floral and fruit colors. Anthocyanins are found in the vacuole.

Antipodal nuclei The three or more nuclei at the end of the embryo sac opposite the egg nucleus (female gamete). They are produced by mitotic divisions of the megaspore and degenerate following sexual fertilization.

Apical Used in reference to tip regions of stems or roots.

Apical dominance The suppression of the development of lateral buds by high concentration of auxins in the shoot apex.

Apical meristem A meristem at the tip of a root or a stem (or a shoot). These cells are undifferentiated and allow the plant to grow in height or depth.

Apomixis Seed formation without sexual fertilization.

Arrow The inflorescence of a sugarcane plant.

Asexual Used in reference to reproduction by any method in which sexual fertilization is absent, or not completed.

Aspirator An air-flow seed-cleaning device relying on the difference in density of seeds of various species, and of foreign material, for seed separations.

Auricle In a grass leaf, the clasplike structure at the juncture of the leaf blade and sheath. Auricles are prominent in barley and absent in oats.

Autotrophic Capable of producing food from nonfood materials. The opposite of heterotrophic.

Auxins An important group of plant-growth hormones, or regulators. Auxins partly control bud development and cell elongation. Artificial auxins, such as 2,4-D, are leading herbicides.

Awn An extension of the mid-rib of a glume or a lemma. Species or cultivars with awns are frequently referred to as having beards or being bearded.

Axil The angle on the upper side of the union of a branch and main stem or of a leaf and a stem. Buds are frequently located in axils.

Bagasse A fibrous by-product of sugarcane processing that is used in the manufacture of some types of paper products.

Beard See Awn.

Benchland A relatively flat area, or plateau, that is less sloping than the surrounding area. Used in reference to a tillable area of farmland.

Berry A fleshy, simple fruit developed from one or more carpels with many seeds and a thin exocarp covering a watery mesocarp; a tomato is a berry.

Biennial A plant that completes its life cycle in two years.

Blade The wide, thin part of a leaf above the petiole or sheath. The blade is generally the site of most photosynthetic activity and of most transpiration.

Bloat Accumulation of excessive gases in the rumen of an animal; often caused by the ingestion of fresh succulent legume forage.

Boll The rounded seed pod of cotton or flax.

Boot The sheath of the last (flag) leaf through which the inflorescence emerges.

Botany The science of plant life.

Brace root A type of adventitious root that grows from aboveground parts of the stem and serves to support the plant much like a guy wire. Brace roots are frequently seen in corn. Also called prop roots.

Bract A leaflike structure that is associated with flowers or groups of flowers but is neither a sepal nor a petal.

Bran The outer layers of a caryopsis removed in milling.

Bud A region of meristematic tissue with the potential for developing into leaves, branches, flowers, or combinations of leaves and flowers.

Budding A form of grafting in which a vegetative bud from one plant is transferred to stem tissue of another plant.

Bulb A modified stem that serves as an underground storage organ; it is composed of closely packed, modified, fleshy leaves.

Bunch type Growth habit of grasses that do not spread by rhizomes or stolons. See Sod type.

Butyric acid An organic acid with the formula $CH_3CH_2CH_2COOH$.

Calcium pectate A calcium compound found in the middle lamella, in which it serves as an intercellular cement.

Calorie The amount of energy required to raise the temperature of one gram of water one degree Celsius at constant pressure. See also Kilocalorie.

Calvin, Melvin Scientist awarded a Nobel Prize in 1961 for describing the basic processes of photosynthesis.

Calyx The collective term for the sepals, the lowermost series of flower parts.

Capillary capacity The amount of liquid that can be moved through small tubes or channels as a result of surface tension.

Capsule A dry fruit formed from a compound ovary with variable dehiscence and many seeds.

Carbohydrate A molecule composed of carbon, hydrogen, and oxygen, in the ratio one C : two H : one O (CH_2O).

Carbon dioxide fixation The addition of H^+ to CO_2 to yield a chemically stable carbohydrate. The H^+ is contributed by $NADPH_2$, the reduced (hydrogen-rich) form of NADP produced in the noncyclic phase of the light reactions of photosynthesis; the H^+ comes originally from the photolysis of water.

Carpel A leaflike structure with ovules along its margins.

Caryopsis See Grain.

Cash crop A crop produced for direct sale for cash, as contrasted with crops produced as livestock feed.

Catabolism Destructive metabolism: the metabolic processes that break down complex molecules with the release of energy. Respiration is a form of catabolism. *See also* Anabolism.

Cation A positively charged atom or molecule.

Cation exchange capacity A measure of the potential of a soil (usually the clay, or colloidal, part) to give up or exchange positively charged ions or molecules.

Cell membrane The membrane that separates the cell wall and the cytoplasm. The cell membrane, like most membranes, is composed of outer layers of protein with a double layer of phospholipid material between them. The cell membrane regulates the flow of material into and out of the cell.

Cell plate The precursor of the cell wall, formed as cytokinesis starts. It forms in the region of the equatorial plate and is developed from membranes in the cytoplasm.

Cellulose A complex, fibril polysaccharide that is a major component of the cell wall.

Cell wall The outermost, cellulose limit of the cell; the barrier that develops between nuclei after telophase of mitosis.

Center of origin A geographical area in which a species is reported to have evolved through natural selection from its wild ancestors.

Centromere A region of a chromosome at which fibrils from the spindle are associated with or attached to the chromosome. The location of the centromere accounts in part for the shape of the chromosome. Individual chromosomes may be visually identified by the location of their centromeres.

Cereal A grass grown primarily for its seed, which is used for feed or food.

Certified seed Seed that meets rigid standards of purity and germination and is so designated by an authorized agency.

Chiasmata The physical crossing of chromatids from members of a homologous pair of chromosomes. Chiasmata may lead to genetic recombination through crossing over.

Chlorophyll A complex organic molecule that traps light energy and, through a complex series of steps, converts it into chemical energy.

Chloroplast The cellular organelle in which chlorophyll is located. Enzymes necessary for photosynthesis are associated exclusively with the chloroplast in higher plants.

Chlorosis A condition in which a plant or a part of a plant is light green or greenish yellow because of poor chlorophyll development or the destruction of chlorophyll resulting from a mineral deficiency or disease.

Chromatids The duplicated halves of a single chromosome attached by a common centromere; in metaphase of mitosis and in metaphase II of meiosis, chromatids separate to yield sister chromosomes.

Chromosome A specific, highly organized body in the nucleus that contains DNA and behaves in a highly regular fashion during nuclear division. Chromosomes are so named because they stain easily with basic dyes.

Cigar binder A type of tobacco used to wrap cigars.

Cigar filler A type of tobacco used in cigars.

Citric acid cycle *See* Krebs cycle.

Clay Small mineral soil particles having a diameter less than 0.002 mm.

Cold frame A transparent, covered shelter that is unheated, relatively small, and low; it is used to protect plants from low temperatures.

Coleoptile A leaf at the second node of the embryonic stem in the embryo of a grass caryopsis. The coleoptile protects the stem as it grows through the soil and emerges from the soil surface.

Colza oil Oil from rapeseed.

Combine A machine that harvests and threshes seed in one operation.

Companion crop A crop sown with another crop (e.g., a cereal sown with a small-seeded forage).

Compensation point The metabolic stage at which the rates of photosynthesis and respiration are equal.

Complete and perfect flower A flower with all parts of the perianth and pistil and stamen.

Contact herbicide A nonsystemic weed killer—one that does not have to be metabolized by a plant to be effective.

Corm A short, enlarged base of a stem in which food is stored.

Corolla The collective term for all petals of a flower.

Cortex A region in roots and stems between the vascular tissue and the epidermis. The cortex is sometimes a storage area for starches. Most cortical cells are loosely packed and thin-walled.

Cotton gin A machine that separates the cellulose fibrils (cotton) from the seed on which they are produced. The invention of the cotton gin revolutionized the cotton industry.

Cotton staple Used in reference to the length and fineness of cotton fiber.

Cotyledons Leaves at the first node of the primary stem. In dicots, cotyledons are the food-storage organs of the young plant.

Coumarin A white crystalline organic compound with the formula $C_9H_6O_2$; it imparts a characteristic odor and flavor to sweet clover.

Crimper A machine designed to crush the stems of hay, which hastens drying and thus reduces the time that the hay is exposed to weather.

Crop botany A division of botany in which various subdivisions of botany are related to problems of crop production.

Cross-fertilization Follows cross-pollination. A male gamete from one plant fertilizes a female gamete of a different plant.

Crossing over The physical exchange of chromosomal material between nonsister chromatids of homologous chromosomes.

Cross-pollination The condition in which pollen from an anther on one plant ultimately fertilizes a flower on a different plant. Corn, many forage grasses, and alfalfa are cross-pollinated. *See also* Self-pollination.

Crown (Crown region) The region at the base of the stem of cereals and forage species from which tillers or branches arise and in which, in perennial species, carbohydrates are stored.

Crude protein Unrefined, total protein; digestible protein is a fraction of the crude protein.

Culm (Stool; Tiller) The common name for the jointed stem of a grass plant.

Cultivar A cultivated variety within a plant species that differs in some respect from the rest of the species.

Curvilinear Used in describing a nonuniform response of a variable (e.g., population size) in relation to a second variable (e.g., time). A geometric progression yields a nonlinear curve or graph.

Cutin A clear or transparent waxy material on plant surfaces that tends to make the surface waterproof.

Cyclic reaction A part of the light reactions of photosynthesis in which a light-excited, high-energy electron cycles through the chlorophyll molecule and comes to rest at its starting point, but in a low-energy state. The energy from the sun that initially excited the electron has been used to form ATP from ADP, the process of photophosphorylation.

Cylinder separator A seed-cleaning device in which a cylinder with indented discs on its surface separates seeds of different lengths and widths.

Cysteine An essential, sulfur-containing amino acid.

Cytokinesis Division of the cytoplasm to form two new cells. Cytokinesis generally follows telophase of mitosis.

Cytology The study of cells and their components and the relationship of cell structure to function.

Cytoplasm The living material of the cell exclusive of the nucleus. The "assembly plant" of the cell in which essential membranes and cellular organelles are found. The cytoplasm is a complex protein matrix or gel.

Deciduous Used in reference to trees that lose their leaves every year, as distinguished from evergreens.

Decumbent A growth habit in which stems or shoots lie close to the soil surface. The apices or extremities of the stems generally turn upward.

Deferred grazing A management practice in which livestock are withheld from a given pasture or range for an extended period to permit the vegetation to recover and the forage condition to improve.

Dehiscence The splitting open of either fruits (usually dry fruits) or anthers. The term may also apply to the breaking-off of plant parts such as awns.

Deoxyribonucleic acid (DNA) A molecule composed of repeating subunits of ribose (a sugar), phosphate, and the nitrogenous bases adenine, guanine, cytosine, and thymine. DNA in chromosomes contains genes, the fundamental units of inheritance.

Desiccant Any chemical used to dry plant materials; a drying agent.

Determinate Used in reference to the growth of a stem that terminates in a floral bud and thus will not grow in length indefinitely; also used in reference to flowering that occurs during a specific period, regardless of environmental factors.

Dichlorodiphenyltrichloroethane (DDT) One of the earliest insecticides of the chlorinated hydrocarbon family.

2,4-Dichlorophenoxyacetic acid (2,4-D) A very effective selective weed killer or herbicide. 2,4-D kills broadleafed weeds growing in fields of grasses, including cereals.

Dicotyledonae (or Dicots) The subclass of flowering plants that have two cotyledons at the first node of the primary stem.

Differentiated cell A cell that has followed a specific pattern of development.

Diffusion The random scattering of molecules due to their own kinetic activity and to external forces.

Digestion The breakdown of complex foods to simpler foods, which can be more easily respired. Digestion requires energy.

Dioecious Literally, two-housed. The term is used in reference to species in which male and female flowers occur on separate plants. Buffalograss and asparagus are dioecious.

Diploid The condition in which both members of all homologous pairs of chromosomes are present in a nucleus. This term is also used in reference to an organism having both members of all homologous pairs of chromosomes present in its somatic cells.

Disc flower (Disc floret) The flowers in the center of the inflorescence (head) of plants in the Compositae (sunflower) family. *See also* Ray flower.

Dominant Used in reference to the allelic form of a unit of heredity that, when present in a heterozygous condition, is expressed or manifest. *See also* Recessive.

Dormant Used in reference to the stage of growth or development of a plant during which the rate of all metabolic processes is at a minimum. Also used to describe a seed that, for various physical or physiological reasons, will not germinate, even under optimum conditions.

Dormin *See* Abscisic acid.

Double cross Used in reference to double-cross seed—the seed produced from a cross between two single crosses. Much hybrid corn seed is double-cross seed.

Double fertilization The process of sexual fertilization in the angiosperms in which one nucleus from the male gametophyte fertilizes the egg nucleus to form the embryo and a second nucleus from the male gametophyte jointly fertilizes two polar nuclei to form endosperm.

Drupe A fleshy, simple fruit with a single seed, in which a hard endocarp is associated with the seed.

Dry fruit Any fruit formed from an ovary with walls that are dry at maturity.

Emasculate Remove the male reproductive structures.

Embryo The minute plant produced as a result of fertilization and development of the zygote in a seed.

GLOSSARY

Embryo sac The cell that develops from mitosis in the megaspore; it consists of eight or more haploid nuclei, one of which is the female gamete.

Endocarp The inner layer of cells of the pericarp.

Endodermis In roots, a single layer of cells at the inner edge of the cortex. The endodermis separates the cortical cells from cells of the pericycle.

Endosperm Food-storage material in a caryopsis that develops as a result of double fertilization.

Endosperm nucleus One of two generative nuclei in the microspore. The endosperm nucleus fertilizes the two polar nuclei to form endosperm.

Energy The potential to do work. Generally, energy is expressed as light, heat, or chemical energy.

Enzyme A protein that functions as a biological catalyst and regulates cellular functions.

Epidermis The outer layer of cells covering an organ or an entire individual: the skin.

Epigeal germination In dicots, germination in which the cotyledons rise above the soil surface.

Equatorial plate A plane through the center of the nucleus along which chromosomes are aligned at metaphase: it is a region, not a physical structure.

Erosion The wearing away of the surface soil by wind, by moving water, or by other means.

Etiolated Describes a plant that is overelongated (excessively tall) and spindly, usually as a result of insufficient light.

Exocarp The outer layer of cells of the pericarp.

F_1 First-generation offspring after a particular mating.

F-test A statistical test in which the ratio of two variances is evaluated to determine the probability that the difference between the variances is due to random error. The test is frequently used in conjunction with the analysis of variance.

Fallow Land left uncropped for one or more seasons to accumulate moisture, destroy weeds, and allow the decomposition of crop residue.

Fanning mill An air-screen seed cleaner in which screens and air flow separate weed seed and foreign material from seed.

Farmers' Educational and Cooperative Union A large, cooperative organization established to protect and foster the economic interests of farmers and ranchers.

Farming The art and science of crop production. The direct application of the multiple aspects of agronomy.

Fats *See* Lipids.

Fermentation Alternate path of respiration followed when oxygen is absent or limited. This process produces less ATP than the Krebs cycle and its end product may be ethyl alcohol, lactic acid, or other substances.

Fertilization (plant) The union of sperm and egg.

Filament The stalk that supports an anther.

Fire cure A method of curing tobacco leaves that uses open fires in the curing sheds.

Fleshy fruit Any fruit formed from an ovary that has fleshy or pulpy (not dried) walls at maturity. Also, those fruits that include fleshy parts of the perianth or the receptacle.

Floret The floral (reproductive) subunit of a grass spikelet. A floret has a palea and a lemma as its outer or lower limit.

Florigen A hormone or group of hormones thought to affect flowering.

Flue cure A method of curing tobacco leaves in which sheds are heated through ducts or flues.

Flying shuttle Part of a loom.

Foliar On or pertaining to leaves.

Follicle A dry, dehiscent, simple fruit.

Forage Vegetation used as feed for livestock (e.g., hay, pasture, silage, or green chop).

Frost heaving The swelling, bulging, and contracting of soil, and the associated uprooting of plants as a result of the freezing and thawing of soil.

Fructosan A ketose form of a six-carbon sugar that is present as a storage material in many grasses.

Fructose A ketose sugar with the formula $C_6H_{12}O_6$.

Fruit A mature ovary.

Futures markets A system of trading contracts for the delivery of specific kinds and amounts of a particular commodity at a specific time and place.

Gamete A haploid cell or nucleus that may unite with another gamete to produce a zygote, the process of sexual fertilization. A gamete is a sex cell or nucleus. Both egg and sperm nuclei are gametes.

Gametophyte The spore that, as a result of mitosis, ultimately produces the male (sperm) and female (egg) gametes or nuclei.

Gene A fundamental unit of inheritance. The gene has been defined as the code (DNA nitrogen base sequence) for a single enzyme, although this definition is not universally accepted.

Generative nuclei Nuclei that unite to yield an embryo. The egg and sperm nuclei.

Genetic code In a narrow sense, the sequence of nitrogen bases in a DNA molecule that codes for an amino acid or protein. In a broad sense, the full sequence of events from the translation of chromosomal DNA to the final stage of the synthesis of an enzyme.

Genetics The science or study of inheritance.

Geniculate Bent abruptly at an angle, like a bent knee.

Genotype The hereditary (genetic) makeup of a nucleus or of an individual. See also Phenotype.

Geometric (progression) Sequence of terms in which the ratio of each term to the preceding one is the same throughout the sequence (e.g., 2, 4, 8, 16, 32, . . .).

Germ A common name for the embryo, as in "wheat germ."

Germination The sequence of events occurring in a viable seed, starting with the imbibition of water, that leads to the growth and development of an embryo.

Gibberellins A group of hormones that contribute to increases in cell size, fruit size, and fruit yield, among other effects.

Glucose A simple sugar composed of six carbon, twelve hydrogen, and six oxygen atoms ($C_6H_{12}O_6$). In discussing respiration, the initial foodstuff is usually taken to be glucose.

Glume The bractlike structures at the base of the grass spikelet.

Gluten A protein substance in wheat flour that allows dough to rise.

Glycolysis A respiratory process in which glucose is converted anaerobically into lactate or pyruvate with a net gain of two molecules of ATP.

Graftage The process of grafting. See also Grafting.

Grafting The transfer of aerial parts of one plant to the root or trunk material of a different plant. Grafting is important in commercial fruit-tree and vine (grape) production. See also Graftage.

Grain (or Caryopsis) A simple, dry, indehiscent fruit with ovary walls fused to the seed. The "seed" of cereal or grain crops such as corn, wheat, barley, and oats.

Grain drill Any of several devices used to plant grain at a relatively constant depth.

Gram molecule A quantity of a substance equal to its molecular weight expressed in grams; for example, a gram molecule of glucose, $C_6H_{12}O_6$, equals (6 × 12) + (12 × 1) + (6 × 16) = 180 grams of glucose. Also called a mole.

Grange, The A local unit of the fraternal lodge of Patrons of Husbandry or the organization itself. The group was founded in 1867.

Grass Any plant of the Gramineae family.

Green chop Forage that is chopped in the field while succulent and green and fed directly to livestock.

Green manure A crop that is plowed under while still green and growing to improve the soil.

Ground meristem One of the three primary meristematic tissues. Ground meristem cells differentiate into endodermis and cortex.

Growth An irreversible increase in cell size or cell number. An increase in dry weight, regardless of cause.

Growth regulator See Hormone.

Gynophore A structure in peanut plants that leads to the subterranean formation of fruits.

Halum The residue from a peanut crop that may be used as livestock feed.

Haplocorm See Corm.

Haploid Used in reference to either a nucleus or an entire organism in which only one member of each set of homologous chromosomes is present. In plants, the characteristic chromosome number of the gametophyte.

Hay Vegetation that is cut and cured for later use as a livestock feed.

Head inflorescence The inflorescence typical of members of the Compositae, or sunflower, family. A head has a large, flattened receptacle.

Hectare In the metric system, a measure of area equal to 10,000 square meters or about 2.5 acres.

Hedging In the futures market, the execution of opposite sales or purchases to offset purchases or sales of commodities. This practice gives some protection to sellers and buyers of grain against uncertainties that are the result of unstable grain prices.

Heliotropism The tendency of a part of a plant to follow the sun in a diurnal cycle.

Heterozygous Containing unlike alleles at one or more loci. See Homozygous.

Hilum The scar on a dicot seed, such as a bean seed, where the seed was attached to the fruit.

Homologous chromosomes Chromosomes that are members of a distinct morphological pair. Homologous chromosomes physically pair during prophase I of meiosis.

Homozygous Containing like alleles at a given locus, or at all loci—depending on whether the condition pertains to a single locus or to an individual.

Hormone A naturally occurring substance produced in one part of an organism that may move to another part of the organism and exert a specific effect. Plant hormones include the growth regulators, such as auxins and gibberellins.

Hulling In processing legumes, the removal of the hull or pod; in processing grass seeds, the removal of the lemma and palea.

Hydrocyanic acid An organic acid, with the formula HCN, that is highly toxic to livestock; also called prussic acid.

Hydrolysis A chemical reaction in which water participates as a reactant and not as a solvent. Usually, the splitting of a molecule to form smaller molecules that incorporate hydrogen and hydroxyl ions, derived from water, in their structures.

Hypogeal germination In dicots, germination in which the cotyledons do not rise above the soil surface.

Imperfect flower A unisexual flower.

Inbred line A plant or group of plants that is homozygous as a result of continued self-pollination. Many inbred lines come from naturally cross-pollinated species that have been artificially self-pollinated.

Incompatibility A genetic condition in which certain normal male spores are incapable of functioning on certain pistils.

Incomplete flower A flower that is missing all or some of the following parts: sepals, petals, stamens, or pistils.

Indehiscent fruit A fruit that does not split open naturally at maturity.

Indeterminate growth The growth exhibited by a stem that terminates in a vegetative bud and will thus elongate indefinitely. A plant that flowers more or less continuously until climatic conditions become unfavorable is also said to exhibit indeterminate growth.

Inferior ovary An ovary that is imbedded in the receptacle, or an ovary whose base is below the point of attachment of the perianth. See also Superior ovary.

Inflorescence Any structurally organized group of flowers.

Inoculation Introduction of bacteria (Rhizobium) on seed or into the soil.

Inorganic compound A chemical compound that generally is not derived from life processes: salts and noncarbon containing compounds. The mineral nutrients essential to plants are inorganic.

Insecticide Any substance that kills insects. Two broad classes of insecticides are the chlorinated hydrocarbons and the organic phosphates.

Integuments The maternal tissues in the ovary that surround and support the egg nucleus. The integuments mature to form the testa or seed coat.

Intercalary growth The pattern of stem elongation typical of grasses. Elongation proceeds from the lower internodes to the upper internodes through the differentiation of meristematic tissue at the base of each internode. As a result of this pattern of elongation, unelongated, upper internodes are encased in the sheath material from leaves at the lower, elongated internodes.

Intercellular space A space between the walls of adjacent cells.

Internode The region of a plant stem between nodes.

Interphase The so-called resting, nondividing phase of a nucleus. During this phase, the genetic code of chromosomal DNA is transferred to messenger RNA, which in turn directs the synthesis of specific proteins in the cytoplasm.

Interphase I In cells capable of meiotic division, the stage preceding the appearance of chromosomes. Visually similar to mitotic interphase.

Iodine number The amount of iodine absorbed by a fat or oil expressed as a percentage of its molecular weight; a measure of the degree of unsaturation of the fat or oil.

Irrigation Any method of applying water to a crop so that it reaches the rooting zone.

Karyolymph See Nuclear sap.

Ketone An organic compound having the carbonyl group CO attached to two organic radicals (e.g., acetone—CH_3COCH_3).

Ketose Any sugar with a ketone group or its equivalent.

Kilocalorie The amount of energy required to raise the temperature of one kilogram of water one degree Celsius at constant pressure: 1,000 calories.

Kilogram In the metric system, a measure of mass equal to 1,000 grams, or about 2.2 pounds.

Kinins A group of plant hormones that regulate scar development. Also called cytokinins.

Krebs cycle A cyclic series of chemical reactions through which pyruvate is oxidized to CO_2 and H_2O and ATP is produced. The Krebs cycle is thought by some to include the terminal electron-transport phases of respiration.

Lactate The anion of lactic acid, with the formula $C_3H_5O_3^-$. It is the end product of anaerobic glycolysis and is oxidized to CO_2 and H_2O (via pyruvate and the Krebs cycle) in the process of respiration.

Lactic acid An organic acid with the formula $CH_3CHOHCOOH$.

Lax Loose, or not densely packed.

Leach To remove soluble material with water. To wash out or down.

Leaflet An individual subdivision of a compound leaf.

Least significant difference (LSD) The smallest difference that could exist between two significantly different sample means. If a statistical test indictaes significant differences between treatment means, the LSD may be computed and used to determine which means contribute to the significance.

Legume A simple, dry, dehiscent fruit that dehisces along two margins. The fruit type of beans and other members of the pea family. Also, any plant in the legume or pea family, Leguminosae. Alfalfa and beans are important crop plants in this family.

Lemma The outer, larger bractlike structure of a single floret. Frequently, an awn will form as an extension of the mid-rib of the lemma. See also Palea.

Life cycle In annuals and biennials, the sequence of events from the shedding of a seed until a new seed is produced. In perennials, the events from the initiation of vegetative growth until seed is produced during a year or a growing season.

Light reactions The reactions of photosynthesis in which light energy is required: the photo (light) activation or excitement of an electron in the chlorophyll molecule, electron transfers of this electron, and associated reactions. This electron may ultimately contribute to the formation of ATP and DPNH because it is raised to a high energy level.

Lignin An organic substance found in secondary cell walls that gives them strength and hardness. Wood is composed of lignified cells.

Ligule The membranous structure on the inside of the juncture of the leaf blade and leaf sheath. Oats have a large ligule.

Lime Calcium carbonate ($CaCO_3$). Also, the practice of adding $CaCO_3$ to the soil to help correct an acidic condition.

Linters The short fibers remaining on cotton seed after ginning.

Lipids Generally, fats and oils. Lipids are organic compounds with a much lower proportion of oxygen than carbohydrates.

Lodging A condition most frequently observed in cereals; plants bend at or near the soil surface and lie more or less flat on the ground.

Lodicules Two scalelike structures at the base of the ovary in a grass floret.

Lucerne A common European name for alfalfa.

Macronutrient A mineral required in a relatively large amount for plant growth.

Male sterility A condition in which pollen either is not formed or does not function normally, even though the stamen may appear normal.

Malthus, Thomas British clergyman and economist (1766–1834) who speculated that the world's population would grow faster than food productivity and thus man would face mass starvation.

Market class A classification of grain or other crop products, which may be further divided into subclasses, for the purpose of designating grade under the U.S. Grain Standards Act.

Market spread The difference in price at two levels in the marketing chain.

Mating system The method of pollination in a specific plant; for example, self, cross, or both.

Matrix (or Gel) A complex, structured, semisolid substance, such as cytoplasm.

Megaspore The spore that germinates to form the female gametophyte: the end product of meiosis in the pistil.

Meiosis Reduction division: a form of nuclear division that takes place in sexually reproducing organisms. It includes a single DNA duplication and two nuclear divisions. Four haploid cells are formed from one diploid cell.

Meristem A region in which cells are not fully differentiated and are capable of repeated mitotic divisions.

Mesocarp The middle layers of cells of the pericarp.

Messenger RNA RNA produced in the nucleus and capable of carrying parts of the message coded in chromosomal DNA. Messenger RNA moves from the nucleus to the ribosomes—regions of protein synthesis in the cytoplasm.

Metabolism The overall physiological activities of a plant.

Metaphase The stage of mitosis in which individual chromosomes, composed of two chromatids, are arranged on the equatorial plate. The chromosomes are being moved at this stage by their association with fibrils of the mitotic apparatus.

Metaphase I Like metaphase of mitosis, except that pairs of homologous chromosomes rather than single chromosomes are arranged on the equatorial plate.

Metaphase II Individual chromosomes are arranged on the equatorial plate as in mitotic metaphase. Only one member of each homologous pair is found in each metaphase II nucleus.

Methionine An essential, sulfur-containing amino acid.

Metric ton In the metric system, a measure of mass equal to 1,000 kilograms, or 2,204.6 pounds.

Micronaire values Values of the flow of air under specified weight and volume of cotton fiber that indicate fiber fineness.

Micronutrient A mineral required in a relatively small amount for plant growth.

Microspore The male gametophyte, the end product of meiosis in an anther. The haploid microspore undergoes mitosis to produce the male gametes or sperm. The male gametophyte.

Middle lamella The pectic layer lying between the primary cell walls of adjoining cells.

Millimeter (mm) One-thousandth of a meter.

Millimho (mmho) A measure of electrical conductivity (0.001 mho); a mho is the reciprocal of an ohm.

Mitochondrion A cellular organelle with which the enzymes necessary for oxidative phosphorylation (Krebs cycle and terminal electron transport) are associated. Plural: mitochondria.

Mitosis A form of nuclear division in which chromosomes duplicate and divide to yield two nuclei that are identical with the original nucleus. Usually mitosis includes cellular division (cytokinesis). It is the process by which new cells are formed for growth and tissue repair.

Mitotic apparatus A fibril structure formed during mitosis and meiosis, which regulates or guides chromosome movement during nuclear division.

Monocotyledonae (or Monocots) The subclass of flowering plants that have only a single cotyledon or leaf at the first node of the primary culm or stem.

Monoecious Used in reference to plants that have separate male and female flowers on the same plant, such as corn.

Morphology (plant) The study or science of the form, structure, and development of plants.

Mottled Used to describe a blotched or spotted appearance.

Neppy Used to describe cotton fiber that is full of knots and snares that cannot be removed by the spinner.

Nerved Having a pronounced vein or vascular bundle in a leaf, lemma, or glume.

Nicotinamide adenine dinucleotide phosphate (NADP) A hydrogen acceptor molecule. In photosynthesis, this molecule uses energy from electrons in the noncyclic phase of the light reactions to bind hydrogen for eventual use in making carbohydrates.

Nitrate ion NO_3^-; this form of nitrogen, in the soil solution, is readily available to plants.

Nitrogen fixation The conversion of atmospheric nitrogen (N_2) into a stable, oxidized form that can be assimilated by plants. Certain blue-green algae and bacteria are capable of biochemically fixing nitrogen.

Nitrogenous base A nitrogen-containing compound found in DNA and RNA, which, in sequence, specify precise genetic information. The nitrogenous bases in DNA are adenine, thymine, cytosine, and guanine. In RNA, they are adenine, uracil, cytosine, and guanine.

Nobel blade A cultivation implement having a single wide blade that rides below the soil surface, cutting off weeds without turning the soil. It is used to preserve mulch on the surface without exposing moist soil to surface drying, thus preserving moisture.

Nodes Swollen regions of stems that are generally solid, to which leaves are attached and at which buds are frequently located. Stems have nodes; roots do not.

Noncyclic reaction A part of the light reactions of photosynthesis in which a light-excited, high-energy electron travels through a series of electron-acceptor molecules, and, rather than return to its starting point, the electron provides the energy to bind H to NADP, helping to form NADPH.

Nuclear sap The material confined by the nuclear membrane in which chromosomes are embedded and enzymes are found.

Nucleus A cellular organelle surrounded by an envelope of two membranes. Chromosomes are located in the nucleus.

Nut A dry, indehiscent, single-seeded fruit with a hard, woody pericarp.

One-tenth bloom In flowering plants, a stage of maturity reached when approximately 10 percent of the flowers have opened with the showy petals exposed.

Opposite arrangement An arrangement of leaves or buds in pairs on opposite sides of a single node.

Organelle A functional body within the cell (e.g., mitochondrion, or chloroplast).

Organic compound A chemical compound composed of carbon and usually hydrogen and oxygen. Organic chemistry is the chemistry of hydrocarbons.

Osmosis The diffusion of fluids through a semipermeable or selectively permeable membrane until the concentration of fluid on either side of the membrane is equal.

Ovary The basal, generally enlarged part of the pistil in which seeds are formed. The ovary, at maturity, is a fruit. It is a characteristic organ of plants in the class Angiospermae. It protects the ovum and developing seed.

Ovum The female reproductive cell, or gamete; the egg.

Oxidation-reduction reaction A chemical reaction in which one substance is oxidized (loses electrons, or loses hydrogen and its associated electron, or combines with oxygen) and a second substance is reduced (gains electrons, or gains hydrogen and its associated electron, or loses oxygen).

Oxidative phosphorylation Production of ATP through oxidative processes (e.g., glycolysis or Krebs cycle).

Oxidative respiration The chemical decomposition of foods (glucose, fats, and proteins) requiring oxygen as a terminal electron acceptor and yielding carbon dioxide, water, and energy. The energy is commonly stored as the bond energy of ATP.

Palatability Used to describe how agreeable or attractive feed stuff is to livestock or how readily they consume it.

Palea The innermost, smaller, bractlike structure enclosing a single grass floret. See also Lemma.

Palisade parenchyma The cell layer in leaves immediately below the upper epidermis; columnar, with their long axes perpendicular to the leaf surface, and packed with chloroplasts. This tissue is found in dicots, but not in monocots.

Palmate Used in reference to the arrangement of leaflets of a compound leaf or of the veins in a leaf. The arrangement is characterized by subunits arising from a common point much as fingers arise from the palm of the hand. A maple leaf exhibits palmate venation. See also Pinnate.

Panicle An inflorescence, common in the grass family, that has a branched central axis. Oats and bromegrass have a panicle type of inflorescence.

Parenchyma A tissue composed of thin-walled, loosely packed, unspecialized cells.

Parthenocarpy Fruit production without sexual fertilization.

Parthenogenesis A form of reproduction in which an unfertilized egg develops into a new organism.

Pasture An enclosed area in which vegetation is produced to be grazed by animals.

Pathology The science or study of disease, its causes, and its control.

Pearl A process of grinding off the hull, bran, aleurone, and germ of barley to yield a pellet of endosperm.

Pedicel In an inflorescence, a branch that bears or supports a single flower or floret.

Peduncle The top part of a stem that supports an inflorescence.

Perennial A plant that grows more or less indefinitely from year to year and may produce seed more than once.

Perianth The collective term for the calyx and corolla.

Pericarp The walls of the ovary at maturity. The fruit walls.

Pericycle The layer of cells immediately inside of the endodermis. Branch roots arise from the pericycle.

Petal A single unit of the corolla.

Petiole The stalk that attaches a leaf blade to a stem. In rhubarb, the petiole is harvested for food.

pH A measure of acidity or alkalinity, expressed as the negative logarithm (base 10) of hydrogen-ion concentration.

Phenotype The external physical appearance of an organism. *See also* Genotype.

Philadelphia Society for Promoting Agriculture An organization founded shortly after the end of the Revolutionary War to foster the interests of agriculture.

Phloem A tissue through which materials are translocated through the plant. Commonly, the phloem consists of sieve tube cells and companion cells.

Phospholipid A molecule containing both fats and phosphorous compounds. Phospholipids are important components of cell membranes.

Photolysis Splitting of water into H^+ and OH^-, utilizing solar energy in the light reactions of photosynthesis. Photo = light, and lysis = rupturing or splitting.

Photoperiodic response The flowering response of a plant in relation to the relative length of light and dark periods, usually in terms of a 24-hour day.

Photophosphorylation The production of ATP using the energy of light-excited electrons produced in the light reactions of photosynthesis. Photo = light, phosphorylation = adding phosphorus.

Photosynthesis The process of converting water and carbon dioxide into sugar using light energy; accompanied by the production of oxygen. Photo = light, and synthesis = building.

Phototropism A change in the manner of growth of a plant in response to nonuniform illumination. Usually, the response is a bending toward the strongest light. This response is auxin-regulated.

Phylogenic Used in reference to evolutionary relationships.

Physiology The study or science of the life processes of an organism, including its cellular chemistry and energetics.

Pigments Molecules that are colored by the light they absorb. Some plant pigments are water soluble and are found mainly in the vacuole. Others are less soluble and are housed in special structures, the plastids (e.g., chlorophyll in chloroplasts).

Pinnate Used in reference to the arrangement of leaflets of a compound leaf or the branches of veins in a leaf. The leaflets or veins are arranged along both sides of a central axis, much like the barbs of a feather. *See also* Palmate.

Pistil The entire female reproductive structure of a flower or floret—the stigma, style, and ovary.

Pith A region in the center of some stems and roots consisting of loosely packed, thin-walled cells that sometimes serve as storage tissue.

Plumule The part of the embryo above the cotyledons.

Pod *See* Legume.

Polar bodies The two nuclei in approximately the center of the embryo sac that are formed by mitosis in the megaspore and yield endosperm when fertilized.

Pollen *See* Microspore.

Pollen tube A tubelike structure developed by the tube nucleus in the microspore that helps guide the sperm and endosperm nuclei through the stigma and style to the embryo sac.

Pollination The transfer of pollen from an anther to a stigma. Pollination is not the same as sexual fertilization.

Polyploidy A condition in which a plant has somatic cells with more than 2N chromosomes per nucleus.

Pome A compound, fleshy fruit that includes parts of the perianth and receptacle.

Postemergence spray A pesticide or herbicide spray that is applied after the crop has emerged from the soil.

Preemergence spray A pesticide or herbicide spray that is applied after planting, but before the crop emerges from the soil.

Preirrigation Irrigation before final seedbed preparation and planting; a method to insure adequate moisture for the germination of crop seeds. Some fields are preirrigated to make weeds germinate so that they can be destroyed before a crop is planted.

Primary cell wall The cellulose layer formed on both sides of the middle lamella.

Primary plant nutrients Nitrogen, phosphorus, and potassium.

Primary terminal A central marketing institution where products are concentrated for processing.

Procambium One of the three primary meristematic tissues. Cells of the procambium ultimately differentiate into primary vascular tissue: xylem, phloem, and vascular cambium.

Prophase The first identifiable phase of mitosis. During this phase, chromosomes wind or coil and become distinctly visible when stained and viewed with a compound microscope. Prophase chromosomes consist of two chromatids, reflecting the fact that DNA duplication takes place in late interphase.

Prophase I The first identifiable stage of meiotic division. During this prolonged stage, homologous chromosomes (each composed of two chromatids) pair, chiasmata are formed, and genetic recombination may occur.

Prophase II A short phase in meiosis in which chromosomes shorten and thicken as a result of winding or coiling. Each prophase II nucleus contains a single member of each pair of homologous chromosomes, each consisting of two chromatids. (No DNA duplication occurs in interphase II.)

Propionic acid An organic acid with the formula CH_3CH_2COOH.

Prostrate Used in reference to plants lying flat on the ground.

Protein Linked groups of specific amino acids (polypeptides) with a precise order and arrangement. Proteins may be either structural (as in membranes) or functional (as in enzymes).

Protoderm One of three primary meristematic tissues. The protoderm ultimately differentiates into epidermis.

Protoplasm The contents of living cells.

Prussic acid See Hydrocyanic acid.

Pubescence Fine hairs or fuzz.

Quesnay, François French economist who stressed the importance of a strong agriculture to a nation's well-being. His philosophy was adopted by Benjamin Franklin.

Raceme An inflorescence in which flowers on pedicels are borne on a single, unbranched main axis.

Rachilla The central axis of a spikelet; a short rachis. See also Rachis.

Rachis The central axis of a spike type of inflorescence.

Radicle The part of the embryonic axis that grows into the primary root. The first part of the embryo to grow during germination.

Range An extensive natural pasture area.

Ratoon Regrowth of sugar cane, or the crop harvest from regrowth.

Ray A vascular cell that develops from vascular cambium and accounts for the lateral movement of materials in a plant.

Ray flower (Ray floret) The flowers at the edge of the inflorescence (head) of plants in the Compositae (sunflower) family. Many ray flowers (or florets) have large, showy petals. See also Disc flower.

Receptacle The enlarged tip of a stalk on which a flower is borne.

Recessive Used in reference to the allelic form of a unit of heredity that, when present in a heterozygous condition, is not expressed.

Reproductive growth Growth that leads to the development of flowers or other reproductive structures.

Reproductive nucleus See Generative nuclei.

Respiration The oxidation of food to yield energy for cellular activities.

Rhizobium Genus of bacteria that live symbiotically in the roots of legumes and fix nitrogen that is used by plants.

Rhizome A horizontal underground stem.

Rhizosphere The area in the soil in which roots grow or can grow.

Ribonucleic acid (RNA) A molecule related to DNA. RNA occurs in three forms:

Ribonucleic acid (RNA) (continued) messenger RNA, transfer RNA, and ribosomal RNA. Each form plays a precise role in genetic coding.

Ribosomal RNA A form of RNA found in a cellular organelle, the ribosome. Ribosomal RNA aids in the synthesis of proteins.

Root cap A mass of hard cells covering the tip of a root and protecting the root's tip from mechanical injury.

Root hair An elongation of an epidermal cell on a root.

Root stock The trunk or root material to which buds, or scions, are transferred in grafting.

Rosette A crowded cluster of radiating leaves appearing to rise from the ground.

Rotational grazing A management scheme in which livestock are systematically rotated from one pasture to another to maximize forage use and increase productivity.

Ruminant A cud-chewing mammal characteristically having a stomach divided into four compartments, such as cattle, sheep, goats, and deer.

Saline soil A soil containing excessive amounts of soluble salts.

Samara A dry, indehiscent, simple fruit that has winglike appendages on both sides of the ovary. These appendages help carry wind-borne fruit. The fruit of a maple is a samara.

Scalping A method of range renovation in which vegetation is turned over in strips to improve water infiltration, hasten the decay of organic matter, and reduce competition.

Scarify Scratch, chip, or nick the seed coat to enhance the imbibition of water.

Scar tissue The tissue formed at the site of an injury.

Schizocarp A dry, dehiscent fruit that dehisces along the margins of the carpels. The pistil is compound.

Scientific method An approach to a problem that consists of the following steps: awareness of the problem, establishment of one or more hypotheses for solutions to the problem, testing these hypotheses by experimentation or observation, and accepting or rejecting the hypotheses.

Scion (or Cion) The aerial part of a plant that is transferred to a new root stock in grafting.

Scutellum The rudimentary leaflike structure at the first node of the primary (embryonic) culm of a grass plant. The single cotyledon of a monocot.

Secondary cell wall Material deposited on the cytoplasmic side of the primary cell wall. Generally, the deposits strengthen and harden the cell wall and make it waterproof. Many types of cells can be identified by the type and location of secondary cell-wall materials. The secondary cell wall consists of a cellulose matrix that may or may not have other materials embedded in it.

Secondary phloem Phloem cells formed from vascular cambium on the inside of the primary phloem. Secondary phloem is found in biennials and perennials, not in annuals. Secondary phloem cells (and xylem cells) generally have characteristic cell walls.

Secondary xylem Xylem cells formed from the vascular cambium on the outside of primary zylem. The development of secondary xylem accounts for "annual rings" in many trees. See also Secondary phloem.

Seed coat See Testa.

Seed potatoes Pieces of potato that are planted to produce commercial crops.

Self-fertilization The condition in which pollen from an anther of a flower ultimately effects fertilization of the same flower or a flower on the same plant. See also Cross-pollination.

Self-pollination The transfer of pollen from an anther of one plant to a stigma on the same plant. See also Cross-pollination.

Seminal roots Roots that differentiate in the embryo.

Sepals The outermost series of floral parts, usually green, leaflike structures at the base of a flower; collectively, they form the calyx.

Sessile Used in reference to flowers, florets, leaves, or leaflets that are attached directly to an axis, not borne on any type of a stalk or stem.

Sexual fertilization The union of two haploid gametes to yield a single diploid zygote.

Sheath A part of the leaf of a grass plant. The sheath wraps around the culm and supports the leaf blade.

Silage Forage that is chemically changed and preserved by fermentation.

Silique A dry, dehiscent, elongated fruit from a compound ovary with many seeds; common in the Cruciferae (mustard) family.

Silo A structure for making and storing silage.

Simple leaf A leaf composed of a single, undivided blade that may or may not have a petiole.

Single cross A cross between two inbred lines. Frequently used in reference to hybrid corn.

Society for Promoting and Improving Agriculture and Other Rural Concerns, The An organization founded in the southern United States shortly after the Revolutionary War to foster the interests of agriculture.

Sod The top few inches of soil that is filled with root.

Sod type Growth habit of grasses that spread by stolons and rhizomes and yield a dense mat of vegetation. See also Bunch type.

Soilage Forage produced to be chopped green and fed directly to livestock. See also Green chop.

Soil profile A two-dimensional view of soil from which horizon development can be described.

Soil structure The tendency of soil particles to clump together or form aggregates.

Soil texture The relative proportions of sand, silt, and clay particles in surface soil.

Somatic tissue Nonreproductive, vegetative tissue. Tissue developed through mitosis that will not undergo meiosis.

Specific-gravity separator Seed-cleaning device that separates weed seed and foreign material from seed of a desired species according to differences in seed density.

Speculation Trading in futures contracts in which buyers take the risk of price change, hoping for a financial gain. Speculators are people who underwrite the risk for the hedging process. See also Hedging.

Sperm nucleus The haploid nucleus developed in the male gametophyte that unites with the female gamete (egg nucleus) to culminate sexual fertilization and yield a zygote.

Spike An inflorescence, common in the grass family, that has a central axis on which sessile spikelets are borne. See also Rachis.

Spikelet The flowering unit of a grass inflorescence. A spikelet is composed of one or more florets and has as its outer or lower limit a pair of glumes.

Spindle See Mitotic apparatus.

Spinning jenny A device used to spin cotton into yarn.

Spongy parenchyma The cell layer in a leaf located between the palisade parenchyma and the lower epidermis; these cells have thin cell walls and are loosely packed. This tissue is a storage area for products of photosynthesis.

Spore A haploid cell, produced by meiosis, that is capable of living free of the plant that produced it. The spore undergoes mitosis to produce gametes. Thus, a spore may be referred to as a gametophyte or gamete producer.

Spore mother cell Diploid cells, found in either the anther or the ovary, that go through meiosis to yield one or more functional haploid spores.

Sporophytic Used in reference to the plant or life cycle stage (generation) of a plant in which spores are produced. Crop plants (e.g., wheat, corn, and alfalfa) are sporophytic plants (or sporophytes) and are in the sporophytic stage of development in the life cycle.

Staining A technique of applying differential coloring to cell parts so that they may be viewed with various types of microscopes.

Stamen The entire male reproductive structure of a flower. The stamen is composed of a filament on which is borne an anther.

Starch A complex polysaccharide carbohydrate. The form of food most commonly stored by plants.

Statistical significance An attribute that may be assigned to a result obtained by the statistical evaluation of experimental data. It is based on the calculated probability of the occurrence, by random error, of the event in question. If the probability of such an occurrence is less than 5%, then, by convention, the event (e.g., the difference between two averages) is called statistically *significant*. If the probability is less than 1%, the event is called *highly significant*.

Stigma The apical part of the pistil specially modified to provide an ideal site on which pollen may land, germinate, and initiate the processes that lead to fertilization.

Stipule A small structure or appendage at the base of some kinds of leaves. Stipules are commonly leaflike.

Stolon A slender, prostrate aboveground stem. The runners of white clover plants are stolons.

Stoma (plural stomata) An opening or pore; the intercellular space between specialized epidermal cells, the guard cells. Guard cells regulate the size of the stoma. Stomata are sites of gas exchange between leaves and the external atmosphere.

Stool (Culm; Tiller) A common name for the tillers or stems of cereal grasses.

Stooling The stage of development in which tillers are formed.

Stubble mulch Partial incorporation of crop residue into the surface soil to increase water infiltration and reduce soil erosion.

Style That part of the pistil between the stigma and ovary.

Suberin A waxy, waterproofing substance found in cork tissue.

Succinic acid An organic acid with the formula $C_2H_4(COOH)_2$.

Sucker A tiller or shoot arising from an axillary bud.

Sucrose Common table sugar, $C_{12}H_{22}O_{11}$.

Superior ovary An ovary situated above the receptacle; all other floral parts develop below the base of the ovary. *See also* Inferior ovary.

Symbiosis An obligate relationship between two organisms of different species living together in close association for their mutual benefit.

Synergic nuclei Nuclei formed by mitosis in the megaspore, which lie on either side of the egg nucleus in the embryo sac.

Systemic Used in reference to a plant's entire metabolic processes.

Tannin An organic substance with an astringent, bitter taste.

Taxonomy The study or science of identification and naming of organisms (plants or animals).

Telophase The stage of mitosis immediately following anaphase. At this stage, chromatids have become independent chromosomes, which uncoil and become less distinct. A new nuclear envelope is formed around each telophase nucleus. This is the final stage of mitosis.

Telophase I Follows anaphase I of meiosis. Members of homologous pairs of chromosomes separate and move to opposite poles. Each chromosome still consists of two chromatids.

Telophase II The final stage in the complete meiotic cycle. From the initial diploid cell, four haploid, telophase II nuclei are formed. One or more of these nuclei becomes a spore. Each telophase II chromosome consists of a single chromatid.

Tendril Slender, leafless, coiling organs of climbing plants that support and attach the plant to surrounding surfaces. Tendrils are frequently modified leaves or parts of leaves, as in pea plants.

Testa The tissue surrounding the embryo and (in some cases) endosperm. The testa develops from the integuments and is maternal tissue; the seed coat.

Tiller (Culm; Stool) Stem of a grass plant.

Tilth Used in reference to the overall physical condition of a soil, frequently regarding its suitability as a seedbed.

GLOSSARY

Tomato harvester A mechanical device that uproots individual plants and removes the fruits from them. The tomato harvester has revolutionized the tomato industry in California and Arizona.

Toxicity symptom The undesirable physical change in a plant due to its taking-up of an excess of some substance.

Transfer RNA A form of RNA that exists in the cytoplasm. Transfer RNA associates with specific amino acids and carries these molecules to the site of protein synthesis.

Translocated Moved from one place to another. For example, sugars formed in leaves as a product of photosynthesis are moved (translocated) to roots and stems.

Transpiration The loss of water from plant tissue in the form of vapor.

Tricarboxylic acid cycle (TCA cycle) See Krebs cycle.

Trifoliate A stem having three leaves arising from separate buds at one point, that is, from one node. See also Trifoliolate.

Trifoliolate A compound leaf divided into three leaflets, such as the leaf in clover. See also Trifoliate.

Triploid A specific case of polyploidy in which there are 3N chromosomes.

Tube nucleus A nucleus, formed in the microspore by mitosis, that is characteristically found in the pollen tube after germination of the pollen grain. See also Pollen tube.

Tuber An enlarged, fleshy, underground tip of a stem. A common (Irish) potato is a tuber.

Tull, Jethro English gentleman-farmer of the 1730s who invented the grain drill and was an advocate of the continuous cultivation of crops. George Washington read and followed his writings on farming.

Umbel An inflorescence type in which flowers are borne at the end of stalks that arise like the ribs of an umbrella from one point—for example, the inflorescence of leafy spurge.

Vacuole A region, characteristic of plant cells, bounded by a membrane in which various plant products and by-products are stored. The vacuole may occupy the greater part of the volume of a plant cell.

Variety See Cultivar.

Vascular cambium A meristem that may differentiate to produce either secondary xylem or secondary phloem cells. Vascular cambium is found in biennials and perennials and is located between the primary xylem and primary phloem.

Vascular cylinder In roots, the central or core section immediately inside the pericycle. This mass of cells includes phloem, vascular cambium, and xylem cells.

Vector In biology, a carrier. A vector is generally an insect that carries pollen or disease-causing organisms from plant to plant.

Vegetative growth Growth of roots, stems, leaves, and other somatic tissues.

Volatile Evaporating readily or readily passing off in the form of a vapor.

von Liebig, Justus A nineteenth century German chemist who wrote extensively on the application of chemistry to agriculture.

Watson, Elkahana Organized one of the first producer-oriented agricultural fairs in 1811 in New England.

Waxes A class of fats derived from fatty acids and alcohols other than glycerol. Some waxes are solids.

Weed A plant that is located in such a way as to be more harmful than beneficial.

Whorled The arrangement of more than two buds or leaves in a circle at one node.

Windrowing Gathering herbage into a row; cutting and raking herbage into a row.

Winterhardiness The ability of a plant to tolerate severe winter conditions.

Xylem Specialized cells through which water and minerals move upward through a plant. The xylem cells are in the center of the vascular cylinder. Materials from the soil move upward through the plant by means of the xylem.

Index

A horizon, 111
Abscisic acid, 62
Acid-delinted seed, 387
Adventitious root, 57
Advertising, 220–221
Aerial seeding, 283
Aerobic respiration, 32
Agronomic classification, 89–90
Aleurone, 252
Alfalfa, 470–476
   bacterial wilt, 475
   diseases of, 475
   insects of, 475
   weevil, 475
Alkaline soil, 106–107
Allele, 22
Alsike clover, 471–472, 479–481
Alternation of generations, 74
Aluminum toxicity in soils, 106
American Farm Bureau, 11
Analysis of variance, 528–529
Annual precipitation, distribution of, 131
Anther, 63
Antipodal nuclei, 75
Apical meristem, 45
Apomixis, 73–74
Aquatic plants, 122–123
Asexual reproduction, 63, 73
Assembly, 216
Auricle, 55
Auxins, 61, 69
Available moisture and response to fertilizers, 144
Awn, 66, 230
Azonal soil, 113–114

B horizon, 111
Bacteria, life cycles of, 182–183
Bahiagrass, 502
Banded fertilizer, 243

Barberry, 181
Barley, 310–323
   botanical characteristics of, 312
   comparison of winter and spring types, 313
   crop rotations with, 316
   diseases of, 320–321
   distribution of, in North America, 310
   environmental conditions for production of, 312–313
   ergot in, 321
   harvesting of, 315–316
   high-moisture, 315
   history of, 311–312
   insect pests of, 321
   origin of, 311
   planting dates for, 313, 315
   planting rates for, 313
   precipitation required for 312–313
   seedbed preparation for, 313
   six-row, 66, 312
   smut, 320
   soil for production of, 313
   spike, 66
   storage of, 315–316
   temperature for production of, 313
   two-row, 66, 312
   utilization of, 321–322
   weeds in, 321
   world production of, 310
Basin irrigation, 139–141
Bassengault, J. B., 537
Bentgrass, 499
Bermudagrass, 501
Big bluestem, 506
Binomial system of classification, 86

Biological control of insects, 170
Birdsfoot trefoil, 471, 485–487
Black medic, 491
Blasting, 156, 329, 372
Bluebunch wheatgrass, 505
Blue grama, 506
Boll weevil, losses due to, 167
Bolting, 160–162, 415–416
Border strip irrigation, 139
Borlaug, N., 529
Boron, 59
Bran, 252
Branch root, 42
Brewing, 323
Broadcast seeder, 239–240
Budding, 73
Buds, classification of, 48–49
Buffalograss, 506
Bulb, 73
Bunt, 182

C horizon, 111
C_3 plant, 37, 54, 124
C_4 plant, 37, 54, 124
Calcium, function of, 59
Calyx, 64
Canadian crops, increases in yield of major, 525
Carbamate insecticides, 172
Carbon, 58
Cardinal temperatures, 154–156
Carpel, 63
Carpetgrass, 499
Caryopsis, 252
Castor beans, 460
Cation exchange capacity, 103
Cell membrane, 19
Cellulose, 17
Cell wall, 17
Centers of origin, 9, 339

INDEX

Centipedegrass, 502
Central Valley Project (California), 195-196
Cereals
  classification of, 250
  cool-season, 252
  cultivation for weed control in, 255-256
  current importance of, 248-250
  evolution of, 247-248
  factors affecting quality of, 257
  fruit of, 252
  growth stages and response to stress, 254-255
  harvesting of, 255-256
  and human diet, 249-250
  planting of, 255
  planting rates for, 255
  as a protein source, 250-251
  row spacings for, 255
  and social evolution, 247-248
  storage of, 256
  warm-season, 252
Certified seed, 185, 229-230
Chernozem soil, 114
Chinch bug, 277
Chlorinated hydrocarbon insecticides, 172, 173
Chlorine, 60
Chlorophyll, role of, in photosynthesis, 34, 35
Chloroplast, function and structure of, 21
Chromosome number, 21
Chromosomes, 21
Citric acid cycle, 32
Classification. See plant classification
Clay, 99
Climate
  and crop production, 162-163
  and resource evaluation and utilization, 201-206
Climax vegetation, 509
Cold hardiness, 296
Coleoptile, 252
Colza oil, 460
Combine, adjustment of, 244
Common names, uses and problems of, 90
Compensation point, 148
Competition, 151
Complete and perfect flower, 65
Compound fruit, 76
Compound leaves, 51-52, 472-473

Compound pistil, 63
Conservation and cropping systems, 196-200
  pest management, 200-201
  selective herbicides, 197
Consumptive use of water, 127-128
Continental climate, 162
Cool-season crops, 154
Cool-season grasses, 494
Cooperative marketing, 226
Copper, 60
Cork, 47
Corm, 57
Corn, 259-279
  botanical characteristics of, 260
  compared with other cereals, 254
  crop rotations with, 270
  detasseling of, 533
  diseases of, 272-275
  ear of, 260
  earworm, 277
  effect of distribution of precipitation on production of, 263
  effect of soil pH on, 263
  environmental conditions for production of, 263-264
  fertilizer requirements of, 263-264
  flowers of, 65, 260
  grain harvest, 269-270
  history of, 259-260
  insect pests of, 275-277
  and minimum tillage, 265-266
  North American production of, 258
  origin of, 259-260
  picker-sheller, 270, 272
  planting dates for, 268-269
  planting depth for, 266
  planting rates for, 266-268
  rot diseases of, 273-275
  row spacings for production of, 266-268
  seedbed preparation for, 265
  seed treatment, 275
  smut, 272-273
  temperatures for production of, 263
  types of, 260-261
  utilization of, 278
  weeds in, 275
  world production of, 258
Corolla, 64
Corrugation irrigation, 143

Cortex, 42, 47
Cotton, 382-395
  boll, 385
  botanical characteristics of, 385
  cost of weed control in, 392
  crop rotations with, 392
  defoliants, use of, 388
  diseases of, 392
  distribution of (U.S.), 382
  effect of light on production of, 386
  environmental conditions for production of, 385-386
  fertilizer requirements of, 388
  fiber, 385
  flaming for weed control in, 393
  frost-free period required for, 386
  gin, 391
  harvesting of, 388, 391
  and herbicides, 393
  history of, 383-384
  insect pests of, 393-394
  mechanical harvesters, 388, 391
  origin of, 383
  planting dates for, 387-388
  planting depth for, 387
  planting rates for, 387
  precipitation required by, 386
  processing of, 391-392
  production practices for, 386-392
  row spacings for, 386-387
  seedbed preparation for, 386
  soil for production of, 386
  temperature for production of, 385-386
  topping of, 388
  utilization of crop, 394-395
  weeds in, 392-393
  yields of, 391-392
Cottonseed meal, 395
Cottonseed oil, 395
Cotyledon, 227, 228
Coumarin, 484
Crambe, 460
Crested wheatgrass, 502, 503
Crop botany and crop production, 15-16
Cropping systems, 194-195, 207-209, 354
Crop production, basic steps in, 9
Crop rotations, 194

Crop tolerance of salinity, 107
Crop yields per acre, changes in, 12
Cross-pollination, 69, 70-71
Crown, 45, 56
Culm, 55
Cultivar, 251-252
Cyclic reaction, 35
Cylinder separator, 230
Cyme, 68
Cytokinesis, 23
Cytokinins, 61, 69
Cytology, 17
Cytoplasm, 19
Cytoplasmic male sterility, 532-533

Dallisgrass, 501, 502
Damping off, 321
Dark reactions, 36-37
Day-neutral plant, 153
Degree days, 160
Dehiscent fruit, 76-78
Dehydrated forage, 518
Deoxyribonucleic acid, 21
2,4-Dichlorophenoxyacetic acid, 61, 189-190, 525
Diet, human, 5
  composition of (Asian and U.S.), 337, 524
  minimum kilocalories required, 5, 524
Dioecious, 65, 506
Disc flower, 66, 447
Disease cycle, 179-180
Distribution (marketing), 223
Dominant, 530
Dormancy, 61-62, 228-229, 362-363
Double cropping, 354
Double fertilization, 76
Drought tolerance, 466
Dry fruit, 76
Drying oil, 355
Dry matter accumulation, pattern of, 81-82
Dust bowl, 195

Edible legumes
  diseases and pests of, 342
  environmental conditions for production of, 341
  origins of, 338-340
  production practices for, 341-342
  utilization of, 343
Ehrlich, P., 5
Embryo, 227, 252
Endocarp, 64

Endodermis, 42, 45, 47
Endoplasmic reticulum, 18, 21
Endosperm, 76, 227, 252
Endosperm nucleus, 75
Energy, earth's supply of, 3
Epidermis, 42, 45, 47
Essential amino acids, 337-338
Essential minerals, determination of, 60
European corn borer, 167, 275-276
Exocarp, 64
Expeller, 355

F_1, 531
F test, 528
Faba beans, 381, 490
Fallow, benefits from, 133-135
Fanning mill, 230, 231
Farmers' Educational and Cooperative Union, 11
Farming, origin of, 8
Fermentation, 34
Fertile Crescent, 247
Fertilizers, 115-118
  increased use of, 525
Fescue, 496-497
Fibrous root, 41-42, 45
Field beans
  botanical characteristics of, 371-372
  bunch type, 372
  bush type, 372
  classes of, 370-372
  crop rotations with, 377-378
  diseases of, 378-379
  environmental conditions for production of, 372-374
  growth habit of, 372
  harvesting of, 376-377
  history of, 371
  insect pests of, 379-380
  irrigation of, 373
  origin of, 371
  planting dates for, 375
  planting rates for, 374-375
  precipitation required by, 372-373
  production practices for, 374-376
  row spacings for, 374-375
  seedbed preparation for, 374
  soils for, 374
  temperature required by, 372, 375
  uses of, 380
  weeds in, 379
  yields of, 377
Field capacity, 100

Field peas, 371-380, 491
  botanical characteristics of, 372
  crop rotations with, 378
  diseases of, 379
  environmental conditions for production of, 374
  harvesting of, 378
  history of, 371
  insect pests of, 380
  origin of, 371
  production practices for, 376
  uses of, 380
  weeds in, 379
Filament, 63
Finance, market, 219
Financing, 219
Flax,
  botanical characteristics of, 446
  diseases of, 457
  environmental conditions for production of, 450
  harvesting of, 456
  insect pests of, 459
  planting rates for, 453
  production practices for, 453
  origin of, 444
  weeds in, 458
Fleshy fruit, 65, 76
Flooding, rice, 283
Flood irrigation, 139-143
Floret, 66
Floral induction, 68-69
Flower, 63-66, 340
Foot candle, 148
Forages
  palatability of, 466
  quality of, 466
  yields of, 465
Fragipan, 109
Franklin, B., 11
Freight rates, 220
Frost, 156-157
Frost-free period, 158-160
Fruit, classification of, 76-80
Fumigant, 185
Fungicide, 185
Furrow irrigation, 141-142
Futures trading, 218-219

Gametophyte, 74-75
Gametophytic plant, 74
Gene, concept of, 22
Generative nucleus, 75
Genetic code, 21-22
Genome, 533
Genotype, 530
Geotropic, 61, 360

Germination, conditions for, 228–229
Gibberellins, 61, 69
Glume, 66
Glycolysis, 31
Government classification and grades, 90–93
Grading, market, 216
Graftage, 73
Grafting, 73
Graham flour, 252
Grain drill, 239
Gramineae, 88, 493
Grange, The, 11
Grass flower, 65–66
Grasshoppers, damage to range from, 166
Grass plant, stem of, 55
Grass seeder, 239
Great soil groups, 113–114
Green chop, 518
Green needlegrass, 505
Ground meristem, 45, 47
Guard cell, 51
Gynophore, 360

Hard pan, 109
Hay, 464, 514–518
  acres of (in U.S.), 426–463
  baling of, 517
  field chopping of, 517
  grading of, 217
  long loose, 516
  pelleting of, 518
  quality of, 514–515
  storage of, 516–518
  wafering of, 518
Head, 68
Hedging, 218–219
Herbicide, 189–191
Hessian fly, 167, 307
Heterosis, 532
Heterozygous, 21–22, 531
High-moisture barley. See Barley
Homozygous, 22, 531
Homologous chromosomes, 24, 26–28, 532
Hopkins Bioclimatic Law, 162–163, 268
Horsebeans, 381
Hulling, 230
Hybrid corn, 526, 532, 534
Hybrid vigor, 532
Hydrogen, 58
Hydrogen acceptor, 31
Hydrophyte, 122–123
Hypogeal, 360

Hypothesis, 527

Imperfect flower, 65
Incompatibility, 70
Incomplete flower, 65
Indehiscent, 78
Inferior ovary, 65
Infiltration rate, 236
Inflorescence, 66–68
Inoculation
  methods of, 235
  reasons for, 235–236
Inoculation group, 235
Insecticides
  formulations of, 173
  modes of action of, 173
  pollution by, 174
Insects
  biological control of, 170
  chemical control of, 170–174
  classification of, 164–166
  control of, 168–174
  cultural control of, 167–168
  damage caused by, 166–168
  genetic control of, 169
  life cycles of, 165–166
Integuments, 75
Intercalary growth, 56, 252
Intercellular space, 19
Intermediate wheatgrass, 503–504
Internode, 48, 57
Intrazonal soil, 113–114
Irish (common) potato, 57
Iron, 59
Irrigation
  management, 143–144
  water measurements, 135–136
Irrigation, methods of, 137–143
  aerial (sprinkler), 137–139
  surface, 139–143

Jarmo, 247
Johnsongrass, 499, 500
Jointing, 254

Karyolymph, 20
Kentucky bluegrass, 498–499
Kinins. See Cytokinins
Koch's postulates, 184
Krebs cycle, 31–34
Kudzu, 471, 491

Land available for cultivation, 98
Leaf abscission and hormones, 61
Leaf, 49–56
  anatomy of, 51–56

  axil of, 49
  blade, 49
  blade of grasses, 55
  canopy, 127
  sheath, 55
Leaf area index, 83, 126–127, 284
Leaflet, 51
Least significant difference, 528
Legumes
  botanical characteristics of, 340–341
  flowers of, 340–341
  inflorescences of, 340
  inoculation of, 233
  and nitrogen, 58
  photoperiodic response of, 341
Leguminosae, 88, 470, 472
Lemma, 66
Lespedeza, 471, 488
Liebig, J. von, 11
Life cycle (plant), 15–16
Light, 146–154
  duration of, 153
  intensity of, 148–153, 348
  quality of, 146–148
  quantity of, 148–153
  quantity of, for C_3 and C_4 plants, 149
  quantity and planting rate, 149–151
  quantity and weed control, 151
  and temperature relations, 153–154
  visible part of spectrum, 146
  wavelength and photosynthesis, 36–37
Light reactions, 35–36
Lignin, 17
Lima beans, 371
Limestone, 105
Liming, 104–106
Linnaeus, C., 86
Linseed oil, 460
Little bluestem, 506
Lodging, 150–151, 284, 300
Long-day plant, 153
Lower epidermis, 51
Lux, 148
Lygus bug, 166–167

Macronutrient, 58–59
Magnesium, 59
Male sterility, 70, 273–275, 532–533
Malthus, T., 8

Malt, production of, 322
Malting barley, 316-320
   acceptable protein levels, 317
   combine adjustments for, 319-320
   cultivars, 317-318
   fallow, 318
   fertilizers for, 317-319
   harvesting of, 319-320
Manganese, 59
Market communications, 219
Market economy, 215-216
Marketing agencies, 224-225
Marketing functions, 216-223
   assembly, 216
   financing, 219-221
   futures trading, 218-219
   merchantising, 220-223
   standardizing and grading, 216-218
Marketing institutions, 223-226
Marketing, special considerations, 225
Marketing spread, 225
Megaspore, 64, 75
Meiosis, 26-29, 64, 74-75
   compared with mitosis, 29-30
Mendel, G., 530
Merchandising, 220-223
Mercury fungicides, 232
Meristematic tissue, 47
Mesocarp, 64
Mesophyll, 54
Mesophyte, 122-123
Mesopotamia, 247
Micronutrient, 59-61
Middle lamella, 17
Microspore, 75
Minerals
   forms available to plants, 114-115
   used by plants, 117-119
Minimum tillage, 198, 265-268
Mitochondria, 21
Mitosis, 23-25
   compared with meiosis, 29-30
Moisture
   excessive, 129
   seasonal distribution, 127-128
   sources of, 130
Moisture stress, plant response to, 128-129
Molybdenum, 60
Monoecious, 65, 69
Morrill Act, 12
Mustard, 446, 460

Natalgrass, 502

Nematode, 174, 186, 380, 393-394
Nitrogen, 58, 116
Node, 48, 57
Nodulation, effects of pH on, 233
Noncyclic reaction, 35-36
Nuclear envelope, 20
Nucleus, 20-21

Oats, 324-335
   botanical characteristics of, 327-329
   crop rotations with, 332
   diseases of, 332-333
   distribution of, in North America, 324
   environmental conditions for production of, 329-331
   fertilizers for, 331
   harvesting of, 332
   history of, 325-327
   insect pests of, 334
   origin of, 325
   panicle infloresence of, 68
   planting dates for, 368
   planting rates for, 331
   precipitation required by, 329
   row spacings for, 331
   seedbed preparation for, 331
   soil for production of, 331
   temperature for production of, 329, 331
   utilization of, 334-335
   weeds in, 334
   world production of, 324
Oceanic climate, 162
Oil extraction methods, 355-356
One-tenth bloom stage, 474
Orchardgrass, 495, 496, 497
Organic matter, 108, 119-120
Organic phosphate insecticides, 172
Ovary, 63, 75
Overseeding, 512
Ovule, 75
Ovum, 64
Oxidative phosphorylation, 33
Oxygen, 32, 58, 283-284

Packaging, 221-223
Paddy rice, 286
Palea, 66
Palisade parenchyma, 54, 57
Palmate venation, 51
Pangolagrass, 499
Panicle, 68
Paragrass, 502

Parent material, 111
Pasture
   permanent, 510
   renovation, 512
   rotation, 513
   supplementary, 513
   temporary, 513
Pasture calendar, 467
Peanuts, 358-369
   ability of, to extract minerals from soil, 363, 365
   botanical characteristics of, 360-361
   bunch type, 360, 363
   composition of, 368
   crop rotations with, 365
   diseases of, 365-367
   distribution of (U.S.), 358
   environmental conditions for production of, 361-365
   fertilizer for, 363-364
   frost-free period required by, 361
   harvesting of, 364-365
   history of, 359-360
   insect pests of, 368
   moisture requirements of, 361
   origin of, 359
   planting dates for, 362
   planting depth for, 362
   planting rates for, 362-363
   production practices for, 361-365
   production regions, 361
   row spacings for, 362
   runner type, 360, 362
   seedbed preparation for, 361-362
   seed quality, 362
   seed treatment, 362-363
   soil for production of, 361
   temperatures required by, 361, 362
   uses of, 368
   weeds in, 367-368
Ped, 102
Pedalfer, 113
Pedicel, 64, 68
Pedocal, 113
Pedon, 110
Peduncle, 64, 68
Pegging, 360
Perianth, 64
Pericarp, 76
Pericycle, 42, 47
Permanent wilting percent, 101
Petal, 64
Petiole, 49

INDEX 563

Phenotype, 530
Phloem, 42, 46
Phosphorus, 58
Photolysis of water, 58
Photoperiod, 153-154, 346
Photoperiodism, 68
Photosynthesis, 34-38
Phototropism, 61
Pink beans, 380-381
Pinnately compound leaves, 51-52, 472-473
Pinnate venation, 51
Pistil, 63, 66
Pith, 46, 47
Pith ray, 47
Plant breeding, 526, 529-539
Plant classification, scientific system of, 86-89
Plant diseases, 174-186
  control of, 184-186
  effect of environment on, 180-181
  losses due to, 174-177
  types of damage from, 177-179
Plant domestication, 8
Plant-growth regulators, 61-62
Plant pathogens, life cycles of, 181-184
Plant pathology, 15, 174
Plant taxonomy, 15, 86-89
Plasmalemma, 19
Podzol, 113-114
Poisonous plants, 468, 484
Polar nuclei, 75
Pollen, 75
Polyploid, 532
Population density, 7
Population growth, 5
Pop-up fertilizer, 243, 351
Pore space, 100-102
Potassium, 58
Potatoes, 430-441
  botanical characteristics of, 432
  diseases of, 437-438
  distribution of (U.S.), 430, 432
  fertilizers for, 436
  harvesting of, 436
  insect pests of, 439-440
  irrigation requirements for, 433
  planting rates for, 435
  and specific gravity, 440
  storage of, 437
  temperature for production of, 435-436
  vegetative reproduction, 432

  weeds in, 439
Potato scab, 183, 438
Press wheel, 239
Pricing, 220, 230
Primary cell wall, 17
Primary consumer, 4
Primary phloem, 45, 47
Primary plant nutrient, 58
Primary producer, 4
Primary xylem, 45, 47
Procambium, 45, 47
Processing (marketing), 221
Proteins in human diet, 336
Protoderm, 45, 47
Pubescent wheatgrass, 504
Punnet square, 70-71
Pure live seed ratio, 239

Quesnay, F., 11

Raceme, 68
Radicle, 45
Rachilla, 66
Rachis node, 66
Rainfall, effectiveness of, 130-132
Range grass, 502
Range, western, 508-510
Rapeseed, 444-460
  botanical characteristics of, 446
  cultural practices for, 452
  diseases of, 456
  environmental conditions for production of, 449
  harvesting of, 455
  history of, 444
  insects of, 458
  origin of, 444
  planting rates for, 452
  production practices for, 452
  weeds in, 458
Ray flower, 68, 447
Recessive, 530
Red clover, 471, 477
  diseases of, 479
  insect pests of, 479
Redtop, 499
Reed canarygrass, 497, 498
Rescuegrass, 502
Research, 524-525, 526
Resources
  evaluation of, 201-209
  utilization of, 195, 206-207
Respiration, 30-34
  and photosynthesis, 37-38
Rhizobium, 58, 233-234
Rhizomes, 57, 63, 77
Ribosome, 21

Rice, 280-289
  botanical characteristics of, 282
  crops rotated with, 286-287
  diseases of, 287-288
  distribution of (U.S.), 280
  drying of, 286
  environmental conditions for production of, 282-283
  fertilizers for, 285
  florets of, 282
  growing season required by, 282
  harvesting of, 285-286
  history of, 281
  insect pests of, 288
  irrigation of, 283
  milling of, 286
  origin of, 281
  oxygen requirements of, 283-284
  planting dates for, 283
  planting rates for, 284-285
  precipitation required by, 283
  seedbed preparation for, 283
  soil for production of, 283
  storage of, 286
  temperature for production of, 282
  types of, 282
  utilization of, 288-289
  water quality and production of, 283
  weeds in, 288
  world production of, 280
Risk sharing, 220
Root cap, 45
Root hair, 42, 45
Roots, 39-46
  annual compared with perennial, 42-43
  cross sections of, 42-43
  food storage in, 40-41
  functions of, 39-41
  longitudinal sections of, 44
  morphology of, 41-42
  origin of, 45
  pattern of distribution of, in soil, 41-42
Rough rice, 286
Russian wild rye, 505
Rye, 327-335
  botanical characteristics of, 329
  diseases of, 333-334
  distribution of, in North America, 326
  environmental conditions for

Rye *(continued)*
    production of, 331
    harvesting of, 332
    history of, 327
    insect pests of, 334
    origin of, 327
    seedbed preparation for, 331-332
    weeds in, 334

Safflower, 443-459
    botanical characteristics of, 445
    diseases of, 456
    environmental conditions for production of, 449
    harvesting of, 454
    insect pests of, 458
    origin of, 443
    planting rates for, 451
    production practices for, 451
    utilization of, 459
Safflower oil, 459
Sainfoin, 472, 488, 489, 490
Saline seep, 134, 135, 195-196
Saline soil, 107
Salinity, 107-108
Sand, 99
Scalping, 230
Scarification, 230
Scientific method, 13, 526-527
Secondary cell wall, 17
Secondary inoculum, 182
Secondary phloem, 43
Secondary roots, 252, 291-293
Secondary xylem, 43
Seed, 78, 227-228
    certification of, 229-230
    cleaning of, 230
    drying of, 244-245
    harvesting of, 243-244
    quality of, 83-84, 228
    storage of, 231-232
    treatment of, 232-233, 362-363
    uses of, 227
Seed coat, 75, 227-228
Seed-metering devices, 239
Seed potato, 435
Seed production, commercial, 245-246
Seedbed preparation, 236-238
Seeders
    calibration of, 240-242
    row-crop, 242-243
    types of, 238-240
Seeding, 238-240
    and fertilizers, 243
    rate, 238
Self-pollination, 69, 71-72, 531

Seminal root, 45
Sepal, 64
Sesame, 460
Sessile leaves, 51
Shade tolerance, 140-150, 267
Shattering, 243-244
Short-day plant, 153
Sideoats grama, 506
Sieve element, 43
Silage, 518, 519, 520, 521
Silo, 518, 519, 521
Silt, 99
Simple fruit, 76
Simple leaf, 49
Simple pistil, 63
Six-row barley. *See* Barley
Slender wheatgrass, 504
Smooth bromegrass, 495, 496, 497
Smudging, 158
Snow, 132-133, 296
Sodic soil, 106-107
Soil classification, systems of, 112-114
Soil conservation, 121
Soil Conservation Service, 12
Soil, defined, 97
Soil formation, 111-112
Soil horizons, 110
Soil nomenclature, 112-113
Soil profile, 110
Soil taxonomy, 112
Soil testing, 120
Soil triangle, 99-100
Soils and plant nutrition, 114-115
Soils and resource evaluation, 204-206
Soils, properties of, 99-108
    depth, 108
    moisture-holding capacity, 99
    pH, 104-105
    porosity, 101
    structure, 102
    texture, 99, 102-103
Soilage, 518
Solvent extraction, 355-356
Sorghum, 260-278
    botanical characteristics of, 260-262
    crop rotations with, 271
    diseases of, 275
    distribution of, in North America, 265
    drought resistance of, 264
    effect of soil pH on, 264
    environmental requirements for production of, 264-265
    harvesting of, 270-271

    history of, 260
    insect pests of, 277-278
    irrigation of, 270
    origin of, 260
    planting dates for, 269
    planting rates for, 269
    precipitation required by, 264
    seedbed preparation for, 268-269
    temperatures for production of, 264
    types of, 261-263
    utilization of, 261-278
    weeds in, 275
Southern corn leaf blight, 175
Soybeans, 471-491
    botanical characteristics of, 346
    crop rotations with, 353-354
    diseases of, 354
    distribution of (U.S.), 344
    environmental conditions for production of, 346-348
    fertilizers for, 350-351
    frost-free period required by, 346-347
    harvesting of, 351
    history of, 345-346
    inoculation of, 351
    insect pests of, 355
    light intensity required by, 348
    maturity groups, 346, 348
    oil yields, 356
    origins of, 345
    photoperiodic reaction of, 346
    planting dates for, 349, 350
    planting depth for, 350
    planting rates for, 350
    precipitation required by, 348
    production practices for, 348-354
    row spacings for, 348
    seedbed preparation for, 348, 350
    soil for production of, 348
    storage of, 351, 353
    temperatures for production of, 346
    utilization of, 355-356, 471, 491
    weeds in, 355
    world production of, 344
Space for crop production, 7, 97, 201
Specific-gravity separator, 230, 231
Speculation, 218
Sperm nucleus, 75

INDEX

565

Spike, 66
Spikelet, 66
Spongy parenchyma, 54
Spore, 74
Sporidia, 181
Sporogenesis, 64
Sporophyte, 74
Sprinkler irrigation, 137–139
Spring wheat, 295, 296–297
  planting dates for, 296–297
  temperatures required by, 296–297
Square, 385
Sugar beets, 411–429
  botanical characteristics of, 412
  diseases of, 425–426
  distribution of (U.S.), 410
  environmental conditions for production of, 414–416
  fertilizers for, 419–420
  harvesting of, 420–421
  history of, 411–412
  insect pests of, 428
  irrigation for, 417
  monogerm seed, 417
  origin of, 411
  planting dates for, 417
  planting rates for, 417
  processing of, 429
  root storage in, 40
  temperatures for production of, 415–416
  thinning of, 418
  weeds in, 427
Sugarcane, 412–429
  botanical characteristics of, 414
  diseases of, 426–427
  distribution of (U.S.), 410
  environmental conditions for production of, 416–417
  fertilizers for, 423–424
  harvesting of, 424
  insect pests of, 428
  processing of, 429
  row spacings for, 422
  soil for production of, 417
  temperature for production of, 416
Sulfur, 59
Summer annual, 16
Sunflower, 444–460
  botanical characteristics of, 447
  diseases of, 457
  environmental conditions for production of, 450
  harvesting of, 456

insect pests of, 459
origin of, 444
planting dates for, 454
planting rates for, 454
production practices for, 454
Sunflower oil, 460
Superior ovary, 64
Sweetclover, 471, 472, 482, 483, 484
Symbiosis, 58, 115, 233–235, 342, 472
Synergid nuclei, 75
Systems approach, 526–527
Solar energy, 3, 4
Stamen, 63
Stand establishment, 236
Standardization and grading, 216–218
Statistics, use of, in research, 527–529
Stem, 46–49, 55
Stigma, 63
Stinking smut, 182
Stipules, 51
Stolon, 57, 63, 73
Stomata, 51, 54
Stools, 56
Storage organ, 57
Storing (marketing), 218
Strawbery clover, 466, 491
Style, 63
Synergid nuclei, 75

Tall wheatgrass, 504
Taproot, 41
Teliospore, 181
Temperature
  control of, by irrigating, 158
  conversions of, 154
  and crop distribution, 159–161
  and crop growth, 154–163
  and crop management, 157–162
  and planting date, 157–158
Temperature extremes, damage from, 156–157
Tennessee Valley Authority, 195–196
Testa, 75, 361
Test weight, 257
Thickspike wheatgrass, 504
Tiller, 56, 252–254
Tilth, 236
Timothy, 496, 498
Tobacco, 397–409
  botanical characteristics of, 398
  curing of, 407–408
  diseases of, 405

distribution of (U.S.), 396
fertilizers for, 402–403
harvesting of, 403–405
history of, 397
insect pests of, 406
priming of, 403–405
production practices for, 400–405
temperatures for production of, 398
transplanting of, 40
types of, 407–408
weeds in, 406
Topography, 108, 109
Toxic materials in soil, 106, 109
Transpiration ratio, 101, 122–124
Transportation, 220
Trashy fallow, 197
Tricarboxylic acid cycle (TCA), 32
Trifoliolate leaves, 51, 473
Triticale, 533–535
Two-row barley. See Barley
Tube nucleus, 75
Tuber, 57
Tull, 11
Type specimen, 89

Umbel, 68
U.S. Department of Agriculture, 11
Uridiospore, 182
Upper epidermis, 51

Vacuole, 19
Vascular bundle, 46
Vascular cambium, 43, 45
Vascular cylinder, 42
Vavilov, N. I., 338–339
Vegetable oil, 443
Vegetative reproduction, 73
Velvet roll separator, 230
Venation, 51
Vernalization, 162, 296
Vessel element, 43
Vetch, 471, 472, 490
Virus, life cycle of, 183–184

Warm-season crops, 154–155
Warm-season grasses, 495
Water
  function of, in plant, 122
  measurements of, 135–136
Watershed development, 200
Water use
  and climate, 126
  efficiency, 124–125
  and photosynthesis, 126
Watson, E., 11

Weeds, 186-192
  control of, 127, 151-152, 189-192
Western wheatgrass, 503
Wheat, 290-309
  botanical characteristics of, 291
  classes of, 293-295
  coleoptilar node elongation, 291
  composition of kernel, 78, 293
  crop rotations with, 303-304
  diseases of, 304-305
  distribution of, in North America, 290
  environmental conditions for production of, 295-298
  fertilizers for, 301, 303
  grading of, 91-92
  harvesting of, 302-303
  history of, 291
  insect pests of, 306-307
  irrigation of, 297
  milling of, 307-308
  origin of, 291
  planting dates for, 299
  planting depth for, 299
  precipitation required by, 297-298
  row spacing for, 299-301
  rust diseases of, 175, 304-305
  seedbed preparation for, 298-299
  soil for production of, 297
  spike, 66, 291
  stand adjustments, 299-300
  utilization of, 307-308
  weeds in, 305-306
  world production of, 290
  yields of, 303
Wheat stem sawfly, 167, 306-307
White clover, 471, 480, 482
White lupine, 472
Wilting
  permanent, 101
  temporary, 101
Winter annual, 16
Winterhardiness, 468
Winter wheat, 295-296

Xylem, 42-43, 46

Zerophytes, 122-123
Zinc, 59-60
Zonal soil, 113-114